Skeletal Muscle Damage and Repair

Peter M. Tiidus, PhD

Wilfrid Laurier University

EDITOR

Human Kinetics

Library of Congress Cataloging-in-Publication Data

Skeletal muscle damage and repair / Peter M. Tiidus, editor.
    p. cm.
  Includes bibliographical references and index.
     ISBN-13: 978-0-7360-5867-4 (hard cover)
     ISBN-10: 0-7360-5867-2 (hard cover)
  1. Muscles--Wounds and injuries. 2. Muscles--Wounds and injuries--Treatment. I. Tiidus, Peter M., 1955-
  [DNLM: 1. Muscle, Skeletal--injuries. 2. Muscle, Skeletal--metabolism. 3. Soft Tissue Injuries--therapy. WE 500
S6262 2007]
    RD688.S63 2007
    617.4'73044--dc22

                   2007034028

ISBN-10: 0-7360-5867-2
ISBN-13: 978-0-7360-5867-4

The Web addresses cited in this text were current as of November, 2007, unless otherwise noted.

Acquisitions Editor: Loarn D. Robertson, PhD; **Developmental Editor:** Elaine H. Mustain; **Managing Editor:** Melissa J. Zavala; **Copyeditor:** Joyce Sexton; **Proofreader:** Red, Inc.; **Indexer:** Nancy Gerth; **Permission Manager:** Carly Breeding; **Graphic Designer:** Nancy Rasmus; **Graphic Artist:** Carol Smallwood; **Cover Designer:** Keith Blomberg; **Photographer (interior):** © Human Kinetics, unless otherwise noted; **Photo Asset Manager:** Laura Fitch; **Visual Production Assistant:** Jason Allen; **Art Manager:** Kelly Hendren; **Associate Art Manager:** Alan L. Wilborn; **Illustrator:** Tammy Page; **Printer:** Thomson-Shore, Inc.

Printed in the United States of America   10   9   8   7   6   5   4   3   2   1

Human Kinetics
Web site: www.HumanKinetics.com

United States: Human Kinetics
P.O. Box 5076
Champaign, IL 61825-5076
800-747-4457
e-mail: humank@hkusa.com

Canada: Human Kinetics
475 Devonshire Road Unit 100
Windsor, ON N8Y 2L5
800-465-7301 (in Canada only)
e-mail: info@hkcanada.com

Europe: Human Kinetics
107 Bradford Road
Stanningley
Leeds LS28 6AT, United Kingdom
+44 (0) 113 255 5665
e-mail: hk@hkeurope.com

Australia: Human Kinetics
57A Price Avenue
Lower Mitcham, South Australia 5062
08 8372 0999
e-mail: info@hkaustralia.com

New Zealand: Human Kinetics
Division of Sports Distributors NZ Ltd.
P.O. Box 300 226 Albany
North Shore City
Auckland
0064 9 448 1207
e-mail: info@humankinetics.co.nz

This book is dedicated to my family. To my wife, Ann, the love of my life, my companion, my comfort, and my friend—may your constant quest for adventure keep us forever young. To my sons, Erik and Tommi, who have grown into fine young men and who are my greatest source of pride—may you prosper and realize your fondest dreams. And to my late parents, Arvo and Helene Tiidus, who taught me that the combination of perseverance and laughter will lead to success. If the object of life is happiness, mine has been achieved.

CONTENTS

Part II Muscle Damage and Repair in Applied Situations

Part III **Treatments and Interventions in Muscle Damage and Repair**

LIST OF CONTRIBUTORS

Karin Alev, PhD
  Functional Morphology Laboratory
  University of Tartu, Tartu, Estonia

Muhammel Al-Jarrah, PhD, PT
  Department of Physical Therapy
  and Rehabilitation Science
  University of Kansas Medical Center

Mary F. Barbe, PhD
  Physical Therapy Department
  College of Health Professions
  and Department of Anatomy and Cell Biology
  School of Medicine, Temple University

Ann E. Barr, PT, PhD
  Physical Therapy Department
  College of Health Professions
  and Department of Anatomy and Cell Biology
  School of Medicine, Temple University

Susan V. Brooks, PhD
  Department of Molecular and Integrative Physiology
  and Institute of Gerontology University of Michigan

Priscilla M. Clarkson, PhD
  Department of Exercise Science
  University of Massachusetts

Tommy G. Gainer, MS
  Department of Human Nutrition, Foods, and Exercise
  Virginia Polytechnic Institute and State University

Allan H. Goldfarb, PhD
  Exercise and Sport Science Department
  University of North Carolina Greensboro

Robert W. Grange, PhD
  Department of Human Nutrition, Foods, and Exercise
  Virginia Polytechnic Institute and State University

Dawn T. Gulick, PhD, PT
  Widener University Institute for Physical Therapy,
  Chester, Pennsylvania
  Coowner of AquaSport Physical Therapy, PC

Thomas J. Hawke, PhD
  School of Kinesiology and Health Science
  Faculty of Science and Engineering
  York University, Toronto

Monica J. Hubal, PhD
  Department of Exercise Science
  University of Massachusetts

Leesa K. Huguenin, MBBS, MS
  Private medical practice
  Dromana, Victoria, Australia

Adam Johnston, MSc
  School of Kinesiology and Health Science
  Faculty of Science and Engineering
  York University, Toronto

Priit Kaasik, PhD
  Functional Morphology Laboratory
  University of Tartu, Tartu, Estonia

Timothy J. Koh, PhD
  Department of Movement Sciences
  University of Illinois at Chicago

Richard M. Lovering, PhD, PT
  Department of Physiology
  School of Medicine, University of Maryland

Douglas J. Mahoney, PhD
  Apoptosis Research Center
  Children's Hospital of Eastern Ontario
  Research Institute, Ottawa

Ken Nosaka, PhD
  School of Exercise, Biomedical, and Health Sciences
  Edith Cowan University, Perth, Australia

Leigh E. Palubinskas, MPT
  Division of Physical Therapy
  Georgia State University

Ando Pehme, PhD
  Functional Morphology Laboratory
  University of Tartu, Tartu, Estonia

Michael Pierrynowski, PhD
  School of Rehabilitation Science
  McMaster University, Hamilton, Ontario

Francis X. Pizza, PhD
  Department of Kinesiology
  University of Toledo

Stephen P. Sayers, PhD
  Department of Physical Therapy
  University of Missouri-Columbia

Teet Seene, PhD
  Functional Morphology Laboratory
  University of Tartu, Tartu Estonia

Karin Shortreed, MSc
  School of Kinesiology and Health Science
  Faculty of Science and Engineering
  York University, Toronto

Lisa Stenho-Bittel, PhD, PT
  Department of Physical Therapy
  and Rehabilitation Science
  University of Kansas Medical Center

Mark A. Tarnopolsky, MD, PhD
  Department of Pediatrics and Medicine
  McMaster University, Hamilton, Ontario

Maria Umnova, PhD
  Functional Morphology Laboratory
  University of Tartu, Tartu, Estonia

Qiong Wang, MS
  Department of Human Nutrition, Foods, and Exercise
  Virginia Polytechnic Institute and State University

Christopher W. Ward, PhD
  School of Nursing
  University of Maryland at Baltimore

Gordon L. Warren, PhD
  Division of Physical Therapy
  Georgia State University

S. Janette Williams
  Department of Physical Therapy
  and Rehabilitation Science
  University of Kansas Medical Center

PREFACE

Understanding the physiological mechanisms of skeletal muscle damage and repair is important for many health professionals, therapists, kinesiologists, physical educators, and researchers. It is also important to understand how basic mechanisms of skeletal muscle damage and repair pertain to specific populations and how they may be affected by various potentially therapeutic mechanisms. This edited book covers a wide range of both current basic research and applied clinical topics related to skeletal muscle damage and repair mechanisms and their application. In so doing it highlights the inter-relationships among these areas and how knowledge of the physiology of muscle damage and repair can affect our understanding of therapeutic modalities and clinical populations and vice versa. The book is divided into three sections comprising related chapters, each written by an internationally acclaimed researcher or research group; all the authors are experts in their field.

The chapters in part I review current research and cover our present understanding of issues related to the mechanisms of muscle damage, physiological responses to damage, and subsequent muscle repair mechanisms. They also deal with experimental models for muscle damage research and new trends in molecular biological research related to muscle damage and repair. Although the main focus of this book is on exercise- and overuse-induced muscle damage, the basic physiology of muscle damage and repair is generally applicable to muscle damage and subsequent repair, whatever the causation.

The chapters in part II examine muscle damage and repair mechanisms and issues in specific populations including older adults, persons with diabetes, and ath-letic populations. They also cover other applied topics related to muscle damage and repair, such as gender and hormonal influences, effects of muscle damage on gait mechanics, muscle damage and repair in workplace settings, and the issue of "high responders"—individuals who seem extraordinarily susceptible to muscle damage. Many of the specific muscle damage and repair issues discussed in this part of the book can also be more generally applied across various populations.

The authors of the chapters in part III take a critical look at various specific interventions that have been used or advocated for the treatment of muscle damage. They cover our current understanding of the potential physiological mechanisms of these interventions and the evidence, or lack thereof, for their effectiveness. Topics discussed in this section include massage, ultrasound, trigger point therapy, physical therapy, dietary antioxidant interventions, and hyperbaric oxygen treatment.

This book will be helpful to those interested in both basic physiological and applied clinical factors in skeletal muscle damage and repair, including health professionals and clinicians, kinesiologists, physiotherapists, and researchers, as well as graduate and undergraduate students. The text exposes readers to a range of basic and applied issues related to muscle damage and encourages them to consider questions pertaining to muscle damage and repair that are outside their typical range of interest. If this leads to greater cross-fertilization of ideas and communication between those who do research on the basic mechanisms of muscle damage and repair and those who treat muscle injury in various populations, the book will be a success.

ACKNOWLEDGMENTS

I need first to thank all the wonderful scientists who agreed to contribute chapters to this book. It is they and others like them whose curiosity and dedication have advanced our knowledge of muscle physiology to where it is today. I would also like to thank those scientists and teachers who served as mentors and inspiration for me in my career: David Ianuzzo, who taught me scientific rigor and who as my master's supervisor at York University introduced me to the study of muscle damage; Roy Shephard, my boss at the University of Toronto, who showed me what producing a book was all about; and the dedicated teacher and scientist, Mike Houston, my friend and doctoral supervisor at the University of Waterloo, who had confidence in me and allowed me the freedom to succeed. Lastly I want to thank my wonderful colleagues and students at the Department of Kinesiology and Physical Education at Wilfrid Laurier University, who make it a pleasure to come to work each day.

PART I

Physiology of Muscle Damage and Repair

CHAPTER 1

Physiology and Mechanisms of Skeletal Muscle Damage

Timothy J. Koh, PhD

Skeletal muscle is a remarkable tissue. Because of its ability to shorten and produce force, we are able to breathe, walk, and perform all the tasks and activities required in our daily lives. When a skeletal muscle is injured through trauma or unaccustomed exercise, we feel pain and lose some of our ability to perform these important daily tasks. All of us are familiar with the soreness and loss of muscle function that result from hitting our shin on the corner of the coffee table or our first workout at the gym after a long absence. The scientific data indicate that these muscle injuries are initiated by mechanical factors, and that the initial injury may be exacerbated by the ensuing inflammatory process, a loss of calcium homeostasis, or both. Understanding the mechanisms of muscle injury may lead us to design better strategies for reducing the severity of muscle injuries, as well as better approaches for speeding their healing. The purpose of this chapter is to convey current scientific knowledge about the initial phases of muscle injury. Later stages of injury, as well as repair, are the focus of subsequent chapters.

Foundations

As we begin our discussion about the physiology and mechanisms of skeletal muscle damage, we will first provide a definition of muscle injury, and then describe the typical time course of changes that occur in muscle following injury.

Definition of Muscle Injury

In any discussion of muscle injury we must first know what is meant by "injury." Webster's dictionary defines "injury" as "hurt, damage or loss sustained." Similarly, "to injure" means "to inflict material damage or loss." Many different markers are used to characterize injury, including loss of muscle function, altered morphology at the light and electron microscope levels, altered intracellular protein levels and localization, and loss of intracellular muscle proteins to the surrounding environment. Measuring changes in each of these markers sheds light on alterations that occur in muscle following an event that we deem "injurious." However, few authors have explicitly defined what they mean by the term "injury."

Some have argued that measurements of muscle function are the most useful way of quantifying injury (Brooks et al. 1995; Warren et al. 1999). Thus, we could define muscle injury simply as loss of function following a specific injurious event. However, function can be impaired in situations not normally associated with injury (e.g., fatigue and atrophy), and thus loss of function alone may not be adequate to define injury. Others (Morgan and Allen 1999) have defined contraction-induced injury as all the changes that take place within muscle after a bout of eccentric contractions (those in which the muscle is stretched while activated). Although this definition would certainly include everything we might associate with exercise-induced injury, it lacks specificity.

For this chapter, muscle injury is defined as the loss of muscle function caused by the physical disruption of muscle structures involved in producing or transmitting force. This definition is based on the assumption that loss of function in muscle injury is due to physical disruption of muscle structures—an assumption that the remainder of this chapter provides support for. We will refer to the event that causes the injury as an injurious event, and will refer to the injury itself as a process triggered by the injurious event.

Time Course of Muscle Injury

A number of studies have characterized the changes that occur in muscle following exercise-induced injury, particularly exercise that includes high-force eccentric contractions. Fewer studies have characterized the changes that occur after traumatic muscle injury. However, following the initial damaging event, what is known about the subsequent processes of injury and repair appear to be similar in trauma- and exercise-induced injury. The rest of this chapter focuses on exercise-induced injury.

High-force eccentric contractions injure skeletal muscle that is unaccustomed to such exercise. Early changes observed following eccentric contractions include disruption of sarcomeres, disruption of cytoskeletal elements involved in force transmission, damage to the muscle cell membrane, impaired excitation–contraction coupling, and loss of force production (figure 1.1 and table 1.1). Although it is generally accepted that these events occur within minutes after the initiation of unaccustomed eccentric contractions, the precise time course and sequence of events remain to be determined. Following the initial disruption to muscle cells, resting intracellular calcium levels increase, inflammatory cells accumulate, and myofibrillar and other protein is degraded. The latter processes are sometimes associated with a secondary loss

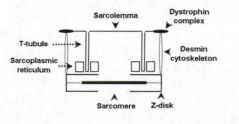

Figure 1.1 Schematic of structures involved in initiation of contraction-induced muscle injury. Within minutes of the initiation of eccentric contractions, disruption has been observed in sarcomeres, Z-discs, other cytoskeletal elements involved in force transmission (the desmin cytoskeleton and the dystrophin complex), structures involved in excitation–contraction coupling (T-tubules and sarcoplasmic reticulum), and the sarcolemma.

in muscle force production (Faulkner et al. 1989). During a subsequent repair period that lasts from days to weeks, the injured muscle is returned to a state not detectably different from the state of a muscle that was never injured. Histological appearance, total protein content, and muscle function then typically return to normal.

An intriguing phenomenon can be observed in skeletal muscle that is subjected to contraction-induced injury and then allowed to recover. A muscle that has been "conditioned" by prior eccentric contractions demonstrates a remarkably reduced susceptibility to injury following later bouts of eccentric contractions. The mechanisms underlying this "repeated-bout effect" are an exciting area of current research, as well as the focus of another chapter in this book. At this point, it is not clear whether the protective effect of muscle conditioning results from protection from the initial mechanical injury, more rapid recovery from this initial injury, or protection from the progression of the injury. Elucidation of the protective mechanisms involved in the repeated-bout effect may shed light on the events of eccentric contraction–induced injury that are critical for the initiation or progression of the injury, as well as the most efficacious time for potential interventions to protect muscle from injury.

Fiber Type and Injury

A number of investigators have reported that fast glycolytic muscle fibers are more susceptible to eccentric contraction–induced injury than slow oxidative muscle fibers. This has been demonstrated in human as well as animal experiments and even in isolated muscle fibers. A research group from Sweden (Friden et al. 1983) had subjects perform an eccentric bicycle exercise and took biopsies from the knee extensor muscles at 18 to 72 h postexercise. Using electron microscopy, they found signs of injury predominantly in fast glycolytic muscle fibers of the quadriceps muscle group. In a similar study, another group of researchers had subjects perform eccentric exercise with the elbow flexors or foot plantarflexors, and found again that fast glycolytic fibers were preferentially damaged in muscle cross sections viewed with a light microscope (Jones et al. 1986). Experiments utilizing sophisticated equipment to ensure that muscles are exposed to the same eccentric contraction conditions have produced results supporting the idea that fast muscle fibers are more susceptible to injury than slow muscle fibers in mice and rabbits. This phenomenon was even observed in single-fiber segments of rat extensor digitorum longus (EDL) muscle and soleus muscle: When exposed to the same eccentric contraction conditions, single fibers from the predominantly fast EDL were more susceptible than fibers from the predominantly slow soleus muscle.

Table 1.1   Summary of Early Changes in Eccentric Contraction–Induced Muscle Injury

Type of disruption	References
Sarcomere disruption	Newham et al. 1983; Friden et al. 1983; Brown and Hill 1991
Cytoskeletal disruption	Lieber et al. 1996; Komulainen et al. 1998; Lovering and De Deyne 2004; Koh and Escobedo 2004
Membrane disruption	McNeil and Khakee 1992; Hamer et al. 2002; Lovering and De Deyne 2004
Loss of calcium homeostasis	Duan et al. 1990; Balnave and Allen 1995; Ingalls et al. 1998
Excitation–contraction coupling impairment	Warren et al. 1993; Ingalls et al. 1998; Takekura et al. 2001; Yeung et al. 2002
Loss of force production	McCully and Faulkner 1985; Faulkner et al. 1989; Warren et al. 1993; Brooks and Faulkner 1995

One hypothesis that has been developed to explain the differential susceptibility of fast and slow muscle fibers to eccentric contraction–induced injury relates to the metabolic differences between these fiber types (Patel et al. 1998). According to this hypothesis, the low oxidative capacity of fast glycolytic fibers may predispose them to injury during repetitive eccentric contractions through depletion of high-energy phosphates and subsequent formation of actin-myosin cross-bridges in the rigor state. The breaking of these rigor cross-bridges through mechanical loading induced by eccentric contractions would then lead to muscle injury. To test their hypothesis, the authors trained rabbit dorsiflexor muscles, which are composed of predominantly fast glycolytic fibers, with isometric electrical stimulation designed to enhance oxidative capacity. Despite increasing the oxidative capacity of the muscle fibers, the training program did not protect these muscles from eccentric contraction–induced injury. Thus, other factors must account for the differing susceptibility to injury between slow and fast muscle.

In addition to metabolic differences, it appears that many other differences distinguish fast and slow muscle fibers. Slow muscle fibers appear to contain higher levels of certain cytoskeletal proteins that may provide structural support for the sarcomeres and the cell membrane and may help to maintain the integrity of these structures in the face of repeated mechanical loading. Thus, the higher levels of these proteins may protect slow muscle fibers from eccentric contraction–induced injury. In addition, the levels of other protective molecules, including a family of "stress proteins," are higher in slow than in fast muscle fibers. These stress proteins, also known as "heat shock proteins," protect different types of cells from a variety of stresses, and may help to protect muscle cells from mechanical stress such as that experienced during eccentric contractions (Koh 2002). More information on the relationship between fiber type

and injury would enhance our understanding of factors influencing susceptibility to contraction-induced injury and may provide insight into how these factors could be altered to make muscle less susceptible to injury.

Mechanisms of Muscle Injury

The injurious event in contraction-induced injury appears to be mechanical in nature. A number of researchers have probed the specific mechanical factors involved in contraction-induced injury.

Contraction Type

The following studies compared the susceptibility of muscle to injury through isometric, concentric, and eccentric contractions.

• A research group at the University of Michigan (McCully and Faulkner 1985) developed a model of exercise-induced injury that involved an anesthetized mouse, an electrical stimulator to control muscle activation, and a servomotor to control the length of the EDL. With use of this system, the EDL could be exposed to bouts of isometric, concentric, or eccentric contractions for which the number of contractions, level of muscle activation, muscle length, and muscle velocity could be controlled. The investigators found that a bout of isometric or concentric contractions did not produce a deficit in maximum isometric muscle force or overt histological damage three days post-exercise. However, they found that a bout of eccentric contractions produced force loss and histological damage three days after the exercise bout, even when the peak force of eccentric contractions was matched to that of isometric contractions. These findings are consistent with data from a number of other studies, in both rodent and human muscle, using controlled eccentric contractions or using exercise requiring repetitive eccentric contractions (e.g., downhill running).

• Interestingly, a recent study (Pizza et al. 2002) demonstrated that a bout of isometric contractions, while not inducing overt fiber damage or a force deficit, can induce an inflammatory response, albeit of lesser magnitude than that induced by eccentric contractions. Isometric contractions may cause minor damage that results in the release of chemoattractants for inflammatory cells, but is not severe enough to induce overt structural damage or loss of function. In short, the scientific data indicate that eccentric contractions are particularly injurious for skeletal muscle.

Mechanical Factors in Muscle Injury

Other investigators have examined the precise mechanical factors that are important in producing injury during eccentric contractions.

• In a systematic investigation of the mechanical factors associated with the loss in muscle force production (Warren, Hayes et al. 1993), soleus muscles isolated from rats were exposed to five eccentric contractions during incubation in a tissue bath. The researchers varied mechanical factors such as muscle force, muscle lengthening, and lengthening velocity during the eccentric contractions and measured the force deficit produced by these eccentric contractions. Loss of force production immediately after the eccentric contractions was most strongly related to peak force during the first eccentric contraction; the larger the peak force, the larger the functional impairment.

• A similar study was performed with EDL muscles from mice exposed to a single in situ eccentric contraction with different amounts of muscle lengthening (Brooks et al. 1995). These investigators found that the work performed on the muscle was the best predictor of the loss of force immediately following the eccentric contraction; the more work done on the muscle, the larger the force loss. In addition, both the amount of lengthening and the average force during lengthening were strongly associated with the loss of force.

• Other investigators have found that muscle length is a key factor in the injury produced by eccentric contractions: The longer the initial muscle length and the greater the amount of stretch during the contraction, the greater the injury (Talbot and Morgan 1998).

Thus, the scientific data indicate that muscle tension and muscle length are key mechanical factors in determining the magnitude of contraction-induced injury.

Nature of Exercise–Induced Muscle Injury

While much remains to be discovered about the nature of muscle damage caused by exercise, there are a number of areas in which a good deal is now understood about this process. These include disruption to sarcomeres, other cytoskeletal elements, and the cell membrane; impairments in the excitation–contraction coupling process; loss of calcium homeostasis; and free radical damage.

Disruption of Muscle Structures

Damage to muscle structures that occurs early during the induction of contraction-induced muscle injury provides insight into the causes of injury. These early events include disruption to sarcomeres, other cytoskeletal elements, and the cell membrane.

Sarcomeric Disruption

Immediately after a bout of eccentric contractions, overt damage in muscle cross sections is generally difficult to detect at the light microscope level. Alterations at this level tend to be limited to a few swollen fibers and perhaps some edema. At two to three days following the exercise, overt damage is obvious with the accumulation of inflammatory cells, disrupted sarcoplasm, and numerous swollen fibers. In longitudinal sections, focal areas of damage have been observed, with areas of overstretched sarcomeres apparent.

When observed under the electron microscope, focal myofibrillar damage is evident in muscle immediately following a bout of eccentric contractions (figure 1.2), with such damage increasing over time for up to three days following the injury. Immediately following eccentric contractions of human quadriceps muscles, single disrupted sarcomeres, or even disrupted half-sarcomeres, were observed that were surrounded by normal-appearing sarcomeres (Newham et al. 1983). Disruption of Z-line architecture was also present, with "streaming" or widening of Z-lines apparent; and sometimes these structures appeared to be completely absent. In a separate study, damage following eccentric contractions included disruption of sarcomeres, Z-disc streaming, and even complete Z-disc disruption that seemed to peak three days after exercise and then to decrease with time (Friden et al. 1983). Other investigators have confirmed that early focal damage to sarcomeric structures is a hallmark of eccentric contraction–induced injury.

The underlying cause of sarcomeric disruption immediately following eccentric contractions has been under scrutiny for decades. Particularly revealing experiments utilized single frog muscle fibers exposed to a single eccentric contraction. The contraction was followed by rapid chemical

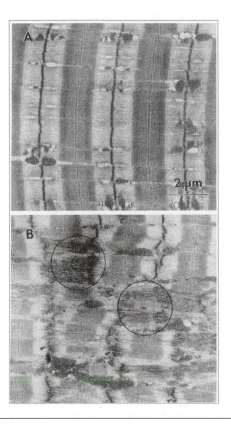

Figure 1.2 Sarcomere disruption following eccentric contractions. *(a)* Sarcomeres from normal muscle show excellent alignment and regular banding patterns; *(b)* sarcomeres from muscle exposed to eccentric contractions show regions of Z-disc streaming and frank sarcomere disruption next to sarcomeres that appear normal.

Reprinted, by permission, from R.L. Lieber, T.M. Woodburn and J. Friden, 1991, "Muscle damage induced by eccentric contractions of 25% strain," *J Appl Physiol.* 70(6): 2498-2507.

fixation of the fiber to preserve the morphology of the sarcomeres immediately after the contraction (Brown and Hill 1991). This process minimized any potential changes that may occur between the injurious event (i.e., the eccentric contraction) and observation under the microscope, and allowed the investigators to more directly associate morphological changes in the muscle fiber with the mechanical loading during the contraction. When these muscle fibers were viewed under the microscope, the investigators observed sarcomeres in which the thin filaments were pulled out of their normal position interdigitating between the thick filaments, either on one side or on both sides of the sarcomere. These data provided direct evidence that sarcomere disruption could be induced by a single eccentric contraction.

Similar experiments were performed with whole toad muscle; in this later study (Talbot and Morgan 1998), eccentric contractions on the descending limb of the force–length relationship produced more disrupted sarcomeres than either

eccentric contractions on the ascending limb or isometric contractions on the descending limb. These data supported the notion that muscle length plays an important role in injury produced by eccentric contractions; the longer the muscle length, the greater the damage.

Experiments further supporting the link between muscle length and injury have used single permeabilized rat muscle fiber segments exposed to eccentric contractions (Macpherson et al. 1997). In these experiments, regions of the fiber with the longest sarcomeres showed the greatest amount of sarcomere disruption.

Taken together, the in vivo and in vitro studies indicate that focal sarcomeric disruption may be caused by inhomogeneities in the strengths of neighboring sarcomeres, with strong sarcomeres pulling weaker sarcomeres apart. When sarcomeres are on the descending limb of the force–length relationship, any further lengthening of weak sarcomeres moves them farther down the descending limb and makes them even weaker. This exacerbates the strength discrepancy between short and long sarcomeres, making eccentric contractions performed at long muscle lengths particularly damaging to muscle fibers (Morgan 1990).

Disruption of Other Cytoskeletal Elements

In addition to overt disruption of sarcomeres, recent studies have demonstrated that cytoskeletal elements involved in transmitting force from the sarcomere through the membrane are disrupted early following eccentric contractions (figure 1.3). These cytoskeletal elements include those that may be important in maintaining Z-disc structure, sarcomeric organization, and cell membrane integrity.

Two approaches have been used to examine cytoskeletal disruption following eccentric contractions: immunostaining of muscle cross sections, which provides information about the localization of specific proteins; and western blotting of muscle homogenates, which provides information about the relative quantity of these proteins in the entire muscle.

Desmin is part of a structural scaffolding that is thought to play a role in maintaining the structure of Z-discs as well as helping to maintain the proper alignment of sarcomeres within and between myofibrils. Immunostaining of muscle cross sections revealed that a number of muscle fibers lost staining for desmin within minutes after the initiation of eccentric contractions in rabbit dorsiflexor muscles, and the number of fibers that lacked desmin staining increased with time following the exercise bout (Lieber et al. 1996). Some of the fibers that lost desmin staining demonstrated accumulation of plasma fibronectin, indicating a loss of membrane integrity in these fibers. The investigators concluded that disruption of the desmin cytoskeleton is an early event in eccentric contraction–induced injury, which leads in some fibers to loss of membrane integrity. Although this

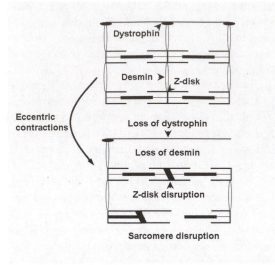

Figure 1.3 Schematic of potential sites of cytoskeletal disruption following eccentric contractions. Arrowheads indicate reported sites of disruption following eccentric contractions (dystrophin complex, desmin intermediate filaments, Z-discs, sarcomeres).

Adapted, by permission, from T.J. Koh, 2002, "Do small heat shock proteins protect skeletal muscle from injury?," *Exercise and Sport Science Reviews* 20(3): 117-121.

may be true, early loss of desmin following eccentric contractions in muscles of species other than rabbit remains to be demonstrated.

Dystrophin is a large cytoskeletal protein associated with the cell membrane and is thought to help maintain the integrity of the membrane during the repeated mechanical loading that muscle cells experience every day through their own contractions. Loss of the dystrophin gene in Duchenne muscular dystrophy leads to a severe progressive muscle degeneration that is thought to result from the loss of this membrane-stabilizing function. Using immunohistochemical procedures, Komulainen and colleagues (1998) investigated loss of dystrophin in cross sections of rat dorsiflexor muscles that had been exposed to eccentric contractions. Immediately after the eccentric contractions, dystrophin staining at the cell membrane of certain muscle fibers became discontinuous, in contrast to the continuous peripheral staining present in normal fibers. Six hours after exercise, dystrophin staining was completely missing in some fibers, and this was accompanied by loss of desmin staining and disorganized actin staining. The number of affected fibers increased for up to two days after exercise, again indicating a progression of damage over time.

Researchers in two other laboratories also found loss of dystrophin staining immediately after eccentric contractions in muscle cross sections from mice and rats (Koh and Escobedo 2004; Lovering and De Deyne 2004). Both

groups found that loss of dystrophin was detectable in western blots of muscle homogenates, indicating that the loss was substantial. In one of these studies, loss of dystrophin was associated with loss of another membrane-associated protein, beta-spectrin, which is thought to play a membrane stabilization role similar to that of dystrophin. In human muscle exposed to eccentric contractions, western blots have been used to demonstrate loss of alpha-sarcoglycan immediately following eccentric contractions (Feasson et al. 2002). Alpha-sarcoglycan is a member of the complex of proteins associated with dystrophin, and genetic mutations in this protein also result in a form of muscular dystrophy called limb-girdle muscular dystrophy. Thus, loss of members of the membrane-associated dystrophin complex of proteins appears to be an early event in eccentric contraction–induced injury. The rapid loss of these proteins is striking, and loss of their membrane-stabilizing function may render the cell membrane more susceptible to damage by further contractions.

Evidence for Membrane Disruption

Early studies of exercise-induced muscle injury used an increase in plasma levels of intracellular muscle proteins (e.g., creatine kinase) as a marker of injury. Increased plasma levels of muscle proteins were taken to indicate disruption of the cell membranes of muscle fibers from which the intracellular proteins could leak into the blood. Loss of intracellular proteins is also commonly used to indicate cardiac muscle injury; and currently, elevated plasma levels of troponin isoforms are the gold standard for detecting heart attack. Although an elevated level of creatine kinase is sometimes still used to detect skeletal muscle injury, more sensitive and precise methods have been developed to detect membrane damage.

Recent studies have used the appearance of extracellular proteins and dyes in the intracellular space of muscle fibers as markers of membrane damage following skeletal muscle injury. The idea behind these methods is that membrane damage would allow entry of macromolecules that are normally excluded from the intracellular space by the cell membrane. With use of these methods, the specific muscle fibers that have incurred membrane damage and the time of wounding following eccentric contractions can be investigated. In an initial study of exercise-induced membrane damage, McNeil and Khakee (1992) performed experiments using rats running downhill, a form of exercise that has been associated with eccentric contraction–induced injury. These researchers localized serum albumin in muscle cross sections using immunohistochemical techniques and found that in noninjured muscle, albumin is confined to the extracellular space. Immediately following downhill running, approximately 20% of the muscle fibers in the medial head of the triceps brachii muscle stained positively for serum albumin, suggesting that albumin had leaked into these muscle fibers through membrane disruptions. The appearance of albumin within the muscle fibers was also associated with an increased plasma level of creatine kinase, both markers indicating muscle cell membrane damage immediately after downhill running.

Another research group injected Evans blue dye into rats and mice and used entry of the dye into muscle fibers following exercise as a marker of membrane disruption. This dye binds to serum albumin and thus provides essentially the same information as the study just described. With use of dye injection 24 h before exercise, dye-positive fibers could be observed within minutes following eccentric contractions, whereas muscle not exposed to exercise did not demonstrate uptake of dye (Hamer et al. 2002).

Other investigators (Lovering and De Deyne 2004) probed further to test the relationship between the loss of dystrophin at the cell membrane and the uptake of dye into muscle fibers following eccentric contractions. They found that fibers with normal staining for dystrophin did not take up dye, and that fibers containing dye showed loss of dystrophin staining (figure 1.4), suggesting that contraction-induced disruption of dystrophin at the cell membrane was associated with membrane disruption and dye uptake.

Interestingly, in mice that lack dystrophin due to a genetic mutation, muscle fibers take up Evans blue dye with a much greater frequency than muscle fibers from normal mice, both in nonexercised muscle and in muscle subjected to exercise. Together, these data support a role for dystrophin in maintaining cell membrane integrity.

When dystrophin is missing at the membrane as a result of either exercise or genetic defect (e.g., muscular dystrophy), membrane integrity appears to be compromised.

Loss of membrane integrity is generally perceived as a negative event, compromising the homeostasis of the muscle fiber by disrupting the barrier that helps to maintain a desirable balance of intracellular and extracellular molecules. However, small transient membrane disruptions may provide a normal pathway for the release and uptake of certain molecules, especially in tissues that are exposed to repeated mechanical stresses (McNeil and Khakee 1992). Muscle cells appear to have the ability to rapidly repair small disruptions in the cell membrane, so any negative effects of these small disruptions may be limited. One molecule whose export from muscle fibers could be regulated by transient membrane disruption is basic fibroblast growth factor (bFGF). This growth factor can act to stimulate proliferation of a variety of cells that contribute to muscle repair, as well as a chemotactic agent for these cells; thus bFGF is thought to play a critical role in muscle repair and adaptation following exercise. Hence, transient membrane wounding could be seen as a signaling event in the response of skeletal muscle to exercise, releasing growth factors and other molecules from muscle that may be important in repair and adaptation. Whether membrane disruption results in repair and adaptation or cell death may depend on the magnitude of the injury and on other factors that contribute to the processes of injury and repair.

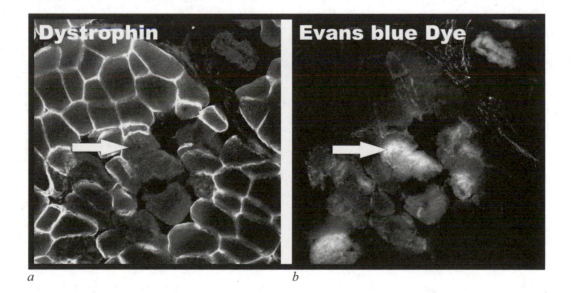

Figure 1.4 Loss of dystrophin associated with membrane disruption immediately after eccentric contractions. *(a)* Immunostaining for dystrophin. *(b)* Uptake of Evans blue dye; normal peripheral immunostaining for dystrophin was lost in certain muscle fibers after eccentric contractions (e.g., arrow in *a*). Fibers that demonstrated intact dystrophin staining did not take up Evans blue dye, and fibers that did take up the dye showed loss of dystrophin (e.g., arrow in *b*).

Adapted, by permission, from R.M. Lovering and P.G. De Deyne, 2004, "Contractile function, sarcolemma integrity, and the loss of dystrophin after skeletal muscle eccentric contraction-induced injury," *Am J Physiol Cell Physiol*. 286(2): C230-238.

Other Exercise-Induced Disruptions

Disruptions in other muscle processes also provide insight into the causes of contraction-induced injury. These other disruptions include impaired excitation–contraction coupling, loss of calcium homeostasis, and free radical–induced damage.

Disruption of Excitation–Contraction Coupling

Most studies of eccentric contraction–induced injury show a large deficit in force production immediately after the exercise bout but only scattered focal structural damage within the muscle. The small volume of tissue affected by this structural damage would not seem to account for the large force deficit observed. Over time, the structural damage becomes more evident while the force deficit tends to either remain unchanged or decrease. This disassociation between structural disruption and loss of function has led to the idea that much of the early loss of function following eccentric contractions results from an impairment of excitation–contraction coupling. Impairment in excitation–contraction coupling could occur at any point in the chain of events between depolarization of the muscle cell membrane and the release of calcium from the sarcoplasmic reticulum that is required for force production in the myofibrils. Impaired excitation–contraction coupling would reduce calcium activation of myofibrils and subsequent force production.

Investigators have used caffeine as a way to bypass sites of impaired excitation–contraction coupling to explore the role of such impairment in the force deficit following eccentric contractions. Caffeine induces calcium release in muscle fibers without the need for electrical activation, and thus can be used to test whether the failure in force development occurs prior to or following the calcium release step in muscle contraction. Initial experiments were performed on isolated mouse soleus muscles incubated in a tissue bath (Warren, Lowe et al. 1993). These experiments demonstrated that eccentric contractions caused a greater force loss than isometric contractions, as previously noted. In contrast, caffeine-elicited muscle forces were not different between muscles exposed to eccentric versus isometric contractions; this suggests that the force impairment immediately following eccentric contractions is not due to disruption of myofibrillar force-generating structures, but due primarily to a disrupted excitation–contraction coupling apparatus. Further experiments by the same investigators and by other laboratories confirmed that calcium release during electrical activation was indeed impaired both in single muscle fibers and in whole muscles (Balnave and Allen 1995; Ingalls et al. 1998). In addition, eccentric contractions have been found to alter the structure of T-tubules, and such structural changes may be responsible for the impairment in excitation–contraction coupling (Takekura et al. 2001; Yeung et al.

2002). In summary, much of the force deficit in the early hours of eccentric contraction–induced injury appears to result from impaired excitation–contraction coupling.

A question that arises at this point is how eccentric contractions may result in disruption of the excitation–contraction coupling apparatus. Remember that as a result of eccentric contractions, it appears that strong sarcomeres pull weaker sarcomeres apart. In these areas of sarcomeric disruption, the excitation–contraction coupling apparatus may be subjected to damaging loading conditions. Indeed, experiments designed to test this idea (Ingalls et al. 1998) indicate that the disruption in excitation–contraction coupling lies in the junctions between the T-tubules and sarcoplasmic reticulum. This junction would seem to be susceptible to damage through the progressive development of length discrepancies between neighboring sarcomeres, and would thus be a likely site for eccentric contraction–induced excitation–contraction coupling failure.

Loss of Calcium Homeostasis

Free calcium concentrations outside of muscle cells are typically in the millimolar range, while intracellular free calcium levels are in the submicromolar range. Disruptions of the cell membrane would allow calcium to move down its concentration gradient, with the potential for greatly increasing calcium concentrations in the cell. Likewise, calcium concentrations are in the millimolar range within the sarcoplasmic reticulum, and disruption of the membrane boundary of the sarcoplasmic reticulum would result in an influx of calcium into the cell. A number of studies have demonstrated that calcium concentrations are increased in skeletal muscle cells following eccentric contractions. Mitochondrial calcium concentration was reported to be significantly increased following downhill running (Duan et al. 1990). Since mitochondria are known to take up calcium when intracellular levels are increased, this observation indicated that cytosolic calcium was increased in these muscles. Using calcium-sensitive dyes, other investigators reported that the resting level of intracellular calcium was increased immediately following controlled eccentric contractions but that the calcium levels attained during tetanic contractions were decreased (Balnave and Allen 1995; Ingalls et al. 1998). These results are consistent with the ideas that damage to the sarcolemma or to the sarcoplasmic reticulum leads to increased cytosolic calcium at rest and to impaired calcium release upon electrical activation.

Increased calcium has the potential for activating many different molecular pathways in skeletal muscle, including the phospholipase—prostaglandin pathway and the calpain proteolytic pathway. Activated phospholipase A_2 may play a role in muscle cell damage by contributing to breakdown of the cell membrane and subsequent loss of intracellular molecules (Duncan and Jackson 1987). Transient membrane disruption caused by eccentric contractions could allow calcium influx and activation of phospholipase A_2, which in turn could lead to more severe membrane damage. Calcium could also activate the calpains, which are proteolytic enzymes thought to be responsible for initiating the breakdown of myofibrils. It is known that calpains cleave many of the cytoskeletal elements reported to be disrupted following eccentric contractions, including desmin, dystrophin, and spectrin. Calpains may specifically degrade Z-discs in skeletal muscle and may thus be responsible for the Z-disc disruption following eccentric contractions. In summary, increased levels of intracellular calcium could trigger many of the degradative events observed in skeletal muscle fibers following eccentric contractions, both immediately following an injurious event and during the progression of the injury.

To support the role of calcium in contraction-induced injury, experiments have utilized calcium ionophore, an agent that increases intracellular calcium by transporting it across the cell membrane or across the membrane of the sarcoplasmic reticulum.

In nonexercised muscle treated with the ionophore, Z-discs are disrupted, sarcomeres are misaligned, and myofibrils show signs of degeneration. The similarity in the structural damage between muscles treated with ionophore and muscles exposed to eccentric contractions indicates that elevated calcium is sufficient to produce much of the damage observed following eccentric contractions. Although elevated calcium is sufficient to produce such damage, whether elevated calcium actually plays a key role in early events of contraction-induced injury remains to be determined.

Free Radical Damage

Free radicals are produced in noninjured skeletal muscle predominantly as a by-product of aerobic metabolism, during the reduction of molecular oxygen to water in the mitochondria. Following injury, the accumulation of inflammatory cells is accompanied by increased production of free radicals and increased oxidative modifications of lipids and proteins. Repeated

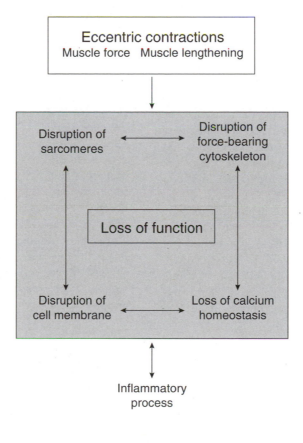

Figure 1.5 Schematic of skeletal muscle injury following eccentric contractions. Early responses include disruption of sarcomeres, force-bearing cytoskeletal elements, and the cell membrane. Loss of calcium homeostasis may contribute both to the initial injury and to the progression of the injury. The inflammatory process may also exacerbate the initial injury. Both the initial injury and later events contributing to its progression may cause impaired muscle force production.

isometric contractions have been reported to increase the production of superoxide radicals in skeletal muscle (McArdle et al. 2001).

If free radical damage is an important initiating event in eccentric contraction–induced injury, one would expect increased free radical production and increased free radical damage immediately following eccentric contractions compared with isometric contractions. However, there is little to no evidence that eccentric contractions result in increased free radical production over and above levels produced by isometric contractions. In fact, it is generally agreed that less oxygen is consumed during eccentric contractions than during isometric contractions, reducing the potential for free radical production from mitochondria. Experiments designed to detect free radical damage immediately after eccentric contractions have generally shown no elevations compared to values in nonexercised muscle, suggesting that free radical production is not an important early event in eccentric contraction–induced injury.

The aging of living systems is generally associated with oxidative stress. All of the data just discussed were obtained from muscle in animals that would be considered young or adult. Experiments on eccentric contraction–induced injury in young, adult, and old mice (Zerba et al. 1990) were performed to determine whether the contribution of oxygen free radicals to muscle injury varied with age. The investigators used supplementary injections of superoxide dismutase in an attempt to protect the muscle from contraction-induced injury; superoxide dismutase is an antioxidant enzyme that reduces superoxide radicals. If superoxide production were an important component of contraction-induced injury, superoxide dismutase would be expected to reduce the magnitude of the injury. Interestingly, superoxide dismutase reduced the force deficit immediately following the eccentric contractions only in old mice. These data suggest that superoxide radical or one of its by-products contributes to the immediate injury in old mice but not young or adult mice. In mice of each age group, the exogenous superoxide dismutase reduced the force deficit

observed at three days following exercise, indicating that superoxide radicals contribute to the progression of the injury with time. The source of the radicals at the later time point was not investigated, but these radicals are likely produced by the inflammatory cells that accumulate in the damaged muscle following eccentric contractions. In summary, free radicals may contribute to the initial contraction-induced injury in muscle of old animals, but not young or adult animals. Free radicals may also contribute to the progression of damage following injury.

Summary

When skeletal muscle is injured through unaccustomed exercise, mechanical factors are important in determining the extent of damage (figure 1.5). Eccentric contractions can induce severe injury compared with isometric contractions, particularly in fast glycolytic muscle fibers, and especially when the contractions are performed by muscles not previously conditioned with eccentric contractions. Early events in eccentric contraction–induced injury include disruption of sarcomeres, damage to other cytoskeletal elements involved in transmitting force from the sarcomeres through the cell membrane, loss of cell membrane integrity, and impaired excitation–contraction coupling. Although largely ignored in the literature, disruptions in the connective tissue may also play an important role in the pathogenesis of muscle injury (Fritz and Stauber 1988). Loss of calcium homeostasis results from damage to the cell membrane or to the sarcoplasmic reticulum and may contribute to early events in contraction-induced injury as well as to the progression of injury. Free radicals may play a role in the initiation of contraction-induced injury in old individuals and during the inflammatory process. Better understanding of the mechanisms involved in the early stages of skeletal muscle injury should lead us to improving strategies for protecting muscle from injury, as well as approaches for speeding their healing.

CHAPTER 2

Human and Animal Experimental Muscle Injury Models

Gordon L. Warren, PhD, and Leigh E. Palubinskas, MPT

To best understand the causes, mechanisms, and prevention of skeletal muscle injury, as well as the means by which an injured muscle recovers, one needs an experimental model of muscle injury. Because of the numerous human and animal models available, the selection of an appropriate model may be a daunting task to the novice researcher or to someone new in the field. The objective of this chapter is to systematically review the experimental models of injury in use over the last three decades, a period coinciding with a heightened interest in muscle injury. We describe the various models, discuss their strengths and weaknesses, and provide insight into the model or models preferable for a given set of conditions or biological question.

Our discussion of the models is not all encompassing but rather includes models described in research publications within the authors' personal files and those resulting from a PubMed search conducted on January 31, 2005. The following terms were used in that search: skeletal muscle AND (injur* [ti] OR damag* [ti]) AND (eccentric OR pliometric OR "lengthening contraction" OR down* OR trauma* OR freeze OR crush OR blunt OR contusion OR penetrat* OR toxin OR ischemia OR reperfusion) NOT ("nerve injury" OR "neuronal damage"). After exclusion of review articles and articles not written in English, this search yielded 984 articles. These were screened further to exclude articles in which a marker of muscle injury was not assessed, the injury was accidental, or the injury was secondary to a disease or impairment (e.g., muscular dystrophy, disuse atrophy, denervation, transplantation). Additionally, we restricted our analysis to models that had been used at least twice in the research literature. In the end, a total of 252 research articles were analyzed in detail, of which 91 and 161 utilized human and animal models, respectively.

Human Muscle Injury Models

Human muscle injury models are limited in scope primarily because of ethical considerations. For this reason, experimental models have focused on inducing muscle injury through exercise, a normal activity of daily life. In general, exercise-induced injury in human subjects requires the performance of repetitive eccentric contractions that are novel to the subject. Numerous studies have unequivocally demonstrated that this type of exercise can produce injury to the involved muscles (e.g., Komi and Viitasalo 1977; Clarkson et al. 1992; Gibala et al. 1995; Michaut et al. 2001; Nosaka et al. 2001). It is reasonable, then, that most researchers have chosen to induce muscle injury using models employing a bout of eccentric contractions or exercise biased to the performance of eccentric contractions. Because the "eccentric injury" models have been almost exclusively used in human research, they are the only ones considered in this part of the review.

Our discussion of the eccentric injury models is subdivided into two groups: (1) injury resulting from an acute bout of eccentric contractions and (2) injury induced by whole-body exercise with an eccentric contraction component or bias (e.g., running downhill or downward stepping exercises). Within the two categories, we describe the commonly used models and protocols for inducing injury. Following this is a comparison across studies of the ability of the major models to induce injury as assessed by strength loss and the peak blood level of creatine kinase (CK) activity.

Acute Bouts of Eccentric Contractions

In general, models involving acute bouts of eccentric contractions induce injury to muscle groups that function about a uniaxial joint, such as the elbow flexors, knee extensors, or ankle dorsiflexors. Movement at these joints is typically easier to control or evaluate because the joint can be easily isolated, compensatory movements can be readily controlled, and the available degrees of freedom are limited by joint biomechanics. Consider, for example, the knee joint, which has only flexion and extension movements available for control or analysis; shoulder movement would be much more difficult to control given the number of directions in which the joint can move.

This category of models can be further divided into (1) isokinetic models, in which the joint angular velocity is

kept constant throughout the eccentric contraction; and (2) isotonic models, in which the external load is kept constant during the eccentric contraction. Isokinetic injury models most often involve the use of a commercially available isokinetic dynamometer, whereas isotonic eccentric models typically use a protocol of lowering weights such as dumbbells or barbells. By far the most commonly studied muscle groups using these injury models have been the elbow flexors and knee extensors; although other muscle groups have been studied, these models are not as well developed and are not discussed in detail here.

Elbow Flexor Models

In terms of the sheer number of research studies done to date, the muscle group most commonly injured through experimental means has been the elbow flexors (table 2.1). Injury to this muscle group has been induced using an isokinetic dynamometer, controlled lowering of free weights as the elbow is extended (isotonic eccentric exercise), or

Table 2.1   Human Muscle Injury Models Using Eccentric Contractions Done by the Elbow Flexors

Study	Contraction type	Load	No. reps	Angular excursion (degrees)	Angular velocity (°/s)	Isometric strength loss (%MVC)	Peak blood CK level (IU/L)
Kilmer et al. 2001	Isokinetic	Maximal	16	90 to 180	30	–	1800
Paddon-Jones et al. 2001	Isokinetic	Maximal	24	40 to 170	30	35	–
Chen and Hsieh 2001	Isokinetic	Maximal	30	50 to 180	60	58	14,000
Paddon-Jones et al. 2000	Isokinetic	Maximal	36	?	100	50	4,750
Paddon-Jones and Abernethy 2001	Isokinetic	Maximal	36	40 to 170	30	45	4,030
Eston and Peters 1999	Isokinetic	Maximal	40	?	33	25	400
Bloomer et al. 2004	Isokinetic	Maximal	48	100 to 180	20	45	820
Evans et al. 2002	Isokinetic	Maximal	50	50 to 170	120	41	2,110
Gleeson et al. 2003	Isokinetic	Maximal	50	?	60	13	–
Michaut et al. 2001	Isokinetic	Maximal	50	?	60	30	–
Michaut et al. 2002	Isokinetic	Maximal	50	?	60	11	–
Donnelly et al. 1992	Isokinetic	Maximal	70	?	105	60	6,500
Stauber et al. 1990	Isokinetic	Maximal	70	?	120	–	–
Childs et al. 2001	Isokinetic	Submaximal	30	?	?	–	1,300
Lambert et al. 2002	Isokinetic	Submaximal	125	?	?	25	1,920
Nosaka et al. 2001	Quasi-isokinetic	Maximal	2 6 24	100 to 180	–	20 33 55	490 3,230 2,890
Nosaka and Clarkson 1997	Quasi-isokinetic	Maximal	12	50 to 170	–	20	3,110
Nosaka et al. 2002	Quasi-isokinetic	Maximal	12 24 60	90 to 180	–	42 53 76	5,550 15,280 16,230

(continued)

Table 2.1 *(continued)*

Study	Contraction type	Load	No. reps	Angular excursion (degrees)	Angular velocity (°/s)	Isometric strength loss (%MVC)	Peak blood CK level (IU/L)
Gulbin and Gaffney 2002	Quasi-isokinetic	Maximal	24	30 to 180		37	10,220
Nosaka and Clarkson 1992	Quasi-isokinetic	Maximal	24	50 to 170		48	3,100
Nosaka and Clarkson 1996a	Quasi-isokinetic	Maximal	24	50 to 170		55	11,930
Nosaka and Clarkson 1996b	Quasi-isokinetic	Maximal	24	40 to 170		27	12,770
Nosaka and Sakamoto 2001	Quasi-isokinetic	Maximal	24	50 to 130 100 to 180		34 36	2,890 10,640
Nosaka and Newton 2002a	Quasi-isokinetic	Maximal	24	100 to 180		50	6,350
Nosaka and Newton 2002b	Quasi-isokinetic	Maximal	30	50 to 180		53	7,710
Savage and Clarkson 2002	Quasi-isokinetic	Maximal	50	?		50	2,210
Sayers, Knight, and Clarkson 2003	Quasi-isokinetic	Maximal	50	?		54	–
Sayers and Clarkson 2003; Sayers, Peters et al. 2003	Quasi-isokinetic	Maximal	50	?		43	2,700
Thompson et al. 2002	Quasi-isokinetic	Maximal	50	?		–	500
Clarkson and Ebbeling 1988	Quasi-isokinetic	Maximal	70	?		50	2,700
Rinard et al. 2000	Quasi-isokinetic	Maximal	70	?		65	–
Sbriccoli et al. 2001	Quasi-isokinetic	Maximal	70	30 to 170		37	6,210
Newham et al. 1987	Quasi-isokinetic	Maximal	80	?		50	10,900
Foley et al. 1999	Isotonic	15% body mass	50	?		–	21,000
Teague and Schwane 1995	Isotonic	60% MVC	10	70 to 155		18	–
Howell et al. 1993	Isotonic	90% MVC	45	?		35	–
Chleboun et al. 1998	Isotonic	90%, 80%, 70% MVC	18	?		47	–
Howatson and Van Someren 2003	Isotonic	70% 1RM	30	?		–	800
Phillips et al. 2003	Isotonic	80% 1RM	30	?		–	800
Gibala et al. 1995	Isotonic	80% 1RM	64	?		37	–
Clarkson et al. 1986	Isotonic	100% 1RM	60	?		–	110
Paddon-Jones and Quigley 1997	Isotonic	110% 1RM	64	?		26	–
Jones et al. 1986	Isotonic	?	Until exhaustion	?		–	2,700
Rawson et al. 2001	Isotonic	?	50	?		50	4,500
Folland et al. 2001	Isotonic	?	60	40 to 160		15	1,500

No. reps: number of eccentric contractions; MVC: maximal voluntary contraction isometric strength; CK: creatine kinase; 1RM: 1-repetition maximum; 180° = elbow completely extended. Isometric strength losses and peak blood CK levels represent the mean values reported for a study; the values were extracted from figures in many studies and thus are approximate values. Some of the variation among studies in peak blood CK levels may be attributed to changes in the temperature at which the CK assay was run. In 2002, the international standard temperature was changed from 30° to 37° C (Siekmann et al. 2002). Because of the Q_{10} effect, the CK activity of a sample measured at 37° C would be ~75% greater than that measured at 30° C.

a custom-designed device mimicking a dynamometer. A dynamometer, as already mentioned, enables an eccentric movement at a constant angular velocity regardless of the applied load. Provided that the subject is maximally activating the elbow flexors, the force imposed on those muscles can be the maximum possible throughout the entire range of movement for that angular velocity. On the other hand, the external load used in isotonic eccentric exercise is based on the maximum torque that the muscle group can exert at its weakest joint angle. This load will always be submaximal during the eccentric movement at joint angles other than the weakest one.

In the past, two highly productive research groups, those of Priscilla Clarkson and Kazunori Nosaka, have induced injury using a custom-designed device that is quasi-isokinetic (figure 2.1). With this device, as for most others inducing injury to the elbow flexors, the subject is seated in front of a padded "preacher curl" bench that is used to support the axilla and upper arms. The device requires that the investigator apply the external load manually using a lever. During the eccentric movement, the researcher adjusts the external load so as to maintain a nearly constant angular velocity. This quasi-isokinetic type of eccentric exercise is as effective as the true isokinetic and isotonic eccentric models in inducing injury, as evidenced by comparison across the 45 studies in table 2.1 that measured the peak isometric

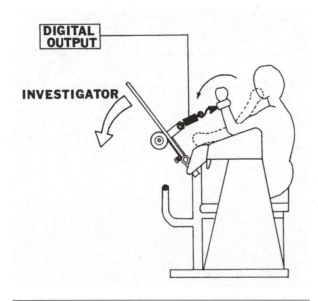

Figure 2.1 Custom-designed apparatus used for inducing injury to the elbow flexor muscles as originally described by Clarkson and colleagues (1986). A bar is attached to the lever of an arm curl machine. The investigator moves the bar down, and the subject is encouraged to resist the motion into elbow extension. Digital force output is measured by a strain gauge load cell connected to the machine.

Reprinted, by permission, from P.M. Clarkson, K. Nosaka and B. Braun, 1992, "Muscle function after exercise-induced muscle damage and rapid adaptation," *Medicine and Science in Sports and Exercise* 24(5): 512-520.

strength loss or peak blood CK level (or both) following a bout of eccentric contractions.

Most studies have used one arm for the eccentric contraction bout and the other as an unexercised control for comparison. Aside from this, there has been considerable variability in the exercise conditions among studies (e.g., load, number of repetitions, angular excursion, and angular velocity). In most studies, the angular excursion for the contractions has been described as going from full flexion to full extension. In those studies giving specific angular excursions, these have varied from a movement from 50° to 130° of extension (Nosaka and Sakamoto 2001) to a movement from 100° to 180° of extension (Nosaka and Sakamoto 2001; Nosaka et al. 2001). For the studies in table 2.1, the angular excursion used cannot explain a significant portion of the variance among studies in the injury induced, though there is a trend toward greater strength losses when the eccentric contractions ended at more extended joint angles ($r = 0.40$; $p = 0.06$). However, *within* a study, the range of angular excursion at the elbow has been shown to have an appreciable effect on the location and extent of muscle injury. Through ultrasound and magnetic resonance imaging analyses, Nosaka and Sakamoto (2001) found that eccentric contractions with an elbow angular excursion over a more extended range (from 100° to 180°) preferentially induced injury to the brachialis muscle. Over a more flexed range (from 50° to 130°), injury was induced in both the brachialis and biceps brachii muscles. Furthermore, Newham and colleagues (1988) found that greater strength losses and pain resulted when the elbow flexors performed eccentric contractions at more extended joint angles.

The number of eccentric contractions used in the elbow flexor models has been quite variable, from as few as six (Nosaka et al. 2001) to as many as 125 (Lambert et al. 2002). Though one would presume that the degree of injury incurred in a study would be influenced by the number of contractions, the data do not support this. For the studies in table 2.1, we analyzed the extent to which contraction number could explain the between-study variability in strength loss and peak blood CK level. There was no correlation whatsoever between the number of contractions used in a study and either index of injury ($r \leq 0.09$). In contrast, the number of eccentric contractions performed by the elbow flexors has been demonstrated to affect the degree of injury observed *within* a study, with greater injury associated with the performance of more contractions (Nosaka et al. 2001).

Knee Extensor Models

The second most frequently tested muscle group has been the knee extensors (table 2.2). As with the elbow flexors, one may induce injury to the knee extensors by having the subject perform a bout of eccentric contractions

Table 2.2   Human Muscle Injury Models Using Eccentric Contractions Done by the Knee Extensors

Study	Contraction type	Load	No. reps	Angular excursion (degrees)	Angular velocity (°/s)	Isometric strength loss (%MVC)	Peak blood CK level (IU/L)
Komi and Viitasalo 1977	Isokinetic	Maximal	40	?	?	40	130
Day et al. 1998	Isokinetic	Maximal	50	?	?	43	–
Serrao et al. 2003	Isokinetic	Maximal	60	110 to 40	5	52	150
Brown et al. 1999	Isokinetic	Maximal	100	?	60	24	490
Byrne et al. 2001	Isokinetic	Maximal	100	140 to 40	90	24	810
Deschenes et al. 2000	Isokinetic	Maximal	100	?	30	13	2250
Eston et al. 1996	Isokinetic	Maximal	100	?	30	–	410
Babul et al. 2003	Isokinetic	Maximal	300	110 to 35	30	–	–
Child et al. 1998	Isokinetic	Submaximal	75	160 to 80 120 to 40	90	15 40	1000 1970
Byrne and Eston 2002a	Isotonic	70% body mass	100	180 to 90		17	540
Mair et al. 1992	Isotonic	110% MVC	70	150 to 90		22	1440
Sorichter et al. 1995	Isotonic	110% MVC	70	180 to 90		–	320
Wojcik et al. 2001	Isotonic	120% MVC	100	?		24	1270
Stupka et al. 2001	Isotonic	120% MVC	136	165 to 90		–	900
Mair et al. 1995	Isotonic	150% MVC	70	?		–	280
Sorichter, Mair, Koller, Gebert et al. 1997	Isotonic	150% MVC	70	?		–	200
Sorichter, Mair, Koller, Muller et al. 1997	Isotonic	150% MVC	70	?		–	250
Sorichter et al. 2001	Isotonic	150% MVC	70	?		–	450
Kaufman et al. 2001	Isotonic	75% 1RM	100	180 to 90		15	–
Ploutz-Snyder et al. 2001	Isotonic	75% 1RM	100	?		10	–
Byrne and Eston 2002b	Isotonic	80% 1RM	100	180 to 90		30	810
Hortobagyi et al. 1998	Isotonic	80% 1RM	100	180 to 90		37	7600
Blais et al. 1999	Isotonic	110% 1RM	50	?		–	320
Chen et al. 2003	Isotonic	?	300	?		33	–

MVC: maximal voluntary contraction isometric strength; No. reps: number of eccentric contractions; CK: creatine kinase; 1RM: 1-repetition maximum; 180° = knee completely extended.

using an isokinetic dynamometer or an isotonic exercise device. The body position is similar for the two exercise modes (figure 2.2). The subject is placed in a sitting or semireclined position, and lap, chest, and thigh belts are employed to minimize unnecessary movement. The lower leg is strapped to the machine's rotating arm, and the arm's axis of rotation is aligned with the lateral femoral condyle. An external load is applied to the lower leg in an attempt to bend the knee into flexion. The subject is encouraged to resist the external load maximally (in the case

Figure 2.2 Isokinetic dynamometer configuration for inducing injury to the knee extensor muscles. The subject is secured onto the machine using a waist belt. The arm of the dynamometer moves downward at a constant angular velocity, pushing the leg into flexion as the subject attempts to maximally resist.

of isokinetic dynamometry) or in a controlled manner (in the case of isotonic exercise). Modified leg press machines have also been used to induce injury to knee extensors. However, with such machines, injury is not restricted to the knee extensors because eccentric contractions are performed by the hip extensors as well as the knee extensors. As shown in table 2.2, the injury elicited by the isotonic models is comparable to that elicited by the isokinetic ones.

As with studies using the elbow flexors, studies using the knee extensor models have varied considerably with regard to load, number of repetitions, angular excursion, and angular velocity used during the eccentric contraction bout. Joint angular excursions have ranged from a movement from 165° to 90° of knee extension (Stupka et al. 2001) to a movement from 110° to 35° (Babul et al. 2003). Again, the variation in angular excursion across the studies in table 2.2 could not explain a significant portion of the variance in either of the injury indices, though there was a trend toward greater strength losses when the eccentric contractions ended at joint angles corresponding to longer muscle lengths ($r = 0.59$; $p = 0.10$). Conclusions are similar when the knee angular excursion varies *within* a study. Child and colleagues (1998) found greater injury when the eccentric contractions ended at 40° of extension as opposed to 80°. Studies also differed substantially in the number of repetitions completed. The smallest number of contractions was 40 (Komi and Viitasalo 1977), and the largest was 300 (Babul et al. 2003; Chen et al. 2003). However, the number of contractions performed could not explain a significant portion of the variability in the injury indices among studies ($r \leq 0.30$).

Whole-Body Exercise With Eccentric Contraction Component or Bias

The models in the next category induce muscle injury consequent to whole-body exercise, often prolonged, that has a strong eccentric contraction component or bias. In

contrast to the models described in the previous section, these models utilize exercise that involves multiple muscle groups encompassing movement at several joints (e.g., as in running). Thus the injury is not restricted to one muscle group and is more difficult to quantify. The eccentric contraction component of the exercise is usually not as intense as in the studies just discussed, necessitating longer-duration exercise to elicit comparable injury.

Downward-Stepping Model

Injury models using downward-stepping exercise protocols were among the first to be used with human subjects and are also the simplest to implement (Newham et al. 1983, 1986; Duarte et al. 1999). Beginning at the top of a step, the subject steps down with one leg in a controlled manner. Lowering one's body weight results in eccentric contractions of the hip and knee extensors contralateral to those of the initially downward-stepping leg. The muscles of the initially downward-stepping leg then perform concentric contractions to raise the subject back up to the starting point atop the step. It has been shown that minimal injury is induced in the concentrically exercising leg (Newham et al. 1983). The cycle of upward and downward stepping is repeated for a specified number of repetitions (300-900) or until exhaustion. Typically, the step height is adjusted to 110% of the subject's lower-leg length, and the speed of stepping is set to 15 cycles per minute so as to maintain a consistent exercise pattern among subjects (Newham et al. 1983, 1986).

Downhill Running Models

Similar to the downward-stepping model are the downhill running and walking models (table 2.3). Locomoting downhill induces muscle injury because of the eccentric contractions done by the hip and knee extensor muscle groups as they lower the body down the slope. This exercise is normally performed on a motor-driven treadmill at downhill grades varying from –5% (Akimoto et al. 2002) to –25% (Balnave and Thompson 1993). The speed at which a subject exercises has varied considerably among studies but is often based on a percentage (70-77%) of the $\dot{V}O_2$max or maximal heart rate. For the studies in table 2.3, the speed, grade, or exercise duration used could not account for a significant portion of the variance in injury across the studies.

Prolonged-Racing Models

Prolonged running, such as marathon or ultramarathon racing, may also be used as a model for inducing skeletal muscle injury (table 2.4). A standard marathon is defined as 42.2 km long, and an ultramarathon is any race exceeding that distance. Subjects capable of running such distances are usually well-trained athletes who

Table 2.3 Human Muscle Injury Models Using Downhill Running or Walking

Study	Exercise mode	Treadmill speed	Grade (%)	Duration (min)	Knee extensor isometric strength loss (%MVC)	Peak blood CK level (IU/L)
Sorichter, Mair, Koller, Gebert et al. 1997	Run	70% $\dot{V}O_2$max	−16	20	–	310
Koller et al. 1998	Run	9 km/h	−25	25	–	2,610
Shave et al. 2002	Run	70% max run velocity	−15	30	–	380
Feasson et al. 2002	Run	?	−12	30	–	160
Thompson et al. 2003	Run	77% max HR	−10	30	–	110
Akimoto et al. 2002	Run	9.6 km/h	−5	30	–	300
Donnelly et al. 1990	Run	70% max HR	?	45	6	380
Yu et al. 2002	Run	Subject preference	?	45	–	940
Cannon et al. 1994	Run	77% max HR	−16	45	–	290
Sacheck et al. 2003	Run	75% $\dot{V}O_2$max	−16	45	–	390
Rowlands et al. 2001	Run	10.5 km/h	−8.5	45	27	–
Byrnes et al. 1985	Run	170 bpm HR	−10	60	–	360
Balnave and Thompson 1993	Walk	6.4 km/h	−25	40	17	340

HR: heart rate; bpm: beats per minute; $\dot{V}O_2$max: maximal oxygen uptake; MVC: maximal voluntary contraction isometric strength; CK: creatine kinase.

Table 2.4 Human Muscle Injury Models Using Marathon or Ultramarathon Racing

Study	Race type	Average or range of completion times (min)	Distance (km)	Peak blood CK level (IU/L)
Cummins et al. 1987	Marathon	195	41.3	860
Akimoto et al. 2002	Marathon	251	42.2	3880
Apple and Rhodes 1988	Marathon	192	42.2	2250
Ide et al. 1999	Wheelchair marathon	?	42.2	730
Noakes and Carter 1982	Ultramarathon	251-351	56	960
Koller et al. 1998	Ultramarathon	530	67	430
Matin et al. 1983	Ultramarathon	?	80.5-160	8640
Noakes et al. 1983	Ultramarathon	337-615	88	640

CK: creatine kinase.

have leg and hip musculature that is relatively resistant to injury. As seen in table 2.4, neither completion time nor the distance covered relates very well to the degree of injury observed among the studies.

Comparisons of Human Muscle Injury Models

We sought to compare the various muscle injury models on the basis of the amount of injury induced as reported in the literature. It was practical to do this on the basis of only two indices of injury, that is, peak isometric strength loss and the peak blood level of CK activity, because most studies measure one or both of these indices. Though muscle soreness is also commonly measured, it is not feasible to compare across studies on this measure.

We initially compared the elbow flexor (table 2.1), knee extensor (table 2.2), downhill running (table 2.3), and prolonged-racing models (table 2.4) on the basis of the peak blood CK level. (Minimal data are available on the running models for a comparison based on strength loss.) An analysis of variance was run on the CK data in tables 2.1 through 2.4, followed by Tukey post hoc tests. This analysis indicated that the studies employing elbow flexor injury models yielded a mean (±SE) blood CK value (5440 ± 850 IU/L) that was 5- and 10-fold higher than those induced by the knee extensor (1080 ± 370 IU/L) and downhill running (550 ± 200 IU/L) models, respectively. The mean blood CK values for the knee extensor, downhill running, and prolonged-racing (2300 ± 990 IU/L) models were not significantly different from each other.

Using a *t*-test, we then compared the isometric strength loss data from the elbow flexor and knee extensor studies in tables 2.1 and 2.2. The analysis indicated that average strength loss across the elbow flexor model studies (40% ± 2%) was significantly greater than that observed for the knee extensor model studies (27% ± 3%). We interpret these findings as suggesting that the elbow flexor muscles may be more easily injured than the knee (and possibly hip) extensor muscles following a bout of eccentric contractions or eccentric-biased exercise. This conclusion is supported by a recent study demonstrating that the injury susceptibility of the elbow flexors was greater than that of the knee extensors (Jamurtas et al. 2005). It is particularly interesting that eccentric injury elicits higher blood CK levels for the elbow flexors compared to the knee extensors even though the knee extensors have a ≥10 times greater mass and thus CK content. It is also interesting that a few eccentric contractions done by a single muscle group (the elbow flexors) can elicit a higher blood CK level on average than can prolonged eccentric-biased exercise involving multiple, large muscle groups.

We attempted to explain the apparent greater susceptibility of the elbow flexors to eccentric injury in comparison to the knee extensors. We postulated that the elbow flexor injury models might be associated with a particularly damaging contraction parameter (e.g., relatively high eccentric loads). It is difficult to compare the two categories of injury models on the load, angular excursion, and angular velocity used. However, it was possible to compare the elbow flexor and knee extensor studies listed in tables 2.1 and 2.2 on the number of eccentric contractions performed. Contrary to what one might predict, the average number of contractions performed in the elbow flexor model studies (41 ± 3) was less than half the average number performed in the knee extensor model studies (100 ± 13). Thus, the greater injury incurred in the elbow flexor model studies on average cannot be explained by a greater number of eccentric contractions.

Another possible explanation for the muscle group difference in injury susceptibility is that compared to the elbow flexors, the knee extensors are subjected to relatively greater chronic use involving the performance of eccentric contractions and thus are better trained and protected from injury. One might think that because people routinely perform eccentric contractions of the knee extensors while descending stairs, lowering themselves into a chair, and so on, the eccentric contractions in the knee extensor model studies would be a less novel exercise than those for the elbow flexors. However, the available data do not support this hypothesis. The relative activities of the elbow flexors (biceps brachii) and knee extensors (vastus lateralis and vastus medialis) as assessed by electromyography (EMG) over the course of a day are not significantly different (Kern et al. 2001). However, these EMG measurements cannot distinguish eccentric contraction activity from other contractile activity, so the possibility still remains that knee extensors perform relatively more eccentric contractions throughout the day.

Because fast-twitch muscle fibers are more readily injured by eccentric contractions than are slow-twitch fibers (Friden et al. 1983; Jones et al. 1986; Friden et al. 1988; Warren et al. 1994), it has been hypothesized that the elbow flexors have relatively more fast-twitch fibers than the knee extensors. Several studies have contrasted the fiber type composition of the human elbow flexors and knee extensors, though no one study has assessed all muscles in both muscle groups (Edstrom and Nystrom 1969; Johnson et al. 1973; Edgerton et al. 1975; Clarkson et al. 1982; Elder et al. 1982; Staron et al. 2000). On average, the knee extensor and elbow flexor muscle groups are both composed of 50% to 60% fast-twitch fibers, and none of the six studies reviewed showed a substantial difference in fiber type composition between the two groups. Thus, a difference in fiber type composition is not a plausible explanation for the difference in eccentric injury susceptibility.

To summarize, the elbow flexors appear to be more susceptible to eccentric injury than the knee extensors, but the explanation for this remains elusive. Future research directed toward probing this question would seem warranted.

Animal Muscle Injury Models

Research on experimentally induced muscle injury using animal models began in earnest in the early 1980s, at about the same time research using human models picked up speed. However, compared to human models, the animal models used over the last 25 years have been much more diverse, including conscious in vivo; anesthetized in vivo or in situ; and isolated (in vitro) muscle, fiber bundle, or fiber models. These models have made use of a number of species, including monkey, cat, chicken, rabbit, rat, and mouse; among these, the rodent models are more popular than the others. For these various models, the injury may be categorized as either (1) contraction induced, resulting from strenuous exercise or high-force eccentric contractions; or (2) trauma induced, from exposure of the muscle to a toxin, extreme temperatures, crushing, laceration, blunt impact, or ischemia. This section focuses on the contraction-induced injury models for two reasons: First, this type of injury is more common than are traumatic injuries (Warren, Lowe, and Armstrong 1999); secondly, the trauma-induced models are often used not to study the injury per se but rather to study the muscle's regenerative capacity following frank destruction of its fibers.

Contraction–Induced Injury Models

Before discussing the contraction-induced models, we need to provide definitions for in vivo, in situ, and in vitro that we use for categorization purposes. We define an in vivo model as one in which the muscle of interest is left unperturbed in the animal's body. The in situ model differs only in that the distal tendon for the muscle to be injured is surgically exposed and connected to either a servomotor, a materials testing machine, or another injury-inducing device. The in vitro model is one in which the muscle is surgically removed from the animal and then injured while bathed in a medium approximating extracellular fluid. In some in vitro models, a fiber bundle or single fiber is isolated from an excised muscle and then injured. We will discuss four broad categories of contraction-induced injury models ranging from whole-animal or in vivo models to the in vitro models at the muscle or single-fiber level. Within each category, we describe commonly used models and protocols for inducing injury, and then discuss the advantages and disadvantages of the different model categories.

Conscious In Vivo Models

Conscious in vivo models include ones in which animals, usually rodents, perform forced exercise on a motorized treadmill (table 2.5). The earliest model in this category was that of prolonged and intensive running done by rodents on a level treadmill (e.g., Vihko et al. 1978). However, the most popular model has been downhill walking or running performed by rats. Armstrong, Ogilvie, and Schwane (1983) introduced this model in their classic paper. Walking or running downhill is thought to predispose the antigravity (or extensor) muscles to injury because the exercise is biased toward the performance of eccentric contractions.

Rats typically walk on a decline of $-16°$ to $-18°$ (-29% to -32% grade) at a speed of 15 to 17 m/min. An exercise duration of 90 min was originally used by Armstrong and colleagues and has been mimicked in most studies, although durations of up to 200 min have also been used. Because the rats used in these studies are normally naïve to treadmill exercise and relatively unfit, most protocols include frequent rest breaks. Five-minute bouts of exercise interspersed with 2 min rest intervals can enable most rats to complete the entire exercise trial. The muscles showing the greatest degree of injury are the predominately slow-twitch extensor muscles, soleus and vastus intermedius (Armstrong, Ogilvie, and Schwane 1983; Ogilvie et al. 1988; Komulainen et al. 1994; Smith et al. 1997). The medial head of the triceps brachii (Armstrong, Ogilvie, and Schwane 1983) and red portions of the quadriceps femoris muscles (Komulainen et al. 1994) have also been reported to be injured in this exercise model. The degree of injury sustained by the soleus muscle appears to be related to the animal's size, age, or both (Kasperek and Snider 1985a). Soleus muscles in larger, older rats are injured more easily and to a greater degree than are the muscles in smaller, younger rats.

The downhill walking or running model has also been used with mice, but not as often and with less success. The exercise conditions in the mouse studies have been comparable to those for the rat, and in some cases more stressful. Despite this, the degree of injury sustained by the soleus muscle in the mouse studies has been minimal to modest. Carter, Kikuchi, Abresch and colleagues (1994), as well as Lowe and colleagues (1993), reported no histological injury or strength reduction in the soleus muscle after 2.2 to 5 h of downhill running at higher speeds or steeper grades (or both) than those typically used for rats. One study with mice did show histological injury and modest strength reductions in the soleus muscle following downhill running, but the strength loss was less than that observed for the dorsiflexor muscle, extensor digitorum longus (EDL) (Lynch et al. 1997). However, the explanation for EDL's functional deficit is most likely related to fatigue, because the EDL muscle does not appear to perform eccentric contractions during rodent locomotion (Gambaryan 1974). A plausible explanation for the lesser injury to the soleus in mice compared to rats may be the effect of body size, as mentioned earlier

Table 2.5 Conscious Animal In Vivo Injury Protocols (Rodent Downhill and Uphill Treadmill Running)

Study	Species	Incline (degrees)	Speed (m/min)	Duration (min)	Comments
Komulainen et al. 1994	Rat	−13.5	17	90	5 min bouts with 2 min rest periods
Komulainen et al. 1999	Rat	−13.5	17	130	Same
Smith et al. 1997	Rat	−16	12-15	30	
Armstrong, Ogilvie, and Schwane 1983; Schwane and Armstrong 1983; Snyder et al. 1984; Ogilvie et al. 1988; Shimomura et al. 1991; McNeil and Khakee 1992; Kyparos et al. 2001; Sotiriadou et al. 2003	Rat	−16	16	90	5 min bouts with 2 min rest periods except in Protocol I of Armstrong, Laughlin et al. 1983
Tsivitse et al. 2003	Rat	−16	17	90	5 min bouts with 2 min rest periods
Takekura et al. 2001	Rat	−16	18	90	Same
Duan et al. 1990	Rat	−17	15	130	Same
Warren et al. 1992	Rat	−17	25	150	Same
Kasperek and Snider 1985a	Rat	−18	24	200	5 min rest after first 100 min
Kasperek and Snider 1985b	Small rat	−18	16	120	Greater injury in larger rats
	Med. rat	−18	25	200	
	Large rat	−18	16	90	
Armand et al. 2003	Mouse	−14	30-40	150	Speed progressed up to 40 m/min
Lynch et al. 1997	Mouse	−16	13	60	
Lowe et al. 1994	Mouse	−16	20	300	
Carter, Kikuchi, Abresch et al. 1994	Mouse	−20	21	130	30 min bouts with 5 min rest periods
Carter, Kikuchi, Horasek, and Walsh 1994	Mouse	−20	21	155	Same
Komulainen and Vihko 1994	Rat	+5.5	17	240	
Kuipers et al. 1983	Rat	+10	21	60	
Tiidus et al. 2001	Rat	+12	21	60	
Armstrong, Ogilvie, and Schwane 1983	Rat	+16	16	90	
Komulainen and Vihko 1995	Mouse	+6	13.5	360	
Salminen and Kihlstrom 1987	Mouse	+6	13.5	540	
Carter, Kikuchi, Abresch et al. 1994	Mouse	+20	21	130	

in relation to differences in injury susceptibility among rats. For rats running downhill (−16°) at 15 to 30 m/min, the energetic cost is 25% to 30% less than for running uphill (+16°) at the same speed (Armstrong, Laughlin et al. 1983); and the difference has been attributed to the bias during downhill running toward the more energy-efficient eccentric contraction. In contrast, the energetic cost of running downhill in mice is the same as that for running uphill (Taylor et al. 1972). These data suggest less reliance on eccentric contractions during downhill running in mice compared to rats.

Though downhill walking or running has been a popular model for inducing injury, an acute bout of uphill running by rodents has been found to be almost equally effective (table 2.5). Treadmill speeds used in these bouts have been comparable to those used in downhill walking or running, with inclines varying between +6° and +20°. Exercise durations have also been comparable to those used in downhill walking or running except in those protocols using relatively low inclines (~+6°); in those, the animals ran for 4 to 9 h (Salminen and Kihlstrom 1987; Komulainen and Vihko 1994, 1995). The soleus and quadriceps femoris muscles exhibit histopathology following uphill running (Armstrong, Ogilvie, and Schwane 1983; Kuipers et al. 1983; Komulainen et al. 1994; Komulainen and Vihko 1995). The degree of histological injury incurred by the soleus muscle during uphill running is comparable to, if not greater than, that incurred during downhill running (Armstrong, Ogilvie, and Schwane 1983; Komulainen et al. 1994). A plausible explanation for this is not readily obvious because presumably uphill running biases the exercise toward the performance of concentric contractions, which are not associated with inducing injury. Armstrong and colleagues (1983) suggested that upon hindfoot strike in uphill running, the ankle is more dorsiflexed and thus the soleus muscle longer than in the same phase of the gait cycle during running on the level or downhill. Thus, during the hindfoot's acceptance of the body's weight, the resulting eccentric contraction performed by the plantarflexors, including the soleus, is done at a longer muscle length during uphill running. Injury resulting from eccentric contractions has been shown to be greater when the contractions are initiated at longer muscle lengths (Newham et al. 1988; Gosselin and Burton 2002); thus this combination of factors might explain the apparent anomaly.

Anesthetized In Vivo Models

All models in the anesthetized in vivo category induce injury by eliciting electrically stimulated eccentric contractions in the ankle dorsiflexors or plantarflexors of anesthetized rodents or rabbits (table 2.6). Models injuring the ankle dorsiflexor muscles have been more popular because injury is more easily induced in these muscles than in the plantarflexor muscles (Warren et al. 1994; R.G. Cutlip, personal communication). This is attributed to the plantarflexor muscles' relatively higher proportion of slow-twitch fibers, which are associated with greater injury resistance, and to the fact that the plantarflexors are chronically more active and thus better trained because of their role as weight-bearing muscles (Friden et al. 1983; Jones et al. 1986; Friden et al. 1988; Warren et al. 1994).

Table 2.6   Anesthetized Animal In Vivo Injury Protocols (Electrically Stimulated Eccentric Contractions of the Ankle Dorsi- or Plantarflexors)

Study	Species	Muscle group	No. contractions	Contraction interval (s)	Angular velocity (°/s)	Angular excursion (degrees)	Starting angle (degrees)	Comments
Lovering and De Deyne 2004	Rat	Dorsiflexors	1	–	900	90	90	
Peters et al. 2003	Rat	Dorsiflexors	30	120	95	38	90?	
Yasuda et al. 1997	Rat	Dorsiflexors	50	4	150	150	?	
Sakamoto et al. 1996	Rat	Dorsiflexors	60	10	37.5	150	?	
Cutlip et al. 2004	Rat	Dorsiflexors	70	0	500	50	70 or 90	7 bouts of 10 stretch–shortening cycles

(continued)

Table 2.6 *(continued)*

Study	Species	Muscle group	No. contractions	Contraction interval (s)	Angular velocity (°/s)	Angular excursion (degrees)	Starting angle (degrees)	Comments
Geronilla et al. 2003	Rat	Dorsiflexors	150	0	500	50	70	15 bouts of 10 stretch–shortening cycles
Komulainen et al. 1998	Rat	Dorsiflexors	240	3	500	50	80	
Warren, Hermann et al. 2000	Mouse	Dorsiflexors	50	10	2000	40	70	
Barash et al. 2004	Mouse	Dorsiflexors	50	60	312	78	60	
Newton et al. 2000	Mouse	Dorsiflexors	120	5	333	~100	?	
Lowe et al. 1995; Warren, Lowe et al. 1996; Ingalls, Warren, and Armstrong 1998; Ingalls, Warren, Williams et al. 1998; Warren, Ingalls et al. 1999; Warren, Fennessy, and Millard-Stafford 2000; Rathbone et al. 2003; Ingalls et al. 2004	Mouse	Dorsiflexors	150	10	2000	40	70	
Sacco et al. 1992	Mouse	Dorsiflexors	240	5	333	~100	?	
Faulkner et al. 1989	Mouse	Dorsiflexors	360	5	333	~100	Full flexion	
Lieber et al. 1996	Rabbit	Dorsiflexors	150-900	2	75	30	~100	
Lieber et al. 1994; Mishra et al. 1995; Friden and Lieber 1998; Patel et al. 1998	Rabbit	Dorsiflexors	900	2	75	30	~100	
Willems and Stauber 2002a	Rat	Plantarflexors	15	180	50	50	90	

(continued)

Table 2.6   *(continued)*

Study	Species	Muscle group	No. contractions	Contraction interval (s)	Angular velocity (°/s)	Angular excursion (degrees)	Starting angle (degrees)	Comments
Willems and Stauber 2002b	Rat	Plantarflexors	20	120	90	50	90	
Willems et al. 2001	Rat	Plantarflexors	20	180	~250	50	90	
Willems and Stauber 2000	Rat	Plantarflexors	30	180	50 or 600	50	90	
Stauber and Willems 2002	Rat	Plantarflexors	30	40 or 180	~475	50	90	
McBride 2000; McBride et al. 2000	Rat	Plantarflexors	24-192	20	?	?	?	Stimulated sciatic nerve, forcing the dorsiflexors to perform eccentric contractions against load of the stronger plantarflexors
Fritz and Stauber 1988; Stauber et al. 1996	Rat	Plantarflexors	50	0	~150 or ~375	?	?	5 bouts of 10 stretch–shortening cycles, done as the researcher manually moved the foot while the muscles were stimulated
Tomiya et al. 2004	Mouse	Plantarflexors	100	4	~475	50	180	
Schneider et al. 2002	Mouse	Plantarflexors	450	?	85	17	?	
St Pierre Schneider et al. 1999	Mouse	Plantarflexors	450	?	180	17	?	

180° = foot is completely plantarflexed.

Most models in this category utilize servomotors to have the muscles perform injurious high-force eccentric contractions. The hindfoot of the rodent or rabbit is secured in a shoelike structure that is attached to the shaft of a servomotor (figure 2.3). The ankle is positioned such that its axis of rotation coincides with that of the servomotor shaft. With this configuration, the torque produced about the ankle during a muscle contraction can be measured by the servomotor and, while at a given angle, is directly proportional to the forces produced by the contracting muscles. With the lower leg stabilized, eccentric contractions are created as the animal's foot is moved at a constant angular velocity about the ankle. An eccentric contraction of the ankle dorsiflexors, for example, would entail a plantarflexion movement of the foot as the ankle dorsiflexors contract in an attempt to resist the movement. Normally, a brief (100-500 ms) isometric contraction immediately precedes the eccentric contraction. This ensures that the muscles are maximally active at the initiation of the movement and that the highest possible peak eccentric forces are attained.

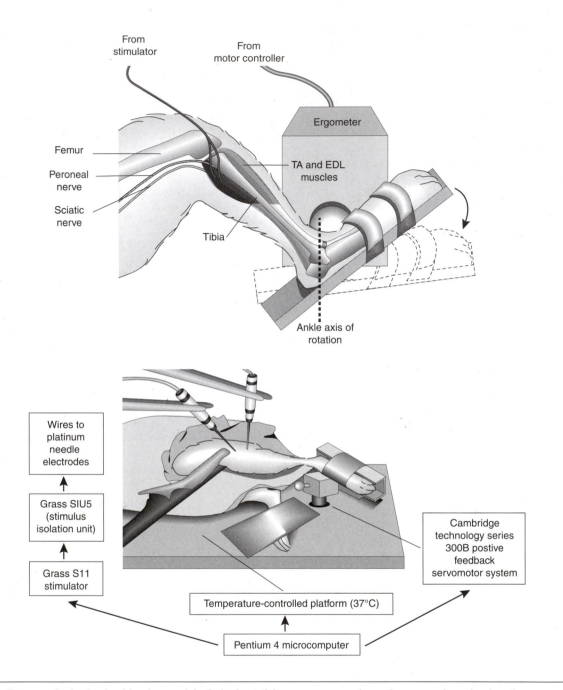

Figure 2.3   Two anesthetized animal in vivo models. In both models, percutaneous electrodes are used to stimulate the common peroneal nerve during forced ankle plantarflexion (see arrow in top figure). The ankle dorsiflexor muscles are injured as a result.

Reprinted, by permission, from R.L. Lieber et al., 1994, "Contractile and cellular remodeling in rabbit skeletal muscle after cyclic eccentric contractions," *J Appl Physiol*. 77(4): 1926-1934.

The angular excursions and angular velocities used in these models have been quite variable (table 2.6)—sometimes within the range attained during normal locomotion, sometimes not. Rat ankle angular velocity during unforced ambulation averages 300° to 350°/s, but peak instantaneous velocities are ~600°/s; ankle angle during the stance phase ranges from 72° (18° of dorsiflexion) to 111° (21° of plantarflexion) (Varejao et al. 2002). Mouse ankle angular velocities during moderate-speed

(20-30 m/min) treadmill running average 700° to 1650°/s, though peak instantaneous velocities can easily exceed 2000°/s; ankle angle ranges from 64° to 114° (D. Lowe, unpublished observations). The ideal angular excursion and velocity for a model would be those within the physiological range, along with an angular velocity that is also on the plateau of the eccentric torque: velocity relationship. For eccentric contractions of the mouse ankle dorsiflexors, this would require an angular velocity of at least 600°/s (Ingalls et al. 1996; Warren, Fennessy, and Millard-Stafford 2000).

The number of eccentric contractions used in these injury models has varied from as few as one to as many as 900 (table 2.6). The ideal number of contractions is the minimum number necessary to attain a substantial, reproducible injury; and it is clear that this number is dependent on the mechanical parameters associated with the contractions. Eccentric contractions of maximally activated muscles that are initiated at relatively long muscle lengths, and conducted over relatively large angular excursions at high angular velocities, appear to elicit the most injury for a given number of contractions. The time between contractions, another important parameter, has varied from 2 s to 3 min (table 2.6). When the interval is too short, the strength loss and histopathology induced by injury may be confounded by those induced by fatigue (Lannergren et al. 2002; Stauber and Willems 2002). The optimum interval between contractions has not been determined for most models but is most certainly species and muscle group dependent. For our mouse model using the ankle dorsiflexors, we have found that 10 s between contractions of 120 ms duration results in no fatigue (Warren, Ingalls et al. 1999).

Performance of eccentric contractions by anesthetized in vivo (and in situ) models requires electrical stimulation of the muscle either directly (with electrodes on the skin over the muscle) or indirectly (with electrodes near or on the motor nerve). Indirect stimulation can be accomplished in one of three ways. The most common way is to surgically expose a segment of the nerve and place the stimulating electrodes in direct contact with the nerve, one placed more distally than the other. A second way is to percutaneously insert two needle electrodes such that the nerve passes between the two electrodes but is not in direct contact with either electrode (figure 2.3). This technique is less likely to injure the nerve than the first but is technically more difficult because insertion of the electrodes is done in a blind fashion. The third means of stimulating the nerve is to chronically implant a stimulating cuff on the motor nerve and allow the animal four weeks or more to recover before inducing injury (Warren et al. 1998). To stimulate the nerve, wires connected to the nerve cuff are externalized in the dorsal cervical region and connected to a stimulator. This third means is preferable when it is necessary to make repeated functional measurements (e.g., strength, EMG) on an animal following injury. As a final comment on nerve stimulation, because of the associated difficulties it is essential that the investigator monitor muscle-generated torque continuously throughout an injury protocol so as to be able to confirm the adequacy of stimulation.

Anesthetized In Situ Models

Models in the anesthetized in situ category, made popular by two of the classic studies in the field (McCully and Faulkner 1985, 1986), have been widely used. These models induce injury by eliciting electrically stimulated eccentric contractions in individual muscles or motor units of the ankle dorsiflexors or plantarflexors of anesthetized rodents, rabbits, or cats (table 2.7). For reasons mentioned in the previous section, models injuring one of the dorsiflexor muscles, the tibialis anterior or EDL, have been more commonly used than others. In these models, the lower leg is stabilized by clamping or pinning of the tibia, femur, knee, or some combination of these. The distal tendon of the muscle to be injured is then exposed and attached via suture to the lever arm of a servomotor or to a materials testing machine (e.g., Instron). In acute experiments, the tendon is cut; but in experiments in which the animal is to recover, the tendon is normally left intact.

Table 2.7   Anesthetized Animal In Situ Injury Protocols (Electrically Stimulated Eccentric Contractions of Individual Ankle Dorsi- or Plantarflexor Muscles)

Study	Species	Muscle	No. contractions	Contraction interval (s)	Lengthening velocity (L_f/s)	Length change (% L_f)	Starting length (% L_f)	Comments
Brooks and Faulkner 2001	Mouse	EDL	1	–	1-16	30-50	100	
Brooks and Faulkner 1996	Mouse	EDL	1	–	2	10-50	100	
Brooks et al. 1995	Mouse	EDL	1	–	2	10-60	100	
Hunter and Faulkner 1997	Mouse	EDL	1	–	2	50-60	90, 100, or 120	

(continued)

Table 2.7 (*continued*)

Study	Species	Muscle	No. contractions	Contraction interval (s)	Lengthening velocity (L_f/s)	Length change (% L_f)	Starting length (% L_f)	Comments
Koh and Escobedo 2004; Pizza et al. 2005	Mouse	EDL	75	4	1	20	100	
Brooks and Faulkner 1990	Mouse	EDL	225	4	1	20	90	
McCully and Faulkner 1986	Mouse	EDL	450	2-10	0.2-1	20	90	
McCully and Faulkner 1985	Mouse	EDL	450	2	1	20	90	
McArdle et al. 2004	Mouse	EDL	450	2	1.5	20	100	
Black and Stevens 2001	Mouse	EDL	Variable	45	~0.8*	10*	95*	
Mair et al. 1996	Rabbit	EDL	1	–	1, 10, or 50 cm/min	Until rupture	?	Used materials testing machine
Strickler et al. 1990	Rabbit	EDL, TA	1	–	10 cm/min	Until rupture	?	Same
Safran et al. 1988	Rabbit	EDL, FDL, TA	1	–	10 cm/min	Until rupture	?	Same
Hasselman et al. 1995	Rabbit	EDL, TA	1	–	10 cm/min	Until near rupture	?	Same
Garrett et al. 1987	Rabbit	EDL	1	–	100 cm/min	Until rupture	?	Same
Benz et al. 1998	Rabbit	Soleus	180	10	1.67	25	100	
Best et al. 1998; Corr et al. 2003	Rabbit	TA	1	–	?	?	?	Length change specified as 90° ankle plantarflexion movement; distal tendon shortened prior to contraction
Lieber and Friden 1988; Friden et al. 1991; Lieber et al. 1991	Rabbit	TA	1800	1	0.63	25	100	
Parikh et al. 2004	Cat	MG	2-50	40	2.5	30	100	

(*continued*)

Table 2.7 *(continued)*

Study	Species	Muscle	No. contractions	Contraction interval (s)	Lengthening velocity (L_f/s)	Length change (% L_f)	Starting length (% L_f)	Comments
Brockett et al. 2002	Cat	MG	10	40	~2.5	~15	100	Individual motor units studied
Morgan et al. 2004	Cat	MG	10	30	2.5	30	100-130	Groups of motor units studied
Whitehead et al. 2003	Cat	MG	50	40	2.5	30	85-145	Entire muscle or groups of motor units studied
Whitehead et al. 2001	Cat	MG	150	20	2.5	30	85	
Uchiyama et al. 2001	Rat	Plantaris	1	–	?	?	?	Heavy, falling load imposed on distal tendon during muscle stimulation

EDL: extensor digitorum longus; TA: tibialis anterior; FDL: flexor digitorum longus; MG: medial gastrocnemius
*Expressed relative to muscle length instead of relative to fiber length.

Eccentric contractions are elicited via stimulation of the motor nerve for the muscle of interest as the servomotor or materials testing machine pulls distally on the distal tendon, thus lengthening the muscle. As is the case for the anesthetized in vivo models, a brief isometric contraction usually precedes the eccentric contraction. The degree to which the muscle is stretched during the eccentric contraction has been highly variable (table 2.7). Even the way in which the degree of stretch is specified has been variable. Most commonly, the stretch has been specified as a percent of muscle fiber length (L_f), but it has also been specified as a percent of muscle length (L_o) and in absolute units (e.g., centimeters). Stretches expressed as percent of fiber length have ranged from 10% to 60%. Usually, these stretches have been initiated from the so-called physiological L_o, which is specified as the muscle length yielding the highest active force production. However, stretches have sometimes been initiated at muscle lengths corresponding to short (i.e., 85-95% L_f) or long (i.e., 120% L_f) fiber lengths.

For many in situ injury models, the physiological nature of the stretch imposed during the eccentric contraction protocol is subject to question. First, the ability to precisely control fiber length by pulling distally on the muscle's distal tendon is doubtful (Butterfield and Herzog 2005). Fibers in most muscles do not run from end to end of the muscle and thus are in series with connective tissue (e.g., aponeuroses), other fibers, or both. In order to lengthen a fiber by a certain percentage of its length, one would need to know the relative stiffnesses of the fiber and the elements in series with it. Such stiffness values are not known for muscles during eccentric contractions. Furthermore, it is not clear for most in situ injury models whether the investigators have even considered a relative stiffness correction estimate in an attempt to control fiber length.

Second, assuming that it is possible to precisely control fiber length, it is likely that in many of the in situ models the fibers are pulled to lengths not anatomically possible in the intact animal. Such extreme lengthening may predispose the muscle to a type of injury (e.g., disruption of certain cytoskeletal or extracellular structures) that would not occur if the muscle was injured in vivo. As table 2.7 shows, the mouse EDL muscle is often injured in situ via stretching of its fibers to lengths greatly exceeding those attainable in the intact animal (i.e., 120% L_f) (Burkholder and Lieber 2001). The case is similar for the performance of eccentric contractions in cat medial gastrocnemius muscle. In that muscle in vivo, fiber length does not exceed 99% L_f (Burkholder and Lieber 2001). Simply setting the muscle length to the one that is optimal for force production is setting it to one that is not attainable in the intact cat.

Third, the stretch velocities used in these studies have been either of unknown physiological relevance or too slow to be physiologically relevant. Velocities used in the studies employing materials testing machines have ranged from ~0.1 to ~10 muscle lengths per *minute,* limited by the maximum speed at which the machine head can move. Such slow speeds coupled with very long stretches (i.e., until the muscle ruptures or is close to rupturing) make for a very prolonged eccentric contraction—that is, tens of seconds. In the studies employing faster and apparently more relevant stretch velocities (i.e., 0.2-16 fiber or muscle lengths per second), it is still not known for most muscles how these velocities compare to those occurring in vivo during locomotion.

In Vitro Models

In the category of in vitro models, neither muscles nor muscle fibers are injured while in the animal but after excision from the animal. Muscles or fibers from rodents and amphibians have been injured in vitro. In vitro models from the 1980s typically induced injury by subjecting the isolated muscle to noxious agents or to repetitive isometric contractions with little recovery time in between contractions (e.g., Publicover et al. 1978; Jackson et al. 1984; Jones et al. 1984; Duncan 1988; McCall and Duncan 1989; Shamsadeen and Duncan 1989).

Today, injury is most often induced through elicitation of electrically stimulated eccentric contractions of the muscle, fiber bundle, or fiber (table 2.8). In most in vitro preparations, the muscle or fiber is bathed in a solution with an ionic composition similar to that of extracellular fluid. This solution, referred to as a Ringer's, Krebs, or Krebs-Ringer solution, may also contain glucose and insulin to support energy metabolism, branched chain amino acids to promote nitrogen balance, and a protein such as albumin to slow the inhibition of muscle enzymes (e.g., CK) whose activities in the bathing solution are used as injury markers (Warren, Lowe et al. 1993). The bathing solution is normally bubbled with a 95% O_2:5% CO_2 gas while being maintained at temperature of 20° to 37° C. The CO_2 in the gas is necessary to maintain pH at ~7.4 in bicarbonate-buffered solutions. Because O_2 delivery to the fibers deep within the muscle is dependent on O_2 diffusion from the bathing solution, muscle size is restricted to 30 mg or less for experiments done at 37° C (Segal and Faulkner 1985). This excludes the use of limb muscles (e.g., soleus and EDL) from adult rats and larger animals for in vitro studies at 37° C. Muscles larger than 30 mg can be studied in vitro, but the temperature of the preparation must be reduced below 37° C so as to decrease the muscle's metabolic rate. The viability of in vitro muscle preparations is limited to 3 to 4 h; the smaller, more oxidative muscles (e.g., mouse soleus) are the most stable when kept at cooler temperatures.

In the in vitro preparation, one tendon of the muscle, fiber bundle, or fiber is attached to a fixed support while the other is attached to the arm of a servomotor. The servomotor lengthens or shortens the muscle or fiber while at the same time measuring force production. Eccentric contractions are done by lengthening the muscle or fiber(s) as the muscle is stimulated directly. The stimu-

lating electrodes are usually wires or plates made of a noble metal (e.g., platinum) and are positioned so that the muscle is directly in between the electrodes but not in contact with them. During stimulation, the ions in the bathing solution complete the circuit and allow the current to pass through the muscle or fiber.

The parameters associated with the eccentric contraction protocols implemented in vitro are shown in table 2.8. The degree and velocity of lengthening for intact muscles during eccentric contractions are often specified as a percent of muscle length and muscle lengths per second, respectively, as opposed to being normalized to fiber length. The length changes for intact muscles have been in the 10% to 30% L_o range and have been done at velocities of 0.5 to 4 L_o/s. Just as for the in situ models, there is potential for stretching in vitro muscles to anatomically impossible lengths unless precautions are taken.

The greatest differences between the in vitro injury protocols and the anesthetized in vivo and in situ protocols are in the number of eccentric contractions done and the time between contractions. Typically, 5 to 20 eccentric contractions are performed in vitro whereas 10 times that many or more must be performed in vivo or in situ to elicit comparable injury. It is not known why the in vitro muscle is relatively more susceptible to injury. The unnatural environment and the likelihood for overstretching the muscle probably contribute. The time between eccentric contractions for isolated muscles is on the order of minutes (e.g., 2-4) as opposed to seconds for the in vivo and in situ models. The longer interval is used to avoid metabolic fatigue and is tied to the dependence of the isolated muscle on O_2 diffusion from the bathing medium as opposed to O_2 delivery via the vasculature. At 25° C, a 10 mg mouse EDL muscle requires ~9 min to recover energetically from a ≥2 s contraction (Warren, Williams et al. 1996). The recovery time is certainly shorter at higher temperatures and for more oxidative muscles, but still must be considered.

In vitro studies of intact single fibers and fiber bundles (three to five fibers) from rodents are technically demanding. Such preparations are easily damaged. Eccentric contraction injury protocols for these preparations are comparable to those for isolated muscles, with two exceptions. First, the temperature of the preparation is typically below the physiological range, that is, at room temperature, so as to enhance the stability of the preparation. Second, the interval between contractions can be shorter since the O_2 diffusion distance is greatly reduced compared to that for a whole muscle. Investigators can skirt the difficulty associated with these preparations by using either frog muscle fibers or skinned fiber segments taken from rodent muscle. In a skinned fiber, the plasmalemma has been permeabilized so that the fiber is no longer electrically excitable. An eccentric contraction in this preparation is done via exposure of the fiber to a

Table 2.8 Animal In Vitro Injury Protocols (Eccentric Contractions of Isolated Muscles, Fiber Bundles, or Single Fibers)

Study	Species	Muscle	No. contractions	Contraction interval (s)	Lengthening velocity (L_f/s)	Length change (% L_f)	Starting length (% L_f)	Temperature (°C)
Warren, Hayes, Lowe, Prior, and Armstrong 1993	Rat	Soleus	1-10	240	1.5*	25*	90*	37
Warren, Hayes, Lowe, and Armstrong 1993	Rat	Soleus	5	240	0.5-1.5*	10-30*	85 or 90*	37
Gosselin 2000; Gosselin and Burton 2002	Rat	Soleus	20	5	1*	20*	80 or 100*	26
Hughes and Gosselin 2002	Rat	Diaphragm strip	20	5	1*	20*	80 or 100*	26
Yeung, Bourreau et al. 2002	Rat	Soleus fiber bundle	10	4	2	30	100	22-24
Yeung et al. 2003	Rat	Soleus fiber bundle	10	4	4	40	100	22
Macpherson et al. 1997	Rat	Skinned EDL fiber	1	–	0.5	40	100	15
Lynch and Faulkner 1998	Rat	Skinned EDL fiber	1	–	0.5-4	5-20	100	15
Brooks and Faulkner 1996; Macpherson et al. 1996	Rat	Skinned EDL fiber, skinned soleus fiber	1	–	0.5	5-20 10-40	100	15
Childers and McDonald 2004	Rat	Skinned psoas fiber	3	?	2	10-20	100	?
Warren, Ingalls, and Armstrong 2002	Mouse	EDL	5	180	1.5*	20*	90*	15, 20, 25, 30, 33.5, or 37
Sam et al. 2000	Mouse	EDL (fifth digit belly)	10	180	2	15	115	25
Warren, Williams et al. 1996	Mouse	EDL	15	180	1.5*	20*	90*	25
Warren et al. 1994	Mouse	EDL, soleus	15	240	1.5*	20*	90*	37
Warren, Lowe et al. 1993	Mouse	Soleus	10-20	240	1.5*	25*	90*	37

(continued)

Table 2.8   *(continued)*

Study	Species	Muscle	No. contractions	Contraction interval (s)	Lengthening velocity (L_f/s)	Length change (% L_f)	Starting length (% L_f)	Temperature (°C)
Lowe et al. 1994; Warren et al. 1995	Mouse	Soleus	20	120	1.5*	20*	90*	37
Yeung, Balnave et al. 2002	Mouse	Flexor brevis fiber	10	4	2.5 or 4	25 or 40	100	22-24
Balnave and Allen 1995; Balnave et al. 1997	Mouse	Flexor brevis fiber	10	4	5	25 or 50	100 or ~113	22
Lynch et al. 2000	Mouse	Skinned EDL fiber	1	–	0.5	10-30	100	15
Talbot and Morgan 1998	Toad	Sartorius	5-60	≥180	3-4*	10-13*	70-115*	20-22
Wood et al. 1993	Toad	Sartorius	10, 20, 40, or 60	≥180	~3.35*	10*	~106*	17-20
Patel et al. 2004	Frog	TA fiber bundle	10	180	2	10-35	~100	25
Morgan et al. 1996	Frog	TA fiber	7-26	120	0.45	7.5	100	3

EDL: extensor digitorum longus; TA: tibialis anterior; *expressed relative to muscle length instead of relative to fiber length.

high-calcium–containing solution that otherwise approximates intracellular fluid, followed by stretching of the fiber. However, activating the fiber in this manner bypasses the entire excitation–contraction coupling process. This is important because it has been demonstrated that a failure in excitation–contraction coupling is the main cause of the strength reductions resulting from eccentric contractions (Warren et al. 2001). Thus, data derived from the skinned fiber injury model may be of limited value.

Advantages and Disadvantages of Contraction-Induced Injury Models

No one animal injury model is inherently better than another. Table 2.9 summarizes the advantages and disadvantages of the four main categories of contraction-induced injury models. The conscious in vivo model's main advantages are that it is relatively easy to implement and is physiologically the most relevant model. The injurious contractions are conducted within the anatomical range of muscle length and are initiated using motor unit recruitment from the central nervous system rather than from an electrical stimulator. This latter point is important for studies of the "repeated-bout" effect because neural adaptations are thought to play a role in that effect (McHugh 2003). A main disadvantage

of the model is that the forces, length changes, and degree of activation imposed upon the muscle are totally unknown. Furthermore, compared to that in other injury models, the degree of histological injury is relatively small in the conscious in vivo model (e.g., ~5% of the fibers or 0.2% of the soleus muscle volume) (Armstrong, Ogilvie, and Schwane 1983; Ogilvie et al. 1988). Finally, with this model, the ability to draw conclusions about the cellular and subcellular mechanisms of injury is poor.

The main advantage of the anesthetized in vivo model over the conscious in vivo model is that the investigator controls the degree of muscle activation—usually maximal—during the injurious contractions. In addition, one can restrict the activation to the muscles that one intends to injure. As with the conscious in vivo model but in contrast to what occurs with the in situ and in vitro models, one constrains muscle length to lie within the anatomical range during the eccentric contractions. Furthermore, by ensuring that the joint angular excursion and velocity approximate those achieved during locomotion, one can ensure that the muscle's length excursions and lengthening velocity are physiologically relevant. A major disadvantage of the anesthetized models is that most anesthetics impair muscle contractility (Ingalls et al. 1996), and if the level of anesthesia varies either within or across experiments, the degree of injury induced will also

Table 2.9 Summary of the Advantages and Disadvantages of the Animal Contraction-Induced Injury Models

Category of models	Physiological relevance of model	Difficulty of implementing model	Ability to probe cellular-level mechanisms of injury	Knowledge of biomechanical parameters (e.g., forces, length changes) imposed upon the muscle
Conscious in vivo	Highest	Relatively easy	Lowest	None
Anesthetized in vivo	Moderate to high	Moderate	Moderate	Low
Anesthetized in situ	Low to moderate	Moderate	Moderate	Low to moderate
In vitro muscle or fiber(s)	Lowest	Moderate to hard	Highest	Highest

vary. In addition, because the anesthetized in vivo models induce injury to an entire muscle group, the effect on an individual muscle's function is unknown unless the muscle can be excised and studied in vitro.

The main advantages of the anesthetized in situ model over the anesthetized in vivo one are that the investigator can restrict the injury to the muscle of interest and can precisely control the muscle length excursions during the eccentric contractions. The precise control of muscle length can also be a disadvantage if this leads to a lengthening of the muscle outside of its anatomical range. In the in situ preparation, the muscle's distal tendon is normally connected to a force-measuring device during the injury protocol. However, the force measured by this device most likely does not accurately reflect that being produced within the muscle. Peter Huijing and colleagues over the last eight years have demonstrated in an elegant series of experiments that muscle-generated forces can be easily transferred out of one muscle onto surrounding connective tissue or to another muscle (e.g., Huijing 1999; Huijing and Baan 2001; Maas et al. 2001; Huijing et al. 2003). For the rat ankle dorsiflexors, the force measured at the distal tendon of the EDL muscle can differ by up to ±25% from that measured proximally in the muscle. This means that substantial force transfer must occur between the EDL and tibialis anterior muscles or between the EDL muscle and the surrounding connective tissue. Furthermore, the "force transfer" relationship between the EDL and tibialis anterior muscle changes throughout an injury protocol done in situ as one muscle, the muscle undergoing injury, becomes weaker and weaker. The implications are that the measured changes in eccentric forces occurring during an injury protocol, as well as the change in isometric force from before to after injury, may not be physiologically meaningful.

An advantage of the in vitro model over the others is that the forces imposed upon muscles or fibers during the injury protocol can be more accurately determined. Furthermore, in the in vitro models, muscle or fiber length and lengthening velocity can be more precisely controlled, and there are no or minimal anesthetic effects on contractility. The in vitro models are the ones best suited for probing the cellular and subcellular mechanisms of injury since in these, it is possible to measure intracellular and extracellular ion, metabolite, and protein concentrations. In this preparation, one has precise control over the muscle's environment and can add drugs or compounds to the bathing solution. The main disadvantage of the in vitro models is the non-physiological nature of the muscle's or fiber's environment. Because there is no vascular perfusion, only relatively small rodent muscles can be studied, and then for only a relatively short time. This means that one can use the in vitro preparation to study only the initial injury events, not those occurring during the inflammatory and regenerative phases.

Trauma-Induced Injury Models

With the widespread interest in how stem cells and the stem cell–like satellite cells participate in or may be induced to enhance muscle regeneration, trauma-induced injury models are quite popular. The objective of these models is the frank destruction of muscle fibers, which stem or satellite cells may be induced to replace. Table 2.10 provides a sampling of these models. The most popular technique is to inject a toxin or long-lasting local anesthetic in the hindlimb muscle of a rodent. Two toxins are commonly used. One is notexin, which is a myotoxic phospholipase A_2 derived from the venom of the Australian tiger snake. The second is cardiotoxin, which is a set of small proteins found in cobra venom that causes skeletal muscle contracture and interferes with neuromuscular transmission. The local anesthetics bupivacaine (Marcaine) and lidocaine are also effective in inducing fiber degeneration. A problem with these models is that the volume of toxin or anesthetic injected is often too large. Five of the studies listed in table 2.10 used an injection volume equal to 50% or more of the

Table 2.10 Trauma-Induced Muscle Injury Models Using Animals

Study	Species	Muscle	Mode of injury	Comments
Fink et al. 2003	Rat	Soleus	Notexin injection	Injected 200 µl of 100 µg/ml solution
Vignaud et al. 2003	Rat	Soleus	Notexin injection	Injected 50 µl of 80 µg/ml solution
Ullman and Oldfors 1991	Rat	Soleus	Notexin injection	Injected 3 µl of 200 µg/ml solution
Kirk et al. 2000	Rat	Biceps femoris	Notexin injection	Injected 10 µl of 200 µg/ml solution
Zhao et al. 2002	Mouse	Gastrocnemius	Cardiotoxin injection	Injected 100 µl of 10 mM solution
Goetsch et al. 2003	Mouse	Gastrocnemius	Cardiotoxin injection	Injected 150 µl of 10 µM solution
Gregorevic et al. 2000	Rat	EDL	Bupivacaine injection	Injected ~1000 µl of 0.5% solution
Shiotani et al. 2001	Rat	TA	Bupivacaine injection	Injected 200 µl of 0.5% solution
Gillette and Mitchell 1991	Rat	TA	Lidocaine injection	Injected 100 µl of 2% solution
Soeta et al. 2001	Rat	Femoral	Hypertonic saline injection	Injected 500 µl of 20% NaCl solution
Fink et al. 2003	Rat	Soleus	Crush	Used forceps to crush muscle
Zhang and Dhoot 1998	Chicken	Peroneus longus, gastrocnemius	Crush	Same
Li et al. 2000	Rat	TA	Laceration	
Menetrey et al. 1999; Fukushima et al. 2001; Chan et al. 2003; Foster et al. 2003	Mouse	Gastrocnemius	Laceration	
Zhang and Dhoot 1998	Chicken	Peroneus longus, gastrocnemius	Laceration	
Crisco et al. 1994; Crisco et al. 1996	Rat	Gastrocnemius	Blunt impact	Weight dropped onto muscle
Pavlath et al. 1998; Kuang et al. 1999	Mouse	TA	Freeze induced	5 s application of steel probe cooled to −79° C
Warren, Hulderman et al. 2002; Rathbone et al. 2003; Summan et al. 2003; Warren et al. 2004; Warren et al. 2005	Mouse	TA	Freeze induced	10 s application of steel probe cooled to −79° C
Qiu et al. 2000; Pachori et al. 2004	Mouse	Hindlimb	Ischemia-reperfusion	Tourniquet applied for 3 h
Fish et al. 1993	Rat	Hindlimb	Ischemia-reperfusion	Tourniquet applied for 2 h
Hayashi et al. 1998	Dog	Hindlimb	Ischemia-reperfusion	Tourniquet applied for 1, 3, or 6 h
Beiras-Fernandez et al. 2003	Monkey	Limb	Ischemia-reperfusion	Blood flow to limbs stopped for 1 h

EDL: extensor digitorum longus; TA: tibialis anterior.

muscle's volume. In these instances, the excess injected volume leaks out of the muscle and can cause unintended systemic effects.

Other popular models involve direct physical damage to the muscle, usually the hindlimb muscles of rodents (table 2.10). The investigator may crush a muscle using forceps, cut a muscle using a scalpel blade, freeze a muscle using a metal probe cooled to dry ice temperature, or drop a weight upon a muscle. A popular traumatic model that does not immediately induce fiber degeneration is the ischemia-reperfusion model, in which a tourniquet is applied to the hindlimb of an anesthetized animal for 1 to 6 h. The tourniquet is then released and the tissues are allowed to reperfuse. It is during the reperfusion phase that most of the muscle injury occurs, in association with oxidative damage and neutrophil invasion (Huda, Solanki, and Mathru 2004; Huda, Vergara et al. 2004). The ischemia-reperfusion model elicits a qualitatively different muscle injury than the other traumatic models. However, among the other trauma-induced models, no one model appears better or worse than another.

Summary

Studies evaluating experimentally induced skeletal muscle injury have focused primarily on exercise-induced injury involving relatively few high-force eccentric contractions or resulting from prolonged exercise biased to the performance of eccentric contractions. Such exercise can be performed by individual muscles or fibers or by multiple muscle groups during whole-body eccentrically biased exercise. Within each category of injury models, investigators can adjust several variables (e.g., number of contractions, angular excursions) to enhance the degree of injury or to meet the needs of the study. When these variables are held constant, the muscle groups yielding the greatest degree of injury are the elbow flexors for the human models and the ankle dorsiflexors for the animal models. Within a given model, eccentric contractions of maximally activated muscles conducted over relatively long muscle lengths at relatively high velocities appear to elicit the most injury for a given number of contractions. Within the animal in situ and in vitro models, it is imperative that muscle length changes be constrained to lie within those that are anatomically possible in the intact animal. An approach for avoiding this issue would be to always injure muscles in vivo in either the conscious or the anesthetized animal and then study the muscles or fibers in vitro if the intent is to probe a biological question at the cellular or subcellular level.

CHAPTER 3

Histological, Chemical, and Functional Manifestations of Muscle Damage

Stephen P. Sayers, PhD, and Monica J. Hubal, PhD

Eccentric (or lengthening) muscle actions commonly occur during a variety of everyday activities such as running, lowering oneself into a chair, or walking down a flight of stairs. Under normal circumstances, eccentric muscle actions are important components of the complex strategy of successful human movement. However, strenuous activities may cause excessive forces on lengthening muscle, resulting in muscle damage. In this chapter we examine damage caused by such excessive forces during eccentric contractions, focusing on the following:

- **Histological evidence** of muscle damage, particularly direct evidence of muscle damage observed from muscle biopsies using light and electron microscopy and indirect evidence of damage observed using magnetic resonance imaging (MRI)

- **Biochemical and histochemical evidence** of muscle damage, including elevated proteins and enzymes in the blood, inflammatory markers in the tissues and blood, alterations in calcium homeostasis, and impaired glucose metabolism

- **Functional evidence** of muscle damage, including strength loss and recovery, low-frequency fatigue (LFF), neuromuscular disturbances, changes in the range of motion and stiffness of muscles and joints, swelling, and soreness

Within each section, we discuss the limitations of these direct and indirect measures. Table 3.1 presents a comparison of the time course of the changes in these measures after exercise.

Histological Evidence of Muscle Damage

In 1902, Hough first suggested that soreness in muscles could be attributed to muscle fiber injury and necrosis. However, direct evidence of disturbances in myofibrillar organization was not demonstrated in humans for another 80 years, when light and electron microscopic techniques revealed focal (localized) disruptions to the striated banding patterns and ultrastructural features of eccentrically exercised muscle (Friden et al. 1981, 1983; Newham, McPhail, Mills, and Edwards 1983) (see figure 3.1). In this section we describe and interpret the direct evidence of muscle damage after eccentric exercise as seen in muscle biopsy samples, as well as the indirect evidence obtained from assessment of whole muscle using MRI.

Cellular and Subcellular

Using light microscopy, Friden and colleagues (1981) reported that focal disturbances to the muscle fiber two

Table 3.1   Time Course of Changes in Histological, Bio- and Histochemical, and Functional Indices After Eccentric Exercise and Their Values as Indicators of Muscle Damage

Time postexercise	1-12 h	24 h	48 h	3-5 days	5-7 days	7+ days	Value of measure
Histological							
Ultrastructural changes	+ +	+ + +	+ + +	+ +	+	+ +	Low/Moderate
T2 signal intensity		+	+	+ +	+ + +	+ +	Low/Moderate
Bio- and histochemical							
CK[a]		+	+ +	+ + +	+ +	+	Low
Inflammation (tissue)							Low/Moderate
*Acute	+ + +	+	+				
*Chronic					+ + +		
Inflammation (blood)							Low
*Acute	+ + +	+					
*Chronic				+ +	+ + +		
Myoglobin	+	+	+ +	+ +	+	+ +	Low
LDH[b]		+	+ +	+ +	+	+ +	Low
AST[c]			+	+ +	+		Low
MHC[d]			+	+ + +	+ +	+ + +	Low
Functional							
Strength	– – –	– – –	– –	– –	–	–	High
Low-frequency fatigue	+ + +	+ + +	+ +	+ +	+ +	+	High
Range of motion							High
*RANG[e]	– –	– –	– – –	– – –	–	–	
*FANG[f]	+ + +	+ + +	+ +	+ +	+·	+	
Soreness		+ +	+ + +	+ +	+		Low
Swelling		+	+ +	+ + +	+ +	+	Low

CK[a] = creatine kinase; LDH[b] = lactate dehydrogenase; AST[c] = aspartate aminotransferase; MHC[d] = myosin heavy chain; RANG[e] = relaxed arm angle; FANG[f] = flexed arm angle. +/– = minor increase/decrease; + +/– – = moderate increase/decrease; + + +/– – – = large increase/decrease.

days after subjects repeatedly descended stairs were three times larger than in control muscle and muscle obtained seven days postexercise. Subsequent studies using eccentric cycling and stepping exercise showed that focal disruptions were evident immediately post-exercise and worsened between 30 h and three days post-exercise (Friden et al. 1983; Newham, McPhail, Mills, and Edwards 1983). These studies also used electron microscopy to allow a more detailed examination of the changes occurring in the myofibrils after muscle-damaging exercise, and showed Z-line streaming and,

in some instances, complete absence of Z-line material between 30 h and three days postexercise. Taken together, these studies suggested that changes in the muscle become worse one to three days after eccentric exercise and are evidence of myofibrillar damage. However, ultra-structural changes have been observed two to three days (Stupka et al. 2000; Yu et al. 2004), seven to eight days, and as long as 10 days after eccentric exercise (O'Reilly et al. 1987; Manfredi et al. 1991).

Disruption of the Z-line is the principal ultrastructural abnormality reported in eccentric exercise studies (Friden

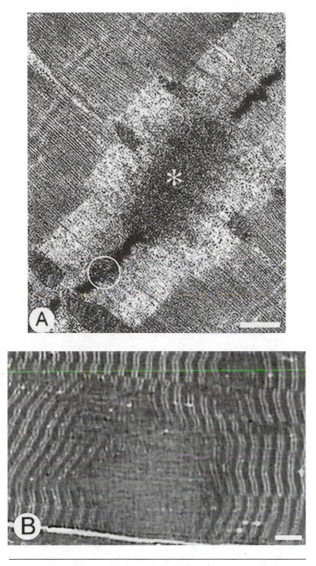

Figure 3.1   (*a*) Electron and (*b*) light microscopy of human muscle following eccentric exercise. Note Z-line streaming in *a* and myofibrillar disruption in *b*.

Reprinted, by permission, from J.G. Yu, L. Carlsson and L.E. Thornell, 2004, "Evidence for myofibril remodeling as opposed to myofibril damage in human muscles with DOMS: An ultrastructural and immunoelectron microscopic study," *Histochem Cell Biol*. 121(3): 219-227.

and Lieber 1992); thus this structure may represent the weak link in the contractile chain of the myofibril (Friden et al. 1981; Evans and Cannon 1991). In addition to Z-line disruption, A-band filament disturbances (Friden et al. 1983; Manfredi et al. 1991; Nurenberg et al. 1992), sarcoplasmic reticulum (SR) disruption (O'Reilly et al. 1987), autophagic vacuoles (Friden et al. 1983), central nuclei (Gibala et al. 1995), disrupted T-tubules and triads (Takekura et al. 2001), swollen or missing mitochondria (Friden et al. 1983; Manfredi et al. 1991), displacement of organelles and randomly oriented myofilaments

(Newham, McPhail, Mills, and Edwards 1983), disrupted I-bands (Nurenberg et al. 1992; Manfredi et al. 1991), cytoskeletal changes (Friden and Lieber 1996; Yu et al. 2002), extracellular matrix disturbances, and capillary disruption (Stauber et al. 1990) have all been reported. Although changes have been observed in both type I and type II fibers after eccentric exercise, myofibrillar disruption appears predominantly in type II fibers (Friden et al. 1983, 1988).

Damage Versus Remodeling

It is theorized that high mechanical forces associated with eccentric exercise result in increased intracellular calcium (Ca^{2+}), which activates highly destructive calcium-activated neutral proteases (CANP) such as calpain. Two substrates for calpain are desmin and α-actinin, cytoskeletal proteins involved in maintaining the integrity of the myofiber (Lieber et al. 1996; Friden and Lieber 1996). Because desmin and α-actinin are components of the Z-line, degradation of these proteins could make the Z-line susceptible to damage from eccentric contractions. Several animal studies (Leiber et al. 1996; Friden and Lieber 1998) have suggested that there is a loss of desmin staining within the first hour after eccentric exercise, indicating rapid structural damage to the muscle fiber.

Yu and colleagues have questioned the interpretation that Z-line streaming and cytoskeletal disruption are indeed evidence of muscle damage from eccentric contractions, arguing instead that these changes are evidence of myofibrillar remodeling. Yu and colleagues (2002) reported that, in contrast to results from animal studies, there was no loss of desmin staining in human muscles 1 h, two to three days, or seven to eight days following eccentric exercise. Instead, there were focal increases in desmin staining and some longitudinal desmin strands that were later shown to extend from one Z-line to another (Yu and Thornell 2002). Thus, it may be that cytoskeletal damage is not the initiating damaging event in the muscle, but rather that desmin is acting as a mechanical linkage in repair and remodeling of the sarcomere. This remodeling is further supported by the findings of supernumerary sarcomeres (additional sarcomeres in series) and increased staining of actin and other cytoskeletal proteins involved with sarcomerogenesis (α-actinin, titin, and nebulin) in some myofibrils (Yu et al. 2003, 2004) (see figure 3.2).

A limitation of the biopsy procedure is that quantitative assessment of whole-muscle damage (or remodeling) using biopsy samples could be under- or overestimated due to the limited amount of muscle tissue extracted during a typical biopsy. In addition, the biopsy procedure itself might have contributed to the ultrastructural damage previously observed in skeletal muscle fiber studies. Roth and colleagues (2000) reported similar hypercontracted and

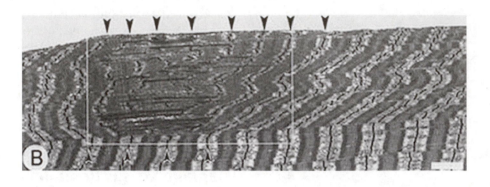

Figure 3.2   Evidence of myofiber remodeling after eccentric exercise. Arrows indicate supernumerary sarcomeres (the addition of one or more sarcomeres in a myofiber) in the days after eccentric exercise.

Reprinted, by permission, from J.G. Yu, L. Carlsson and L.E. Thornell, 2004, "Evidence for myofibril remodeling as opposed to myofibril damage in human muscles with DOMs: an ultrastructural and immunoelectron microscopic study," *Histochem Cell Biol.* 121(3): 219-227.

minced fibers in subjects who did not undergo eccentric exercise compared to those who did. Thus, the results of studies using muscle biopsy samples must be interpreted with caution.

Magnetic Resonance Imaging Observation

A more comprehensive assessment of changes to whole muscle after eccentric exercise has been achieved using MRI. The signal intensity of MRI (T2 relaxation time) is dependent on the amount of water in the tissue and appears to reflect intracellular edema. Thus, acute injury and an increase in edema or hemorrhage will prolong T2 relaxation times. Nurenberg and colleagues (1992) found a high negative correlation between areas of signal intensity increase and normal Z-lines in human muscle ($r = -0.88$) after downhill running; however, MRI has not always correlated well with other indices of muscle damage (Rodenburg et al. 1994; Mair et al. 1992). Moreover, peak T2 relaxation times (three to seven days postexercise) have been shown to occur later than peak functional impairment (Warren et al. 1999). It may be that changes in MRI are more reflective of edema associated with increased levels of muscle proteins in the extracellular space, such as creatine kinase and myoglobin, which are also indicative of muscle damage.

Magnetic resonance imaging may be quite useful in identifying the specific muscles damaged during eccentric exercise, as well as the duration of injury. Nosaka and Clarkson (1996a) reported different effects of eccentric exercise on the biceps brachii and brachialis muscles, with some subjects demonstrating increased signal intensity in either the biceps or brachialis or both. Takahashi and colleagues (1994) observed that damage to the quadriceps muscles was evident only in the vastus lateralis, vastus intermedius, and vastus medialis, not in the rectus femoris. While very

little direct muscle damage has been reported after 10 days following eccentric exercise (O'Reilly et al. 1987), several studies using MRI have shown prolonged changes in signal intensity that remained elevated at 15, 31, 56, and 75 days after eccentric exercise of the elbow flexors (Harrison et al. 2001; Nosaka and Clarkson 1996a; Foley et al. 1999; Shellock et al. 1991). Thus, MRI may be sensitive enough to detect muscle damage long after most conventional methods can and may provide some explanation for the prolonged functional losses observed after eccentric exercise (Sayers et al. 1999; Sayers and Clarkson 2001).

Biochemical and Histochemical Evidence of Muscle Damage

Some of the most commonly reported measures of muscle damage after eccentric exercise feature biochemical analyses of bodily fluids and histochemical analysis of tissue. This section focuses on the bio- and histochemical evidence of muscle damage after strenuous exercise, specifically, elevated protein levels and enzyme activity in the blood; inflammatory markers in the tissues, blood, and urine; alterations in calcium homeostasis; and impaired glucose metabolism.

Elevated Proteins and Enzyme Activity in the Blood

Blood levels and enzymatic activity of several different muscle proteins are routinely studied after strenuous exercise to examine the time course and appearance of muscle damage. Some of the more commonly examined proteins that have been shown to increase are creatine kinase (CK), myoglobin, lactate dehydrogenase (LDH), aspartate aminotransferase (AST), aldolase, troponin-I, myosin heavy chain (MHC), and β-glucuronidase. Little is currently known about the relationship among the various markers of muscle injury with respect to appearance or time course in the blood, as no study has systematically addressed them together. Moreover, blood concentration of a protein is a reflection of what is released by the tissue and what is cleared from circulation. Because CK is cleared by the reticuloendothelial system and myoglobin is cleared through the kidney, differences in rates of clearance may make comparisons among these proteins, and their relationships to damage, dubious. Table 3.1 compares the time course and magnitude of response among several common biochemical markers of muscle damage. In this section we limit our discussion to examination of the strengths and limitations of CK activity as an indirect measure of muscle damage because it is the most thoroughly studied of these proteins (Clarkson and Hubal 2001).

After damage to the muscle, CK is released to the interstitial spaces, where it is taken up by the lymphatic system to then be slowly returned to circulation via the thoracic duct. The time course of CK activity in the blood is dependent on the type of exercise used to induce muscle damage. The two most commonly used means of inducing mild forms of muscle damage are downhill running and high-force eccentric exercise that involves forced lengthening of the muscle using weights, isokinetic dynamometers, or lever systems. Plasma CK activity peaks approximately 12 to 24 h after downhill running and may reach only 300 to 600 IU (Byrnes et al. 1985). After high-force eccentric exercise, however,

CK activity does not increase until 24 to 48 h postexercise and peaks four to five days postexercise (Clarkson et al. 1992). Peak CK activity following high-force eccentric exercise regularly reaches 2500 IU or more (Clarkson et al. 1992) and demonstrates the largest change from baseline compared to other commonly used biochemical markers of muscle damage (see figure 3.3).

Because high-force eccentric exercise results in greater damage to the muscle than downhill running, this greater damage could contribute to the greater CK activity observed; but there is no clear explanation for the differences in time course. Interpretation of the CK response to exercise is also complicated by the large intersubject variability in similarly exercised individuals. Researchers have reported CK activity after high-force eccentric exercise ranging from 500 IU to 34,500 IU (Newham, McPhail, Mills, and Edwards 1983) and from 236 IU to 25,244 IU (Nosaka and Clarkson 1996b). While age, gender, race, muscle mass, and activity level have each been examined in attempts to explain differences in CK response among individuals, the reason for the large intersubject variability in CK remains poorly understood.

Studies in both humans and animals have called into question the utility of CK activity as a measure of quantifiable muscle damage by showing that CK activity does not necessarily reflect the extent of myofiber damage. Manfredi and colleagues (1991) observed large differences between older and younger men in ultrastructural damage to the

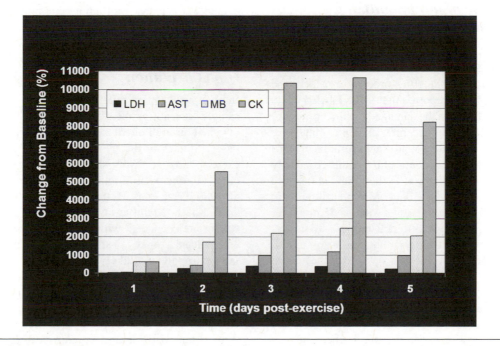

Figure 3.3   Time course and magnitude of change from baseline in several of the most commonly used biochemical markers of muscle damage, lactate dehydrogenase (LDH), aspartate aminotransferase (AST), myoglobin (MB), and creatine kinase (CK) activity.

Data modified from P.M. Clarkson and K. Nosaka, 1996. Myoglobin data modified from S.P. Sayers and P.M. Clarkson, 2003.

vastus lateralis. However, CK activity was not significantly different between the groups. Van der Meulen and colleagues (1991) reported that actual muscle damage in rats, determined through histological assessment after downhill running, was significantly less than damage calculated based on CK appearance and disappearance rates. In a similar animal model, Kuipers and colleagues (1985) reported that histological assessment of muscle damage correlated poorly ($r = 0.04$) with serum CK activity. Conversely, Nosaka and Clarkson (1996b) reported greater abnormalities in the T2 relaxation times of the elbow flexors in subjects who exhibited the greatest CK response after eccentric exercise. Mair and colleagues (1992) and Evans and colleagues (1998) also reported that the time course of elevated CK activity and increases in MRI signal intensity were similar, suggesting that the two measures may be related to the same aspect of muscle damage.

The Inflammatory Response

An increased inflammatory response following eccentric exercise has been documented in the literature with use of muscle biopsy and in vivo radiolabeling techniques, as well as serum and urine assays. The time course of the inflammatory response to eccentric exercise is dependent on the mode, duration, and intensity of exercise; the muscle group examined; and whether muscle samples are examined directly or muscle damage is measured indirectly via blood analysis.

Direct Evidence of Inflammation From Skeletal Muscle Tissue

Several human studies have presented evidence of inflammation in skeletal muscle tissue after eccentric exercise. Using an in vivo radiolabeling procedure, MacIntyre and colleagues (1996) reported that ^{99m}Tc-labeled white blood cells were found in the damaged quadriceps tissues between 4 and 20 h after eccentric exercise. MacIntyre and colleagues (2000) later found that infiltration of these inflammatory cells (likely neutrophils) occurred as early as 2 h after eccentric exercise. Fielding and colleagues (1993) found direct evidence of neutrophil infiltration in muscle tissue of the quadriceps as soon as 45 min after downhill running. Despite different modes of damage-inducing contractions, it appears that eccentric exercise results in a rapid migration of neutrophils into the muscle tissue, which likely act as chemoattractants for additional inflammatory mediators.

Like neutrophils, macrophages are actively phagocytic (Abrams 1997) and may contribute to damage to the muscle in the days after eccentric exercise because they give rise to proinflammatory cytokines including interleukin (IL)-1, IL-6, and tumor necrosis factor (Evans and Cannon 1991). Cannon and colleagues (1989) reported

increased IL-1β in the muscles of the quadriceps up to five days after downhill running. Fielding and colleagues (1993) reported that IL-1β was elevated 135% by 45 min after eccentric exercise, suggesting a robust immediate response of these cytokines in the muscle, and 250% after five days. Interestingly, other studies suggest a chronic inflammatory phase occurring from several days to weeks after eccentric exercise. Jones and colleagues (1986) and Round and colleagues (1987) observed an increase in mononuclear cells (hypothesized to be macrophages and T-lymphocytes) infiltrating muscle tissue 9 to 14 days following eccentric exercise. Jones and colleagues suggested that these cells were not exacerbating damage but were likely involved in the repair process. Indeed, subpopulations of macrophages have been shown to be involved in removal of cellular debris (ED1$^+$) as well as muscle repair (ED2$^+$) (Tidball 1995).

Again, the biopsy procedure itself, failure to use a control group, or other variations in experimental design might be potential limitations in these studies. The findings of the Malm et al. (2000) study suggest that the biopsy procedure itself may generate a measurable inflammatory response. For example, Malm and colleagues (2000) found that changes occurring in IL-1β, neutrophils, and macrophages up to seven days after eccentric exercise were similar to a control group, and a control group was not used in the studies by Fielding and colleagues (1993) and Cannon and colleagues (1989). Still, the findings of MacIntyre and colleagues (1996, 2000) with use of whole-muscle imaging provide strong evidence of increased inflammation in the tissues stemming from eccentric exercise.

Indirect Evidence of Inflammation From Circulation

Indirect measures of inflammation using blood or urine samples are not as clearly defined as those measures obtained through direct examination of skeletal muscle tissue. Studies of the neutrophil response to eccentric exercise have generally shown significant increases in circulation between 2 and 12 h postexercise (Smith et al. 1989; Pizza et al. 1995; Malm et al. 1999) but not at 24 and 48 h (Malm et al. 1999). Taken together, these studies are consistent with those using tissue bioassays and indicate a rapid, but short-lived, increase in circulating neutrophils that likely precedes infiltration into the damaged muscle tissue. Furthermore, the time course of circulating monocytes and macrophages is fairly consistent with that reported in muscle tissue. Malm and colleagues (1999) observed that circulating monocytes were elevated at 6 h after eccentric exercise but not at 24 h, again preceding the appearance of macrophages in the tissues reported at 12 h after exercise (Tidball 1995; Abrams 1997). In a second study, Malm and colleagues

(2000) observed the monocyte response at approximately four days after eccentric exercise, preceding the response seen in muscle tissue 9 to 14 days postexercise (Jones et al. 1986; Round et al. 1987).

The results of studies on the cytokine response to eccentric exercise are less clear. While circulating levels of cytokines in the blood increase after strenuous exercise, often the time course and magnitude of the change depend on the type of eccentric exercise imposed. For example, in studies using eccentric exercise with an endurance component (eccentric cycling or downhill running), Cannon and colleagues (1991) reported increases in IL-1β and tumor necrosis factor-alpha (TNF-α) as early as 6 h postexercise while Evans and colleagues (1986) reported increases in IL-1 at 3 h postexercise. Nosaka and Clarkson (1996a), however, using a high-force eccentric exercise model, failed to show any changes in cytokines. Some high-force eccentric exercise models show a different time course compared to endurance-type eccentric exercise models, with increases in some cytokines occurring at 24 and 72 h postexercise (Smith et al. 2000). Other high-force eccentric exercise studies have shown decreases in IL-1β and P-selectin (Smith et al. 2000) and TNF-α and IL-8 (Hirose et al. 2004). Thus, eccentric exercise without an endurance component generally results in a less robust cytokine response and a more prolonged time course compared to strenuous endurance exercise. Conflicting results among studies could also be attributable to the different amounts of muscle mass exercised during the eccentric exercise protocols and subsequently damaged.

Disruption of Calcium Homeostasis

The effect of intracellular calcium (Ca^{2+}) accumulation on muscle tissue has long been implicated in cell necrosis (Busch et al. 1972; Duncan 1978). It is likely that the mechanical forces associated with eccentric exercise damage the cell membrane and increase the permeability of the cell to high concentrations of intracellular Ca^{2+} (Armstrong 1984). Increased Ca^{2+} then gives rise to CANP, which have been shown to preferentially degrade Z-lines and muscle proteins (Busch et al. 1972). Lynch and colleagues (1997) reported that free intracellular Ca^{2+} concentration was elevated 24 to 48 h after downhill running in mice. Duan and colleagues (1990) reported increased mitochondrial Ca^{2+} concentration after downhill walking, while Warren and colleagues (1995) observed increased free cytosolic Ca^{2+} concentration after stimulated eccentric contractions. Hill and colleagues (2001) were the first to report in humans that Ca^{2+} uptake and release were impaired immediately after eccentric exercise, indicating potential exercise-induced damage to the SR. In the animal model, Yasuda and colleagues (1997) reported SR membrane degradation 12 h after electrically stimulated eccentric contractions; however, Warren and colleagues (1993) showed that force deficits in mice after eccentric contractions were not due to a damaged SR.

Impaired Glucose Metabolism

Several studies have indicated that glycogen concentrations in skeletal muscle following eccentric exercise are depressed. O'Reilly and colleagues (1987) examined biopsy samples following eccentric cycling and reported that glycogen concentrations immediately postexercise had dropped 61% from baseline and were still 44% depressed at 10 days postexercise. Kuipers and colleagues (1985) reported that reductions of glycogen were evident 24 h after compared to immediately after eccentric exercise. These data overall suggested an impairment of glycogen resynthesis with eccentric exercise. There is also evidence that glucose disposal rates are significantly reduced by eccentric exercise (as compared to both nonexercised control and concentric exercise conditions [Kirwan et al. 1992]) and that plasma insulin levels are higher (King et al. 1993). Taken together, these studies point to dysfunction in the glucose transport system following eccentric exercise.

The muscles participating in eccentric contractions may also be more resistant to certain aspects of insulin action. Asp and colleagues (1995) reported depressed glucose transporter-4 (GLUT4) concentration one to two days following eccentric knee extension exercise, with concomitant decreases in glycogen concentration, both of which returned to baseline by four days postexercise. Asp and colleagues (1996) linked the decrease in GLUT4 protein content with reduced insulin sensitivity, reporting that GLUT4 content of the eccentrically exercised thigh was 39% lower compared to that in resting conditions, despite high levels of insulin. Further studies point to a defect in insulin signaling following eccentric exercise. For example, Kristiansen and colleagues (1997) reported a decrease in GLUT4 transcription rate and a reduction in GLUT4 messenger RNA (mRNA) at 48 h following eccentric exercise. Del Aguila and colleagues (2000) reported that insulin-stimulated insulin receptor substrate (IRS-1) phosphorylation was 45% lower than in controls following downhill running, while insulin-stimulated phosphatidylinositol (PI3) kinase (−34%), protein kinase B (Akt) serine phosphorylation (−65%), and Akt activity (−20%) were also depressed.

Functional Evidence of Muscle Damage

Other commonly reported indirect indicators of muscle damage include muscle function measures. This section focuses on functional evidence of muscle damage, specifically strength loss and recovery, low-frequency fatigue (LFF), neuromuscular disturbances, changes in range of motion and stiffness of muscles and joints, swelling, and soreness.

Strength Loss and Recovery

Prolonged strength loss after eccentric exercise is one of the most commonly used indicators of exercise-induced muscle damage. In a review of the eccentric exercise literature, Warren and colleagues (1999) estimated that 50% of human studies measured maximal voluntary contraction torque, while 67% of the animal studies under review tested maximal isometric tetanic force capabilities. Loss of strength is also considered one of the most valid and reliable indirect markers of exercise-induced muscle damage (Warren et al. 1999).

Immediately following exercise, strength losses may be attributed primarily to metabolic fatigue, and such losses occur following all modes of exercise (concentric, isometric, and eccentric). However, strength losses following eccentric exercise outlast changes in metabolite levels (Edwards et al. 1977) and are likely due to the effects of muscle damage. The persistence of strength losses into the days (and possibly weeks) following exercise is dependent not only on mode (Jones et al. 1989), but also on the intensity and duration of the exercise protocol. Isometric and concentric protocols in humans produce a wide range of strength losses (fatigue) immediately postexercise; a typical concentric, dynamic isotonic exercise produces a voluntary strength loss of ~10% to 20% immediately following exercise, with strength returning to baseline over the next several hours. The pattern of strength loss and recovery following eccentrically biased exercise protocols depends on the model used. Downhill running or eccentric cycling typically results in immediate strength losses from 10% to 30%. Strength recovery following downhill running takes longer than that following concentric protocols, with a return to preexercise baseline levels about 24 h after exercise. Moderately intensive eccentric exercise protocols typically elicit 30% to 50% strength losses immediately postexercise (see figure 3.4), while intense protocols can produce 50% to 70% average strength losses. Recovery times for these more intense protocols are usually 7 to 10 days but can extend to several weeks, especially in animal models using maximal stimulated eccentric contractions.

Evidence from the literature suggests that excitation–contraction (EC) coupling failure plays an important role in force reductions observed after eccentric exercise. Warren and colleagues (1993) reported a 43% reduction in tetanic force after eccentric contractions in the mouse soleus; but when caffeine was used to potentiate Ca^{2+} release from the SR, force reductions were eliminated. Ingalls and colleagues (1998) also potentiated Ca^{2+} release from the SR to examine force loss at various time points up to 14 days following eccentric exercise; they concluded that impairment of EC coupling is the primary mechanism responsible for tetanic force reductions from zero to five days after eccentric contractions, accounting for 57% to 75% of this reduction (Ingalls et al. 1998).

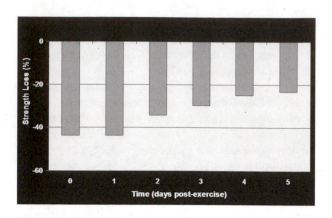

Figure 3.4 Percent strength loss following eccentric exercise.

Data modifed from Clarkson and Tremblay, 1988.

Factors associated with strength loss and recovery following exercise have also been investigated. The exacerbation of strength loss following exercise at longer muscle lengths has been documented (Child et al. 1998), as have the protective effects of either a single bout of eccentric exercise or eccentric exercise training (Clarkson and Tremblay 1988; Clarkson et al. 1992; Balnave and Thompson 1993). Treatment modalities for facilitating strength recovery that demonstrated positive findings include immobilization (Sayers et al. 2003) and compression therapy (Kraemer et al. 2001), while some therapies such as antioxidant supplementation and hyperbaric oxygen treatment have produced conflicting findings. Use of nonsteroidal anti-inflammatory drugs has demonstrated positive effects in both younger and older humans (Sayers et al. 2001; Baldwin et al. 2001), but animal studies have indicated either no effect (Pizza et al. 1999) or enhanced early recovery but delayed long-term recovery of strength (Mishra et al. 1995).

Low-Frequency Fatigue

One of the functional changes consistently demonstrated following eccentric exercise is the relative impairment of force produced at low stimulation frequencies, first described by Edwards and colleagues (1977). After exercise, the ability of muscle to generate force with use of tetanic stimuli at frequencies such as 10 to 20 Hz is impaired to a greater degree than the ability to generate force with use of higher-frequency stimulations (50+ Hz) (Jones 1996). This has been demonstrated in several different muscle groups, including the elbow flexors (Jones et al. 1989) and quadriceps (Newham, Mills, Quigley, and Edwards 1983; Martin et al. 2004).

Edwards and colleagues (1977) demonstrated that LFF persisted longer than the reductions of high-energy phosphates or changes in central activation, suggesting that LFF is caused by a reduction in contractile activation or peripheral failure of EC coupling. This proposition is supported by work from single-fiber experiments (Westerblad et al. 1993; Chin et al. 1997) implicating altered calcium kinetics within the muscle fiber as the cause of LFF. In humans, Hill and colleagues (2001) reported decreased torque production at lower frequencies and decreases in calcium release from muscle, which suggests possible EC coupling failure. There is also evidence that LFF may be caused by physical damage to sarcomeres and remodeling of the tissues, in particular a shift in the length–tension relationship toward greater muscle lengths (Jones 1996). However, recent evidence suggests that the shift in the length–tension relationship may be the result of damage to the sarcomeres and postexercise fatigue (Butterfield and Herzog 2006).

Neuromuscular Disturbances

The inability to generate muscle force following muscle-damaging exercise is considered the best indirect indicator of muscle damage. However, the capacity to generate force is a product of both neural drive and muscle contractile function. Greater neural drive impairments could theoretically play a role in the inability to generate maximal muscle force output after eccentric exercise.

Decreases in voluntary activation can also impair muscle function through central fatigue (Kent-Braun and Le Blanc 1996; Kent-Braun 1999; Loscher and Norlund 2002), in which decreased motoneuron discharge rate or decreased motor unit recruitment can lead to force loss. While central fatigue resolves quickly after exercise (usually within minutes) (Bigland-Ritchie et al. 1986), disturbance of neuromuscular variables in the period following eccentric exercise has been reported. Avela and colleagues (1999) demonstrated a biphasic reduction in reflex sensitivity following stretch–shortening cycles, the first phase potentially attributable to disfacilitation and presynaptic inhibition and the secondary decline potentially attributable to changes from inflammatory processes. Bulbulian and Bowles (1992) described decreased motoneuron pool excitability following downhill running, while Deschenes and colleagues (2000) described a decrease in neuromuscular efficiency (the ratio of torque to integrated electromyography [iEMG]) that lasted until 10 days postexercise—longer than changes in all other symptoms of exercise-induced muscle damage. Several papers have also demonstrated alterations in neuromuscular variables such as force matching (Proske et al. 2003, 2004), tracking (Pearce et al. 1998), and position sense (Saxton et al. 1995; Walsh et al. 2004) following eccentric exercise.

On the other hand, some recent studies have provided evidence against any centrally mediated effects on force production. Behm and colleagues (2001) showed no correlation between muscle activation and force output following exercise with both eccentric and concentric components. Furthermore, Loscher and Norlund (2002) reported no

association of motor cortex excitability with levels of muscle activation during eccentric exercise. While these studies do not rule out contributions to force losses postexercise, the preponderance of evidence suggests that central processes do not play a significant role.

Range of Motion

Many studies have documented decreases in the voluntary range of motion of a joint (i.e., the distance and direction through which a joint can move) following eccentric exercise (see figure 3.5). Eccentric exercise results in an increased arm angle when subjects attempt to flex the arm and a decreased arm angle when the arm is hanging passively by the side. These two arm angles are referred to as the flexed arm angle (FANG) and the relaxed arm angle (RANG), respectively. By subtracting the smaller angle (FANG) from the larger angle (RANG), one can obtain a measure of the overall range of motion (ROM) (Nosaka and Sakamoto 2001). Warren and colleagues (1999) estimated that 19% of the human studies surveyed at the time quantified range

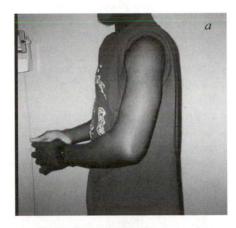

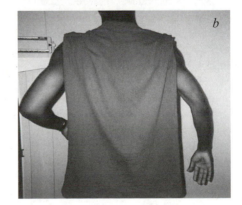

Figure 3.5 *(a)* Swelling and *(b)* changes in range of motion (stiffness) after eccentric exercise of the elbow flexors. This individual performed 50 maximal eccentric contractions with his left arm. Note the swelling response of the left arm *(a* and *b)* as well as the reduction in arm angle compared to the right arm in *b*.

of motion following eccentric exercise, and this measure has been reported to be one of the most valid and reliable indicators of exercise-induced muscle damage (Warren et al. 1999).

Immediately after eccentric exercise, the increase in FANG is maximal and subjects are unable to fully flex the arm. Recovery begins within 24 h and is not fully achieved until 10 days postexercise (Clarkson et al. 1992). In normal individuals prior to eccentric exercise, FANG has been reported to be between 36° and 40°, and changes of greater than 20° immediately after maximal eccentric contractions have been reported (Clarkson and Tremblay 1988; Ebbeling and Clarkson 1990). Although the mechanism explaining the increase in arm angle has not been determined, FANG and reductions in maximal isometric force show a similarity in time course. Immediately after eccentric contractions, there is a decrease in RANG that becomes more acute in the days after exercise. At three days postexercise, the greatest decrease in RANG is observed, followed by a gradual return to baseline by 10 days postexercise. In normal individuals prior to eccentric exercise, RANG has been reported to be between 155° and 170° (Clarkson and Tremblay 1988; Ebbeling and Clarkson 1990), and changes of up to 45° after eccentric contractions have been reported (Jones et al. 1987). Despite shortening of the muscle after eccentric exercise, these changes have not been associated with an increase in neural activity (Jones et al. 1987; Bobbert et al. 1986). RANG has been used as a measure of muscle "stiffness" due to an increased resistance of the muscle to passive lengthening (Howell et al. 1993; Chleboun et al. 1998).

Some studies have also quantified muscle stiffness using devices that provide force feedback information at given joint angles. Chleboun and colleagues (1995, 1998) documented increased muscle stiffness in the elbow flexors up to five days postexercise. These investigators (1998) also documented a disconnect between the time course of stiffness change following exercise and the time course for muscle volume increases, indicating that swelling in the muscle does not account for changes in muscle stiffness. In a rabbit model, Benz and colleagues (1998) documented a 30% reduction in muscle stiffness during eccentric exercise but no change in stiffness during an isometric protocol. Joint stiffness has also been assessed during passive and active conditions. Leger and Milner reported no changes in passive joint stiffness or joint stiffness at low muscle activation levels (i.e., 10-15%) in either the wrist (2000b) or the first dorsal interosseus (2000a) following exercise, but significant decreases in joint stiffness at higher activation levels (50-65%).

Swelling

Swelling occurs within the muscle following exercise-induced injury (see figures 3.5 and 3.6). Common methods used to test for the presence of edema include

anthropometrics (muscle group circumference), ultra-sonography (muscle thickness) and MRI (T2 [trans-verse or spin-spin] relaxation time) (see figure 3.6). Circumference increases, which have been noted in many studies, typically peak at four to five days post-exercise, similar to the time course seen for changes in the ultrasound images (Nosaka and Clarkson 1996a; Chleboun et al. 1998). Studies demonstrating increased T2 following damaging exercise (Ploutz-Snyder et al. 1997; Foley et al. 1999) have indicated the presence of both an acute phase increase (from 0 to 1 h post-exercise) and a gradual phase increase from one to six days postexercise. Typically, the T2 signal peaks around 3 to 10 days postexercise (Shellock et al. 1991; Nosaka and Clarkson 1996a; Foley et al. 1999), which is a bit later than peaks in circumference or ultrasound thickness (Nosaka and Clarkson 1996a).

The location of swelling in the muscle appears to change over time (Nosaka and Clarkson 1996a). Using ultrasound, Howell and colleagues (1993) reported that 65% of the swelling of the elbow flexors after eccentric exercise was located in the muscle compartment, with the rest located in the subcutaneous spaces, from 1 to 10 days postexercise. However, Nosaka and Clarkson (1996a), using MRI, found that most of the fluid accumulation began in the intracel-lular space of the muscle fibers and remained there for

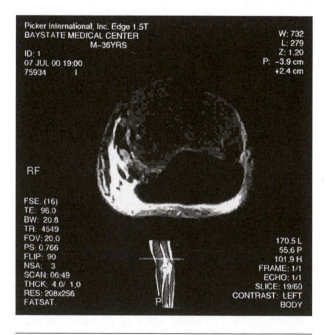

Figure 3.6 Magnetic resonance image (MRI) of the swollen left arm of a 36-year-old volunteer nine days after 50 maximal eccentric contractions of the elbow flexors. Cross section was obtained just proximal to the elbow. Note fluid accumulation outside the muscle that appears to have pooled toward the back of the arm after a prolonged period of time spent supine in the magnet.

approximately five days. The intracellular fluid gradually moved toward the outside of the perimysium over 10 days postexercise. Over the course of several days, the effects of gravity result in fluid accumulation around the elbow, forearm, and hand.

Muscle Soreness

Delayed-onset muscle soreness (DOMS) is a hallmark manifestation of exercise-induced muscle damage. This connection was first hypothesized by Hough (1902), who suggested that mechanical stress during strenuous exercise caused muscle damage, resulting in delayed muscle soreness. Delayed-onset muscle soreness is unrelated to fatigue, following a different time course of change after exercise than muscle function measures such as strength loss (see figure 3.7).

Warren and colleagues (1999) reported that subjectively and objectively determined soreness levels were measured in 63% and 12% of the reviewed human studies of exercise-induced muscle damage, respectively. Of all the measurements documented in Warren and colleagues' (1999) review, soreness was the most commonly assessed in human muscle damage studies. Subjectively determined DOMS is most often assessed via questionnaire. A visual analog scale (VAS) or numerical scale is typically used to describe soreness levels. In this test, a subject is asked to mark on a fixed-length written scale the level of soreness felt in the muscle according to verbal instructions from the investigator. Objective measures include measurement of force manually applied to various points on the muscle that evokes pain or tenderness (Newham, Mills, Quigley, and Edwards 1983; Newham et al. 1986).

Typically, DOMS develops between 24 and 48 h postexercise and peaks between 24 and 72 h postexercise (Clarkson et al. 1992). By five to seven days postexercise, DOMS has usually abated and pain or soreness levels return to baseline (Armstrong 1984; Clarkson and Tremblay 1988). Peak values of soreness are often protocol dependent. On a soreness scale from 1 (no soreness) to 10 (very sore), tasks such as downhill running typically produce soreness levels of 4 to 5, while maximal tasks such as maximal eccentric contractions of the elbow flexors often evoke average peak soreness levels of 7 to 8. One of the limitations in using DOMS to indicate muscle damage is the incidence of low responders or nonresponders. Sayers and colleagues (2001) reported that almost one-third of subjects undergoing maximal eccentric contractions reported no "meaningful" muscle soreness (at least 50 mm on a 100 mm VAS).

One study showed that DOMS was localized throughout the entire muscle after eccentric exercise (Newham, Mills, Quigley, and Edwards 1983); others have reported pain and soreness at the distal portion of the muscle at the myotendinous junction (Edwards et al. 1981; Newham, Mills, Quigley, and Edwards 1983). The reason for this pain localization may be the architecture of the muscle at this site. Friden and colleagues (1986) suggested that the oblique orientation of muscle fibers just proximal to

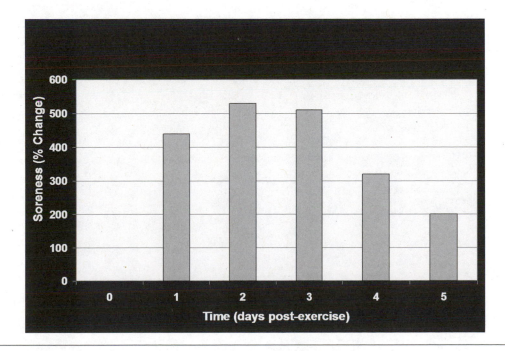

Figure 3.7   Percent increase in delayed-onset muscle soreness (DOMS) after eccentric exercise.

Data modifed from Clarkson and Tremblay, 1988.

the neuromuscular junction makes these fibers vulnerable to the high tensions associated with eccentric exercise. However, others have suggested that pain receptors are most concentrated in the tendons and connective tissue (Newham et al. 1982), thus accounting for pain location in the distal portion of the muscle.

Perception of the dull, aching pain that signals DOMS is thought to be attributable primarily to the type III and type IV afferent neurons. Mechanical nociceptors (pain receptors) are typically type III fibers, while type IV fibers are polymodal (responding to mechanical, thermal, and chemical signals). Two primary mechanisms underlying the development of DOMS sensations have been proposed. The first is that swelling and subsequent increased pressure within the muscle activate resident mechanoreceptors. The second is that chemical changes within the muscle (i.e., increased histamines, bradykinins, and prostaglandins following the infiltration of inflammatory cells and breakdown of myocellular components) activate the polymodal nociceptors that are sensitive to chemical signals. Recent evidence suggests that larger, myelinated afferent fibers might also be responsible for DOMS sensations, as Weerakkody and colleagues (2001, 2003) reported that pressure block of the sciatic nerve (which blocks only large afferent fibers) significantly increased pain thresholds in the triceps surae following backward walking.

Summary

Histological, bio- and histochemical, and functional evidence of muscle damage has been reported after exercise-induced muscle damage. There is some controversy about whether direct myofibrillar disturbances observed with use of light and electron microscopy are truly evidence of muscle damage or instead reflect muscle remodeling. The small amount of muscle tissue observed with the typical biopsy and the invasive nature of the extraction procedure are limitations to the use of histological measures; thus, results should be interpreted with caution. To address these limitations, MRI has been used as a way to quantify whole-muscle damage; however, MRI (T2 relaxation times) reflects only intracellular edema. Creatine kinase is probably the most commonly used biomarker of muscle damage; however, variability in time course among different exercise protocols and the large intersubject variability have called into question the utility of CK activity as a measure of quantifiable muscle damage. Some of the most valid and reliable measures of muscle damage are functional measures such as strength loss and changes in the range of motion. Other functional measures such as muscle swelling and soreness are commonly reported because of the relative ease of data collection using these methods; however, there is large variability in these measures, and they cannot be used to accurately reflect the extent of muscle damage after strenuous exercise.

CHAPTER 4

Neutrophils and Macrophages in Muscle Damage and Repair

Francis X. Pizza, PhD

Clinicians often resort to "anti-inflammatory" treatments (e.g., nonsteroidal anti-inflammatory drugs) for muscle injury. These treatments are intended to alter the inflammation biology of injured skeletal muscle in the hope of preventing further loss of muscle function (e.g., strength and range of motion), reducing muscle soreness and pain, and enhancing muscle repair and regeneration. Many of these treatments, however, are largely ineffective in alleviating muscle injury symptomatology and inflammation or in enhancing muscle repair and regeneration.

This unfortunate outcome is compounded by the lack of information on (a) how injured skeletal muscle initiates the acute inflammatory response and (b) the contribution of a subpopulation of leukocytes, commonly referred to as inflammatory cells (neutrophils and macrophages), to functional, cellular, and molecular events associated with muscle injury and repair and regeneration. If the severity of a muscle injury is to be reduced and the rate of muscle repair and regeneration enhanced, more research is needed on the complex interplay between inflammatory cells and injured skeletal muscle. Once the interplay has been understood, it will be possible to develop targeted approaches to minimize the extent of the injury and to enhance muscle repair and regeneration.

This chapter begins by describing the time course of appearance of neutrophils and macrophages in injured skeletal muscle and then discusses mechanisms for their entry into (diapedesis) and their migration to and within (chemotaxis or haptotaxis) injured skeletal muscle. These topics provide the foundation for understanding how other functional activities of neutrophils and macrophages (e.g.,

phagocytosis, reactive oxygen species [ROS] production, degranulation, and cytokine production) could influence the events associated with muscle injury and repair and regeneration. The chapter concludes with a working model for how injured skeletal muscle orchestrates functional activities of neutrophils and macrophages and how these inflammatory cells contribute to events associated with muscle injury and repair and regeneration.

The Timing and Mechanism of Neutrophil and Macrophage Accumulation

This section discusses when and how neutrophils and macrophages accumulate in injured skeletal muscle. The timing of their accumulation is important because it provides some insight into how neutrophils and macrophages could influence the sequential stages of muscle injury, repair, and regeneration. Understanding how neutrophils and monocytes leave the vasculature and enter into injured skeletal muscle could lead to the development of therapeutic or pharmacological strategies to reduce or enhance their presence within skeletal muscle.

The Timing of Accumulation

The majority of studies that have characterized the time course of neutrophil and macrophage accumulation in injured skeletal muscle have used animal models of muscle injury such as lengthening contractions (Pizza et al. 2002; Tsivitse et al. 2003; Lapointe, Frenette, and Cote 2002), mechanical loading of atrophic muscle (i.e., hindlimb suspension reloading model) (Tidball, Berchenko, and Frenette 1999; Frenette et al. 2002, 2003), or chemical (myotoxin) (Teixeira et al. 2003) or physical (e.g., crush or freeze) trauma (Farges et al. 2002; Warren et al. 2004). Despite differences in the nature, kinetics, and magnitude of the injury, all of these models cause neutrophils and macrophages to accumulate in injured muscle in hours to days after the injury.

The time course of neutrophil and macrophage accumulation in injured skeletal muscle resembles the classical response to an infection. Specifically, we have reported that neutrophils begin to accumulate in skeletal muscle within 2 h, reach their peak concentration at one day, and return to control levels within seven days after injury induced by either lengthening contractions (Koh et al. 2003; Pizza et al. 2002; McLoughlin, Mylona et al. 2003; Tsivitse et al. 2003; Pizza et al. 2005) or mechanical loading of atrophic muscle (Frenette et al. 2002) (figure 4.1). Macrophages, on the other hand, generally appear in injured skeletal muscle after the arrival of neutrophils and remain elevated while neutrophil concentrations are diminishing (Koh et al. 2003; Pizza et al. 2002; McLoughlin, Mylona et al. 2003; Tsivitse et al. 2003; Pizza et al. 2005) (figure 4.2).

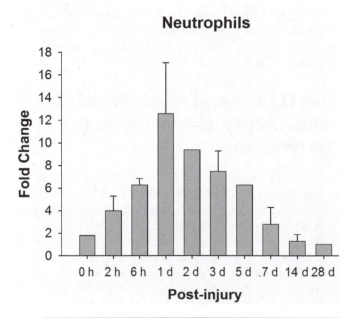

Figure 4.1 Fold changes in neutrophil concentrations after muscle injury. Values were calculated relative to control levels using data derived from studies in which muscle injury was induced by either lengthening contractions or the loading of atrophic skeletal muscle. Neutrophil concentrations in these studies were determined via immunohistochemistry using antibodies specific for mouse or rat neutrophils (Ly6G and HIS48, respectively). Mean ± SEM.

Several studies have examined the accumulation of inflammatory cells in human skeletal muscle after injury. Investigators have reported an elevation (MacIntyre et al. 2001; Hellsten et al. 1997; Stupka et al. 2001) or no change (Peterson et al. 2003; Beaton, Tarnopolsky, and Phillips 2002; Stupka et al. 2001) in the concentration of neutrophils one to four days after exercise-induced muscle injury. Conflicting observations may be attributable to differences in the exercise protocol, sampling time point, techniques used to quantify neutrophils in skeletal muscle, or some combination of these. On the other hand, macrophages have consistently been reported to be elevated in human skeletal muscle one to four days after injurious exercise (Stupka et al. 2001; Peterson et al. 2003; Beaton, Tarnopolsky, and Phillips 2002).

The kinetics of neutrophil and macrophage accumulation in humans after muscle injury is difficult to ascertain experimentally because multiple muscle biopsies are required and the biopsy procedure itself elicits an inflammatory response, confounding the interpretation of when and for how long neutrophils and macrophages accumulate in the injured muscle. Given this and other experimental issues such as ensuring that the muscle sample is obtained from the injured region, we can only assume that the time course for neutrophil and macrophage accumulation in injured human muscle is similar to that for animals after muscle injury.

It is important to emphasize that the mere accumulation of neutrophils and macrophages provides little information on their biological function in injured skeletal muscle. What can be gleaned from their accumulation profile is that inflammatory cell diapedesis and migration occur after muscle injury. Whether neutrophils or macrophages (or both) are performing other functional activities (e.g., phagocytosis, ROS production, degranulation, and cytokine production) in injured skeletal muscle cannot be inferred from their mere presence in injured skeletal muscle.

The Mechanism of Accumulation

Diapedesis, also referred to as transmigration or extravasation, is the movement of neutrophils or monocytes from the postcapillary venules into tissues. Diapedesis of neutrophils or monocytes is preceded by a complex sequence of events that result in their adhesion to vascular endothelial cells. Neutrophil and monocyte tethering, rolling, activation, and firm adhesion compose the classical paradigm for their interaction with endothelial cells prior to diapedesis.

Tethering and rolling are usually mediated by the selectin family of adhesion molecules (P-, E-, L-selectin; CD62P, CD62E, CD62L, respectively) (Patel, Cuvelier, and Wiehler 2002). P- and E-selectin are upregulated

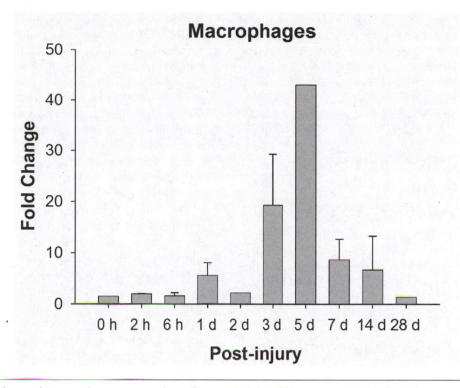

Figure 4.2   Fold changes in macrophage concentrations after muscle injury. Values were calculated relative to control levels using data derived from our studies in which muscle injury was induced by either lengthening contractions or the loading of atrophic skeletal muscle. Macrophage concentrations in these studies were determined via immunohistochemistry using antibodies specific for mouse or rat macrophages (F4/80 and ED1, respectively). Mean ± SEM.

on endothelial cells during inflammation, whereas L-selectin is constitutively expressed on leukocytes. Frenette and colleagues (2003), using the hindlimb suspension reloading model of muscle injury and knockout mice for both P- and E-selectin, reported that P- and E-selectin serve as a mechanism for the early entry (less than one day) of neutrophils (Ly6G+ cells), but not macrophages (F4/80+ cells), into injured skeletal muscle. Baker and colleagues (2004) examined the contribution of P-selectin to inflammatory cell diapedesis after lengthening contractions by injecting (IV) P-selectin glycoprotein ligand-1 (sPSGL-1). The administration of sPSGL-1 would in theory neutralize P-selectin by binding to P-selectin on endothelial cells and thus prevent circulating leukocytes from binding to the P-selectin. The authors reported that neutralization of P-selectin via sPSGL-1 did not influence inflammatory cell concentrations (CD11b+ cells) throughout seven days of recovery from lengthening contractions. Conflicting observations on the contribution of P-selectin to inflammatory cell accumulation in injured muscle may be attributable to differences in (a) the means by which muscle injury was induced, (b) the antibodies used to detect inflammatory cells, or (c) the method for inhibiting one or more of the selections (Frenette et al. 2003; Baker et al. 2004).

Leukocyte activation and firm adhesion to endothelial cells follow selectin-mediated tethering and rolling. The activation step refers to the increased avidity, which takes place in fractions of seconds, of another group of leukocyte adhesion molecules called *integrins*. A β_2 integrin (Mac-1; $\alpha M/\beta_2$; CD11b/CD18) on neutrophils mediates their firm adhesion to endothelial cells, a prerequisite for diapedesis, via its interaction with vascular ligands such as intracellular adhesion molecule-1 (ICAM-1) (Walzog and Gaehtgens 2000). Firm adhesion of monocytes to endothelial cells can occur via a β_1 (VLA-4; α_4/β_1; CD49d/CD29)- or a β_2 integrin (LFA-1; $\alpha L/\beta_2$; CD11a/CD18)-dependent interaction with several vascular ligands (e.g., vascular cell adhesion molecule-1 [VCAM-1] and ICAM-1, respectively) (Meerschaert and Furie 1995). Because human blood neutrophils do not express VLA-4, their firm adhesion to endothelial cells is largely dependent on CD11b/CD18. After firm adhesion, neutrophils and monocytes migrate between endothelial cells by interacting with adhesion molecules in the intercellular junction between endothelial cells, as well as with extracellular matrix proteins. Neutrophils are also thought to migrate through endothelial cells by traversing their cytoplasm (Kvietys and Sandig 2001).

Using mice deficient in CD18, we reported that CD18 is required for neutrophil (Ly6G+ cells), but not macrophage (F4/80+ cells), accumulation in skeletal muscle injured by lengthening contractions. Because the accumulation of macrophages in injured skeletal muscle was independent of CD18 (Pizza et al. 2005), monocyte diapedesis into injured skeletal muscle may be mediated by a β_1 integrin mechanism.

The Causative Factors in Neutrophil and Macrophage Accumulation

Little is known about the factors that cause neutrophils and monocytes or macrophages to migrate to and within injured skeletal muscle. This section summarizes research in this area and speculates on the source and identity of factors released from injured skeletal muscle that promote migration of inflammatory cells.

Causes of Neutrophil Accumulation

Neutrophil accumulation in injured skeletal muscle is most likely the result of factors that promote their migration to and within skeletal muscle via chemotaxis or haptotaxis. Chemotaxis is the migration of cells toward or away from a concentration gradient of a soluble chemical attractant (chemoattractant); haptotaxis is the migration of cells toward or away from an immobilized chemoattractant. After neutrophil migration, neutrophils in injured skeletal muscle could remain elevated due to the presence of cytokines (e.g., granulocyte-macrophage colony-stimulating factor, interleukin [IL]-1β and IL-6) that prevent or delay their apoptosis (programmed cell death) (Simon 2003).

Research in this area has been directed toward determining the cellular origin and the identity of chemoattractants for neutrophils after exercise or muscle trauma. Because different cell types within injured skeletal muscle (e.g., skeletal muscle cells, endothelial cells, fibroblasts, neutrophils, and macrophages) are capable of releasing a bewildering number of neutrophil chemoattractants, the cellular source of chemoattractants for neutrophils cannot be identified using in vivo muscle injury models. We recently developed a cell culture model to test whether human skeletal muscle cells are a source of neutrophil chemoattractants after simulated exercise and after muscle trauma (Tsivitse et al. 2005). We exposed human multinucleated skeletal muscle cells (myotubes, immature myofibers) to mechanical strain or trauma in vitro and assayed for neutrophil chemotaxis. Both mechanical strain and traumatic injury caused myotubes to release one or more factors that

resulted in neutrophil chemotaxis (Tsivitse et al. 2005). Chemotaxis induced by mechanical strain occurred in the absence of muscle injury and progressively grew with incremental increases in the magnitude of strain. The higher neutrophil chemotaxis observed after injurious strain relative to noninjurious strain agrees with our prior findings of higher neutrophil concentrations in rodent skeletal muscle after injurious relative to noninjurious exercise (Pizza et al. 2002). Together, our results demonstrate that skeletal muscle cells are an important source of neutrophil chemoattractants after exercise and traumatic injury.

The identity of the skeletal muscle–derived chemoattractants that cause neutrophils to migrate after exercise or trauma remains to be determined. Neutrophil migration could be initiated by a large number of skeletal muscle cell–derived products (Grounds and Davies 1996). Specifically, skeletal muscle cells can produce cytokines (e.g., granulocyte colony-stimulating factor, IL-1β, IL-8, and tumor necrosis factor-α [TNF-α]), extracellular matrix proteins (e.g., fibronectin and laminin), complement proteins (C3a and C5a), and lipid derivatives (e.g., leukotriene B4) that could directly or indirectly contribute to neutrophil migration after muscle injury (Nagaraju 2001; Grounds and Davies 1996; Cassatella 1999). Lastly, neutrophil accumulation could also be amplified by the release of ELR+ CXC chemokines from neutrophils that have entered injured skeletal muscle. The ELR+ CXC chemokines (IL-8, neutrophil-activating peptide 2, growth-related gene product α/β, and epithelial cell-derived neutrophil-activating protein 78) are considered the most potent and specific chemoattractants for neutrophils (Matsukawa et al. 2000).

Previous investigators have provided evidence that complement proteins and skeletal muscle protease (calpain) activity cause neutrophils to accumulate in skeletal muscle. Frenette and colleagues (2000) reported that inhibition of the complement system resulted in a modest reduction in neutrophil concentrations at 6 h, but not 24 h, after injury induced by mechanical loading of atrophic skeletal muscle. These observations suggest that cleavage products of complement proteins (e.g., C3a and C5a) contribute to the migration of neutrophils to injured skeletal muscle only during early recovery (<6 h). Raj and colleagues (1998) have reported a significant positive correlation between muscle calpain activity and neutrophils (myeloperoxidase activity) in rat skeletal muscle after exercise. Because calpain activity had been reported to produce peptides that are chemotactic for neutrophils, these investigators proposed that calcium-stimulated proteolysis is a mechanism by which neutrophils accumulate in skeletal muscle after exercise (Belcastro, Shewchuk, and Raj 1998; Raj, Booker, and Belcastro 1998).

Causes of Macrophage Accumulation

The initial elevation of macrophages in injured skeletal muscle is probably the result of factors that promote monocyte chemotaxis (Grounds and Davies 1996), whereas their continued presence could be the result of factors that impair their apoptosis (Sandri et al. 2001). Similar to what happens with neutrophils, monocyte chemotaxis and haptotaxis could be initiated by a bewildering number of factors produced by cells that reside within injured skeletal muscle (e.g., neutrophils, macrophages, skeletal muscle cells, endothelial cells, and fibroblasts).

Because neutrophil accumulation precedes that of macrophages and since neutrophils can produce chemo-attractants for monocytes (e.g., macrophage inflammatory protein-1 [MIP-1] and monocyte chemoattractant protein-1 [MCP-1]) (Cassatella 1999), neutrophils were initially thought to promote monocyte chemotaxis after muscle injury. Recent observations, however, indicate that neutrophils do not contribute to macrophage accumulation in injured skeletal muscle. Specifically, we demonstrated that macrophages accumulate in muscles depleted of neutrophils after muscle injury induced by lengthening contractions (Pizza et al. 2005).

Current evidence indicates that skeletal muscle cells are an important source of chemoattractants for monocytes and macrophages after muscle injury. Chazaud and colleagues (2003) proposed that proliferating myoblasts are an important source of chemoattractants for monocytes and macrophages after muscle injury in vivo. Their proposal was based on experiments in which cultured human myoblasts produced factors that caused monocyte chemotaxis in vitro. These authors also reported that skeletal muscle cell–derived MCP-1, macrophage-derived chemokine, vascular endothelial cell growth factor, fractalkine, and urokinase-type plasminogen activator all caused monocyte chemotaxis in vitro.

Several investigators have reported increased expression of CC chemokines, which attract mainly monocytes and lymphocytes (Matsukawa et al. 2000), after traumatic muscle injury. Warren and colleagues (2004) reported increased MCP-1 in skeletal muscle after freeze injury in mice. Antibody neutralization of MCP-1 and a genetic deficiency in its receptor (CCR2), however, did not reduce macrophage concentrations after freeze injury (Warren et al. 2005, 2004). Other CC chemokines that have been reported to be elevated after traumatic muscle injury or to be produced by cultured skeletal muscle cells (or both) include MCP-3, macrophage inflammatory proteins (MIP-1α, -3α, and -1β), and RANTES (regulated upon activation, normal T-cell expressed and secreted) (Summan et al. 2003; Nagaraju 2001). Whether these CC chemokines cause monocyte and macrophage chemotaxis in vivo after muscle injury remains to be determined.

Monocyte and macrophage migration can be initiated by fragments of complement proteins and by lipid derivatives (Grounds and Davies 1996). Frenette and colleagues (2000) provided evidence that cleavage products of complement proteins (e.g., C3a and C5a) cause monocyte migration during early recovery from a muscle injury. Bondesen and colleagues (2004) recently demonstrated that lipid derivatives from arachidonic acid contribute to accumulation of inflammatory cells (CD11b+ cells), concluding that the cyclooxygenase-2 enzyme contributes to the accumulation of macrophages after freeze injury in mice.

Neutrophil- and Macrophage-Induced Injury

Because neutrophils and macrophages are accumulating at a time when the muscle shows signs of injury, previous investigators have speculated that these cells may actually contribute to the injury. This section summarizes experimental evidence supporting and refuting this speculation.

Neutrophils as a Possible Cause of Skeletal Muscle Injury

Earlier investigators speculated that neutrophils cause muscle dysfunction and cytoskeletal disruptions after contraction-induced muscle injury. The speculation was largely based on a temporal analysis showing that neutrophils are elevated in skeletal muscle when ultrastructural, histological, and functional signs of muscle injury worsen (2 h to three days; secondary injury) (Faulkner, Jones, and Round 1989; Papadimitriou et al. 1990). Since then, several investigators have tested whether neutrophils cause muscle injury.

We and others have used an in vitro experimental model to explore whether neutrophils cause muscle injury. This approach consists of culturing neutrophils isolated from human blood, or from the peritoneum of rodents, with human or rodent myotubes. Using this model, we demonstrated that human neutrophils injure human myotubes in a concentration-dependent manner (Pizza et al. 2001). The mechanism for injury was dependent on CD18-mediated neutrophil adhesion and ROS production (McLoughlin and Mylona et al. 2003). Of the ROS tested, hydrogen peroxide and the hydroxyl radical were most injurious to myotubes; superoxide anion (O_2^-) was found to be noninjurious (McLoughlin and Mylona et al. 2003). Interestingly, myotube-derived nitric oxide afforded some protection from the injury induced by neutrophils (McLoughlin and Mylona et al. 2003).

We recently tested whether the neutrophil-mediated muscle injury we had observed in vitro would occur in vivo after contraction-induced muscle injury. We exposed normal mice (wild type) and mice deficient in CD18 (C18$^{-/-}$) to injurious lengthening contractions (Pizza et al. 2005). Neutrophil concentrations after lengthening contractions were significantly elevated in wild-type mice and remained at control levels in CD18$^{-/-}$ mice. The elevation in macrophage concentrations after lengthening contractions was not influenced by the CD18$^{-/-}$ deficiency. Histological and functional signs of muscle injury and total carbonyl content (a marker of oxidative damage) were significantly higher in wild-type than in CD18$^{-/-}$ mice three days after lengthening contractions. These data indicate that neutrophils may cause secondary muscle injury by oxidatively modifying skeletal muscle proteins after contraction-induced muscle injury. This interpretation is in agreement with Brickson and colleagues' (2003) findings that administration of a blocking antibody for CD11b reduced blood neutrophil-derived ROS, muscle neutrophil concentrations, and histological signs of muscle injury in rabbits 24 h after a single injurious lengthening contraction.

Few investigators have examined the contribution of neutrophil degranulation and the subsequent release of proteases to skeletal muscle injury after exercise. The most compelling evidence supporting a role of neutrophil degranulation in secondary injury is derived from findings by Morozov and colleagues (2001) and Raj and colleagues (1998). These investigators reported that antibody neutralization of neutrophils, a technique to inhibit blood neutrophils, decreased proteolytic activity in skeletal muscle after exercise. In addition, Lowe and colleagues (1995) reported a significant positive correlation between myeloperoxidase activity in skeletal muscle and protein degradation after lengthening contractions. Whether proteolysis initiated by neutrophils is a negative consequence of their presence in injured skeletal muscle or serves as a mechanism to remove damaged proteins via phagocytosis remains to be determined.

The contribution of neutrophils to histological and functional abnormalities after traumatic injury (e.g., myotoxin, crush, or freeze injury) is not well understood. In contrast to what occurs after injury by lengthening contractions, Teixeira and colleagues (2003) reported that antibody neutralization of neutrophils did not influence histological signs of muscle injury or blood creatine kinase activity (an indirect marker of muscle injury) one day after myotoxin injection in mice.

The inability of neutrophils to cause skeletal muscle injury after myotoxin injection may be the result of differences in the magnitude and time course of histological abnormalities after myotoxin injection compared to lengthening contractions. Specifically, ~50% of the myofibers exhibited signs of injury one day after myotoxin injection (Teixeira et al. 2003), whereas histological signs of muscle injury are not prevalent until three days after lengthening contrac-

tions when ~20% of the myofibers are classified as injured (Pizza et al. 2005). Muscle dysfunction also reaches its nadir within one day after traumatic injury (Warren et al. 2002); in contrast, after lengthening contractions, muscle function is partially restored during early recovery (less than one day) and then declines for up to three days of recovery (Faulkner, Jones, and Round 1989). Thus, secondary injury appears to be a phenomenon that occurs only after contraction-induced muscle injury. Whether neutrophils cause muscle injury in vivo may therefore be dependent on the nature, mechanism, and the extent of the initial injury.

Macrophages as a Possible Cause of Skeletal Muscle Injury

Few investigators have tested whether macrophages are capable of causing muscle injury. An initial investigation by Tidball and colleagues (1999) indicated that macrophages do not cause muscle injury in vivo. However, recent in vitro work from the same laboratory suggested that macrophages do injure skeletal muscle cells (Nguyen and Tidball 2003). Specifically, Nguyen and Tidball (2003) reported that rat peritoneal macrophages injured rat myotubes in vitro. The mechanism for the macrophage-mediated injury was attributed to macrophage-derived nitric oxide production. Because human monocytes and macrophages (and neutrophils) do not produce nitric oxide (Padgett and Pruett 1992), the applicability of these observations to muscle injury in humans remains to be determined.

The contribution of inflammatory cell–derived proteases to the loss of muscle proteins after traumatic muscle injury has been investigated by Farges and colleagues (2002). They reported that crush injury to skeletal muscle increased transcription of several lysosomal proteases (cathepsin B, L, H, and C) and the activity of cathepsin B at two days postinjury. Increased levels of cathepsin B were found within inflammatory cells that were presumed to be macrophages. Additional experiments revealed that lysosomal proteases caused muscle proteolysis after crush injury. The authors speculated that increased activity of lysosomal proteases within macrophages during phagocytosis, as opposed to their release of proteases during degranulation, is the mechanism by which macrophages contribute to muscle proteolysis after crush injury.

Neutrophil- and Macrophage-Assisted Muscle Repair

Neutrophils remain elevated in skeletal muscle when the muscle is beginning to show signs of muscle repair. Macrophages, on the other hand, are substantially elevated throughout the repair process. This section summarizes the contribution of neutrophils and macrophages to events associated with muscle repair or regeneration.

Possible Contribution of Neutrophils to the Resolution of a Muscle Injury

The contribution of neutrophils to the restoration of normal structure and function to injured skeletal muscle is largely unknown. On the basis of their ability to perform phagocytosis, neutrophils are thought to contribute to muscle repair by removing necrotic or injured tissue. Support for this belief was provided by Papadimitriou and colleagues (1990), who reported qualitative electron microscopy observations of neutrophils internalizing remnants of injured skeletal muscle after traumatic injury. Unfortunately, quantifying the phagocytic performance of neutrophils (or macrophages) is technically difficult for several reasons, including the vast number of potential phagocytic targets after muscle injury.

In theory, neutrophils could also influence muscle regeneration (i.e., myogenesis) by producing factors that are known to either directly or indirectly influence myogenesis. For example, Seale and Rudnicki (2000) hypothesized that neutrophils contribute to the activation and proliferation of muscle precursor cells (satellite cells or myoblasts) via the interaction of a neutrophil β_1 integrin (VLA-4) with VCAM-1 expressed on satellite cells or myoblasts. The role of VCAM-1 in myogenesis and the expression of the β_1 integrin on neutrophils after diapedesis, however, are controversial areas. Neutrophils could also promote myoblast proliferation by producing IL-6 or hepatocyte growth factor (or both), which have been reported to cause myoblast proliferation in vitro (Cassatella 1999; Hawke and Garry 2001; Seale and Rudnicki 2000).

A similar line of logic argues that neutrophils impair muscle regeneration. For example, neutrophils are capable of producing other cytokines (e.g., TGF-β1, TNF-α, IL-1β, and interferon) that have been reported to inhibit myoblast proliferation, differentiation, or both (Cassatella 1999; Hawke and Garry 2001; Seale and Rudnicki 2000). Neutrophils could also delay the rate of muscle regeneration by injuring myotubes, as we have previously reported (Pizza et al. 2001; McLoughlin et al. 2003). Thus, in theory, neutrophils could aid or impair muscle injury resolution via multiple mechanisms. (TGF is the transforming growth factor.)

We have begun to explore the contribution of neutrophils to the restoration of muscle structure and function after contraction-induced muscle injury. In these experiments we compared markers of muscle repair and regeneration in normal muscle (wild-type mice) and muscle depleted of neutrophils (CD18$^{-/-}$ mice) after contraction-induced muscle injury (Pizza et al. 2005). To our surprise, muscle function after lengthening contractions was restored sooner in the absence of neutrophils. Specifically, muscle function was restored within 7 and 14 days of the injury in CD18$^{-/-}$ and wild-type mice, respectively. The faster rate of restoration of muscle func-

tion in CD18$^{-/-}$ mice was not attributable to differences in the magnitude of the initial or secondary injury. Our functional observations were corroborated by a higher myofiber expression of embryonic myosin heavy chain (a marker of regeneration) and a larger cross-sectional area of centrally nucleated myofibers (an indicator of growth of regenerating myofibers) in CD18$^{-/-}$ compared to wild-type mice. The percentage of centrally nucleated myofibers (an indicator of the number of regenerating myofibers), however, was not influenced by the absence of neutrophils. Together, our preliminary observations indicate that neutrophils may delay some, but certainly not all, of the events associated with restoring structure and function to skeletal muscle injured by exercise (Pizza et al. 2005).

Few studies have directly addressed the contribution of neutrophils to muscle repair and regeneration after muscle trauma. In contrast to our observations on injury following lengthening contractions, Teixeira and colleagues (2003) reported that antibody neutralization of neutrophils after myotoxin injection decreased the percentage of central nucleated myofibers. Qualitative observations of muscles depleted of neutrophils revealed a reduced diameter of central nucleated myofibers and an increased amount of necrotic tissue seven days after the injury. These observations indicate that neutrophils may enhance muscle repair by removing necrotic tissue and may promote muscle regeneration by increasing the number and growth of regenerating myofibers.

Conflicting observations on the contribution of neutrophils to muscle repair and regeneration after contraction-induced muscle injury (Pizza et al. 2005), as well as traumatic injury (Teixeira et al. 2003), indicate that the nature and magnitude of the initial injury may also dictate how neutrophils influence the restoration of normal structure and function to injured skeletal muscle.

In theory, the influence of neutrophils on skeletal muscle is dependent on which products (e.g., ROS, proteases, cytokines) they are producing or releasing in skeletal muscle. Because functional activities of neutrophils are agonist specific, the presence or absence of agonists for ROS or for specific cytokines produced by neutrophils in injured muscle may be the key determinant of whether and how neutrophils influence the resolution of muscle injury.

Possible Contribution of Macrophages to the Resolution of a Muscle Injury

The seminal research on the contribution of macrophages to muscle repair and regeneration was conducted by Miranda Grounds and her colleagues. In a series of studies, they observed a temporal relationship between inflammatory cell accumulation, the removal of necrotic tissue, and muscle regeneration after muscle trauma (Grounds and McGeachie 1989; Robertson, Grounds, and Papadimitriou 1992; Mitchell, McGeachie, and Grounds 1992; Roberts, McGeachie, and Grounds 1997). They also noted that replication of satellite cells and myotube formation were closely associated with the accumulation of macrophages. Their observations initiated research efforts to determine whether and how macrophages influence muscle regeneration.

The majority of research on the contribution of macrophages to muscle regeneration has used a coculture model in which macrophages were cultured with myoblasts. These studies demonstrated that peritoneal macrophages cause myoblast proliferation in vitro (Robertson et al. 1993; Cantini et al. 2002; Merly et al. 1999; Massimino et al. 1997). Because macrophages may cause muscle injury (Farges et al. 2002; Nguyen and Tidball 2003), and because skeletal muscle cells can dedifferentiate and reenter the cell cycle (Echeverri and Tanaka 2002), macrophage-induced myoblast proliferation could result from a direct action of macrophage-derived products or to a secondary response that could be initiated by the macrophage-mediated injury. Cantini and colleagues (Cantini et al. 2002, 1995; Cantini and Carraro 1995) have provided evidence that the mechanism by which monocytes or macrophages cause myoblast proliferation in vitro is via the release of soluble agonists. This was shown with use of a cell culture medium that was free of human monocytes or rat macrophages and that contained soluble factors released from monocytes or macrophages after stimulation (monocyte/macrophage-conditioned medium). Monocytes or macrophages were stimulated for several days with lipopolysaccharide (a cell wall component of gram negative bacteria), a potent agonist for the release of some, but not all, inflammatory cell–derived cytokines (Cassatella 1999). The authors reported that monocyte/macrophage-conditioned medium caused myoblast proliferation in vitro (Cantini et al. 2002, 1995; Cantini and Carraro 1995).

Few investigators have directly tested the contribution of macrophages to muscle regeneration in vivo. Cantini and colleagues (2002) reported that macrophage-conditioned medium promotes muscle regeneration after surgical removal of a large fraction of a rodent skeletal muscle (muscle ablation). Specifically, qualitative observations after muscle ablation revealed a greater number and larger size of regenerating fibers in muscles treated with macrophage-conditioned medium than in nontreated controls. Muscles treated with macrophage-conditioned medium also had a greater mass. These findings corroborate studies using macrophage and myoblast cultures by demonstrating that macrophage-derived products promote muscle regeneration in vivo.

The mechanism by which macrophages promote muscle regeneration has yet to be determined. Investigators have speculated that macrophages promote muscle regeneration via the release of cytokines that are known to cause myoblast proliferation in vitro (e.g., leukemia inhibitory factor [LIF], IL-6, insulin-like growth factor-1 [IGF-1], and platelet-derived growth factors [PDGFs]) (Robertson et al. 1993; Hawke and Garry 2001; Seale and Rudnicki 2000; Grounds and Davies 1996). Macrophages are also known to produce other cytokines that have been reported to inhibit myoblast proliferation or differentiation (or both) in vitro and to induce muscle catabolism in vivo (e.g., TNF-α and IL-6, IL-1, allograft inflammatory factor-1, IFN, IL-1) (Kuschel et al. 2000; Hawke and Garry 2001; Seale and Rudnicki 2000; Grounds and Davies 1996). Thus, further work is needed to determine which cytokines macrophages are actually releasing into the milieu of injured skeletal muscle and to identify the agonist for their production.

An interesting mechanism by which macrophages were thought to promote muscle regeneration is via fusion with regenerating myofibers. This was thought to be possible because bone marrow–derived stem cells can differentiate into skeletal muscle cells or cells of monocyte or macrophage lineage. Comprehensive experiments conducted by Doyonnas and colleagues (2004), however, clearly showed that macrophages do not contribute nuclei to regenerating myofibers.

The evidence to date indicates that the accumulation of macrophages in injured muscle enhances muscle repair and regeneration via the release of growth-promoting cytokines. Sustained elevations of macrophages in injured muscle, on the other hand, may impair regeneration. Sandri and colleagues (2001) reported that inhibition of Fas-mediated apoptosis caused a long-lasting elevation in inflammatory cells (CD11b+ cells) and impaired the rate of muscle regeneration in mice after myotoxin injection. Warren and colleagues (2004, 2005) also reported that a persistent elevation in macrophages was associated with impaired regeneration in mice after freeze injury. Thus, the time between macrophage accumulation and their subsequent elimination from injured muscle may be a key determinant of the rate of muscle repair and regeneration.

Summary

Injurious exercise or trauma causes neutrophils and macrophages to accumulate in skeletal muscle at a time when the muscle shows signs of both injury and repair and regeneration. The mechanisms by which skeletal muscle initiates the inflammatory response, as well as the contribution of neutrophils and macrophages to muscle injury and repair and regeneration, are beginning to be discovered.

We propose a model for how injured skeletal muscle initiates the accumulation of inflammatory cells and for the common and separate contributions of neutrophils and macrophages to the events of muscle injury and

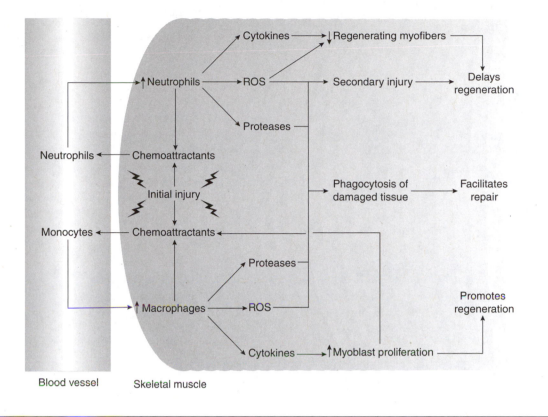

Figure 4.3   A model for how injured skeletal muscle initiates the accumulation of neutrophils and macrophages and for their contribution to the events of muscle injury and repair and regeneration.

repair and regeneration (figure 4.3). The foundation of this model is the idea that the environment of injured skeletal muscle dictates when and how many neutrophils and macrophages appear in skeletal muscle, as well as whether and how they influence skeletal muscle.

In our model, neutrophil accumulation is initiated by chemoattractants released from skeletal muscle cells after the initial injury. Neutrophil chemotaxis and diapedesis after the initial injury are likely enhanced by the release of ELR$^+$ CXC chemokines from neutrophils that have invaded the injured muscle. The environment of injured muscle probably contains soluble or insoluble factors (or both) that initiate the removal of damaged or necrotic tissue by promoting phagocytosis. During phagocytosis, however, the release of ROS from neutrophils (and also to some extent from macrophages) appears to cause secondary injury by damaging healthy regions of injured fibers, adjacent uninjured fibers, or both. Because it has been shown that neutrophils impair some of the events associated with restoring normal structure and function to skeletal muscle injured by exercise, neutrophil-derived ROS, cytokines, or both could serve as a mechanism for this impairment. Specifically, neutrophil-derived ROS or cytokines (e.g., TGF-β1, TNF-α, IFN-γ, and IL-

1β) could delay muscle regeneration by decreasing the number or growth of regenerating myofibers.

Macrophage concentrations generally reach their peak at a time when neutrophil concentrations are diminishing and when the injured muscle is in an intermediate stage of muscle repair and regeneration. The initial stimulus for monocyte chemotaxis and diapedesis may originate from chemoattractants released from injured skeletal muscle cells, from proliferating myoblasts, or both. Macrophage accumulation could be amplified via the release of CC chemokines from macrophages within the injured skeletal muscle. Macrophages appear to promote muscle repair by removing damaged or necrotic tissue via ROS production during phagocytosis. Furthermore, macrophages aid muscle regeneration presumably by releasing cytokines (e.g., LIF, IL-6, IGF-1, and PDGFs) that have been reported to enhance myoblast proliferation in vitro.

Further study is needed to support our model and to reveal the underlying mechanisms for the complex interplay between inflammatory cells and skeletal muscle after injury. Once the mechanisms have been revealed, targeted treatments could be developed to minimize any negative consequences of neutrophils and to enhance the apparent beneficial role of macrophages in injured skeletal muscle.

CHAPTER 5

Muscle Soreness and Damage and the Repeated-Bout Effect

Ken Nosaka, PhD

Unaccustomed exercise consisting of repeated or forced lengthening (eccentric) contractions induces muscle damage. The most noticeable symptom of this damage is the muscle soreness that we experience after performing such exercise, which is often referred to as delayed-onset muscle soreness (DOMS). Eccentric exercise–induced muscle damage is also characterized by morphological changes such as disruption of contractile and noncontractile proteins and the plasma membrane, increases in muscle proteins in the blood, prolonged loss of muscle function, swelling, and abnormality detected by ultrasound and magnetic resonance images. These are used as markers of muscle damage, but it is not clear how they are related to each other and are associated with DOMS. A bout of unaccustomed exercise confers a protective effect against DOMS and muscle damage in subsequent bouts of the same or a similar exercise. This effect, referred to as the repeated-bout effect, is a unique feature of eccentric exercise–induced muscle damage. This chapter focuses on the physiology of muscle soreness, the relationship between DOMS and muscle damage, and the repeated-bout effect.

Muscle Soreness

Pain originating from skeletal muscle is caused by an acute or overuse injury or by chronic syndromes such as fibromyalgia and muscular rheumatism. Muscle soreness refers to muscle pain felt after exercise when the muscle is palpated or moved, and is most com-

monly experienced after the performance of unaccustomed exercise consisting of lengthening (eccentric) contractions. The development of such muscle soreness is generally delayed. Although DOMS is an extremely common symptom, its underlying mechanisms are not clearly understood, nor are the reasons for the delay.

Understanding Pain

The International Association for the Study of Pain (IASP) defines pain as "an unpleasant sensory and emotional experience associated with actual or potential tissue damage, or described in terms of such damage" (Merskey and Bogduk 1994, pg. 210). We experience various kinds of pain in association with injuries, illnesses, and diseases. We also experience pain in many other nonpathological conditions (e.g., when we hold a heavy object or climb up stairs). Pain is a crucial signal that informs us of an abnormality in the body or a potential risk, but it is difficult to appreciate its value when we are suffering from it. In some cases, pain has no physiological value and yields only suffering. Pain perception is influenced by many factors, such as age, gender, and social or cultural norms about acceptable behavior in relation to pain (Unruh et al. 2002). In addition, given the same magnitude of tissue damage, the magnitude of pain perception ranges widely among individuals, since pain is a subjective symptom. These factors add to the complexity of understanding pain.

Function of Pain

Pain is generally considered a warning signal of actual or perceived tissue damage, and one of its major functions is to protect the organism from injurious stimuli (Millan 1999; Unruh et al. 2002). In order for this to occur, (1) pain should be felt before injurious stimuli induce irreversible damage to tissue, (2) the magnitude of the pain should reflect the severity of tissue damage, and (3) the duration of the pain should match the process of tissue damage and recovery. However, pain often occurs without clear evidence of tissue damage; and its onset, magnitude, and duration do not necessarily correspond to tissue damage (Melzack 1982). In cancer, for example, pain does not exist in the developmental phase, but cancer patients often suffer from severe pain in the phase when the disease is no longer treatable.

There are two clinical states of pain: physiological (nociceptive) pain and neuropathic (intractable) pain (Kingsley 2002). The former results from the direct stimulation of pain receptors due to injury of tissue and inflammatory responses, and the latter results from injury to the nervous system that causes permanent changes in central nervous system connections (Kingsley 2002). Chronic pain may be associated with changes in the nervous system (Unruh et al. 2002). Neuropathic pain does not appear to have a useful physiological function (Millan 1999), and is not necessarily functioning as a warning signal.

Exercise and Muscle Pain

Muscle pain occurs in many situations in sport and exercise, ranging from stretching a stiff muscle to holding a heavy object to incurring a muscle cramp or muscle tear. The onset of muscle pain varies depending on the cause. Figure 5.1 shows changes in the magnitude of muscle pain relative to its peak in five different situations: The same person experienced a muscle tear of the biceps femoris while playing tennis, a muscle cramp in the knee flexors on a different occasion, pain in the knee extensors from running a marathon (42.195 km), pain in the knee extensors from playing soccer for 120 min, and pain from performing maximal eccentric exercise of the elbow flexors. Pain can occur either (a) only during exercise, (b) both during and after exercise, or (c) only after exercise.

Muscle pain is elicited by sustained or rhythmic muscle contractions, with occlusion of blood flow accelerating its onset and increasing the intensity; but it subsides quickly once muscle activity is terminated and normal blood flow is restored (Miles and Clarkson 1994). It seems likely that pain substances produced during muscle contractions, as well as increased intramuscular pressure, are associated with the pain.

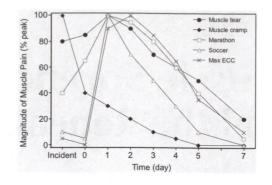

Figure 5.1   Changes in the magnitude of muscle pain relative to its peak (100%) in five different incidents. The pain level when the muscle tear and the muscle cramp occurred or during performance of exercises is shown as "Incident"; "0" is either 30 min after the occurrence of the muscle tear and the muscle cramp or immediately after the exercise. The peak pain level differed among the incidents, and the magnitude of pain was normalized relative to the peak value.

A muscle cramp—a sudden involuntary shortening of muscle—elicits sharp pain that appears to result from stimulation of mechanical receptors by excessive muscle tension (Miles and Clarkson 1994). It appears that muscle cramps can also be a consequence of prolonged low-intensity exercise, and many factors (e.g., fatigue, dehydration, electrolyte abnormality, environmental conditions) appear to be associated with the cause in this context. The pain disappears rapidly if the muscle stops cramping, but a residual pain is present after stretching of the cramping muscle (figure 5.1). Stretching of a cramping muscle is a forced lengthening (eccentric) muscle action that induces the "muscle damage" explained later in the chapter.

If muscle pain occurs during exercise, it may be a signal to stop or slow down. If the cause of the pain is a serious injury such as a laceration or contusion, we have no choice but to stop moving. Pain from these injuries is sustained and often aggravated in the days after the incident. Pain that continues after the cessation of exercise may indicate the need for treatment. Pain is generally one of the classic signs of inflammation (Hargreaves et al. 1989); therefore, anti-inflammatory treatment would be the first choice for relief.

As shown in figure 5.1, pain may develop during endurance events such as the marathon and triathlon, then continue and sometimes become aggravated the next day. In contrast, with other types of exercise, little or no muscle pain develops during performance but begins to develop several hours later. We experience this type of delayed-onset muscle pain especially when performing an unaccustomed exercise. This type of muscle pain is the main focus of this chapter.

Neurophysiology of Muscle Pain

Muscle pain has been less frequently documented than pain originating from cutaneous or visceral tissue. Most of the pain receptors appear to have a common physiological basis, but it appears that there are some differences between skeletal muscle nociceptors and others.

Pain Receptors (Nociceptors)

Pain receptors, referred to as nociceptors, respond to a noxious (tissue threatening) stimulus. Most of the data on the neuropathology of pain are from studies of cutaneous nociceptors; less information is available for muscle nociceptors. Muscle pain differs from cutaneous pain in that pain associated with muscle lesions is described as aching and cramping, while cutaneous pain is characterized by sharp, pricking, stabbing, or burning sensations (Mense 1993).

A large portion of the afferent nerve terminals lack encapsulated endings; these unencapsulated endings are called free nerve endings. The free nerve endings are nociceptors, which are highly branched and have large areas of sensitivity. This is why our idea of the place of origin of a pain sensation is often vague, particularly for muscle pain (Mense 1993). When we feel muscle pain, it is difficult to tell exactly where in the muscle the pain is originating without careful palpation of the muscle. It appears that the pain spreads beyond its actual site of origin.

Skeletal muscles contain four types of afferent fibers: types I (Aα), II (Aβ), III (Aδ), and IV (C). The free nerve endings of the latter two respond to noxious stimuli such as mechanical pressure, heat, cold, and algesic substances (e.g., bradykinin, potassium, serotonin, histamine). Aδ fibers have thin myelinated axons with a relatively fast conducting velocity (5-30 m/s); they respond to muscle stretch, contractions, and innocuous pressure and are sensitized by thermal and chemical stimuli (Millan 1999). In contrast, C fibers are thin and unmyelinated and transmit signals more slowly (0.5-2 m/s). Like Aδ fibers, C fibers respond to thermal stimuli, ischemia, and hypoxia and are sensitized by chemical stimuli (Millan 1999). Stimulation of C fibers in muscle elicits dull, aching pain and cramping pain. The pain sensation from muscle is thought to be mainly mediated by C fibers and secondarily by Aδ fibers (Hargreaves et al. 1989).

Muscle Nociceptor Locations

In skeletal muscle, the free nerve endings of Aδ and C fibers are located along the walls of arterioles and in the surrounding connective tissue (Mense 1993). It is important to note that there are no pain receptors on the muscle plasma membrane. Therefore, even if some muscle fibers are damaged and stop functioning, we may not feel muscle pain if nociceptors located far from the damaged area are not affected. Interestingly, devastating muscle dystrophies such as the Duchenne type are not painful (Marchettini 1993).

It seems that the sensation of muscle pain is activated by changes in the chemical environment surrounding muscle tissue, or by stimulation of fascia, rather than by primary muscle cell damage (Marchettini 1993). Weerakkody and colleagues (2001) reported that muscle mechanoreceptors (primary endings of muscle spindles) are involved in DOMS and argued against the likelihood that a simple sensitization process is responsible for DOMS. It has also been documented that the sympathetic nervous system contributes to the sensation of pain by augmenting or modifying the nociceptors (Hargreaves et al. 1989).

Muscle Nociceptor Stimuli

As already described, muscle nociceptors are polymodal, responding to mechanical, thermal, and chemical stimuli. Effective stimuli for nociceptors in skeletal muscle are strong mechanical forces and endogenous algesic substances such as bradykinin, serotonin, and potassium ions (Mense 1993). The nociceptors are sensitized by prostaglandin E_2 (PGE_2) (Mense 1993) and hydrogen ions (not lactic acids) (O'Connor and Cook 1999). It appears that muscle nociceptors respond to weak mechanical stimuli when they are sensitized.

An increased sensitivity of nociceptors to a stimulus is termed hyperalgesia, and allodynia refers to the situation in which pain is induced by a stimulus that does not normally provoke pain (Calzá 2001). In normal conditions, palpating, stretching, or contracting muscles does not induce muscle pain; however, the same stimulus evokes pain when muscle damage has occurred. This indicates that muscle damage changes innocuous stimuli to noxious stimuli; muscle nociceptors become hyperalgesic, and the injured person experiences allodynia. It seems that in general, endogenous substances produced by muscle damage and inflammation do not stimulate muscle nociceptors directly (though some of them may to some extent) but instead increase the sensitivity of the nociceptors so that muscle contraction or stretching or palpation pressure becomes painful (figure 5.2). Swelling may also contribute to hyperplasia of the nociceptors.

Pain Pathways

Aδ and C fibers bring signals from skeletal muscle to the spinal cord (figure 5.2). Most Aδ and C fibers enter the dorsal root ganglion and synapse primarily in the dorsal horn of the spinal cord (Guyton 1994). From the spinal cord, pain signals take the spinothalamic pathways to the medulla, the midbrain, and the cerebral cortex. There are two different spinothalamic pathways: the neospinothalamic (lateral spinothalamic) tract and the paleospinothalamic (anterior spinothalamic) tract (O'Connor and Cook 1999). The neospinothalamic tract transmits signals primarily from Aδ fibers and appears to be responsible for sharp, fast pain.

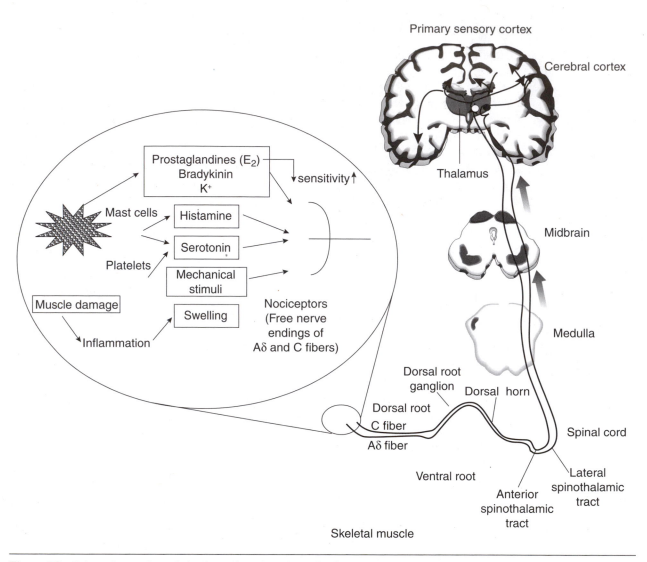

Figure 5.2   Pain pathways from skeletal muscle and a schematic view of the events at the site of skeletal muscle damage.

In contrast, pain signals from C fibers take the paleospinothalamic tract and terminate widely in the brainstem and thalamus. It seems that dull, aching pain is conveyed by this tract. The thalamus is the terminal of the spinothalamic pathways and transfers sensory information to the primary somatosensory areas of the cerebral cortex designated Brodmann's areas 3, 1, and 2 (Guyton 1994). In addition to these areas, multiple cortical areas such as the secondary somatosensory cortex, the anterior cingulate cortex, the insula, the prefrontal cortex, and the supplementary motor area are activated by pain stimuli (Kingsley 2002).

Modulation of Pain and Analgesia

The transmission of pain sensation from the nociceptors to the spinal cord to the cerebral cortex is modulated (Millan 1999). Stimulation of nociceptors does not necessarily reach consciousness and elicit pain. For example, it is often reported that athletes do not feel pain during

a match even if they have a serious injury. The nervous system can interact with the pain pathways to palliate the perception of pain under some conditions. In this pain control process, often called analgesia, not only the brain but also the spinal cord is involved (Millan 1999). Substances involved in analgesia include enkephalins, endorphins, and other opiate neuropeptides. Enkephalin is believed to cause presynaptic and postsynaptic inhibition of type Aδ and C fibers in the dorsal horns (Koltyn 2000). It is also known that multiple areas of the brain have opiate receptors, and opiates such as endorphins and enkephalins suppress pain signals (O'Connor and Cook 1999). Thus, the magnitude of pain does not directly reflect the magnitude of the stimulus to nociceptors.

It has been documented that exercise increases pain thresholds and pain tolerance; this phenomenon is often referred to as exercise-induced analgesia (Koltyn 2000; O'Connor and Cook 1999). Exercise-induced analgesia has been found to occur following running, cycling, and swimming, but little is known about the effect from resistance exercise (Koltyn 2000). More than 100 years ago, Hough (1902) reported that performing a second bout of exercise on the day after an activity that induced muscle soreness caused excessive pain for the first 2 to 3 min, which disappeared over the course of 5 to 10 min of exercise. Armstrong (1984) stated that exercising a sore muscle appeared to provide the most effective way of reducing the soreness; however, the nature of exercise-induced analgesia for DOMS has not been investigated systematically. A later section of the chapter provides further details on this issue.

Delayed-Onset Muscle Soreness

Following unaccustomed or severe exercise, we experience the discomfort of a dull, aching pain, combined with tenderness and often stiffness, for several days. The "delay" in DOMS appears to vary among exercises or individuals, but the pain normally increases in intensity in the first 24 h after exercise, peaks from 24 to 72 h, then subsides and disappears by five to seven days postexercise. There is little or no discomfort at rest; the sensation of pain is elicited when mechanical stimuli such as pressure, stretching, or contraction are imposed on the affected muscles.

Muscle soreness develops during exercise such as long distance running or performance of a marathon, and often gets worse over the days of recovery (figure 5.1). This type of muscle soreness is different from that occurring only after exercise. It may be that the cause of the muscle soreness felt during exercise is not the same as that responsible for the muscle soreness after exercise. However, it seems likely that the cause of the soreness in the days following exercise is similar in the two scenarios. Muscle damage induced by lengthening (eccentric) contractions is associated with the muscle soreness felt over the subsequent days (Armstrong 1984).

Mechanism of DOMS

A number of theories have been proposed to explain DOMS; among these are theories involving lactic acid, muscle spasm, muscle damage, connective tissue damage, and inflammation (Cheung et al. 2003). Theories pertaining to lactic acid and muscle spasm are unlikely to explain DOMS, and there is evidence to refute these explanations (Cleak and Eston 1992; Miles and Clarkson 1994).

Delayed-onset muscle soreness was first described by Hough (1902), who concluded that DOMS was "fundamentally the result of ruptures within the muscle." Although "ruptures" of muscle fibers are not associated with DOMS, ultrastructural disruptions of myofilaments, especially at the Z-disc, characterized by broadening, streaming, or smearing of the Z-disc structure as observed under electron microscope, have been reported to accompany DOMS (Friden et al. 1981, 1984; Yu and Thornell 2002). Connective tissue damage is also indirectly shown by increases in urine hydroxyproline (Abraham 1977) or plasma hydroxyproline and serum type 1 collagen concentration (Brown et al. 1999).

Therefore, the "damage" theory proposed by Hough is still valid with some modification, and it is most likely that muscle or connective tissue damage (or both) and subsequent inflammatory responses are associated with DOMS (Cheung et al. 2003). It may be that changes in connective tissue (endomysium or perimysium), rather than damage to muscle fibers, relate directly to DOMS. Jones and colleagues (1987) suggested that damage and shortening of muscle connective tissue would increase the mechanical sensitivity of muscle nociceptors and cause pain with stretching or palpation. One postulate is that the inflammatory response process leading to sensitization of muscle nociceptors takes time, and this would explain the delay (Smith 1991).

Eccentric Exercise and DOMS

Eccentric exercise, exercise mainly consisting of lengthening (eccentric) contractions, produces greater DOMS than exercise involving mainly shortening (concentric) or static (isometric) contractions (Newham et al. 1983; Talag 1973). It has been reported that isometric contractions at long muscle lengths induce greater DOMS than those at short muscle lengths; however, the magnitude of DOMS following isometric exercise is much less than that after eccentric exercise (Jones et al. 1989).

It appears that although pure concentric contractions do not induce DOMS, when people perform repeated concentric contractions in training they also unintentionally perform eccentric contractions, especially when muscles are fatigued. For example, when lifting a dumbbell using the elbow flexors, if we fail to produce a force larger than the dumbbell, our elbow flexors are lengthened while producing force. This may result in DOMS and muscle damage. In fact, concentric exercise of the elbow flexors with a 10 kg dumbbell may induce minor muscle soreness after exercise. However, the magnitude of the soreness is much less than it would be if the same person at the same level of training had engaged in eccentric exercise with the same weight (see figure 5.3f).

Unaccustomed eccentric exercise is known to induce greater damage to the internal membrane system, intermediate filaments, Z-discs, and contractile proteins than other types of exercises (Lieber and Friden 2002). This provides strong support for the damage and inflammation theory of the mechanism of DOMS. However, alteration of the Z-disc structure occurs to some degree even after concentric exercise (Gibala et al. 1995). The magnitude of ultrastructural alterations may not necessarily determine the magnitude of muscle damage (Nurenburg et al. 1992). It has also been documented that morphological changes at the muscle fiber level (e.g., mononuclear cell infiltration) do not correspond with muscle pain (Jones et al. 1986; Newham 1988). Moreover, Yu and colleagues (2002) did not find muscle fiber degeneration or an inflammatory response in human skeletal muscle with DOMS. Thus, the exact relationships between damage and inflammation in muscle or connective tissue and DOMS are still not clearly understood.

Measurement of DOMS

There is no generally accepted single best measure of pain (O'Connor and Cook 1999). However, there are several methods for evaluating pain, including the pressure pain threshold and tenderness, and several ways to quantify the soreness level using a questionnaire (MacIntyre et al. 1995; O'Connor and Cook 1999; Ohbach and Gale 1989). Types of scales for assessing muscle soreness include visual analog scales (VAS), numerical rating scales, verbal rating scales, descriptor differential scales, and the McGill pain questionnaire. Ohnhaus and Adler (1975) reported that a VAS consisting of a line (50-200 mm) with "no pain" at the left end and "unbearable pain" at the right end reflected a subject's muscle-associated pain more precisely than a verbal rating scale.

A number of difficulties exist with use of the VAS to quantify soreness, although this type of scale has been used in many studies. The difficulties include questions regarding the sensitivity of the instrument. For example, subjects who mark "50," or "unbearably painful," for the peak DOMS level cannot indicate a greater pain level even if they experience greater soreness in subsequent days. This is a disadvantage of using a closed-ended scale. A better way of quantifying pain may be to use an open-ended scale in which subjects can choose any number to represent the pain associated with some intermediate level of stimulation and then scale all subsequent tests in relation to the initial reference stimulus (Jones and Round 1990).

It is also important to note that the pain sensation can vary among subjects. For example, some subjects may mark "40" even if their soreness level is medium; others may mark "10" even if the pain level is very severe. Because of the subjective and individual nature of the pain sensation, the question arises whether it is possible to compare levels of soreness between subjects, or even changes in DOMS over days in the same subject. Perception of a noxious stimulus may differ greatly between individuals and may also vary with a given person's mood, health, or hormonal status. Melzack (1982) stated that "pain is not simply a function of the amount of bodily damage alone, but is influenced by attention, anxiety, suggestion, and other psychological variables (pg. 148)."

For these reasons, it is important to be cognizant of the limitations in quantitatively assessing DOMS in experimental studies.

DOMS and Muscle Damage

As already discussed, DOMS is a symptom particular to eccentric exercise–induced muscle damage. Other symptoms of muscle damage include muscle weakness, stiffness, and swelling. Muscle damage is detected by histological changes, increases in muscle proteins in the blood, and abnormality shown by ultrasound and magnetic resonance images. It is important to note that DOMS does not necessarily represent muscle damage and that the level of DOMS and changes in muscle detected by the different methods are not necessarily correlated.

Defining Muscle Damage

Muscle is damaged when it receives a harmful physical, chemical, or biological stimulus. In relation to exercise and sport, muscle damage occurs as a result of physical trauma, which is also referred to as muscle injury. In relation to exercise, Safran and colleagues (1989) suggested that muscle injury can be divided into three major types based on clinical presentation. A Type I injury is characterized by muscle soreness that occurs 24 to 48 h after unaccustomed exercise (DOMS). A Type II injury is characterized by an acute disabling pain from a muscle tear, ranging from a tear of a few fibers, with fascia remaining intact, to a complete tear of the muscle and fascia. A Type III injury is associated with muscle soreness or cramping that occurs during or immediately after exercise. It may not be accurate to include the Type III injury, because actual injury to muscle does not occur in the case of muscle cramping, although muscle damage may be a part of the process of treating a muscle cramp (e.g., stretching the cramping muscle).

Again, the Type I injury is peculiar to eccentric exercise. Eccentric exercise–induced muscle damage is evident by morphological changes in the intracellular structure and extracellular matrix (Armstrong et al. 1991; Lieber and Friden 2002; Stauber and Smith 1998). The earliest events associated with the muscle damage are mechanical, and later events indicate muscle remodeling (Friden and Lieber 2001). Damage to muscle and connective tissue is followed by an inflammatory response that is necessary for regeneration (Kuipers 1994). During this process, neutrophils and macrophages infiltrate damaged muscle fibers and degrade damaged proteins (MacIntyre et al. 1995). However, it seems possible that this degradation process does not necessarily take place if the damage is not severe (Proske and Morgan 2001). The relationship between ultrastructural changes and the inflammatory process has not yet been clarified.

Strictly speaking, muscle damage should be verified only through morphological examination. However, the highly focal nature of damage to specific sarcomeres in individual fibers makes quantitative evaluations of the magnitude of muscle damage difficult (Faulkner et al. 1993). This is why symptoms are more frequently used as markers of muscle damage.

Indirect Markers of Muscle Damage

Two typical symptoms of eccentric exercise–induced muscle damage are muscle soreness and loss of muscle function, and these effects have been used as markers of muscle damage (Bär et al. 1997). Muscle damage is also assessed via increases in muscle-specific proteins in the blood such as creatine kinase (CK) or myoglobin (Clarkson et al. 1992). Swelling of muscles, detected by increases in circumference or magnetic resonance or ultrasound images, is often included among the markers of muscle damage (Howell et al. 1993; Nosaka and Clarkson 1996). Among indirect markers, muscle function measures such as muscle strength and range of motion (ROM) are considered the best tools for quantifying muscle damage (Warren et al. 1999).

Figure 5.3 shows changes in some indirect markers of muscle damage after eccentric and concentric exercise of the elbow flexors. Nonresistance-trained subjects performed a bout of eccentric exercise (ECC) of the elbow flexors with one arm and a bout of concentric exercise (CON) with the other arm; bouts were four to six weeks apart. In both exercises, six sets of five muscle contractions with 2 min rest between sets were performed with a dumbbell set at 40% of each arm's maximal isometric strength measured at an elbow joint angle of 90°. The average weight of the dumbbell used for both exercises was approximately 10 kg; the range of motion was 90° for both, with ECC starting from an elbow joint angle of 90° and CON from a fully extended position (180°). The time taken for each movement was 3 to 4 s for ECC and 2 to 3 s for CON. Significantly larger changes in maximal isometric strength, ROM, and upper arm circumference were evident following ECC compared to CON. A prolonged loss of muscle strength and ROM, profound swelling, DOMS, and increases in plasma CK activity and myoglobin concentration were peculiar to eccentric exercise.

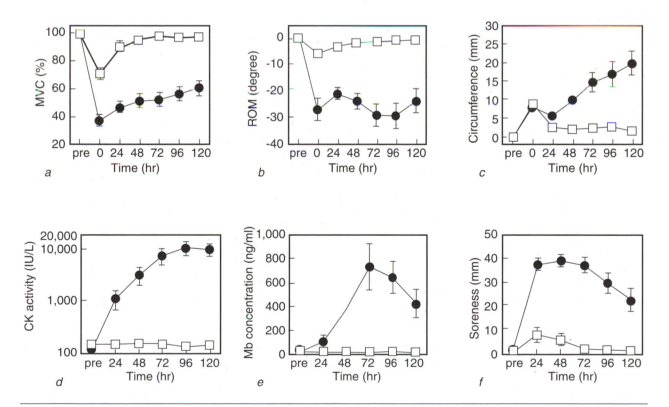

Figure 5.3 Comparison between eccentric (circles) and concentric (squares) exercise of the elbow flexors for changes in markers of muscle damage before (pre), immediately after (0), and 24 to 120 h after exercise. *(a)* Normalized changes in maximal isometric strength; *(b)* absolute changes in range of motion from baseline (pre); *(c)* absolute changes in upper arm circumference from baseline (pre), *(d)* plasma creatine kinase activity; *(e)* plasma myoglobin concentration; *(f)* muscle soreness as indicated on a VAS of 50 mm (0: no pain; 50: very painful).

Data from Lavender and Nosaka, 2006.

Connections Between DOMS and Muscle Damage

As already discussed, although DOMS is one of the symptoms associated with muscle damage, it may not be a direct reflection of muscle damage. It is possible for severe DOMS to develop with little or no indication of muscle damage, and severe muscle damage does not necessarily result in severe DOMS. The dissociation between DOMS and other indicators of muscle damage has been documented (Newham 1988; Rodenburg et al. 1993; Nosaka et al. 2002a).

Time Course

Although unaccustomed eccentric exercise results in DOMS and a number of functional, structural, and biochemical changes, it is important to note that the time courses of these changes differ (figure 5.4). After maximal eccentric exercise of the elbow flexors, muscle soreness does not develop immediately whereas muscle strength shows its largest decrease at this point. Range of motion is more affected a couple of days after exercise but recovers more quickly than muscle strength. When muscle soreness subsides, swelling of the upper arm peaks, and abnormality in magnetic resonance or ultrasound images is greatest around this time period. Plasma CK activity peaks around five days after exercise and returns to baseline two to three weeks later. Not only do the time courses of changes in the markers of muscle damage differ from one another, but also none of them match the time course of muscle soreness (Newham 1988). All this clearly shows that muscle soreness is not a cause of loss of muscle function, swelling is not a direct stimulus for muscle soreness, and muscle soreness does not peak when plasma CK peaks.

Correlations Between DOMS and Other Markers Among Subjects

It has been reported that the level of DOMS correlates poorly with the magnitude of changes in other indicators of muscle damage (Nosaka et al. 2002a; Rodenburg et al. 1993). As shown in figure 5.5, the peak muscle soreness score does not correlate strongly with other markers. It is generally thought that the larger the decrease in muscle strength following eccentric exercise, the greater the magnitude of muscle damage. However, subjects who show large decreases in maximal isometric strength do not necessarily have severe soreness.

It is also interesting that resistance-trained subjects and untrained subjects reported a similar magnitude of peak muscle soreness after performing maximal eccentric exercise of the elbow flexors, although changes in other markers of muscle damage such as maximal isometric strength and plasma CK activity were significantly larger for untrained compared to trained subjects (figure 5.6). This suggests that the magnitude of muscle damage is less for resistance-trained than for untrained individuals after eccentric exercise, but that resistance-trained individuals may feel a similar magnitude of DOMS with significantly less muscle damage.

The Magnitude of DOMS and Muscle Damage in the Same Subject

The examples shown so far are based on groups of subjects. It may be that findings of little or no relationship between DOMS and muscle damage are due to differing pain perception among subjects. However, there is evidence to support the idea that DOMS does not reflect

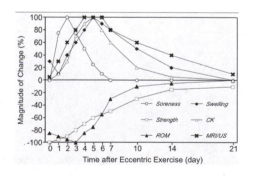

Figure 5.4 Time course of changes in markers of muscle damage after maximal eccentric exercise of the elbow flexors: muscle soreness, upper arm circumference (swelling), maximal isometric strength (strength), creatine kinase (CK), range of motion (ROM), and magnetic and ultrasound images (MRI/US).

Data from Nosaka and Clarkson, 1996.

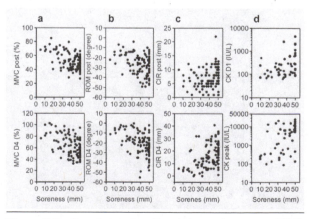

Figure 5.5 Correlations between peak muscle soreness (VAS, 0-50 mm scale) and other markers of muscle damage. (a) Percent changes in maximal isometric strength immediately or four days after exercise; (b) changes in range of motion immediately or four days after exercise; (c) changes in upper arm circumference from baseline immediately or four days after exercise; (d) plasma creatine kinase activity at one day postexercise and peak.

Data from Nosaka and Clarkson, 1996.

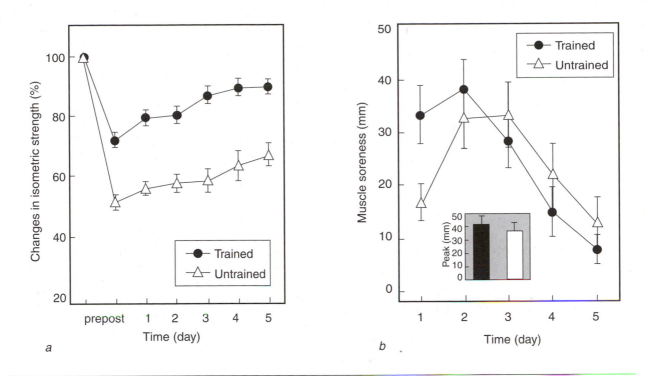

Figure 5.6   Comparison between trained (circles) and untrained (triangles) subjects on changes in *(a)* maximal isometric strength and *(b)* muscle soreness after eccentric exercise of the elbow flexors. The muscle soreness graph also shows the peak soreness value for each group.

Data from Newton et al., 2007.

the magnitude of muscle damage even within a given individual, depending on the type and intensity of exercise as well as on which muscles are being used.

- **Type and intensity of exercise.** When the same subject performed two different intensities of eccentric exercise using the same muscle group, all of the indirect markers of muscle damage showed larger changes for maximal intensity than for 50% intensity. However, no significant difference in muscle soreness was observed between the exercises (Nosaka and Newton 2002c). Moreover, two different types of exercise of the elbow flexors (maximal eccentric exercise in which 24 forced eccentric muscle actions were performed under maximal force generation; endurance exercise in which elbow flexion and extension movements were repeated for 30 min) performed by the same subject induced a similar magnitude of muscle soreness, but the magnitudes of changes in other markers of muscle damage such as muscle strength and plasma CK activity were significantly different (figure 5.7). These findings also support the notion that the magnitude of DOMS does not necessarily reflect the magnitude of muscle damage.

- **Difference between arm and leg muscles.** The responses to eccentric exercise differ between the leg and arm muscles. Having subjects perform both arm and

leg exercises, Jamurtas and colleagues (2005) compared effects of an eccentric exercise of the elbow flexors with those of an eccentric exercise of the knee extensors, matching the number of eccentric actions (six sets of 12 reps) and the relative intensity using 75% of predetermined maximal eccentric torque. The arm eccentric exercise induced larger decreases and slower recovery of strength, as well as larger increases in blood markers of muscle damage (CK, myoglobin), than the leg exercise. However, DOMS did not differ between the two exercises (figure 5.8). These findings again support the idea that the magnitude of DOMS does not represent the magnitude of muscle damage.

The Message of DOMS

Delayed-onset muscle soreness may play a protective role by acting as a warning to reduce muscle activity and prevent further injury. However, we may find that further activity, although causing more pain initially, in fact then alleviates muscle soreness. Hough (1902) reported that performing a second bout of exercise with a sore muscle caused excessive pain for the first 2 to 3 min but that the pain disappeared over the course of 5 to 10 min. Saxon and Donnelly (1995) investigated the effect of submaximal (50%) concentric exercise performed one to

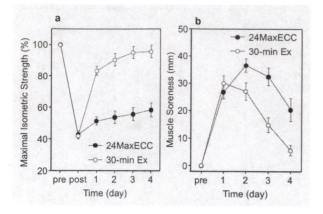

Figure 5.7 Comparison between two different types of exercise of the elbow flexors for changes in *(a)* maximal isometric strength and *(b)* muscle soreness. Each subject performed the two types of exercise using a different arm: 24 maximal eccentric actions of the elbow flexors (24MaxECC, filled circles) and repeated elbow flexion (1 s) and extension (1 s) movements for 30 min (30 min Ex: open circles) using a light dumbbell.

Nosaka et al. Unpublished data.

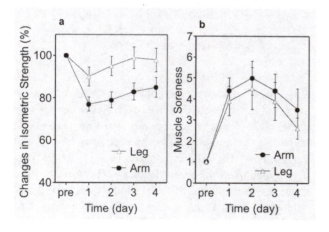

Figure 5.8 Comparison between eccentric exercise of knee extensors (leg: triangles) and elbow flexors (arm: circles) for changes in *(a)* maximal isometric strength and *(b)* muscle soreness after exercise.

Modified from Jamutas et al., 2005.

four days following a bout of maximal eccentric exercise on DOMS. They reported that DOMS was significantly reduced immediately after performing the concentric exercise on day 2, but no significant effect on DOMS was found on the other days.

A recent study by Zainuddin et al. (2006) demonstrated that muscle soreness was palliated significantly (≈40%) immediately after performance of light concentric exercise on each of days 1 through 4 following eccentric exercise that induced muscle

soreness and damage (figure 5.9*a*). This suggests that the light concentric exercise was effective for attenuating DOMS. However, the palliative effect of the light exercise was temporary and did not influence changes in overall muscle soreness (figure 5.9*b*). No adverse effects of the light concentric exercise on recovery of muscle function were evident, either. Thus, it seems unlikely that DOMS is a warning sign not to use the affected muscle.

Several studies have shown that performing eccentric exercise in the early recovery days after the initial bout (two to three days) with sore muscles does not exacerbate muscle damage or retard the recovery process (Chen 2003; Nosaka and Newton 2002c, 2002d). These results suggest that muscle soreness is not necessarily a warning signal not to use the muscle. Such muscle pain may be telling us that we should not use the muscle, but we find that the pain is reduced when we use the sore muscle anyway. Thus it is still unclear how we should treat this type of muscle pain. We often face a situation in which we need to consider whether to ignore and overcome pain or accept and try to remove it.

Repeated-Bout Effect

Once an individual has experienced severe DOMS after performing an unaccustomed exercise, the exercise is no longer unaccustomed. After the person performs a similar exercise a couple of weeks later, severe DOMS no longer develops. It is as if muscles adapt to exercise rapidly to

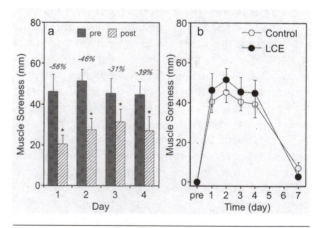

Figure 5.9 *(a)* Changes in muscle soreness with extension of the elbow joint before (pre) and immediately after (post) light concentric exercise performed on days 1 to 4 after maximal eccentric exercise; *(b)* changes in soreness following eccentric exercise for control and light concentric exercise (LCE) conditions. Asterisk indicates significant ($p < 0.05$) difference from the pre value. The left graph also shows changes from pre to post.

Based on Zainuddin et al., 2006.

protect themselves from muscle damage. This phenomenon of a protective effect against muscle damage has been termed the repeated-bout effect (McHugh 2003).

Along with attenuated DOMS after the exercise, the repeated-bout effect is characterized by faster recovery of muscle strength and range of motion, reduced swelling and muscle soreness, and smaller increases in muscle proteins (e.g., CK, myoglobin) in the blood following the second bout of a given eccentric exercise compared to the first (figure 5.10). It has also been demonstrated that immune responses are attenuated in the second bout (Pizza et al. 1999). Fewer abnormalities in ultrasound or magnetic resonance images are evident after the subsequent bout as well (Foley et al. 1999).

It is important to note that the protective effect conferred by the initial bout of eccentric exercise does not necessarily "prevent" muscle damage, but attenuates the magnitude of changes in markers of muscle damage or enhances the recovery process. It appears that the adaptation is more specific to the muscles involved in the exercise, as evidenced by a study in which subjects performed the second eccentric exercise bout with a different arm from the initial bout. No significant difference in the changes in isometric strength is evident between the right and left arm bouts separated by two weeks, but the first and second bouts performed by the same arm show a distinct difference in strength recovery (figure 5.11). Howatson and van Someren (2007) reported that changes

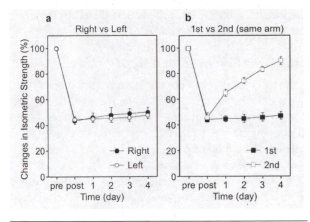

Figure 5.11 Comparison between (a) right-(filled circles) and left-(open circles) arm bouts and (b) first (filled squares) and second (open squares) bouts, using the same arm, of changes in maximal isometric strength following eccentric exercise of the elbow flexors. The two exercise bouts were separated by two weeks.

Nosaka et al. Unpublished data.

in maximal isometric strength, serum CK activity, and muscle soreness were attenuated when a second bout of eccentric exercise of the elbow flexors was performed by the contralateral arm two weeks later; however, the magnitude of protection in the contralateral arm was much less than that shown in the ipsilateral limb. Newton et al. (2007) compared changes in the markers of muscle damage between arms after maximal eccentric exercise of the elbow flexors separated by four weeks by counterbalancing the use of dominant or nondominant arm for the first bout among subjects. Changes in maximal isometric strength, range of motion, upper arm circumference, plasma CK activity, and muscle soreness measures were not significantly different between arms, but a significant difference between the bouts was evident for maximal isometric torque, circumference, and plasma CK activity, such that the changes were significantly smaller after the second bout compared with the first bout. These suggest that some effect is transferred from one arm to the other, but the effect is weak, and the repeated bout effect appears more strongly for the muscle that previously performed the same eccentric exercise.

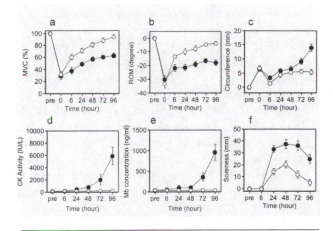

Figure 5.10 Comparison between first (filled circles) and second (open circles) bout of exercise of the elbow flexors for changes in markers of muscle damage before (pre), immediately after (0), and 24 to 120 h after exercise. (a) Normalized changes in maximal isometric strength; (b) absolute changes in range of motion from baseline (pre); (c) absolute changes in upper arm circumference from baseline (pre); (d) plasma creatine kinase activity; (e) plasma myoglobin concentration; and (f) muscle soreness as indicated by a VAS (50 mm line: 0, no pain; 50, very painful).

Based on Hirose et al., 2004.

Characteristics of Repeated-Bout Effect

The repeated-bout effect occurs whenever unaccustomed exercise is repeated within a certain period of time. However, many factors influence the magnitude of the effect, such as time between bouts, the number of eccentric contractions, muscle length, and exercise mode. Additionally, it seems that intersubject variability exists in relation to this effect, and the magnitude of the effect differs among the markers of muscle damage.

Effect of Time Between Bouts

If the time period between the unaccustomed and subsequent exercise bout is too long, the repeated-bout effect does not occur. However, it appears that the repeated-bout effect lasts for at least several weeks and that its length is dependent on markers of muscle damage (Nosaka et al. 2005a). It was reported that changes in indirect markers of muscle damage following exercise were suppressed more when the interbout interval was 6 weeks compared to 10 weeks (Nosaka et al. 1991). As shown in figure 5.12, faster recovery of strength and range of motion, reduced swelling, less development of muscle soreness, and smaller increases in muscle proteins in the blood following a second eccentric exercise bout compared with the initial bout persisted for more than six months for eccentric exercise of the elbow flexors. However, the magnitude of the protective effect appears to decrease gradually as the time between bouts increases, and the time course of attenuation of the protective effect varies among the measures. No protective effects seem to last more than a year.

Several studies showed that when the second eccentric exercise was performed two weeks after the first bout, prior to full recovery of muscle function, prolonged decreases in muscle function and development of muscle soreness, but no increases in CK activity, occurred (Clarkson and Tremblay 1988; Newham et al. 1987; Nosaka et al. 2005b). When the second bout was performed within a week of the initial exercise, in the early recovery phase,

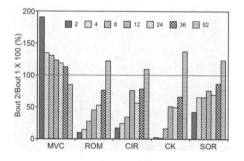

Figure 5.12 Magnitude of repeated-bout effect for maximal isometric strength (MVC), range of motion (ROM), upper arm circumference (CIR), plasma creatine kinase activity (CK), and muscle soreness (SOR) when the interval between bouts was 2, 4, 8, 12, 24, 36, or 52 weeks. For MVC, ROM, and CIR, the values at four days postexercise in relation to the preexercise values were compared between bouts. For CK and SOR, peak values after exercise were compared between bouts.

Data from Nosaka et al., 2001a, Nosaka et al., 2001b, Nosaka et al., 2005a, and Nosaka et al., 2005b.

no adverse effects on markers of damage were observed, although acute decreases in muscle function occurred immediately after exercise (Chen 2003; Ebbeling and Clarkson 1989; Nosaka and Newton 2002c, 2002d).

Effect of Number of Eccentric Contractions

Muscles do not appear to require the same exercise stimuli in the two bouts in order to show the repeated-bout effect. It has been reported that performing an initial eccentric bout with a relatively small number of eccentric actions produced the repeated-bout effect. An initial bout of 24 maximal eccentric repetitions reduced plasma CK activity and the magnitude of the strength loss and DOMS when a 70-repetition bout was performed two weeks later (Clarkson and Tremblay 1988). It has also been demonstrated that 10, 30, or 50 eccentric actions provided equal protection for a bout of 50 eccentric actions three weeks later, in which increases in plasma CK activity were attenuated and the magnitude of isometric force loss was reduced (Brown et al. 1997).

Nosaka and colleagues (2001b) investigated whether a small volume of an initial eccentric exercise bout could still confer the repeated bout effect on a second bout of a larger volume eccentric exercise that was performed two weeks later. The volume for the initial exercise bout was less than 10% (two maximal eccentric actions: 2ECC), of the number of contractions to be performed in the second bout (24 maximal eccentric actions: 24ECC). All variables changed significantly after 2ECC, but the amount of change in isometric strength and muscle soreness after for the 2ECC was significantly smaller than that for 24ECC (figure 5.13). After the second bout, the group that performed 24ECC initially showed a profound repeated-bout effect that was indicated by a faster recovery of isometric strength and less development of muscle soreness. The group that initially performed 2ECC (2-24ECC) also demonstrated the repeated-bout effect, although the magnitude of the effect was not as strong as that of the 24-24ECC. These results suggest that the repeated-bout effect can be produced by two maximal eccentric actions, and it is not necessary to perform a high number of eccentric actions in the first bout to elicit a repeated-bout effect.

Effect of Muscle Length

It is known that changes in markers of muscle damage are significantly smaller following eccentric exercise of the elbow flexors in which the elbow joint is not fully extended compared to when it is fully extended (Nosaka and Sakamoto 2001). An interesting question is whether eccentric exercise without full extension movements can protect against muscle damage induced by eccentric exercise with full extension.

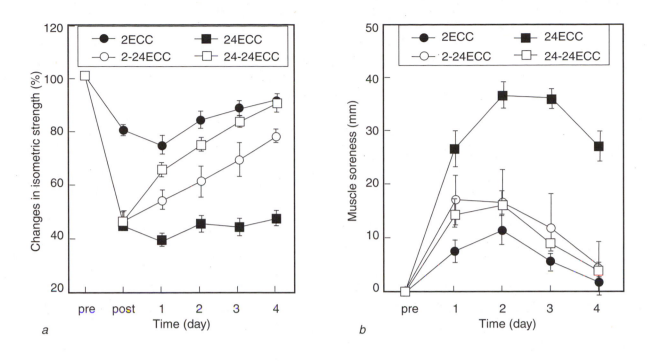

Figure 5.13   Changes in (a) maximal isometric strength and (b) muscle soreness following two (2ECC: filled circles) or 24 (24ECC: filled squares) maximal eccentric actions of the elbow flexors and in the subsequent bout of exercise (in which 24 maximal eccentric actions of the same muscle group were performed by the arm that previously performed 2ECC [2-24ECC: open circles] or 24ECC [24-24ECC: open squares]).

Modified from Nosaka et al., 2001b.

To address this question, Nosaka and colleagues (2005b) compared two groups of subjects. One group (L-L) performed eccentric exercise at a long starting length for both a first and a second bout separated by four weeks; the other group (S-L) performed eccentric exercise at a short starting length followed four weeks later by eccentric exercise at the long starting length (figure 5.14). The eccentric exercise at a long muscle length induced greater muscle damage than an equivalent bout performed at a short muscle length, as shown by the study mentioned in the preceding paragraph. In addition, results showed that eccentric exercise at the short muscle length produced a partial protective effect against muscle damage induced by exercise at the long muscle length. Although the short muscle length exercise induced an appreciable degree of protection against the effects of the more demanding bout, the magnitude of protection varied among the criterion measures. In general, this amounted to a "partial" protection of around 50% for most of the criterion variables. A possible explanation for this result is that the adaptation of the brachialis was more pronounced following the short muscle length exercise, resulting in an enhancement of the protective effect for that muscle compared to that for the biceps brachii.

Strategies to minimize the degree of damage associated with eccentric exercise should include consideration of not only the number of actions, but also the range of motion of the activity, with particular regard to the portion of the range of motion associated with full muscle extension. However, this may not generalize, since McHugh Pasiakos (2004) did not find such an effect for the knee extensors. It may be that arm muscles are different from leg muscles not only in their susceptibility to eccentric exercise–induced muscle damage (Jamurtas et al. 2005) but also with regard to the repeated-bout effect.

Effect of Exercise Mode

Whitehead and colleagues (1998) reported that the susceptibility of muscle to damage from eccentric exercise (downhill running) was increased after subjects performed concentric exercise (450 plantarflexion movements with a 10 kg weight) for five days prior to the eccentric exercise. These authors speculated that the concentric exercise reduced the number of sarcomeres in series and shifted the length–tension relationship in the direction of shorter muscle lengths, which makes the muscle more susceptible to eccentric exercise–induced muscle damage. Conversely, Nosaka and Newton (2002a) showed that eight weeks of concentric training of the elbow flexors using a submaximal dumbbell weight did not exacerbate muscle damage induced by maximal

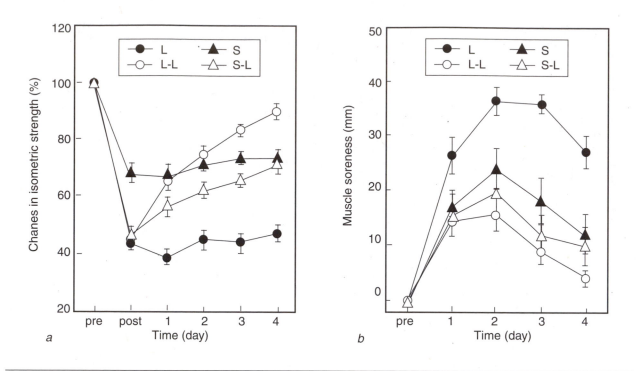

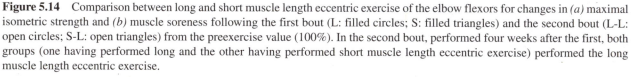

Figure 5.14   Comparison between long and short muscle length eccentric exercise of the elbow flexors for changes in (a) maximal isometric strength and (b) muscle soreness following the first bout (L: filled circles; S: filled triangles) and the second bout (L-L: open circles; S-L: open triangles) from the preexercise value (100%). In the second bout, performed four weeks after the first, both groups (one having performed long and the other having performed short muscle length eccentric exercise) performed the long muscle length eccentric exercise.

Modified from Nosaka et al., 2005b.

eccentric exercise of the same muscle. They suggested that to prevent or reduce eccentric exercise–induced muscle damage, it seems necessary to stimulate the muscles using the same muscle actions and intensity as in the damaging exercise.

Eston and colleagues (1996) showed that muscle soreness, strength loss, and increases in plasma CK activity after a downhill run were reduced when 100 maximal isokinetic eccentric actions had been performed two weeks earlier. This suggests that a different mode of exercise of the same muscle group confers the repeated-bout effect. It is not clear yet how much of a protective effect is conferred by a different mode of exercise of the same muscle.

One unpublished study addressed the extent of protection conferred by a submaximal elbow flexor endurance task against the effects of maximal eccentric exercise of the same muscle group. A group of subjects flexed (1 s) and extended (1 s) their elbow joint rhythmically for 30 min (900 actions) with a wristband load set at 10% of their maximal isometric strength and then, four to six weeks later, performed maximal eccentric exercise of the elbow flexors consisting of 24 forcible extensions of the elbow joint from a flexed (90°) to an extended position (180°). Another group

performed the two bouts of maximal eccentric exercise with the same arm.

Changes in indicators of muscle damage were significantly smaller following the endurance exercise compared with the maximal exercise (figure 5.15). After the maximal eccentric exercise, the subjects who had performed the endurance exercise initially showed smaller changes in the indicators of muscle damage compared with those observed after the first bout of maximal eccentric exercise performed by the other group of subjects. However, the magnitude of the protective effect against the effects of maximal eccentric exercise in the subjects who had initially performed the endurance exercise was not as strong as that shown by the subjects who repeated the maximal eccentric exercise. This suggests that protection against the effects of maximal eccentric exercise can be partially conferred to the elbow flexors using submaximal endurance exercise that results in minor damage.

Protective Effect Induced by Isometric or Nondamaging Exercise

It seems that severe muscle damage is not necessary in order to confer a protective effect on subsequent exercise.

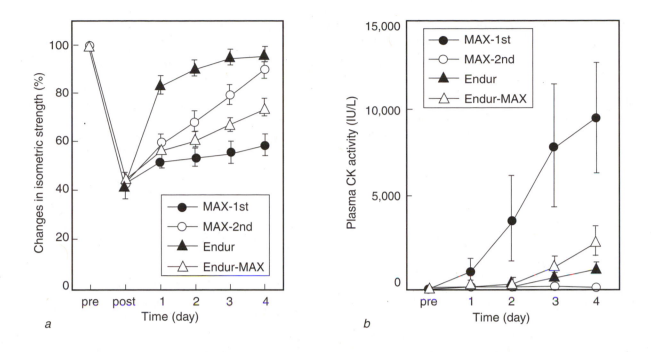

Figure 5.15   Effect of endurance exercise of the elbow flexors (Endur) on maximal eccentric exercise of the same muscles (Max) in comparison to the effect when the maximal eccentric exercise was performed in both the first (Max-1st) and the second bout (Max-2nd). Changes in *(a)* maximal isometric strength and *(b)* plasma CK (creatine kinase) activity following the first (filled circles) and second (open circles) bout of the maximal eccentric exercise and after the endurance exercise (filled triangles) and then the maximal eccentric exercise (open triangles).

Nosaka et al. Unpublished data.

As already mentioned, low-intensity (40%) eccentric exercise could confer protection of 20% to 60% on the indices of muscle damage following a subsequent 100% exercise bout performed two to three weeks later (Chen et al. 2007). Lavender and Nosaka (2007) have shown recently that light eccentric exercise (with a dumbbell set at 10% of maximal isometric strength) induces some protection against the effects of a subsequent bout of eccentric exercise with a heavier weight (a dumbbell set at 40% of maximal isometric strength) carried out two days later. However, whether the protective effect conferred by 10% eccentric exercise lasts longer, say for two weeks, has not been confirmed. Some animal studies have shown that isometric exercise can produce protective effects against the effects of eccentric exercise. Koh and Brooks (2001) reported that maximal isometric contractions or passive stretches of the extensor digitorum longus (EDL) muscles in mice did not cause degeneration of muscle fibers but induced protection against muscle damage from maximal eccentric actions performed three days later. McArdle and colleagues (2004) also reported that nondamaging isometric contractions of the soleus and EDL muscles via electrical stimulation, conducted 4 or 12 h prior to a damaging protocol, reduced CK

release from the muscles of mice. However, no human studies have yet shown the extent of the protective effect conferred by isometric exercise against the effects of eccentric exercise.

Mechanisms of Repeated-Bout Effect

The mechanisms underlying the repeated-bout effect have yet to be fully elucidated, although several potential mechanisms have been addressed. McHugh (2003) reviewed studies associated with the repeated-bout effect and categorized the potential mechanisms as neural, mechanical, and cellular adaptations (table 5.1). He concluded that "there may be several mechanism's for the repeated bout effect and those mechanisms may compliment each other or operate independently of each other (McHugh 2003, p. 96)."

The neural adaptations include more efficient recruitment of motor units, increased synchrony of motor unit firing, better distribution of the workload among fibers, improved usage of synergist muscles, and increased slow-twitch fiber recruitment. To investigate any neural adaptations associated with the repeated-bout effect,

Table 5.1   Potential Mechanisms for the Repeated-Bout Effect

Neural adaptations (spinal cord, CNS)	Mechanical adaptations (noncontractile elements)	Cellular adaptations (contractile machinery)
Increased recruitment of slow-twitch motor units Activation of larger motor unit pool Increased motor unit synchronization Learning effect	Increased passive or dynamic muscle stiffness Remodeling of intermediate filament system to provide mechanical reinforcement (desmin, titin, etc.) Increased intramuscular connective tissue	Longitudinal addition of sarcomeres (shift in the length–tension relationship) Adaptation in inflammatory response Adaptation to maintain excitation–contraction coupling Strengthened plasma membrane Increased protein synthesis Increased stress proteins (heat shock proteins, etc.) Removal of stress-susceptible fibers and replacement by stronger fibers

Nosaka and colleagues (2002b) compared two bouts of stretching of muscles receiving percutaneous electrical stimulation. Since the electrical stimulation bypasses the involvement of central motor drive, the expectation was to examine whether the central nervous system is involved in the repeated-bout effect. The results showed that the repeated bout of exercise resulted in less damage than the first bout, with all of the indirect markers of muscle damage showing smaller changes, and that recovery was significantly faster following the second bout (figure 5.16). If neural adaptations were primarily responsible for the repeated-bout effect, similar effects on the criterion measures would have been observed following the two eccentric exercise bouts. The findings thus suggest that the repeated-bout effect did not result from changes in the motor output of the central nervous system. Some evidence exists to support the neural adaptation theory. Howatson and van Someren (2007) reported a significant attenuation of muscle damage in the second bout of eccentric exercise performed by the contralateral arm two weeks later. Newton et al. (2007) also showed that changes in maximal isometric torque, upper arm circumference, and plasma CK activity were significantly smaller after the second bout performed by the opposite arm than the first bout. A possible explanation for how the attenuation of the changes in some of the criterion measures was conferred by the first bout performed by the contralateral arm may lie in the phenomenon of cross education effect or some learning effect (Hawatson and van Someren 2007; Newton et al. 2007).

According to another theory, suggested by Proske and Morgan (2001), increases in sarcomere number in series are associated with the protective effect. The increases in sarcomere number in series are indirectly assessed by a shift in optimum angle toward a longer muscle length (Proske and Morgan 2001). Philippou and colleagues (2004) recently reported a shift in the optimum angle of the elbow joint for producing maximal force by approximately 16°. If the shift could last for more than several weeks, this theory is attractive; however, the duration of this adaptation has yet to be determined. Chen and colleagues (2007) compared the effect of four different intensities (100%, 80%, 60%, and 40%) of initial eccentric exercise (ECC1) on optimum angle shift and the extent of muscle damage induced by subsequent maximal eccentric exercise (ECC2) performed two to three weeks later with a 100% load. A rightward shift of the optimum angle following ECC1 was significantly greater for the 100% and 80% than for the 60% and 40% exercises, and decreased significantly from immediately to five days postexercise. By the time ECC2 was performed, only 100% exercise retained a significant shift (4°). Although the magnitude of the repeated-bout effect following ECC2 was significantly smaller for the 40% and 60% groups, all groups showed significantly reduced changes in criterion measures following ECC2 in comparison to the ECC1 100% bout. This suggests that the repeated-bout effect is not dependent on the shift in optimum angle. Thus, it seems unlikely that the protective effect can be explained solely by increases in the number of sarcomeres in series.

Ingalls and colleagues (2004) recently showed that the enhanced strength recovery of mouse foot dorsiflexor muscles with repeated lengthening exercise could be attributed to elevated rates of protein synthesis. This could explain the faster recovery of muscle function after a second bout compared with an initial bout as seen in human studies. Newham and colleagues (1987) have postulated that muscle fibers become more resilient and are able to withstand a given eccentric exercise after stress-susceptible fibers are removed and replaced by regenerated fibers. This theory

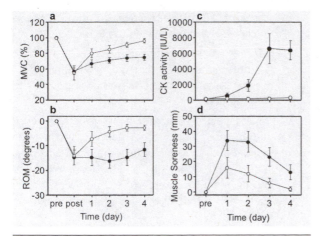

Figure 5.16 Comparison between first (filled circles) and second (open circles) bout of stretching of electrically stimulated muscles of the elbow flexors for changes in *(a)* maximal isometric strength, *(b)* range of motion, *(c)* plasma creatine kinase activity, and *(d)* muscle soreness.

Modified from Nosaka et al., 2002b.

appears to explain the repeated-bout effect very well if the newly regenerated fibers become susceptible again to eccentric exercise in 8 to 12 weeks. Figure 5.17 presents a graphic image of the stress-susceptible fiber hypothesis, which proposes six stages in the repeated-bout effect:

- **Stage 1:** Before performing the first eccentric exercise bout (ECC1), some of the muscle fibers are stress-susceptible fibers.

- **Stage 2:** These fibers are likely to be damaged by ECC1 and to degenerate, and severe muscle damage is induced, but other fibers may survive.

- **Stage 3:** After ECC1, the damaged fibers are regenerated and may be remodeled and become "strong" fibers.

- **Stage 4:** When the second bout of eccentric exercise (ECC2) occurs at this stage, the number of stress-susceptible fibers is small, and less muscle damage is produced.

- **Stage 5:** No stress-susceptible fibers exist and little muscle damage is induced when eccentric exercise is performed at this stage. When eccentric exercise is performed regularly, it may be that muscles are in the Stage 5 condition.

- **Stage 6:** Because of protein turnover, some muscle fibers become stress-susceptible fibers again.

Understanding the mechanisms underlying the repeated-bout effect may be the key to understanding eccentric exercise–induced muscle damage.

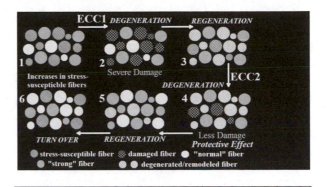

Figure 5.17   Stress-susceptible fiber hypothesis.

According to Koh (2002), heat shock proteins (HSPs) may be involved in protecting skeletal muscle fibers from eccentric exercise–induced muscle damage. A microwave diathermy treatment that increased muscle temperature to over 40° C, 16 to 20 h prior to exercise, resulted in a significantly faster recovery of muscle strength, a smaller change in ROM, and less muscle soreness; however, the protection afforded by the diathermy treatment was significantly less effective than in the second bout, performed four to six weeks after the initial bout (Nosaka et al. 2007). McArdle and colleagues (2004) proposed that activation of the heme oxygenase-1 (HO-1) gene resulting from increased reactive oxygen and nitrogen species (ROS) production was associated with the protective effect. Mikkelsen and colleagues (2006) have recently shown that stimulation of the Na^+-K^+ pump with β_2-adrenoceptor agonists improved force recovery in rat EDL muscles by 40% to 90% following a 30 min electrical stimulation protocol. It is possible that these proteins are associated with protective effects. Further studies are necessary to advance understanding of the mechanisms underlying the repeated-bout effect.

Summary

This chapter has focused on DOMS, muscle damage, the relationship between DOMS and muscle damage, and the repeated-bout effect. Despite advances in our understanding of eccentric exercise–induced muscle damage, we still do not have a complete picture of the phenomenon. More than a century ago, Hough (1902) carefully observed "muscular soreness" and thought creatively in an attempt to explain the cause of DOMS. He stated, "The abnormal condition of the muscle frequently escapes notice, unless attention is specially directed to it by making it work or by over extension." We still need to pay more attention to the condition of muscles so as not to miss what they can tell us. The key points of this chapter can be summarized as follows:

- Muscle soreness is sensed in the brain as a signal from muscle; however, the stimuli that prompt nociceptors or other receptors to evoke muscle soreness have not been fully elucidated.

- Delayed-onset muscle soreness is associated with eccentric exercise–induced muscle damage; however, it is still unclear how a sequence of events in the process of muscle damage induces DOMS.

- Physiological changes used as indicators of eccentric exercise–induced muscle damage include decreases in muscle function, swelling, increases in muscle proteins in the blood, and abnormalities shown by ultrasound and magnetic resonance images.

- The magnitude of DOMS does not necessarily reflect the extent of muscle damage, and the time course of DOMS does not represent the time course of changes in indicators of muscle damage.

- Muscle adapts rapidly after eccentric exercise to prevent muscle damage, and this effect (repeated-bout effect) lasts for several weeks to several months.

- The repeated-bout effect is produced even if the initial bout is less demanding than the second bout in terms of the number of eccentric actions, muscle lengths, and the force generated during eccentric exercise; and conferral of the protective effect does not necessarily require muscle damage.

- Neural, mechanical, and cellular adaptations are likely involved in the mechanisms underlying the repeated-bout effect; however, a unified theory explaining the mechanisms remains elusive.

CHAPTER 6

Satellite Cells and Muscle Repair

Karin Shortreed, MSc; Adam Johnston, MSc; and Thomas J. Hawke, PhD

The skeletal muscle of adult mammals displays a remarkable ability for growth and repair throughout the life span. This adaptability is largely the result of a population of stem cells, termed myogenic satellite cells, resident within the skeletal muscle itself. This chapter focuses on the capacity of skeletal muscle for growth and repair and the role of these unique cells in the regenerative process. The regulation of myogenic satellite cells in health and disease, as well as the role of various extrinsic factors in affecting myogenic satellite cell function, is also discussed.

Skeletal Muscle Stem Cell Populations

Adult skeletal muscles contain various cell populations that display stem cell–like characteristics, including the capacity for self-renewal, proliferation, and plasticity (capacity to become multiple lineages). In particular, the myogenic satellite cell and the muscle side-population (SP) cell are the most well characterized of the skeletal muscle stem cell populations to date.

Myogenic Satellite Cells

The myogenic satellite cell was named on the basis of its location at the periphery of the adult muscle fiber. Although these undifferentiated stem cells were identified over 40 years ago (Mauro 1961) and are the most thoroughly characterized of the resident muscle stem cell populations, a great deal of attention has recently been refocused on these cells as we begin to further appreciate their stem cell–like capacities.

In unperturbed muscle, the quiescent myogenic satellite cell resides outside of the muscle fiber plasma membrane but underneath the overlying basal lamina (figure 6.1, a & c). In this resting state, the nuclei of these cells comprise approximately 2% to 5% of all muscle nuclei and display dense heterochromatin (genetically inactive region of chromosomes), reduced organelle content, and high nuclear to cytoplasmic volume ratio, consistent with their low transcriptional activity (figure 6.1b). Satellite cells exit their quiescent state and enter a proliferative phase in response to stressors such as trauma (figure 6.1d). Studies of rodent skeletal muscles have demonstrated a relationship between myogenic satellite cell content and muscle fiber types (see Hawke and Garry 2001 for review), with oxidative muscle fibers demonstrating a five to six times greater myogenic satellite cell content than glycolytic muscle fibers (table 6.1). As human skeletal muscle displays a more heterogeneous fiber type composition, this phenomenon is less observed in humans.

In response to cellular and extracellular cues associated with intense exercise or muscle damage, the myogenic satellite cells exit their quiescent state (become "activated"), proliferate, and migrate to the site of injury to repair or replace damaged muscle fibers by fusing together, fusing to existing muscle fibers, or both. Although it has been suggested that other stem cell populations contribute to skeletal muscle regeneration, the evidence to date suggests that myogenic satellite

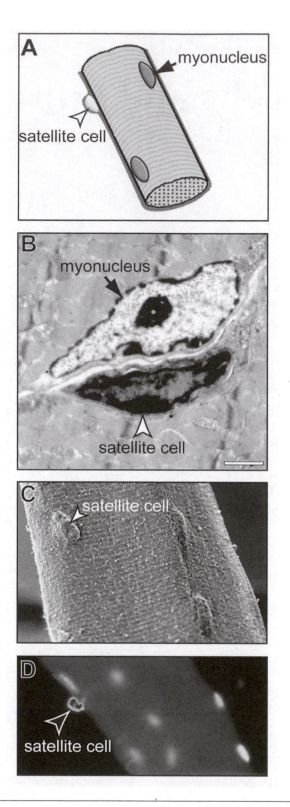

Figure 6.1 Localization of the myogenic satellite cell. *(a)* Illustration of a quiescent satellite cell at the periphery of the adult myofiber. *(b)* This electron micrograph image demonstrates a quiescent satellite cell in comparison to an adjacent myonucleus. Note the separation between the satellite cell and the myofiber as well as the dense heterochromatin compared to that in the myonucleus. *(c)* A scanning electron micrograph image displaying the peripheral location of a satellite cell on the myofiber. The striations of the muscle fiber are very apparent in this image. *(d)* An isolated single skeletal muscle fiber with a myogenic satellite cell in a proliferative state. A satellite cell (arrow) was stained with propidium iodide to label all nuclei and costained with a marker of proliferation, BrdU.

Table 6.1   Percentage of Satellite Cells in Young and Old Skeletal Muscles

Species	Muscle	Young	Old	References
Rodent	Levator ani	1.91	1.15*	Nnodim 2000
	Soleus	6.60	4.70*	Gibson and Schultz 1983
	Extensor digitorum longus	2.90	1.90*	Gibson and Schultz 1983
Human	Biceps brachii	4.17	1.44*	Renault et al. 2002
	Masseter	5.89	1.77*	Renault et al. 2002
	Vastus lateralis	2.80	1.70	Roth et al. 2000
	Tibialis anterior	6.90	3.90*	Kadi et al. 2004

*Denotes a significant ($p < 0.05$) decrease in satellite cell percentage.

cells are fully capable of complete myofiber regeneration (Zammit et al. 2002).

Importantly, myogenic satellite cells are self-renewing, such that a residual pool of these cells is reestablished after each discrete episode of muscle injury and is therefore available for additional rounds of regeneration. While the myogenic satellite cell displays some similarities to other adult stem cell populations (such as self-renewal and a limited plasticity), it is largely assumed that these cells are "committed" to the skeletal muscle lineage.

Other Muscle Stem Cells

Recently, additional cell populations with stem cell–like characteristics have been identified within skeletal muscle. These stem cells are harvested from adult skeletal muscle using either a multiple preplating technique or dual-wavelength flow cytometric analysis (FACS). Harvesting of muscle stem cells using FACS is based on their ability to efflux DNA dyes (e.g., Hoechst 33342). These mononuclear stem cells are termed muscle SP cells as they appear as a "Side Population" on the FACS profile. Muscle SP cells are smaller and far more rare than myogenic satellite cells within resting adult skeletal muscle (<0.2% vs. ~2-5% of all muscle nuclei). Furthermore, these cells appear to reside in the interstitial spaces between muscle fibers, in contrast to the myogenic satellite cell, which resides under the basal lamina. While the muscle SP cell displays heterogeneous expression of various cell surface proteins and differential capacities for self-renewal, plasticity, proliferation, and differentiation in comparison to the myogenic satellite cell, it is still unclear whether these cells are the precursor to the myogenic satellite cell or represent a distinct stem cell population within skeletal muscle (Meeson et al. 2004; Hawke 2005). To date, the contribution of endog-

enous muscle SP cells to skeletal muscle hypertrophy and regeneration appears minimal; however, from a therapeutic perspective, it has been demonstrated that various perturbations (e.g., irradiation, muscle injury, increasing local growth factor levels) increase their overall contribution (Grounds et al. 2002). As the role and regulation of the muscle SP cell population are still largely unknown, this chapter focuses on the myogenic satellite cell and its role in the adaptability and repair of skeletal muscle.

Muscle Repair and the Myogenic Satellite Cell

The capacity for skeletal muscle to repair following extensive injury, as well as to adapt to stressors such as strenuous exercise, is mediated by a complex array of cellular and extracellular cues. The adaptation of skeletal muscle to strenuous exercise (via muscle hypertrophy) relies on the fusion of myogenic satellite cells to existing muscle fibers as postulated by the myonuclear domain theory. In essence, this theory suggests that nuclei within the skeletal muscle fiber control the production of messenger RNA (mRNA) and proteins for a finite volume of cytoplasm, such that increases in fiber size (hypertrophy) must be associated with a proportional increase in muscle nuclei. These new muscle nuclei come from the myogenic satellite cell population. Regeneration from more extensive muscle damage, such as myotrauma, may result in the fusing together of myogenic satellite cells to generate a new muscle fiber. Repaired and newly generated muscle fibers can be identified by their centrally located nuclei.

Initially, skeletal muscle repair is characterized by an inflammatory response and the associated necrosis of the damaged muscle. During this degenerative phase, myogenic satellite cells become activated and begin to proliferate. As the inflammatory response decreases, proliferation and differentiation of myogenic satellite cells proceed, ultimately reconstituting the healthy myofiber population. Figure 6.2 illustrates this process as demonstrated in a study in which injury was induced in the tibialis anterior muscle of mice using cardiotoxin injection. Muscles were harvested at 0 days (no injury) and at 5, 10, 14, or 21 days following injection. By five days of regeneration, the inflammatory response was reduced, the muscle was less edematous, macrophages had invaded damaged myofibers, and myogenic satellite cells were in a highly proliferative state. Numerous newly regenerated myofibers were visible as small basophilic, centronucleated myofibers. At 10 days postinjury, regeneration had largely occurred, with many newly regenerated myofibers present as evidenced by the centrally located nuclei. By 14 days postinjury, regeneration was essentially complete, and the progression to 21 days was characterized by myofiber enlargement.

Although skeletal muscles may be exposed to numerous perturbations that would necessitate a myogenic satellite cell response (hypertrophic stimuli, myotrauma,

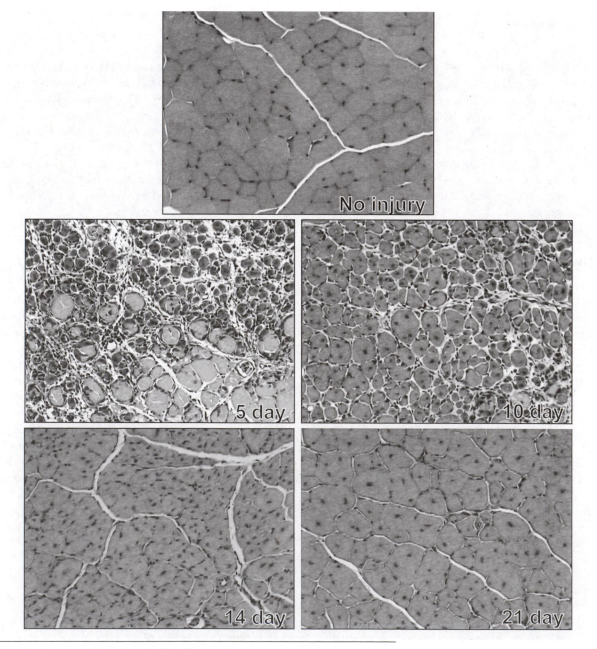

Figure 6.2   Hematoxylin and eosin (H & E) staining of regenerating skeletal muscle.

eccentric exercise), myogenic satellite cell activation, proliferation, and differentiation show similarities in response to these perturbations.

Inflammation and Myogenic Satellite Cell Activation

Myotrauma rapidly (within hours) increases the infiltration of neutrophils and macrophages to the damaged skeletal muscle. These cells release cytokines that further attract neutrophils, monocytes, and myogenic satellite cells. In addition, an increase in blood vessel permeability allows for protein transition and fluid accumulation within the damaged muscle. These inflammatory cells are responsible for phagocytosing cellular debris in the damaged muscle. Macrophages not only are needed to remove cellular debris and attract the myogenic satellite cells to the site of injury; they are also essential in the orchestration of the repair process as they secrete a collection of cytokine factors that regulate the myogenic satellite cell pool (Nathan 1987). The importance of this degradation phase in skeletal muscle regeneration was illustrated by Lescaudron and colleagues (1999), who observed that muscle regeneration was significantly attenuated in the absence of the macrophage response. Furthermore, from a therapeutic perspective, chronic use of nonsteroidal anti-inflammatory drugs (Cox-2 inhibitors) may impair muscle regeneration by inhibiting prostaglandin synthesis, which has been shown to be required during the early stages of muscle regeneration (Bondesen et al. 2004).

Activation is the first step in the response of the myogenic satellite cell to muscle injury and is responsible for preparing the cell for the proliferative phase. The activation phase is characterized by alterations in myogenic satellite cell morphology, increased cytoplasmic to nuclear ratio, increased cytoplasmic organelles, reduced heterochromatin, and changes in adhesion characteristics to the mature muscle fiber. Work from the labs of Allen and Anderson (Tatsumi et al. 2002) has identified a link between mechanical stress on the muscle and myogenic satellite cell activation. In their model, muscle damage leads to the bolus release of nitric oxide, which mediates the release of bound hepatocyte growth factor (HGF) from the extracellular matrix and surrounding muscle fibers. Release of HGF in response to muscle injury occurs very rapidly (within minutes) and is proportional to the degree of muscle injury. Active HGF released following muscle damage binds to its receptor (c-met) located on the myogenic satellite cell plasma membrane and initiates a cascade of signaling events that promote cell proliferation and changes in focal adhesion.

Recently, an alternatively spliced isoform of insulin-like growth factor, termed mechanogrowth factor (MGF), has been identified and shown to be released early (within minutes to hours) following increased loading or stretch of the muscle (Yang et al. 1996; McKoy et al. 1999). The expression pattern and role of MGF appear to diverge from those of insulin-like growth factors in that MGF may be involved in promoting transcriptional changes associated with preparing the cell for increased cellular proliferation. Although the exact role of MGF is still being investigated, its early release following muscle damage and its similarities to potent growth factors (insulin-like growth factors) suggest that it is integral to the activation of the myogenic satellite cell and ultimately the repair process.

Proliferation, Differentiation, and Self-Renewal

Proliferation of the myogenic satellite cell, as with other eukaryotic somatic cells, is a highly organized process involving DNA synthesis and division of cellular contents into two daughter cells (mitosis). The average time of progression of a myogenic satellite cell through the cell cycle is approximately 16 h (Hawke et al. 2003). Depending on the cellular and extracellular cues, the progeny of the myogenic satellite cells (often termed myoblasts) continue to proliferate to increase their numbers or undergo differentiation to reconstitute the healthy muscle fiber population. Numerous factors have been implicated in the regulation of myogenic satellite cells during the proliferative and differentiation phases subsequent to muscle injury. In the following sections we outline some of the extracellular factors that have been demonstrated to be critical for myogenic satellite cell regulation.

The differentiation of myogenic satellite cells appears to be intimately tied to the myogenic basic helix-loop-helix (bHLH) family of transcription factors, in particular MyoD, myf5, myogenin, and MRF4. A number of exquisite studies have demonstrated a role for myf5 and MyoD in the determination of the myogenic satellite cell to a muscle lineage, while myogenin and MRF4 appear to be associated with the promotion of muscle differentiation (Hasty et al. 1993; Rudnicki et al. 1993; Patapoutian et al. 1995; Rawls et al. 1995, 1998; Cornelison et al. 2000; Valdez et al. 2000). These transcription factors promote the expression of numerous genes important for the differentiation and fusion of myogenic satellite cells into (or with) muscle fibers.

As myogenic satellite cells are capable of self-renewal, at least a subpopulation of the activated or proliferating myogenic satellite cells must exit the cell cycle to repopulate the quiescent myogenic satellite cell population. Two key players in the self-renewal process have recently been identified: the transforming growth factor-beta family member (TGF-β) myostatin and the paired homeobox transcription factor, Pax-7. Myostatin appears to be critical for maintaining the myogenic satellite cell in a state of quiescence, as gene disruption studies have shown that the absence of myostatin leads to robust muscle hypertrophy and increased myogenic satellite cell numbers (McCroskery et al. 2003). For example, myostatin-deficient mice display muscle mass increases of 200% to 300%; and, as shown in figure 6.3, *a* through *c,* significant increases in muscle mass result from mutations in the myostatin gene in other species. It has been suggested that increased expression of Pax-7 is one of the critical signals for activated myogenic satellite cells to re-acquire a quiescent, undifferentiated state (figure 6.3*d*) (Olguin and Olwin 2004). Future studies will aid in further delineating the extracellular cues and cellular pathways necessary for myogenic satellite cells to remain in, and reacquire, quiescence.

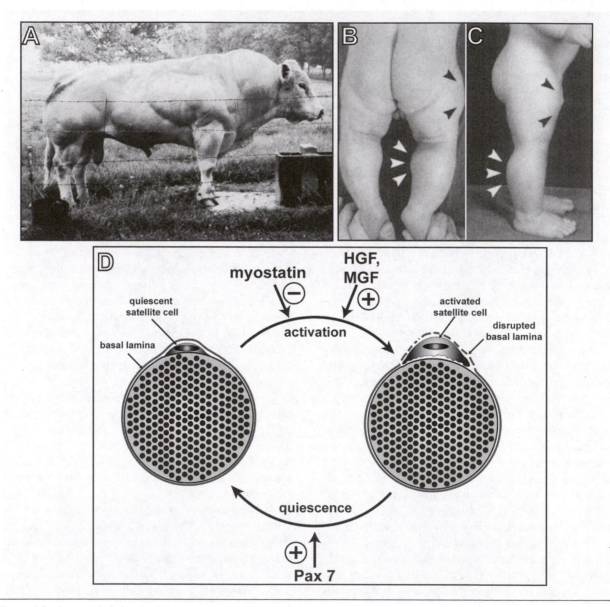

Figure 6.3 Myostatin is involved in myogenic satellite cell quiescence. *(a)* Mutations in the myostatin gene in cattle result in 20% to 25% increases in muscle mass as demonstrated by this image of a full-blood Belgian Blue bull. Myostatin mutations are also observed in humans. Note the gross hypertrophy of the thigh and calf muscles in a child with a myostatin mutation at *(b)* 6 days and *(c)* 7 months. *(d)* A number of factors may affect myogenic satellite cell quiescence and activation. MGF: mechano growth factor; HGF: heapatocyte growth factor.

Extrinsic Regulation of Myogenic Satellite Cells

How myogenic satellite cells are activated in response to skeletal muscle damage was illuminated by a series of elegant experiments by Bischoff (1986, 1989). Using isolated single muscle fibers in culture, these studies demonstrated that myogenic satellite cells located on the single fibers would remain in a quiescent state when exposed to normal muscle extract or crushed extract from other (nonmuscle) tissues but would become activated and proliferate in response to crushed muscle extract. Bischoff concluded that myogenic satellite cells, and ultimately muscle regeneration, are regulated by multiple growth factors (figure 6.4).

In response to stressors such as intense exercise or trauma, myogenic satellite cells become activated, proliferate, and repopulate the myofiber population by fusing together or fusing with existing myofibers. This process is mediated largely through inflammatory cytokines and growth factors that are produced and released locally in response to the injury. Figure 6.4 illustrates the expression profile and proposed role of a number of growth factors and cytokines during each phase (plus sign denotes a positive effect; minus sign denotes an inhibitory effect).

Growth factors are peptides produced within various tissues of the body that function to communicate and coordinate cellular activity. In particular, growth factors are highly effective in stimulating cellular proliferation, differentiation, or both. In response to skeletal muscle

damage, regulation of the myogenic satellite cell population is primarily mediated through inflammatory cytokines that infiltrate the damaged muscle and by growth factors that are produced and released locally in response to the injury. The timing and availability of these growth factors, as well as their receptor density on or within the myogenic satellite cells, are critical in mediating the overall regenerative process (table 6.2). Growth factors involved in myogenic satellite cell regulation include the following:

- **Hepatocyte growth factor.** Hepatocyte growth factor was first identified as a mitogen for hepatocytes within the liver (Thaler and Michalopoulos 1985). It has since been shown to be an integral regulator of multiple myogenic satellite cell functions during skeletal muscle regeneration and hypertrophy (see Hawke and Garry 2001 for review). In addition to its role in activating myogenic satellite cells, HGF has been associated with promoting myogenic satellite cell chemotaxis (migration) to the site of injury, selectively promoting myogenic satellite cell proliferation, and inhibiting differentiation through the transcriptional downregulation of muscle regulatory factors, MyoD and myogenin (Allen et al. 1995; Bischoff 1997; Gal-Levi et al. 1998; Tatsumi et al. 2002).

- **Fibroblast growth factor.** Fibroblast growth factor (FGF) has significant regulatory effects on myogenic satellite cell activity. At least 10 currently identified FGF isoforms are robustly expressed in various tissues throughout the body. While the FGF-6 isoform exhibits high expression levels that appear to be limited to the

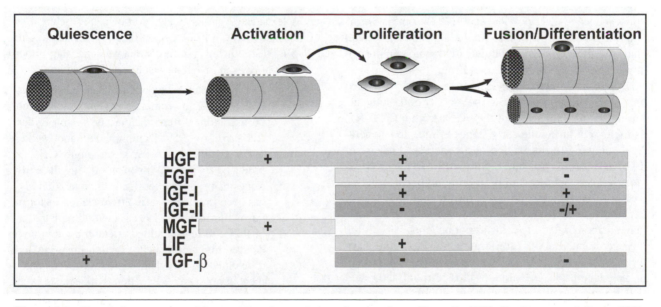

Figure 6.4 Growth factors and cytokines mediate the regulation of myogenic satellite cells. FGF: fibroblast growth factor; IGF: insulin-like growth factor; LIF: leukemia inhibitory factor; TGF-β: transforming growth factor beta.

Table 6.2   Growth Factors Affecting Satellite Cell Activity

Growth factor	Activation	Proliferation	Differentiation	References
In vitro				
HGF	↑	↑	↓	Allen et al. 1995; Miller et al. 2000
FGF		↑	↓	Allen and Boxhorn 1989
IGF-1		↑	↑	Allen and Boxhorn 1989
IGF-2		↑	↑	Haugk et al. 1995; Florini et al. 1991
MGF	↑			McKoy et al. 1999
LIF		↑		Austin and Burgess 1991
TGF-β		↓	↓	Allen and Boxhorn 1989
In vivo				
HGF	↑			Tatsumi et al. 1998
FGF		↑	↓	Armand et al. 2005
IGF-1		↑	↑	Chakravarthy, Davis, and Booth 2000
IGF-2		↓	↓	Kirk et al. 2003
MGF	↑			Hill and Goldspink 2003
LIF		↑		Kurek et al. 1997
TGF-β		↓		Kirk et al. 2000

Up arrow denotes increased expression; down arrow denotes decreased growth factor expression. HGF: Hepatocyte growth factor; FGF: fibroblast growth factor; IGF: insulin-like growth factor; MGF: mechano growth factor; TGF-β: transforming growth factor beta; LIF: leukemia inhibitory factor.

muscle lineage (deLapeyriere et al. 1993), some isoforms are found in multiple tissues (including skeletal muscle) and have the capacity to modulate myogenic satellite cell proliferation (Floss, Arnold, and Braun 1997). One such FGF family member, FGF-2, is released from damaged muscle fibers in proportion to the severity of the damage. Furthermore, expression of the FGF receptor (FGF-R1) is increased during the early stages of muscle regeneration, allowing for utilization of the increased release of FGF in response to injury. Similarly to HGF, FGF is positively associated with promoting myogenic satellite cell proliferation and inhibiting differentiation (Clegg et al. 1987; Sheehan and Allen 1999). While in vitro studies have demonstrated the clear importance of FGF during myogenic satellite cell proliferation, in vivo studies have been less successful in elucidating a critical role for FGF in skeletal muscle regeneration (Floss, Arnold, and Braun 1997; Fiore, Sebille, and Birnbaum 2000).

• **Insulin-like growth factors.** Insulin-like growth factors (IGF-1 and IGF-2) are hormones that are produced in various tissues throughout the body (e.g., liver, skeletal

muscle) and are significant regulators of insulin metabolism (Allen and Boxhorn 1989; LeRoith et al. 1992; Vierck et al. 2000). The following is some of what we understand about the role of IGFs in myogenic satellite cell regulation:

- In vitro studies have demonstrated the ability of IGF-1 and -2 to promote myogenic satellite cell proliferation, differentiation, and fusion (Allen and Boxhorn 1989; Doumit, Cook, and Merkel 1993; Haugk et al. 1995). Although IGF-2 is produced by skeletal myoblasts during the early stages of myogenic terminal differentiation and has been implicated as a differentiation factor in vitro (Florini et al. 1991), local administration of IGF-2 to regenerating skeletal muscle in vivo delayed myogenic satellite cell proliferation and differentiation but enhanced muscle fiber enlargement during late regeneration (Kirk et al. 2003).

- Some very interesting studies have demonstrated the key role played by IGF-1 in the regeneration of skeletal muscle following injury. These in vivo studies, utilizing superfusion or overexpression of IGF-1, showed skeletal muscle hypertrophy and prevention of age-related sarcopenia (Coleman et al. 1995; Chakravarthy, Davis, and Booth 2000). These studies are supported by findings that skeletal muscle overload and eccentric exercise increase local IGF-1 levels within the exercised muscle. The increased IGF-1 in these muscles was correlated with significant increases in DNA content and muscle hypertrophy (Yan, Biggs, and Booth 1993; Owino, Yang, and Goldspink 2001).

- The demonstrated ability of IGF-1 to mediate both myogenic satellite cell proliferation and differentiation may be attributable to its capacity to mediate two distinct signaling cascades. Until recently the proliferative effects of IGF-1 were attributed to the mitogen-activated protein kinase (MAPK) pathway while the phosphatidylinositol kinase (PI3K) signaling pathway was thought to be involved in differentiation (Machida and Booth 2004). However, recent evidence has demonstrated that the PI3K pathway is also upregulated during myogenic satellite cell proliferation, suggesting that at least some cross-talk must occur between these two signaling pathways to promote myogenic satellite cell proliferation and differentiation (Chakravarthy et al. 2000).

- **IL-6 family of cytokines.** Among the array of cytokines associated with muscle damage, the interleukin-6 family of cytokines appears to play an integral role in the repair process following myotrauma. Two members of this family, leukemia inhibitory factor (LIF) and interleukin-6 (IL-6), are critical in the repair process; and it appears that their actions may propagate through the same receptor or signaling pathway (Pennica et al. 1995; Hibi, Nakajima, and Hirano 1996):

  - Release of LIF is significantly increased in response to muscle damage and has been shown both in vitro and in vivo to increase myogenic satellite cell proliferation (Austin and Burgess 1991; Kurek et al. 1997). In vivo experiments have demonstrated that exogenous administration of LIF to skeletal muscle could promote myogenic satellite cell proliferation, hypertrophy, and an enhanced regenerative capacity following injury. These results are supported by studies using LIF-deficient mice, which display a significant decrease in regenerative capacity following injury (Kurek et al. 1997).

  - Although IL-6 is capable of stimulating in vitro myogenic satellite cell proliferation, IL-6 administered in vivo has not been shown to be directly involved in the regenerative process (Austin and Burgess 1991; Charge and Rudnicki 2004). It has been hypothesized that elevations in IL-6 expression during regeneration are involved in synchronizing the cell cycle of myogenic satellite cells, inducing apoptosis in macrophages, and promoting the removal of necrotic tissue after injury (Cantini and Carraro 1996).

- **Transforming growth factors.** The transforming growth factor-beta (TGF-β) family of growth factors transduce their signal through the SMAD family of proteins (Whitman 1998). In response to muscle damage, the TGF-β family inhibits myogenic satellite cell proliferation and differentiation primarily through silencing the transcriptional activation of the MyoD family members (Martin, Li, and Olson 1992). During regeneration from muscle injury, TGF-β2 ligand and receptor levels are reciprocally expressed, and this reciprocal expression pattern may result in the initial promotion of cellular proliferation followed by enhanced muscle differentiation (Sakuma et al. 2000). In addition to modulating myogenic satellite cell activity, the TGF-β family is involved in promoting normal skeletal muscle architecture by regulating local collagen synthesis in tendon-related connective tissue (Sakuma et al. 2000).

Special Populations, Therapies, and Myogenic Satellite Cells

As we have described in the preceding sections, myogenic satellite cells and the factors responsible for their regulation are critical for normal skeletal muscle growth and repair following damage. In certain instances, dysregulation of the myogenic satellite cells, or the factors mediating their regulation, can lead to impaired muscle regeneration or even chronic myopathy. In this section we outline some of the disease states in which myogenic satellite cell regulation is impaired and describe some current and possible future therapies that may help to alleviate the myopathic state.

Muscular Dystrophies

The muscular dystrophies are a group of genetic degenerative diseases that primarily affect voluntary muscles. The most common and the most devastating of the muscular dystrophies is Duchenne muscular dystrophy (DMD), which affects approximately 1 in 3,500 boys. This X-linked genetic disorder results in weak, fragile muscle fibers due to a mutation in the structural gene dystrophin. In response to repeated muscle contractions, widespread muscle degeneration occurs. The myogenic satellite cells respond and repair the damaged muscles; unfortunately, as this is a genetic disorder, the newly regenerated muscle fibers also lack functional dystrophin. This continual degeneration–regeneration cycling ultimately exhausts the myogenic satellite cell pool (Cossu and Mavilio 2000; Heslop, Morgan, and Partridge 2000).

The course of DMD is fairly predictable, with the first signs of muscle weakness occurring by 4 to 5 years of age (Bell and Conen 1968; Monaco et al. 1986; Burghes et al. 1987). At this point a patient has already undergone more skeletal muscle regeneration than six normal patients over the age of 60 combined (Decary et al. 2000). Between the ages of 7 and 12, nearly all children with DMD lose the ability to walk, and by their mid-20s they experience failure of the heart and respiratory muscles, which ultimately leads to death.

There is currently no cure for DMD, although a number of therapies have been tested to alleviate the symptoms of the disease. In particular, significant improvements in muscle strength, timed muscle function (such as time to climb stairs), and pulmonary function have been observed with the use of corticosteroids (e.g., prednisone or deflazacort). However, these effects are not observed in all patients, and side effects (e.g., weight gain, Cushingoid appearance, and gastrointestinal problems) also reduce their universality. There is also evidence that glucocorticoid administration reduces expression of IGF-1 and IGF-2 and increases IGF binding protein (IGFBP) expression in skeletal muscles (Gayan-Ramirez et al. 1999). This scenario would reduce the bioavailability of IGF-1, which is important in muscle regeneration, and thus may help to explain the inconsistent results observed with glucocorticoid administration in DMD patients.

Proliferative fatigue (also termed senescence) of the myogenic satellite cell pool has been implicated as the primary factor in the eventual inability of skeletal muscle to maintain regenerative capacity in DMD patients. One therapy that has shown promise in myopathic rodent skeletal muscle is the administration of growth factors such as IGF-1 (Singleton and Feldman 2001). For example, daily injections of IGF-1 into dystrophic mice for eight weeks resulted in improved muscle structure and function (Gregorevic, Plant, and Lynch 2004), suggesting that this type of therapy may be beneficial for human myopathies. Although IGF-1 administration is under investigation as a therapy for other disease states, there is currently little information regarding its effects within human dystrophic muscle. In association with this, a study of the effects of three months of growth hormone treatment on dystrophic human skeletal muscles showed no changes in skeletal muscle function. One reason for this finding may be the altered "environment" of DMD skeletal muscle, which displays elevated levels of IGFBPs and increased expression of TGF-β growth factors, ultimately resulting in a greater degree of fibrosis within the skeletal muscle (Ignotz and Massague 1986; Grande, Melder, and Zinsmeister 1997; Melone et al. 2000; Bernasconi et al. 1999; Hartel et al. 2001).

Gene transfer therapies are also being assessed as a means of improving the prognosis of DMD patients. Gregorevic and colleagues (2004) injected recombinant adeno-associated viruses containing a functional form of the dystrophin gene into the skeletal muscles of dystrophin-deficient *(mdx)* mice. Widespread muscle-specific expression of functional dystrophin in the *mdx* skeletal muscles was observed two months after injection, and these muscles were less susceptible to eccentric contraction–induced injury than before. Although more research needs to be performed, gene transfer therapies have tremendous potential and may provide a reliable, modifiable method of improving many myopathic conditions.

Aging

Aged skeletal muscles, particularly type II muscle fibers, are more susceptible than younger ones to contraction-induced muscle damage and display an impaired regenerative capacity (Schultz and Lipton 1982; Gibson and Schultz 1983; Dodson and Allen 1987; Faulkner, Brooks, and Zerba 1995; Singh et al. 1999). A number of factors have been proposed as contributing to the age-related loss of skeletal muscle and its contractility, a disease state termed sarcopenia (table 6.3).

Myogenic satellite cell nuclei comprise approximately 30% of the total muscle nuclei in the neonate, and this number decreases to approximately 5% by adulthood. This decrease appears to continue throughout the life span of the organism, with those who are elderly showing an even greater decrease from adulthood, although this result is not observed in all studies (see table 6.1). It has been proposed that the decrease in the percentage of myogenic satellite cells with increasing age results from an increased number of myonuclei and a decrease in the number and proliferative capacity of myogenic satellite cells (Schultz and Lipton 1982; Kadi et al. 2004; Sajko et al. 2004). However, recent findings suggest that the phenomenon is much more complex than merely myogenic satellite cell senescence. For example, the proliferation and fusion of myogenic satellite cells do not appear to be significantly altered with aging (Grounds 1998). Nonetheless, changes in the aged cellular and extracellular milieu (table 6.3) appear to reduce myogenic satellite cell activation (Goldspink and Harridge 2004) and possibly subsequent steps, such that new myofibers are thinner and more fragile—helping to explain the increased susceptibility to contraction-induced muscle damage (Renault et al. 2000).

In support of the critical role played by the host environment, Carlson and Faulkner (1989) found that the mass and maximum force of old muscles grafted into young hosts were not significantly different from those of young muscles grafted into the same young hosts. On the other hand, young muscles grafted into old hosts regenerated similarly to old muscle grafted into the same old hosts. On the basis of these findings, the authors proposed that the environment of a muscle is more important than the age of the muscle in determining regenerative capacity; and

Table 6.3   Age-Associated Changes in Skeletal Muscle That May Affect Myogenic Satellite Cell Regulation

Change with increasing age	References
Decreased MGF expression	Goldspink and Harridge 2004
Increased myostatin expression	Yarasheski et al. 2002
Decreased circulating growth hormone, testosterone, and IGF-1 levels	Proctor et al. 1998
Thickening of the basal lamina	Snow 1977
Increased fibrosis within skeletal muscle	Marshall et al. 1989
Reduced capillary density	Coggan et al. 1992
Reduced genomic maintenance	Hasty 2001
Decreased antioxidative capacity	Fulle et al. 2005
Decreased immune response following injury	Danon et al. 1989

IGF: insulin-like growth factor; MGF: mechano growth factor

this hypothesis was recently supported using old–young mouse parabiotic pairings (Conboy et al. 2005).

A number of therapies have been tested as means of preventing or reducing sarcopenia. In particular, resistance training has received significant attention. Resistance training in elderly individuals has many potential benefits including increased strength; increased bone mineral density (Vincent and Braith 2002); increased satellite cell number, particularly in older women (Roth et al. 2001); decreased myostatin levels (Roth et al. 2003); and the potential to elevate bioavailable IGF-1 (Singh et al. 1999; Parkhouse et al. 2000). It should be noted, however, that although elderly subjects do respond well to resistance training, the adaptations in older people are not as pronounced as in younger subjects (Kraemer et al. 1999).

A therapy that is currently receiving significant attention is hormone replacement therapy. As mentioned previously, aging is associated with changes in body composition, function, and metabolism, as well as a reduction in the activity of the growth hormone–IGF-1 axis. These changes are similar to those in younger adults with pathological growth hormone deficiency. Since pathologically deficient patients display improvements in overall fitness and quality of life with supplementation of growth hormone, IGF-1, or both, similar clinical trials focusing on elderly subjects have been proposed. Results from rodent studies support this endeavor, as increases in local IGF-1 expression attenuated the loss of muscle mass and strength and increased myogenic satellite cell proliferation in older animals (Barton-Davis et al. 1998; Chakravarthy, Davis, and Booth 2000). To date, there is no definite evidence that "frail" elderly human subjects benefit from restoration of growth hormone and IGF-1 levels to a young adult range (Taaffe et al. 1996, 2001; Lange et al. 2002). Furthermore, most studies addressing hormone supplementation have used elderly individuals in good health, partly because of

the potential of these therapies to enhance the growth of hidden malignancies.

Although resistance training appears to be beneficial in reducing the effects of sarcopenia, some elderly subjects may be too frail or impaired in their current state for resistance training to be a viable therapy. In this case, hormone replacement therapy may provide a short-term benefit to elderly patients who are preoperative (e.g., hip surgery) and for whom increased muscle mass and strength would expedite recovery.

Atrophic Stimuli

Atrophy of skeletal muscle results in a reduction in muscle mass and a decrease in myogenic satellite cell number (Darr and Schultz 1989; Mozdiak, Pulvermacher, and Schultz 2000). Atrophy can be induced by numerous factors including denervation, immobilization, and malnutrition (Grounds 1999). Upon resumption of mobility from immobilization-induced atrophy, adult myogenic satellite cells are capable of activation and proliferation to repopulate atrophied skeletal muscle (Mozdiak, Pulvermacher, and Schultz 2000; Wanek and Snow 2000). In contrast, atrophic stimuli in the adolescent skeletal muscle result in an irreversible remodeling process whereby myogenic satellite cell content is decreased and the proliferative capacity of these cells is impaired (Darr and Schultz 1989; Mozdiak, Pulvermacher, and Schultz 2000), such that even with the resumption of weight bearing, the developmental program for muscle fibers to accrue nuclei is altered.

Unlike other forms of atrophy, denervation is a pathological rather than a physiological stress. Denervation produces a form of disuse atrophy characterized by muscle fiber degeneration and is accompanied by distinctive changes in the myonuclei and quiescent myogenic satellite cells (Rodrigues and Schmalbruch 1995; Lu,

Huang, and Carlson 1997; Kuschel, Yablonka-Reuveni, and Bornenam 1999). During the initial period following denervation, the percentage of myogenic satellite cells increases (from 3% to 9%); however, after prolonged denervation (e.g., 18 months), a significant decrease in myogenic satellite cell number (3% to 1%) is observed (McGeachie 1989; Viguie et al. 1997). The decreased myogenic satellite cell number may be due to myogenic satellite cell apoptosis (Viguie et al. 1997) or impaired proliferation and self-renewal of the cells (Hawke and Garry 2001). Long-term denervation, for periods of 6 to 18 months, results in an inability of skeletal muscle to reestablish previous functional capacity even if neuronal regeneration occurs (Sunderland 1978). The mechanisms for this phenomenon are unclear, but considerable data support the conclusion that the intact neuromuscular junction and the denervation model mediate positive and repressive influences, respectively, on the myogenic satellite cell pool (Viguie et al. 1997).

In general, skeletal muscle atrophy has also been linked to elevations in myostatin expression (Dasarathy et al. 2004). In these models, elevated myostatin represses myogenic satellite cell activation and ultimately muscle repair, although other growth factors, such as IGF-1 and LIF, may be upregulated to counteract the increased myostatin levels. Alternatively, increases in these pro-proliferation growth factors may promote the fiber type switching that is observed during atrophy (Reardon et al. 2001).

Myogenic Satellite Cell Transplantation

The transplantation of myogenic satellite cells is currently gaining attention as a viable therapy for various myopathic disease states (Menasche 2003; Skuk and Tremblay 2003). The concept of cell transplantation for the treatment of disease states is not novel, as these types of therapies have been utilized for decades (e.g., bone marrow transplants for the treatment of cancer; Appelbaum 1996). Whether myogenic satellite cells are transplanted in hopes of increasing regenerative capacity, improving muscle architecture, or providing a vehicle for normal genes (e.g., in the genetic myopathies), this therapeutic application, although highly promising, is still in its infancy.

The effective use of myogenic satellite cell transplantation for the treatment of DMD has been well defined in numerous animal studies. However, early human clinical trials, although promising, have been less than optimal (Huard et al. 1992; Gussoni et al. 1992; Karpati et al. 1993). Preventing the massive and rapid death of donor myogenic satellite cells by host immune cells through immunosuppression or local irradiation, promoting migration of the injected donor cells, and improving long-term survival of the transplanted cells are just a few of the issues currently facing researchers (Skuk and Tremblay 2003; Hodgetts and Grounds 2003). Using a protocol based largely on their work with nonhuman primates, the Tremblay group reported some promising results in three children with DMD using multiple intramuscular injections of myogenic satellite cells and an immuno-suppression regime involving tacrolimus (Skuk et al. 2004). This study, although preliminary, is likely to open the door to more widespread use of this cell population for the treatment of DMD.

Even with these promising results and the tremendous potential for myogenic satellite cell transplantation in the treatment of various myopathies, caution is warranted. The injection of any cell with stem cell characteristics inherently has the potential for tumorigenesis; and as discussed previously, the environment into which these cells are injected will play a significant role in determining their capacity for proliferation and appropriate differentiation into viable myofibers.

Summary

The adaptability of adult skeletal muscle in response to a range of stressors is extraordinary. The capacity for muscle hypertrophy in response to increased work demands or for repair following myotrauma is largely the result of the resident myogenic satellite cells. Since the initial description of these cells over 40 years ago, a number of anatomical and physiological studies have established the importance of this cellular pool in the growth, remodeling, and regeneration of adult skeletal muscle. Despite this work, the molecular and cellular regulation of the myogenic satellite cell remains ill defined. Future challenges for this field will include defining the development, maintenance, self-renewal, and potentiality of the myogenic satellite cell population. It is only through a more thorough understanding of this unique cell population that we will enhance their safe and effective use in cell transplant therapies for the treatment of devastating myopathies.

ACKNOWLEDGMENTS

The authors would like to recognize the research efforts of all those who have contributed to this field of study. In some cases review articles have been used to direct the reader to original articles or more comprehensive summaries of the literature. Special thanks to Drs. S. Kanatous and A. Meeson for helpful discussions and critical reviews of this chapter. TJH is supported by grants from the Canadian Foundation for Innovation, National Science and Engineering Research Council, Ontario Innovations Trust, Canadian Institutes of Health Research and the Hospital for Sick Kids Foundation.

CHAPTER 7

Emerging Molecular Trends in Muscle Damage Research

Douglas J. Mahoney, PhD, and Mark A. Tarnopolsky, MD, PhD

The age of technology is upon muscle physiologists! The recent "marriage" between muscle physiology and molecular and cellular biology is beginning to flourish, and with this our approach to addressing muscle physiological questions is being transformed by an enormous expansion in the capability and availability of molecular and cellular biology techniques (figure 7.1). Among these are highly sensitive techniques that enable exquisite analysis of protein abundance, modification, and activation; gene expression and activation; protein–protein interactions; and intra- and extracellular signaling events. Genetic and pharmacological approaches are now routinely used to over- and underexpress proteins in muscle cell cultures as well as mouse skeletal muscle. Fine-tuning of these techniques has allowed researchers to temporally control the expression level of these proteins at their discretion. Advances in imaging technology, coupled with these molecular genetic tools, empower researchers to visualize individual proteins in live cells and live animals. Thus, a protein of interest can be "genetically tagged" in a mouse and visualized in real time after a perturbation, such as a damaging exercise bout, without sacrificing the animal. Together, these techniques offer an unprecedented ability to definitely characterize the cellular physiology of skeletal muscle, as well as the cellular pathology of muscle disease. Furthermore, these techniques enable researchers to decipher the regulatory steps and checkpoints integral to muscle physiology and pathology. In short, the field of muscle physiology is "going cellular," and today's

muscle physiologists have unprecedented capacity to investigate some of the outstanding questions in the muscle damage field, such as the following:

- Which muscle proteins are affected by muscle damage, and how are they affected?

- Which proteins are involved in responding to muscle damage and mediating repair *(effector proteins)*, and at which level of regulation is this effect mediated?

- Which proteins "sense" muscle damage within muscle cells *(regulatory proteins)*, and how do they translate damage sensing into a repair response?

- Which proteins are involved in imparting adaptation in response to repeated bouts of muscle damage *(adaptive proteins)*, and how is adaptation imparted within muscle cells?

- Is it possible to genetically or pharmacologically manipulate the effector, regulatory, or adaptive proteins to attenuate, prevent, or treat muscle damage in disease states such as muscular dystrophy or inflammatory myopathies?

Although we are still in the early years of molecular and cellular muscle physiology research, we are beginning to harvest the fruits of this field as we come to understand muscle damage at its very core: the cellular constituents that are damaged and the proteins that engage and execute programs of repair and adaptation. Yet we have just begun to tap the surface of this

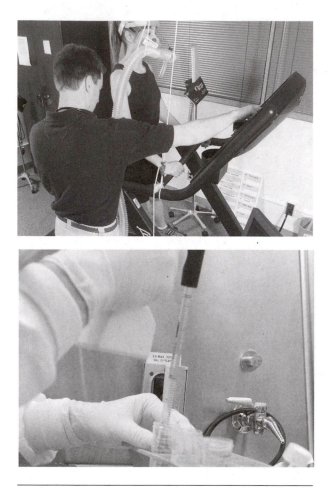

Figure 7.1 The age of technology in muscle physiology. The upper panel depicts several of the classical approaches to studying muscle physiology, such as measuring endurance capacity, voluntary strength and power, and blood markers of muscle damage. The bottom panel depicts some of the newer molecular and cellular approaches to studying muscle physiology, including the use of muscle cell culture models, immunofluorescence analysis of muscle biopsy samples, western blotting for muscle proteins, in vivo imaging of skeletal muscle proteins, genetic modification of muscle proteins in animal models, reverse transcriptase–polymerase chain reaction (RT-PCR) analysis of messenger RNA (mRNA) expression in muscle, and DNA microarray analysis of global gene expression in muscle.

Reprinted, by permission, from J.H. Wilmore and D.L. Costill, 2004, *Physiology of sport and exercise*, 3rd ed. (Champaign, IL: Human Kinetics), 44.

wealth of technology, and the very near future holds many exciting new developments in muscle damage research.

The "Omic" Revolution

Of all the new molecular and cellular techniques available to muscle physiologists, none is more exciting and more promising than "omic" technology (Zhu & Snyder, 2002; Ge et al., 2003; Bilello, 2005). The suffix "omic" is given to a technique or experimental approach that assays an entire level of biological information, or a large part of that level of information. Omic technology was developed primarily for two reasons. Firstly, omic approaches allow scientists to make hundreds, thousands, and even tens of thousands of measurements in a single experiment. The mass of data generated in a single omic-based experiment would take months to years to generate using more conventional techniques. Such mass data generation is beneficial for many reasons, such as discovering a novel member of an existing pathway within a particular biological context or discovering an entirely novel pathway within that biological context. Secondly, omic technology greatly facilitates a "systems biology" approach to the biological sciences (Ge et al., 2003). One of the limitations of conventional molecular and cellular biology techniques is that they are reductionist. Research experiments typically explore one or several aspects of the response to a given biological perturbation. Omic technology empowers scientists to examine *every* aspect of that response, and to integrate all aspects into a network that fully describes the response to the perturbation. In theory, this network could be modeled so that scientists could predict the network-wide effect of a perturbation to any part of the network.

Omic research was inaugurated with the field of *genomics* (McKusick, 1997), which is the study of the entire genetic complement of an organism. The recent completion of the Human Genome Project (Venter et al., 2001), as well as similar genome projects in rats (Gibbs et al., 2004), mice (Waterston et al., 2002), and other lower organisms (Blattner et al., 1997; Goffeau, 1997), laid the groundwork for the field of *transcriptomics* (Hedge et al., 2003), which is the study of the entire mRNA expression profile *(transcriptome)* within a given biological context. DNA microarrays that can measure mRNA expression levels of most genes within the human, mouse, rat, and lower organism genomes are now widely available and routinely used in the biological sciences (Howbrook et al., 2003). These arrays are highly sensitive and reproducible and have become an invaluable resource to biological scientists.

The need for high-throughput analysis of complementary levels of biological information has led to the recent development of other highly parallel molecular techniques. The first of these was *proteomics* (Hedge et al., 2003), which is the study of the entire complement of proteins *(proteome)* within a given biological context. Although not as well developed as transcriptomics, proteomics will undoubtedly become a more powerful and important analytical approach when the technology is refined, as proteins are the ultimate effectors of cellular physiology. Additional omics include *metabolomics* (Goodacre et al., 2004), the study of the entire metabolite profile of a cell, tissue, or system *(metabolome); glycomics* (Shriver et al., 2004), the study of an entire complement of glycan molecules synthesized by a cell *(glycome); localizomics* (Ge et al., 2003), the study of the entire complement of protein localization within a cell, tissue, or system *(localome);* and *interactomics* (Ge et al., 2003), the study of the entire complement of macromolecular interactions in a cell, tissue, or system *(interactome).* More omic techniques are bound to emerge in the coming years.

Of all the "postgenomic omics," the field of transcriptomics is by far the best developed and is the only omic that has made any major inroads into the muscle damage field. Thus this chapter focuses primarily on transcriptomic technology in muscle damage research. A secondary focus is on the future of omics in muscle damage research. In particular, the chapter

- briefly reviews the process of gene expression and addresses its importance in skeletal muscle during recovery from and adaptation to damage,

- introduces the prevailing transcriptomic technologies,

- outlines all current applications of transcriptomics in muscle damage research and what has been learned from these studies, and

- addresses the future applications of transcriptomics and other omics in muscle damage research.

Our goals are to familiarize the reader with the field of transcriptomics and outline how it has been applied to study muscle damage. Furthermore, we hope to convince the reader of the enormous potential stored in transcriptomic and other omic technology, and of ways in which this potential can be harnessed to study muscle damage, while also cautioning about the inherent limitations of these techniques and noting the considerations necessary with their use. Ultimately, we hope that this chapter may be a stepping-stone to future research using omic technology to study muscle damage.

Gene Expression

In the biological community, the term "gene expression" usually refers to the steady state level of the mRNA encoded within a gene and is used interchangeably with the term "mRNA expression" (as is the case throughout this chapter). In reality, a gene is not functionally expressed until the protein that it encodes is translated, as proteins are the functional currency of genes. This section begins with a brief overview of the functional expression of genes within cells.

Gene Expression in All Cells

As we have known for a long time, proteins are encoded for within genes that reside on DNA within nuclei, and mRNA acts as an intermediary that translates the message encoded within the gene into a working protein. In fact, this flow of information is so ingrained that it is often referred to as the "central dogma of cell biology"—genes encode for mRNA; those mRNA encode for proteins; and those proteins are responsible for executing cell function. Although this concept is straightforward, altering protein abundance via gene expression is a very complex process (figure 7.2). Briefly, in response to the appropriate stimuli, cytosolic signal transduction pathways are "turned on" to activate downstream regulatory proteins *(activators, coactivators, repressors, corepressors,* etc.). These regulatory proteins migrate into the nucleus and bind to upstream elements in the promoter (or enhancer) region of genes, either increasing (activators) or decreasing (repressors) the promoter activity of a given gene. For *gene activation,* regulatory proteins act in concert with other members of the transcriptional apparatus (*RNA polymerase, general transcription factors,* etc.) to induce gene transcription, producing a primary mRNA transcript. The primary transcript is processed in the nucleus (*splicing, 5' capping, 3' polyadenylation,* etc.) by the splicesome and is actively transported through a nucleoporin complex into the cytosol, where it is translated into a protein on ribosomes. For *gene repression,* regulatory proteins interact with various members of the transcriptional apparatus to inhibit gene transcription. As mRNA and protein are constantly being

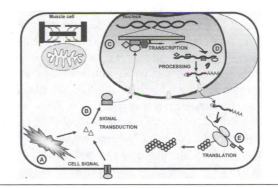

Figure 7.2   Gene and protein expression in cells. This cartoon illustrates a generalized scheme for gene induction and protein synthesis in cells (e.g., skeletal muscle). (*a*) A signal for activating gene expression is generated (e.g., stretch or damage) or received (e.g., cytokines). (*b*) This signal is "sensed" by various cellular kinases and phosphatases and transmitted to regulators of nuclear transcription. (*c*) Activated regulatory proteins are transported into the nucleus, bind to the gene promoter region, recruit other components of the transcriptional apparatus, and induce gene transcription. (*d*) The primary mRNA transcript is processed and exported from the nucleus. (*e*) Messenger RNA is translated into protein on cytosolic ribosomes in the presence of transfer RNA.

turned over, transcriptional inhibition eventually leads to decreased mRNA expression and protein abundance.

In general, the functional *capacity* of a protein is determined by its absolute abundance. Protein abundance can be controlled via regulation of any step along the flow of information from gene to protein, such as mRNA transcription (*transcriptional*); mRNA processing, transport, and stability (*posttranscriptional*); protein translation (*translational*); or protein turnover (*posttranslational*). It is important to recognize that mRNA expression-mediated control is only one regulatory mechanism for controlling protein abundance. It is also important to recognize that the *activity* or *activation state* of a given protein can often be altered by posttranslational modification, such as phosphorylation, glycosylation, or prenylation. Thus, simply expressing a protein does not necessarily mean that it will work properly or maximally, although it has the capacity to do so should the appropriate modifications be made. Excellent reviews of gene transcription and protein translation have been recently published (Lee & Young, 2000; Preiss & Hentze, 2003).

Gene Expression in Skeletal Muscle

Expression of genes in skeletal muscle cells is generally similar to that in other cells. However, muscle cells are unique in that they are multinucleate, containing hundreds of myonuclei along their length. Although all these myonuclei contain identical genomes, the genes and proteins that are expressed may differ among myonuclei as a function of their location within the muscle cell and the environmental cues that they receive. Given the specialized demands at regions such as the neuromuscular and myotendinous junction, "specialized populations" of myonuclei exist that respond to specific local signals within muscle cells. For instance, myonuclei in close proximity to the neuromuscular junction are the only myonuclei within muscle that *express* the acetylcholine receptor (Mejat et al., 2003), although all myonuclei have the gene that *encodes* for it. Such "spatial specialization" adds a layer of complexity to skeletal muscle gene expression that does not exist in mononucleate cells. Unfortunately, myonucleus-specific gene expression is difficult to measure, and most studies of gene and protein expression in muscle measure the combined effect from many myonuclei within a given muscle sample.

Gene Expression in Response to Muscle Damage

A damaging insult (e.g., eccentric exercise) to skeletal muscle is a potent transcriptional stimulus that alters the expression of a large number of genes (Chen et al., 2002, 2003; Barash et al., 2004; Mahoney and Tarnopolsky, 2005). These "damage-responsive" genes are largely involved in two major processes: recovery from and adaptation to the damage (figure 7.3).

- **Recovery** is an imminent process that occurs in the hours and days following the damaging insult, or is an ongoing and ultimately futile process throughout the course of a muscle disease (e.g., muscular dystrophy). Elevated gene expression can induce an *acute* increase in protein abundance that contributes to repairing damaged muscle components and reestablishing muscle cell homeostasis.

- **Adaptation,** thought of primarily in the exercise context, can be broken down into two components: repeated-bout effect and training-induced adaptations.

  - The **repeated-bout effect,** which is addressed in chapter 5, can be observed after a single bout of exercise.

  - **Training-induced adaptations** refer to the hypertrophy and strength increases associated with a resistance exercise program. These adaptations take weeks to months to materialize, and it is thought that repetitive "pulses" of elevated mRNA expression in response to each individual bout of damaging exercise induce *chronic* increases in protein abundance that lead to physiological adaptations.

Although the molecular and cellular details of these two processes are unclear, there is likely a large degree of molecular crossover between them, with a number of proteins being integral to both recovery and adaptation. In contrast, there are also proteins that are likely involved only in either recovery or adaptation. Characterizing these details is currently a subject of much research in the muscle damage field, as it is generally thought that an understanding of the regulatory, effector, and adapter proteins involved in repair and adaptation will provide promising targets for treating muscle damage in disease.

Measuring Muscle Gene Expression in Response to Damage

As proteins are the ultimate effectors of muscle cell function and are regulated at multiple levels, what is the value of measuring gene expression in response to damage? There are two important reasons why muscle physiologists are interested in measuring gene expression in response to muscle damage.

- Firstly, gene expression data often give clues about protein abundance. For example, if a damaging bout of exercise alters the expression level of a gene, there is a good chance that the encoded protein will also be altered at some later time. Although gene expression and protein abundance do not *necessarily* correlate, due to the mul-

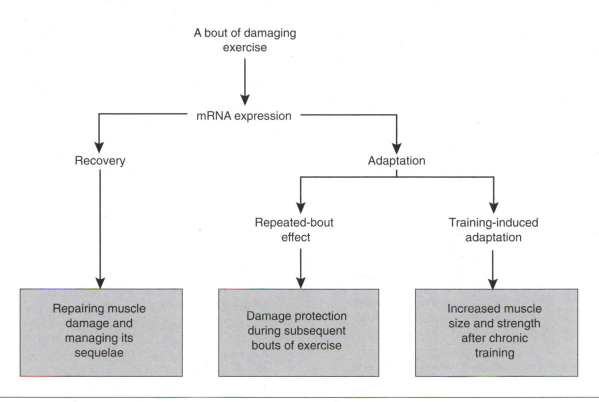

Figure 7.3   The transcriptional response to muscle damage. A single bout of damaging exercise alters mRNA expression in skeletal muscle, primarily during the recovery period. Expression changes may participate in the postexercise recovery process by encoding for proteins that are involved in repairing damaged skeletal muscle and reestablishing muscle homeostasis. Additionally, expression changes may encode for proteins that participate in protecting muscle from subsequent damaging stimuli (repeated-bout effect), as well as for proteins involved in the strength and size gains that occur after weeks to months of resistance training (training-induced adaptation).

tiple levels of regulation that affect protein abundance, the expression level of a given gene in a given cell or tissue is *often* indicative of the abundance of its protein product. Given that it is technically much easier to measure gene expression than protein abundance, in particular when one is measuring the entire transcriptome as compared to the entire proteome, it is often much more practical and cost-effective to examine gene expression. Along these lines, the transcriptome can be used to screen for candidate proteins to investigate.

• Secondly, altered gene expression may provide a mechanism to explain altered protein abundance, particularly in response to an intervention or treatment such as a damaging bout of exercise. For example, if it is known that a bout of eccentric contractions leads to altered protein X abundance in muscle, it may be of interest to ascertain whether or not protein X abundance is regulated at the transcriptional level.

Transcriptomic Technology

Two broad categories of technologies enable rapid and efficient whole-genome transcriptional profiling:

• **Open techniques,** such as differential display (DD), serial analysis of gene expression (SAGE), and massively parallel signature sequencing (MPSS), do not require prior knowledge of the sequence of the genome under study but are expensive and labor-intensive.

• **Closed techniques,** such as complementary DNA (cDNA) and oligonucleotide arrays, are less expensive and labor-intensive but require previous sequence information of the genome under study. Fortunately, with the completion of the major genome projects, sequence information is readily available; thus these closed techniques have unequivocally become the assay of choice for expression profiling.

DNA Microarray Technology

The emergence of DNA microarray technology revolutionized the field of transcriptomics and greatly enhanced research exploring transcriptional regulation and control of cellular events (Heller, 2002; Fehrenbach et al., 2003). DNA microarrays allow parallel analysis of tens of thousands of genes, which is the equivalent of performing tens of thousands of northern blots or reverse transcriptase–polymerase chain reaction (RT-PCR) reactions at the same time. Currently, there are two commonly used variants of the DNA microarray technique: cDNA arrays and oligonucleotide arrays (figure 7.4).

• For **cDNA arrays,** total RNA is extracted from an experimental and a control sample (e.g., skeletal muscle biopsy sample before and after a damaging exercise bout), and the mRNA from both samples are reverse transcribed

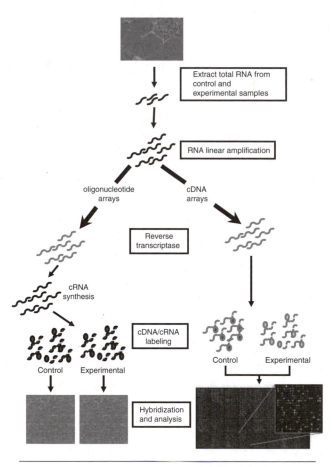

Figure 7.4 Transcriptional profiling using DNA microarrays. Total RNA is extracted from control and experimental muscle samples, amplified, and reverse transcribed to complementary DNA (cDNA). For oligonucleotide arrays, cDNA is converted to cRNA, which is subsequently labeled with biotin. Labeled cRNA is hybridized to short oligonucleotides spotted on a glass array. Control and experimental samples are hybridized to different arrays, and the intensity at each spot is compared between the two arrays. For cDNA arrays, cDNA from the control sample is labeled with a fluorescent dye, while cDNA from the experimental sample is labeled with a different fluorescent dye. Labeled cDNA from both samples is hybridized to a single glass array spotted with long cDNA fragments, and analysis of fluorescent intensity reveals relative mRNA expression in the experimental versus the control sample.

to their cDNA strands. The cDNA from the experimental and control samples *(target)* are subsequently labeled with different fluorescent dyes (e.g., cyanine 3 labeling of control cDNA and cyanine 5 labeling of experimental cDNA, or vice versa). Labeled cDNA from both samples are hybridized to a single glass array that has been spotted with thousands of long (~100-4,000 nucleotides) PCR-amplified cDNA fragments *(probes)*. Each of these probes has been robotically spotted in a unique location on the glass array, and binds a single cDNA target with high specificity. The cDNA targets from both the experimental and control samples will "compete" for binding to each specific probe on the array, and the fluorescent intensity produced at each spot will be representative of the relative difference in mRNA expression between the experimental and control samples.

• For **oligonucleotide arrays,** the process is relatively similar, with a few key differences. Total RNA is extracted from an experimental and a control sample, and the mRNA from both samples are converted to cDNA and subsequently in vitro transcribed to complementary RNA (cRNA). The cRNA targets from both samples are labeled separately with the same tag (e.g., biotin), and then each is hybridized to separate glass arrays that have been spotted with thousands of short (~20-25 nucleotides) oligonucleotide probes. For these arrays, probe synthesis and spotting occur in situ on the array by a process known as *photolithography,* which "builds" the probe nucleotide by nucleotide at very specific locations (spots) on the array. Each probe is highly specific for a single cRNA target; and following hybridization of the target to the probe, the array is scanned and the intensity at each spot is indicative of the amount of target cRNA in the sample. The spot-to-spot difference in intensity from the experimental array versus the control array is representative of the relative difference in mRNA expression between the experimental and control samples.

Both cDNA and oligonucleotide arrays have inherent advantages. Complementary DNA arrays are generally custom-made at "in-house" facilities using a commercially available spotter. As such, cDNA arrays can be specifically fabricated to suit a researcher's needs. For instance, muscle physiologists can generate "muscle-specific arrays" that contain genes expressed only in skeletal muscle. In contrast, oligonucleotide arrays are typically manufactured by companies because of the greater cost of the machinery for synthesis (e.g., Affymetrix GeneChips). Although many different oligonucleotide arrays are available, including a muscle-specific array, the content of the array is not under the complete control of the scientist. However, oligonucleotide arrays are advantageous in that the entire system is highly automated and the arrays have many internal controls, which tends to generate highly reproducible data that can easily be exchanged from lab to lab.

DNA Microarray Data Analysis

Ironically, the greatest asset of gene array technology—the ability to probe the expression level of tens of thousands of mRNAs in a single experiment—has also proved to be one of the greatest challenges in the DNA array field. Statistical analyses of such a large number of data points have proved difficult, and there is no consensus on the statistical approach that should be taken to manage the data. Clearly, multiple *t*-tests would hyperinflate one's *p*-value, and tens to hundreds of statistically significant false-positive results would be generated. Unfortunately, the prevailing *t*-test corrections (e.g., Bonferoni test) tend to generate extremely strict "significance parameters," and many false-negative results ensue. In recent years, microarray-specific statistical tests have been created, such as significance analysis of microarrays (Tusher et al., 2001) and CyberT (Baldi & Long, 2001), which are much preferred over multiple *t*-tests or the use of fold change. However, these tests are still very conservative, and a powerful study design with a large number of biological replicates (i.e., subjects or animals) is needed to extract the majority of the expression changes. *This point cannot be overstated: Given the large amount of data generated and the current limitations in statistical testing, a powerful study design, such as a repeated-measures, nonpooled design with an adequate number of biological replicates, is required in order to realize the full potential of microarray technology.*

Unfortunately, two major drawbacks to DNA microarrays have been that they are expensive and that they require a large quantity of RNA. Thus researchers have often been forced to pool mRNA from multiple samples and interrogate gene expression between the pools rather than individually. This is problematic in that it erodes the statistical power of the experimental approach. Additionally, pooled samples do not allow for analysis of inter-individual differences, which is often of much interest to muscle physiologists. Fortunately, the cost of microarray analysis is becoming more reasonable and techniques are now widely available to amplify RNA (Abmayr et al., 2004), so there should be little need to pool samples in future microarray studies.

Biological Uses of DNA Microarrays

The utility of the DNA microarray technique for examining gene expression is, to say the least, enormous. With regard to muscle damage research, DNA microarrays give researchers unprecedented capacity for

- discovering differentially expressed genes that respond to a damaging insult to muscle, or that are involved in the pathogenesis of muscle disease;
- discovering "programs of gene expression" that are activated in response to muscle damage or disease;
- testing hypotheses regarding the transcriptional response to muscle damage or the efficacy of treatment strategies with muscle disease; and
- defining the entire transcriptional response to a damaging muscle insult or to a muscle disease.

The last point is particularly important, as published expression data are freely accessible to all researchers. Thus, DNA microarray experiments may answer questions outside of the lab in which they were performed.

Transcriptomics and Muscle Damage Research

How has transcriptomics been applied to study muscle damage, and what have we learned from these studies? To date, relatively few studies have applied transcriptomics to study muscle damage, but this research has provided interesting insight into various aspects of the repair and adaptive response. In this section we outline how transcriptomics has been used in muscle damage research and offer some of the important "take-home" messages from these studies.

Over the past three years, several groups have successfully used DNA microarrays to examine global mRNA expression after a bout of damaging eccentric contractions in human vastus lateralis (Chen et al., 2003; Mahoney and Tarnopolsky et al., 2005), rat tibialis anterior (Chen et al., 2002), and mouse tibialis anterior (Barash et al., 2004). Additionally, DNA microarrays have been applied during recovery from a bout of isotonic resistance exercise (Zambon et al., 2003), during compensatory hypertrophy due to synergist ablation (Carson et al., 2002), following cardiotoxin injection in mouse tibialis anterior (Yan et al., 2003), and following a nine-week resistance exercise program (Roth et al., 2002). These studies are outlined in table 7.1. Although the experimental designs of all the studies were different, several very important messages can be extracted when the results are coalesced and interpreted together. Some of these messages confirm old ideas, while others present novel ideas that will undoubtedly form the foundation of much future research.

- **Gene arrays can reproducibly and validly measure gene expression from skeletal muscle samples.** Gene array *reproducibility* is estimated through examination of array-to-array variability using different aliquots of the same muscle sample. Each of these studies, using both oligonucleotide and cDNA arrays, has demonstrated a high degree of array-to-array consistency, proving that both types of arrays can reproducibly measure gene expression in muscle samples. Gene array *validity* is measured through confirmation of the statistically significant array data using a different technique,

Table 7.1 Summary of Skeletal Muscle Array Studies

	Experimental design	Array	Importance and major findings
EIMD studies			
Chen et al., 2002	Stimulated ECC contractions of rat TA muscle; 1 h + 6 h samples; ~8,000 genes	Oligo	Reported first transcriptional profile of muscle after ECC damage; identified *p53* as ECC-responsive protein
Carson et al., 2002	Rat soleus overload via gastroc synergist ablation; 3-day sample; ~8,500 genes	Oligo	Altered genes in metabolism, autocrine/paracrine function, ECM remodeling, intracellular signaling
Zambon et al., 2003	Isotonic RES exercise in human vastus lateralis; 6 h + 18 h samples	Oligo	Identified circadian clock genes (*Cry1, Per2,* and *Bmal 1*) that are altered by a bout of RES exercise
Chen et al., 2003	ECC contractions of human vastus lateralis; 4-8 h samples; ~12,000 genes	Oligo	Demonstrated ~50% homology to expression changes observed in mouse after ECC contractions
Yan et al., 2003	Cardiotoxin injection into mouse TA muscle, harvested at 1, 2, 3, 5, 10, 14 days; 16,267 genes	cDNA	Identified many novel genes activated in vivo during proliferative and differentiative phase of regeneration
Barash et al., 2004	ECC contractions of mouse TA muscle; 48 h sample	Oligo	MLP, MARP1, MARP2—potential mechanical damage sensors in muscle?
Mahoney & Tarnopolsky, 2005	ECC contractions of human vastus lateralis; 3 h + 48 h samples; ~8,400 genes	cDNA	Genes involved in cholesterol and lipid homeostasis—SREBP-2-mediated de novo membrane biosynthesis?
Muscular dystrophy studies			
Tkatchenko et al., 2000	Diaphragm and hindlimb muscles from 12-week-old *mdx* mice; 1,536 genes	*SSH	Altered genes in metabolism, growth, differentiation, Ca^{2+} homeostasis, proteolysis, inflammation, ECM
Chen et al., 2000	6- to 9-year-old DMD ($n = 5$) and 8- to 11-year-old alpha-SG deficiency ($n = 4$) patients; ~6,000 genes	Oligo	Persistent activation of genes involved in development and regeneration
Tkatchenko et al., 2001	11- to 14-year-old DMD ($n = 4$) patients	cDNA	Mitochondrial genes; ↓titin
Porter et al., 2002	Gastroc and soleus muscle from 8-week-old *mdx* mice; ~10,000 genes	Oligo	Inflammatory genes, including cytokine/chemokine signaling, complement activation, leukocyte migration
Rouger et al., 2002	Diaphragm and hindlimb muscle at 12 points during *mdx* mouse life; 1,082 genes	cDNA	Used differentially affected muscles to show that dystrophin deficiency initiates similar pathological cascade but affects different transcriptional circuits
Haslett et al., 2002	Quadricep biopsy from 5-to 7-year-old DMD ($n = 12$) patients	Oligo	Genes in muscle regeneration, ECM remodeling, immune response signals, and growth factors

(continued)

Table 7.1 *(continued)*

	Experimental design	Array	Importance and major findings
Muscular dystrophy studies *(continued)*			
Boer et al., 2002	Hindlimb muscle from 13- to 15-week-old *mdx* mice; ~12,000 genes	Oligo	Explored regenerated *mdx* muscle; ↑genes in muscle development and function, immune response, proteolysis
Bakay et al., 2002	5- to 6-year-old DMD ($n = 5$) and 11- to 12-year-old DMD ($n = 5$) patients	Oligo	First Web-accessible complete transcriptome of DMD muscle; identified IGF family members that are altered
Tseng et al., 2002	Gastroc muscle from 16-week-old *mdx* mouse; 12,488 genes	Oligo	Examined successfully regenerated *mdx* muscle and identified genes that may help explain why
Noguchi et al., 2003	Biceps brachii biopsy from 1- to 5-year-old DMD ($n = 6$) patients; ~3,500 genes	cDNA	HLA-related, myosin light chain, and troponin T proteins altered—markers for necrosis/regeneration?
Haslett et al., 2003	Quadriceps biopsy from 7-year-old DMD ($n = 10$) patients; ~12,500 genes	Oligo	Identified novel differentially expressed genes involved in signaling and cell–cell communication
Porter et al., 2003	Gastroc and extraocular muscle at six points between P7 and P112 in *mdx*; ~10,000 genes	Oligo	Examined affected versus nonaffected *mdx* muscle and identified genes that might help explain the difference
Porter et al., 2004	Diaphragm muscle at time points between P7 and P112 in *mdx* mouse; ~10,000 genes	Oligo	Wild-type diaphragm different expression profile than gastroc, which may explain tissue difference in *mdx*
Myoblast proliferation and differentiation			
Moran et al., 2002	C2C12 cells harvested during proliferation and differentiation; ~11,000 genes	Oligo	First study of its kind; identified genes for contraction, adhesion, metabolism, replication, cell cycle control
Shen et al., 2003	C2C12 cells harvested during proliferation and differentiation; ~10,000 genes	cDNA	Demonstrated coordinate molecular regulation of cell cycle withdrawal and apoptosis
Delgado et al., 2003	C2C12 cells harvested at various time points during first 24 h of differentiation	Oligo	Defined "transcriptional waves" of regulatory proteins activated prior to myogenin
Tomczak et al., 2004	C2C12 cells—proliferation, differentiation, and myotube maturation; ~24,000 genes	Oligo	Defined molecular time course of genes that regulate proliferation, differentiation, and maturation
Sterrenburg et al., 2004	Primary myoblast cells from humans ($n = 3$) harvested over 22 days of differentiation	cDNA + oligo	First human study; data concordant with mouse data, although not identical; many genes in regeneration

*SSH: suppression substractive hybridization

such as northern blotting or RT-PCR. Each of these studies has demonstrated that statistically significant expression changes from both oligonucleotide and cDNA microarrays can be validated using a different technique, proving that both types of arrays generate valid expression data from muscle samples.

- **However, poor study design can lead to low sensitivity and accuracy.** These studies have also demonstrated that low numbers of biological replicates negatively affect the *sensitivity* and *accuracy* of the arrays. With a low *n* it is difficult to observe statistically significant changes in gene expression, and false negatives ensue. Clearly this has occurred in many of the studies to date, as numerous genes that have been shown to increase with use of conventional techniques have not been detected in the microarray experiments. Additionally, with a small sample size or pooled samples, the fold changes do not correlate well with those generated using other techniques (although they are generally still confirmed). For instance, although Barash and colleagues (2004), who used two technical replicates of pooled sample, were able to confirm increased expression of three genes detected by microarray using RT-PCR, the correlation coefficient between the two data sets was very low ($r^2 = 0.06$). In contrast, we recently analyzed global gene expression in four subjects before and after a bout of eccentric exercise (i.e., repeated-measures design) and confirmed the expression of 10 differentially expressed genes using RT-PCR with a correlation coefficient of $r^2 = 0.46$. Furthermore, the latter study generated nearly five times as many genes that were differentially expressed in response to a bout of damaging exercise. To put into perspective the importance of sample size and proper study design (e.g., repeated measures in the same person), we have just completed a study of skeletal muscle global gene expression in 24 human subjects before and after two weeks of leg immobilization, and confirmed 12 genes using RT-PCR with a correlation coefficient of $r^2 = 0.99$ (Tarnopolsky, 2007, unpublished data). Importantly, the power of this design enabled us to detect statistically significant expression changes for >1,700 genes, of which some were altered by as little as 1.3-fold.

- **Contraction-induced muscle damage (CIMD) leads to elevated expression of a relatively large number of genes and to decreased expression of a relatively small number of genes.** In each of these studies, the number of genes whose expression level was elevated was much greater than the number whose expression level was decreased. In broad strokes, this confirms the notion that skeletal muscle recovery from and adaptation to a damaging insult is an *active* process, and part of this process involves gene activation.

- **Contraction-induced muscle damage activates programs of gene expression involved in stress management, structural integrity, protein synthesis, stem or satellite cell activation, inflammation, muscle repair, membrane repair, and cell death, regardless of the species, muscle, or experimental protocol used.** Not surprisingly, gene array studies have consistently demonstrated upregulation of genes involved in the "hallmark" events that ensue after a bout of damaging exercise. Importantly, there were many overlapping genes from study to study, indicating that these responses are generally conserved between muscles (vastus lateralis, tibialis anterior, soleus) and species (human, rat, mouse), irrespective of the experimental damage protocol used (although some differences do exist).

- **Gene array studies can identify differentially expressed genes in response to muscle damage that have not been previously reported (i.e., novel gene expression).** Perhaps the most exciting aspect of this work is that each study clearly demonstrated the power of gene array technology to discover novel differential gene expression in response to muscle damage (figure 7.5).

Several of these findings have led to tantalizing hypotheses about cellular recovery from muscle damage. For instance, muscle LIM protein (MLP) and muscle ankyrin repeat protein 1 and 2 (MARP1 and MARP2) were robustly elevated in most of the array studies. Further investigation by Barash and colleagues (2004) using RT-PCR demonstrated an asynchronous but relatively coordinate induction of these genes in the first 48 h after eccentric contractions in mouse tibialis anterior muscle. Muscle LIM protein is primarily a cytoplasmic protein in adult skeletal muscle that interacts with important muscle structural proteins such as β-spectrin, telethonin, and α-actinin (Flick & Konieczny, 2000; Knoll et al., 2002). Interestingly, during embryonic muscle development, MLP acts as a nuclear transcription factor that activates MyoD, the "master regulator" of muscle differentiation. A major effort in molecular and cellular muscle damage research has been to characterize the putative "mechanical sensors" that respond to structural stress or damage and engage the programs of muscle repair. Given the strategic intramuscular location of MLP, its known ability to activate muscle-specific gene transcription, and its early and sharp increase in response to muscle damage, Barash and colleagues (2004) have speculated that MLP may be a mechanical sensor in skeletal muscle. The same may also be possible for MARP1 and MARP2, which are highly enriched in skeletal muscle, which localize both to the nucleus and to muscle I-bands in the cytoplasm, and which are rapidly and robustly elevated following damaging exercise.

Chen and colleagues (2002) discovered that several *p53* targets were upregulated following eccentric contractions in rat tibialis anterior muscle, including the antigrowth genes PC3 and GADD45. Both PC3 and GADD45 are well-known cell cycle inhibitors and are involved in tissue terminal differentiation. Further investigation using immunoblotting

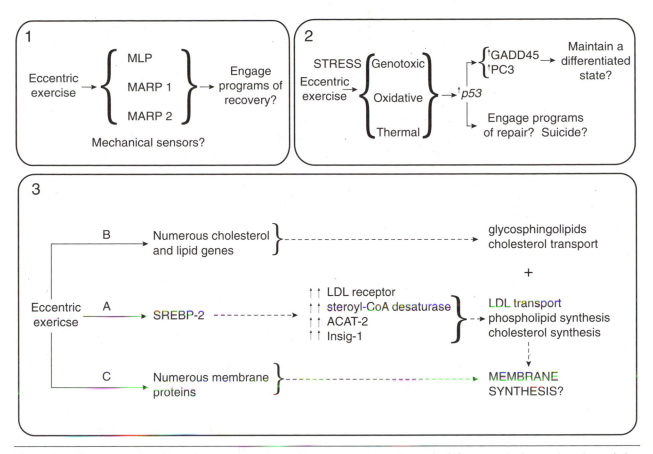

Figure 7.5 Three examples of hypotheses generated regarding recovery, adaptation, or both from muscle damage based on existing array studies. 1. Barash and colleagues (2004) proposed that muscle LIM protein (MLP) and muscle ankryin repeat protein (MARP) respond to mechanical stretch or structural damage and engage programs of recovery from muscle damage. 2. Chen and colleagues (2002) hypothesized that *p53* may be a stress-responsive gene in skeletal muscle that engages programs of repair or that helps maintain muscle in a differentiated state. 3. Mahoney and Tarnopolsky (2007) proposed a model for de novo sarcolemmal synthesis in response to muscle damage, suggesting specific roles for eccentric exercise in activating genes involved in such synthesis.

demonstrated that *p53* protein was elevated in the nuclear compartment following eccentric contractions (Chen et al., 2002). The *p53* pathway is strongly activated by stress, particularly genotoxic stress, and engages a coordinate response that attempts to repair the lesions to the DNA while preventing the cell from cycling. If the stress cannot be appropriately managed, *p53* turns on a suicide response (apoptosis). The total response is intended to prevent the propagation of damaged DNA. To observe this stress response in a fully differentiated tissue that has terminally withdrawn from the cell cycle was particularly interesting. Chen and colleagues (2002) hypothesized that *p53* may be activated after damaging exercise to engage repair programs in response to, or in anticipation of, oxidative, thermal, or genotoxic stress. Furthermore, these authors suggested that *p53*-mediated induction of growth arrest genes may be involved in keeping muscle in a differentiated state.

We recently examined global gene expression in vastus lateralis following a bout of high-intensity eccentric contractions in humans, and observed the coordinate induction of

a very interesting set of cholesterol- and lipid-related genes (Mahoney and Tarnoplsky, 2007, unpublished). In our study, eccentric exercise induced a rapid increase in expression of the transcription factor sterol regulatory element binding protein (SREBP) 2, which was followed by a delayed increase in a number of SREBP-2 gene targets, including the low-density lipoprotein receptor stearoyl-CoA desaturase, acetyl-CoA acetyl transferase 2, and insulin-induced gene 1. These expression changes, together with elevated expression of caveolin-1 and UDP glucose ceramide glucosyl transferase, are characteristic of an established transcriptional program geared toward increasing cholesterol and lipid synthesis, uptake, and modification (Brown & Goldstein, 1997). The expression of several potential upstream regulators of this SREBP-2 program—such as phosphotidyl inositol 3 kinase (PI_3kinase)γ, interleukin (IL)-1 receptor, IL-6 receptor, and peroxisome proliferative activated receptor (PPAR)γ—also rapidly increased after eccentric exercise. Together with a delayed increase in the expression of numerous plasma membrane transport proteins, our results point toward an SREBP-2-mediated program of de novo membrane biosynthesis after eccentric exercise. We speculate that SREBP-2 may act as a "membrane sensor" that responds to sarcolemmal damage or anticipates the need for sarcolemmal growth and engages a transcriptional program of de novo membrane biosynthesis. Although this hypothesis and those mentioned earlier (and those not mentioned) await rigorous evaluation, they offer exciting new leads toward understanding muscle recovery or adaptation (or both) in response to CIMD.

Transcriptomics and Muscular Dystrophy

Over the past five years, no fewer than 12 papers (table 7.1) have described DNA microarray analysis of skeletal muscle in muscular dystrophy (Chen et al., 2000; Tkatchenko et al., 2000, 2001; Bakay et al., 2002; Boer et al., 2002; Haslett et al., 2002; Porter et al., 2002; Rouger et al., 2002; Tseng et al., 2002; Haslett et al., 2003; Noguchi et al., 2003; Porter et al., 2003, 2004). These studies are each unique in that they encompassed both Duchenne muscular dystrophy (DMD) and the mouse *mdx* model; analyzed gene expression in many different muscles and at varying time points; used both cDNA and oligonucleotide arrays; and analyzed the array data in multiple ways. As such, these studies generated *overlapping* but *distinct* data sets. The details of these data are beyond the scope of this chapter; however, the data generally recapitulate the hallmark histopathological features and biochemical dysfunction associated with dystrophin deficiency at a molecular level. Genes involved in muscle regeneration, extracellular matrix remodeling, inflammation, immune signaling, proteolysis, intracellular signaling, growth, metabolism, and muscle development and differentiation are altered. These genes define programs of gene expression secondary to dystrophin deficiency that help explain the pathogenesis of skeletal muscle dysfunction downstream of the loss of dystrophin. Hundreds of genes have been identified that were not previously known to be responsive to dystrophin deficiency, some of which provide exciting new research and potentially therapeutic targets.

Importantly, these array studies have begun to address two particularly outstanding questions regarding dystrophin deficiency: (1) Why does the same primary mutation cause such a disparate response from muscle to muscle? (2) Why is dystrophin deficiency lethal in humans, whereas mice are able to recover?

The first of these questions has been addressed in the *mdx* mouse model, used in several studies of global gene expression in severely affected (diaphragm), moderately affected (hindlimb), and unaffected (extraoccular) muscles. These studies have clearly shown that dystrophin deficiency leads to distinct downstream expression signatures in diaphragm versus hindlimb versus extraoccular muscles. Furthermore,

Porter and colleagues (2003, 2004) have demonstrated that there are large differences in global gene expression between the diaphragm, hindlimb, and extraoccular muscles of *wild-type* mice. These authors speculated that the muscle-specific effect observed in the *mdx* mouse may result from the interaction between dystrophin deficiency and fundamentally distinct "molecular backgrounds" in the different muscles. Identifying one or more differentially expressed genes and proteins in unaffected or lesser-affected muscles that are most involved in mediating protection from dystrophin deficiency is the subject of much current research.

The second question has also largely been addressed in the *mdx* model. Several studies have compared global gene expression in young *mdx* mice undergoing waves of degeneration and regeneration to that of more mature *mdx* mice whose muscle has successfully regenerated. These studies have clearly shown distinct expression changes as a function of the state of the disease, although the goal of identifying the molecular etiology of muscle recovery in mice has not yet been realized. It is hoped that these and future microarray studies will eventually contribute to a full understanding of muscle recovery from dystrophin deficiency in mice and that this knowledge can be translated into treatment targets for DMD.

Myoblast Proliferation and Differentiation

As discussed in chapter 6, muscle satellite or stem cell activation, proliferation, differentiation, and fusion to damaged myofibers constitute an essential and well-documented component of muscle repair. The molecular regulation of these processes, however, is incompletely understood. A number of researchers (table 7.1) have attempted to gain insight into this process by expression profiling both mouse (C2C12) and human (primary) myoblasts during proliferation, cell cycle withdrawal, differentiation, fusion, and myotube maturation in vitro (Moran et al., 2002; Delgado et al., 2003; Shen et al., 2003; Sterrenburg et al., 2004; Tomczak et al., 2004). These studies have generated an enormous amount of novel data; in particular, many novel genes involved in cell cycle regulation, extracellular matrix remodeling, apoptosis, adhesion, replication, and metabolism have been identified. Currently, the data are complex and it is difficult to interpret their potential importance; the hope is that ultimately, a complete understanding of these processes will have therapeutic implications for muscle disease.

Limitations of DNA Microarray Muscle Damage Study

Several limitations to each of these studies must be acknowledged and taken into consideration by the reader. Firstly, skeletal muscle is a heterogeneous tissue, composed not only of myofibers but also of muscle satellite and stem cells, fibroblasts, vascular cells, red blood cells, and resident macrophages. Each of these studies used whole-muscle homogenates, containing a mix of each of these types of cells; thus it is not definitively known whether the observed changes occurred in myofibers (as is generally interpreted in the studies) or another cell type. This having been said, muscle nuclei are the major mRNA contributor to a muscle biopsy sample, and the majority of the expression changes that were observed generally fit into established cellular "programs" in skeletal muscle. However, it is clear that to validate the hypotheses generated in each of these studies, the exact cell type responsible for the observed differential expression will need to be determined. Studies using in situ hybridization methods can be used in the future to highlight the tissue location of the mRNA expression.

Secondly, in the human studies, there is growing awareness that factors other than the damaging exercise bout may be responsible for altered gene expression, such as the damaging effect of repeated biopsies, nutritional status, and the time of day (Zambon et al., 2003; Norrbom et al., 2004; Vissing et al., 2005). None of the human studies employed a nonexercise control group to definitively control for these potentially confounding variables. However, each study did generally control for nutritional status and time of day, and the biopsies were taken in distinct physiological sites (in our study ~6-8 cm apart). These controls, although not perfect, enable the strong suggestion that the observed expression changes were indeed the result of the damaging exercise bout itself.

Thirdly, in the absence of complementary data sets (e.g., protein data), it is very difficult to accurately interpret the biological meaning of DNA microarray data. Thus the most important contribution from the array studies to date is not the author's *interpretation* of the data sets, but rather the data sets themselves. These data sets are deposited in public access databases and are open to interpretation by all scientists. Importantly, these studies have made inroads into a powerful new field and have demonstrated the experimental feasibility and validity of using microarrays to characterize global gene expression in skeletal muscle in response to damage. However, it is clear that microarray data themselves are largely hypothesis generating and that complementary approaches are required to explore these hypotheses in more detail.

Finally, these studies measured between 7,000 and 12,000 genes, yet the human, mouse, and rat genome each contain 20,000 to 40,000 genes. Thus, it is likely that ~75% of genes responsive to muscle damage were not even measured in these investigations. Clearly, with the advent of newer arrays that measure the entire genome, additional studies are needed to fully profile the transcriptome in response to muscle damage.

The Future of Omics in Muscle Damage Research

There is a bright future for omic technology in muscle damage research. First and foremost, only a few studies have applied transcriptomic technology to study muscle damage, and each of these was limited in several important ways. Thus future research is needed to more fully characterize the transcriptome in muscle in response to damage. To maximize the value to the field, researchers conducting these studies should attempt to the following:

- Use arrays that measure the entire genome of interest
- Employ a no-damage control group that is treated *identically* to the experimentally damaged group
- Use a repeated-measures design with a sufficiently large sample size and nonpooled data
- Complete time-course analyses
- Ideally, use or develop techniques to assay gene transcription solely in muscle fibers (e.g., laser capture microdissection, single-fiber analysis)— if the cell type responsible for important expression changes is not definitively known, it should be determined using in situ hybridization

Any hypothesis generated by microarray studies needs to be validated and fully characterized. The ideal way to do this would be to use a combination of omic and non-omic molecular and cellular biology techniques. The "omic arm" of this approach would use many omic techniques to generate entire complements of biological information from multiple levels, such as the proteome or interactome. Integrating these data would allow muscle physiologists to fully characterize muscle damage at the cellular and tissue levels, including the repair, recovery, and adaptive processes that ensue. Whole-cell or whole-tissue models can be generated from such integrated data with the aim of *completely* describing muscle damage. Researchers can test such a model by perturbing different aspects within the model and using the same integrated omic approach to characterize the response to the perturbation (figure 7.6). Although this is the ideal approach, the time, labor, equipment, and computer use required make it very expensive—ultimately the stuff of a large multilab collaborative effort.

At present, a more realistic approach to validating the hypotheses generated from array studies is to use non-omic techniques. To illustrate how this might be done, we will use the hypothesis that muscle LIM protein is a

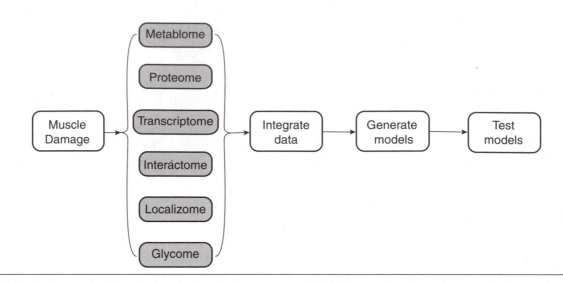

Figure 7.6   The future of omic technology in muscle damage research. To fully appreciate the molecular and cellular response to muscle damage, and to test system-wide hypotheses regarding the etiology of recovery and adaptation, it will be important to use multiple omic techniques. Integrating transcriptomic, proteomic, and other omic data will enable researchers to create cell- or tissue-wide models that fully characterize the damage response in skeletal muscle, including damage resolution and adaptation. Researchers can then test and validate these models by perturbing various aspects of the model and characterizing the response.

mechanical sensor in skeletal muscle that engages repair and recovery programs in response to damage. This hypothesis would initially be explored in a muscle culture or animal model, in which (1) muscle damage can be induced that is representative of that in humans and (2) muscle LIM protein expression can be manipulated (upregulated or downregulated) *exclusively* in skeletal myofibers *at any time* (ideally genetically). Experiments would first be designed to provide proof-of-principle that muscle LIM protein is truly important following muscle damage. This would be done via an examination of the consequence of under- or overexpressing muscle LIM protein on muscle damage, repair, recovery, and adaptation. If these experiments are positive, the next wave of experiments would explore the specific hypothesis that muscle LIM protein is a mechanical sensor that engages repair or recovery responses (or both). This would involve deciphering the upstream mechanism responsible for muscle LIM protein activation (i.e., what does muscle LIM protein actually "sense," and how does it do so?), as well as the important downstream targets responsible for the effect of muscle LIM protein (i.e., after sensing mechanical stress or muscle damage, what does muscle LIM protein do?). Dozens of experimental techniques are now available to muscle physiologists that would enable them to address these questions.

ogy are now widely available. These techniques can empower muscle physiologists to decipher the mechanism of muscle damage, repair, recovery, and adaptation at its very core: the cellular constituents that are damaged and the macromolecules that are responsible for engaging programs of repair, recovery, and adaptation. Most exciting among these are the omic techniques, which promise a comprehensive "systems biology" understanding of muscle damage, repair, recovery, and adaptation. To date, only transcriptomics has made any inroads into the muscle damage field, and it is time for muscle physiologists to move forward and exploit the power of omic technology to its fullest. Full-genome transcriptomic studies, as well as proteomic, localizomic, interactomic, and other omics, need to be undertaken so that models integrating all levels of biological information can be generated and tested. Omic techniques, coupled with non-omic molecular and cellular biology techniques as well as more conventional muscle physiology techniques, now give muscle physiologists unprecedented capacity to understand muscle damage, which will have enormous implications for developing treatments for various muscle diseases.

Summary

Highly sophisticated experimental techniques that explore all aspects of molecular and cellular biol-

PART II

Muscle Damage and Repair in Applied Situations

CHAPTER 8

Changes With Aging

Susan V. Brooks, PhD

The effects of skeletal muscle atrophy, declining strength, increasing fatigability, and increasing susceptibility to injury contribute to the condition of physical frailty. Aging increases susceptibility to contraction-induced muscle injury in rodents and in human beings, and recovery from injury is impaired in old animals. The mechanisms underlying the high susceptibility to injury are not well understood, but impaired recovery may involve altered satellite cell function in old animals. Despite the high susceptibility of muscles in old animals to injury and their decreased ability to recover, exercise training increases resistance to injury in muscles of both adult and old animals. Repeated bouts of exercise reduce the accumulation of inflammatory cells in the muscle associated with contraction-induced injury, the numbers of damaged fibers, and functional deficits. This chapter discusses current research findings and directions related to the ways in which aging alters the mechanisms of muscle damage, repair, and conditioning for protection from damage. Data from both human and animal studies are presented, but whether a specific finding came from human work or a study of another animal will not always be explicitly stated inasmuch as the insights gained are similar.

Physical Frailty

Physical frailty, with the accompanying effects on mobility and increased incidence and severity of falls, is one of the most prominent manifestations of old age and represents a primary factor limiting an elderly person's chances of living independently (Hadley et al., 1993; Holloszy, 1995). Contributing to physical frailty are skeletal muscle atrophy,

declining strength, increasing fatigability, and increasing susceptibility to injury. Between maturity and old age, a wide variety of species, including human beings, show a 30% to 40% decrease in muscle mass and even greater decreases in the development of maximum force and power (reviewed in Brooks and Faulkner, 1994). Muscle mass, force, and power also decrease with reductions in physical activity at any age (Faulkner et al., 1994). Although superficially the deficits in muscle structure and function associated with aging appear similar to those observed with decreased physical activity, highly trained athletes also show significant declines in performance with aging, in particular when competing directly with younger individuals (figure 8.1). Furthermore, in over a century of Olympic competition, the record-setting performances have improved by 20% to 90% in various events requiring maximum or sustained power, while the age of the record holders has remained constant between 16 and 31 years (Schulz and Curnow, 1988). The conclusion is that whereas a portion of the loss in muscle mass, force, and power observed with aging results from decreasing physical activity, intrinsic age-related changes occur in skeletal muscle structure and function that appear to be immutable and irreversible (Faulkner et al., 1995).

For healthy elderly persons, the decrease in muscle mass results from a loss in the total number of fibers per muscle as well as a decrease in the mean cross-sectional area (CSA) of the remaining fibers (Grimby and Saltin, 1983; Lexell, 1995). In addition, some of the large fast fibers become denervated, with subsequent reinnervation by axonal sprouting from small slow fibers (Kanda and Hashizume, 1989; Larsson, 1995). The loss and atrophy of muscle fibers, along

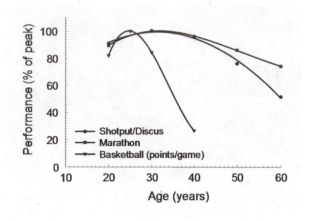

Figure 8.1 Physical performance data plotted against the age of the performer for the marathon run, the shot put and discus throw, and the number of points scored per game by a basketball player. The data for the marathon run and the shot put and discus throw are from Moore (1975), and the data for the basketball player are from the *Official NBA Resister, 1993-94 Edition.*

with the remodeling from large fast to small slow fibers, contribute to muscle atrophy as well as to an increase in the proportion of slow to fast fibers in muscles of old animals (Kadhiresan et al., 1996; Trappe et al., 2003). The loss of muscle fibers with aging represents the portion of the muscle atrophy that is largely irreversible. Conditioning programs that maintain, or even increase, CSA of the remaining fibers can slow the atrophy to some extent (Grimby, 1995; Trappe et al., 2000). In addition to the contribution of muscle atrophy to decreased strength with age, when normalized for the amount of muscle remaining, the muscles of old animals show intrinsic deficits of ~20% in maximum force (Brooks and Faulkner, 1988) and ~30% in maximum power (Brooks and Faulkner, 1991). Consequently, the phenomenon of muscle *weakness* appears to be intrinsic to some aspect of individual muscle cell function (reviewed in Larsson and Ramamurthy, 2000).

Contraction-Induced Injury

Over 100 years ago, during studies of fatigue in the finger flexor muscles of human beings, the phenomenon of *delayed-onset muscle soreness* was observed (Hough, 1902). As described in the latter half of chapter 5 of this text, delayed-onset muscle soreness after exercise involving stretching of muscles was subsequently compared to that after exercise involving primarily shortening contractions (Clarkson et al., 1986; Friden et al., 1983; Newham

et al., 1983). These studies, along with more definitive experiments on individual muscles of mice (McCully and Faulkner, 1985), support the conclusion detailed in chapter 1 that injury is most likely to occur during physical activities that involve lengthening contractions. The utility of various direct and indirect means of identifying and quantifying injury is spelled out in chapter 3, leading to the conclusion that the decrease in force provides the most valid and reliable measure of the totality of the damage to skeletal muscle fibers (Faulkner et al., 1989; Lieber and Friden, 1993; McBride et al., 1995; McCully and Faulkner, 1985; Warren et al., 1993).

In chapter 1, the section "Mechanisms of Muscle Injury" nicely summarizes evidence to support the idea that contraction-induced injury is initiated by the mechanical disruption of ultrastructural elements within or between sarcomeres (figure 8.2). As illustrated in "Histological Evidence of Muscle Damage" in chapter 3, the major disruptions observed immediately following either single or repeated lengthening contractions of isolated skeletal muscles or muscle fibers (Lieber et al., 1991; Macpherson et al., 1996) are highly consistent with those observed for human subjects following injurious bouts of exercise (Friden et al., 1983; Newham et al., 1983). Chapter 1 also points out the importance of both the size of stretches (Lieber and Friden, 1993) and the force generated during stretches (McCully and Faulkner, 1986; Warren et al., 1993) for initiating

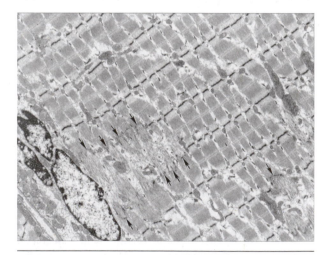

Figure 8.2 Electron micrograph of a longitudinal section of a muscle fiber from an extensor digitorum longus muscle of a young mouse, fixed within 10 min after a protocol of 75 lengthening contractions. Arrows indicate areas of focal damage within single or small groups of sarcomeres. In some sarcomeres, the damage appears to be in the A-band region with Z-lines remaining intact, whereas in others the damage involves the Z-lines. Magnification is 12,900×.

contraction-induced injury, with the work done to stretch a muscle explaining ~80% of variability in the severity of the resultant damage (Brooks and Faulkner, 1996; Brooks et al., 1995; Macpherson et al., 1996).

Mechanical damage to sarcomeres triggers a cascade of events described in detail in previous chapters of this text, including the sections "Other Exercise-Induced Disruptions" in chapter 1, "Biochemical and Histochemical Evidence of Muscle Damage" in chapter 3, and the latter half of chapter 4. These secondary events, which include calcium-dependent proteolytic autodegradation (Beaton et al., 2002; Jackson et al., 1984), the infiltration of the muscle by inflammatory cells (Pizza et al., 2001; Smith, 1991; Stauber et al., 1988; Tidball, 1995), and the generation of reactive oxygen species (ROS) within the injured muscle (Best et al., 1999; McArdle et al., 1999; McCord and Roy, 1982; Zerba et al., 1990), culminate in the degeneration of damaged fibers or portions of a fiber (figure 8.3). The secondary injury peaks between one and five days following the initial injury as evidenced by peaks in the deficit in force-generating capacity and in the number of fibers demonstrating morphological evidence of damage (McCully and Faulkner, 1985). Although the peaks in force deficit and in the number of damaged fibers present in cross sections occur at about the same time, the magnitude of the damage based on force deficit is usually greater than that based on the number of damaged fibers (McCully and Faulkner, 1985; Zerba et al., 1990). The discrepancy is attributed to the focal nature of the injury: Some fibers that appear normal in one cross section are injured in another region of the muscle (McCully and Faulkner, 1985; Ogilvie et al., 1988). Recovery from injury has a time course that varies with the severity of the injury (Brooks and Faulkner, 1990; McCully and Faulkner, 1985). The characteristics of injury and recovery are similar for mice (Faulkner et al., 1989), rats (Van der Meulen et al., 1997), and humans (Faulkner et al., 1993; Friden et al., 1983).

Initial Injury

Evaluations of ultrastructural damage to sarcomeres in muscles from animals of different ages yield varied results. Immediately following 75 lengthening contractions in situ, no differences in the focal damage to sarcomeres were observed among animals of different ages, including old animals (Zerba et al., 1990). In contrast, biopsies of vastus lateralis muscles, obtained immediately following a bout of exercise in which subjects resisted the backward motion of a motor-driven cycle ergometer, showed at least some focal damage to sarcomeres in nearly all of the fibers examined from older subjects compared with only 5% to 10% of the fibers as reported previously for young subjects (Manfredi et al., 1991). Similarly, electron micrographs from biopsies of vastus lateralis muscles, taken within 24 to 48 h following the final bout of a nine-week program of high-load resistance

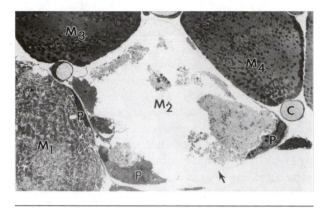

Figure 8.3 Electron micrograph of a cross section of an extensor digitorum longus muscle of a young mouse, fixed three days following a protocol of 75 lengthening contractions. Muscle fibers range from those with intact myofibrils (M3 and M4) to those with degenerating myofibrils (M1) or devoid of cytosolic constituents (M2). Fiber M2 has phagocytes (P) within the basement membrane (arrow). C is a capillary. Magnification is 3800×.

From J.A. Faulkner, S.V. Brooks and E. Zerba, 1995, "Muscle atrophy and weakness with aging: contraction-induced injury as an underlying mechanism," *Journal of Gerontology* 50: B124-B129. Copyright © THE GERONTOLOGICAL SOC OF AMERICA. Reproduced by permission of the publisher

exercise, showed ultrastructural damage in nearly 17% of fibers in older women compared with 2% to 5% in untrained control muscles in both young and older women as well as in the young subjects exposed to the exercise (Roth et al., 2000). In contrast to the greater damage in muscles of old compared with young women, when the same investigators performed the same experiment on young and old men, they found no difference between the age groups for either the types of lesions observed or the numbers of affected fibers, with 7% of fibers in both age groups showing ultrastructural damage compared with ~2% in untrained control muscles (Roth et al., 1999). The explanation for these disparate findings is not known, but the highly dispersed nature of the damage makes quantitative comparisons of ultrastructural damage between subjects or animals of different ages very difficult.

One research group compared amounts of sustained depolarization in the resting membrane potential in young and old animals following protocols of lengthening contractions (McBride, 2000). The amount of the depolarization remaining in the presence of blockers of stretch-activated ion channels was taken as an indicator of physical damage to the sarcolemma (McBride et al., 2000). Despite no difference between control muscles of young and old rats in the initial resting membrane potential, muscles of old rats showed close to 30% greater depolarization immediately following exposure to lengthening contractions than muscles of young rats (McBride, 2000). In addition, the portion of the depolarization that could be restored through blockade of stretch-activated ion channels was less than 20% for the muscles of the old rats, compared with greater than 50% for muscles of young rats. This observation suggested that mechanisms other than the opening of stretch-activated channels made a greater contribution to the depolarization in old rats (McBride, 2000). The conclusion was that compared with the muscles of young rats, those of old rats suffered more extensive membrane damage during lengthening contractions. Despite this indication, the extent to which physical damage to the membrane contributes to the subsequent development of injury has not been established in either young or old animals.

In general, force data support a more severe initial injury in muscles of old compared with young or adult animals. The *force deficit* is the decrease in maximum isometric force developed by a muscle after contraction-induced injury, relative to the maximum force developed before the injury, [1 − (maximum isometric force after injury/maximum isometric force before injury)] × 100%]. Immediately after a protocol of repeated lengthening contractions, the force deficit reflects both fatigue and injury (Faulkner et al., 1989). Recovery from fatigue occurs in minutes to hours, depending on the intensity of the contraction or exercise protocol. In support of a greater initial mechanical injury to muscles of old compared with young mice, the force deficits 10 min after a protocol of 75 lengthening contractions were 57% and 36% for muscles of old and young adult mice, respectively (Zerba et al., 1990). While the 50% greater force deficit observed for muscles of the old mice at 10 min supports the idea of greater susceptibility to initial mechanical damage in muscles of old animals, Zerba and colleagues (1990) did not follow the recovery in force beyond 10 min to ensure that no portion of the force deficit was due to fatigue.

To completely eliminate any effect of fatigue on force deficit, maximally activated muscles and single permeabilized muscle fibers from young, adult, and old animals were exposed to lengthening contractions; see figure 8.4. Figure 8.4a shows the relationship for whole muscles between the work input during single stretches of varying magnitudes and the resultant force deficit 1 min after the stretch. Work is expressed in joules per kilogram and force deficit as a percentage of the isometric force developed just prior to the stretch. The continuous and dashed lines represent the regression relationships for the data from muscles of adult mice and old mice, respectively. For whole muscles, force deficit was well predicted by work. The relationship between work and force deficit for muscles of young and adult mice was not different, but the slope of the relationship was ~40% steeper for muscles of old mice. (figure 8.4a). The steeper work–force deficit relationship for muscles of old mice indicates a higher sensitivity to a given level of work (Brooks and Faulkner, 1996).

To test the hypothesis that the increased susceptibility of muscles in old animals to injury is due to differences at the level of the contractile proteins themselves, single permeabilized fibers from muscles of young and old rats were exposed to single stretches (Brooks and Faulkner, 1996). In figure 8.4b, the data from single fibers are presented as the means ± SEM for the force deficits observed immediately after single stretches of varying strains. Strain is expressed as a percentage of optimum fiber length (L_f) and force deficit as a percentage of the maximum force developed by the fiber segment prior to the stretch. Average force deficits

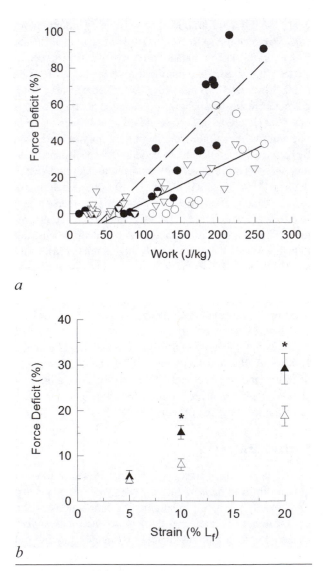

a

b

Figure 8.4   Force deficits following single stretches of maximally activated *(a)* whole extensor digitorum longus (EDL) muscles of young (inverted triangles), adult (open circles), and old (filled circles) mice and *(b)* single permeabilized fibers from EDL muscles of adult (open triangles) and old (filled triangles) rats. Asterisks indicate significant ($p < 0.05$) differences between the age groups.

From S.V. Brooks and J.A. Faulkner, 1996, "The magnitude of the initial injury induced by stretches of maximally activated muscle fibres in mice and rats increases in old age," *Journal of Physiology (London)* 497: 573-580.

immediately after single strains of 10% or greater were approximately twofold larger for fibers from muscles of old compared with adult rats (figure 8.4b). In combination, the whole-muscle and single-fiber experiments indicate that the greater susceptibility of muscles in old animals to injury is due at least in part to a mechanically compromised sarcomeric structure that is less able to withstand stretch.

The precise basis for the mechanical damage is not known. During maximum activation of muscle fibers, heterogeneity in sarcomere lengths develops in series along myofibrils, and the extent of the heterogeneity increases at longer lengths (Julian and Morgan, 1979a, 1979b; Macpherson et al., 1997). These heterogeneities in sarcomere length may arise from slight differences in resting lengths, intrinsic differences in the relative strengths of different sarcomeres, or both. When fibers with imbalances in sarcomere strength are stretched, weak sarcomeres in series with strong sarcomeres are more likely to yield and to be lengthened excessively (Macpherson et al., 1997; Morgan, 1990; Newham et al., 1983). Under these circumstances, weak sarcomeres may be damaged even during relatively small displacements (Brooks et al., 1995). If the hypothesis is supported that damage occurs to weak sarcomeres stretched excessively by adjacent stronger sarcomeres, the greater susceptibility to contraction-induced injury of muscles in old animals would arise from a larger population of weak sarcomeres in muscle fibers of old compared with adult animals.

Secondary Injury

Experiments isolating the effects of a given contraction protocol on the magnitude of the secondary injury in animals of varying ages have not been done definitively. Such an experiment would require following the development of the secondary injury in response to a contraction protocol verified to have produced identical initial injuries in young or adult and old animals. Zerba and colleagues (1990) observed a 40% greater force deficit for muscles of old compared with young or adult mice three days after a protocol of 75 lengthening contractions, but this may have resulted from similar secondary responses to initial mechanical injuries that varied in severity. Similarly, Ploutz-Snyder and colleagues (2001) reported a threefold greater percent reduction in strength for older compared with young women one day following a bout of unaccustomed exercise involving predominantly lengthening contractions. However, this could have resulted from greater susceptibility to the initial mechanical injury with a similar secondary injury response, from a more severe secondary injury in response to a given initial injury, or from some combination of the two.

There are numerous reports of force deficits and morphological evidence of injured fibers that did not differ between adult and old animals at the peak of the secondary injury; but these do not represent experiments that were necessarily

designed to isolate the effects of age on the secondary injury process (Brooks and Faulkner, 1990; Brooks et al., 2001; Devor and Faulkner, 1999; Koh et al., 2003; McArdle et al., 2004; McBride et al., 1995). Rather than demonstrating or failing to demonstrate any effects of age on secondary injury, the similarity in the peaks in the magnitude of the secondary injury for muscles of adult and old animals in these experiments more likely reflects the severity of 15 to 20 min lengthening contraction protocols that overwhelmed any differences between the groups (Brooks and Faulkner, 1990). Supporting this conclusion is the fact that a protocol of 450 lengthening contractions resulted in no differences between the age groups for the force deficit measured at 3 h (McArdle et al., 2004), a time when recovery from fatigue is complete and force deficit reflects the magnitude of the initial mechanical injury. The severe contraction protocol used by McArdle and colleagues (2004) apparently injured all but a population of highly injury-resistant fibers in both adult and old animals (Brooks and Faulkner, 1990) and therefore masked the differences in initial mechanical injury generally observed between the age groups (Brooks and Faulkner, 1996; Zerba et al., 1990). The conclusion is that the ability to resolve differences between the muscles of adult and old animals is limited by the use of an extremely severe lengthening contraction protocol.

Support for a greater secondary injury in muscles of old compared to adult animals comes from data on the accumulation of inflammatory cells in the injured muscles, a key element of the secondary injury. Three days following a 15 min protocol of 225 lengthening contractions, the concentrations of neutrophils and macrophages in the muscles of old mice were 1.5- to 2-fold greater than in the muscles of adult mice subjected to the same contraction protocol (Koh et al., 2003). It has been proposed that oxygen free radical damage contributes to the secondary injury to muscle fibers already injured mechanically by stretches. Although this is an attractive hypothesis, the exact role of oxygen radicals has been difficult to establish. The concentration of free radicals in skeletal muscles is difficult to measure because of their short half-life (Jackson and Edwards, 1988).

As an alternative, one can block the activity of free radicals by increasing the concentration of free radical scavengers. Zerba and colleagues (1990) treated mice with polyethylene glycol-superoxide dismutase (PEG-SOD) for three days before and three days after a protocol of lengthening contractions. Treatment with PEG-SOD virtually eliminated the force deficit at three days for muscles of young mice. In old mice, the force deficit at three days was 30% with treatment compared to 56% without treatment (Zerba et al., 1990). The conclusion was that oxygen free radical damage was a major factor in the secondary injury. The design of this study did not permit resolution of the cause of the 30% force deficit that remained with PEG-SOD treatment in the muscles of old mice, but the twofold higher numbers of inflammatory cells in the muscles of old compared with adult mice three days following lengthening contractions (Koh et al., 2003) suggest that PEG-SOD treatment may have resulted in an incomplete block of a significantly greater oxygen free radical injury in old animals.

Recovery

A highly consistent finding is that recovery from injury is either delayed or impaired in muscles of old animals. Following a severe protocol of 225 lengthening contractions, which resulted in similar force deficits for extensor digitorum longus (EDL) muscles of adult and old mice at three days, muscles in the adult mice recovered completely by 28 days, but those in the old mice showed a remaining 16% force deficit and a decreased number of fibers in the muscle cross section (figure 8.5). Similarly, EDL muscles of old mice demonstrated a 44% force deficit remaining at 28 days following 450 lengthening contractions, while muscles of adult mice generated forces not different from those for uninjured controls (McArdle et al., 2004). For muscles of adult rats, force deficits were eliminated within five days following 24 lengthening contractions, whereas a delay in recovery to 14 days was observed for muscles of old rats (McBride et al., 1995); and recovery of the resting membrane potential to control levels required twice as long (four days vs. two days) for muscles of old compared with adult rats (McBride, 2000).

For human beings as well, after exposure of the knee extensors to an exercise protocol with lengthening contractions, young subjects regained control levels of strength within three to four days compared with seven to nine days for older subjects (Ploutz-Snyder et al., 2001). Additionally, compared with young subjects, older subjects showed a slower recovery of strength of the elbow flexors following a protocol of 24 lengthening contractions (Dedrick and Clarkson, 1990). Even two months after severe contraction-induced injury in old mice, gastrocnemius muscles of both males (Rader and Faulkner, 2006b) and females (Rader and Faulkner, 2006a) displayed persistent force deficits of 35% and deficits in muscle mass of 20%, while the injury was completely resolved for the muscles of adult mice. One should bear in mind that in the experiments of Rader and Faulkner (2006a, 2006b), the old mice were ~30 months old at the final evaluation, an age that represents about 25% survival of the cohort, making it reasonable to assume that the deficits were irreversible.

The irreversible losses of mass and force following severe injury raise questions regarding the factors limiting the recovery process. The decreased capacity for recovery of muscles in old animals following contraction-induced injury is consistent with earlier reports of impaired regeneration after whole-muscle transplantation in old animals (Carlson and Faulkner, 1989). Despite the clear impairment of muscle

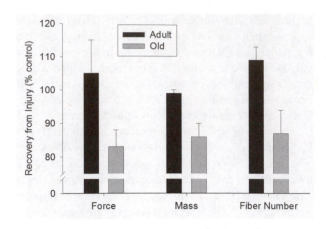

Figure 8.5 Postinjury maximum isometric force, muscle mass, and number of fibers in a cross section for extensor digitorum longus muscles of adult (black bars) and old (gray bars) mice 28 days following a protocol of 225 lengthening contractions. All data are presented as means ± SEM and are expressed as a percentage of values for age-matched uninjured control muscles. Values for adult and old mice are significantly ($p < 0.05$) different for all three variables.

regeneration in old animals, cross-age transplantation experiments demonstrated equally effective regeneration following transplantation of muscles from either a young or an old donor into a young host (Carlson and Faulkner, 1989). The observation of no effect of the age of the donor to reduce muscle regeneration in young hosts supported the conclusion that the intrinsic regenerative capacity of skeletal muscle was not impaired with aging (Carlson and Faulkner, 1989). The complementary demonstration of equally poor regeneration of muscles from either young or old donors transplanted into an old host indicates that the environment of the old host either lacks factors critical for muscle regeneration or contains large amounts of factors that inhibit muscle regeneration. Muscle regeneration requires the activation, proliferation, and differentiation of muscle satellite cells, suggesting the reasonable hypothesis that the factors affecting muscle regeneration influence satellite cell function (Conboy et al., 2005).

A clear impairment was observed in the ability of satellite cells of old mice to become activated in response to injury and to produce myoblasts necessary for muscle regeneration (Conboy et al., 2003; Schultz and Lipton, 1982). The decline in satellite cell activity appears to be due to a loss in old animals of Notch signaling (Conboy et al., 2003)—a signaling pathway of critical importance to numerous aspects of development, including proliferation, differentiation, and tissue formation (reviewed in Artavanis-Tsakonas et al., 1999). Despite the loss of Notch signaling in old animals, forced activation of Notch restored satellite cell function, indicating that the intrinsic regenerative capacity of satellite cells in old animals remains intact (Conboy et al., 2003). Furthermore, consistent with the cross-age transplantation studies of Carlson and Faulkner (1989), showing no deficit with age in the intrinsic regenerative capacity of skeletal muscle, recent experiments very elegantly demonstrated that exposure of old animals to as yet unidentified circulating factors from young animals reestablished Notch activation, enhanced satellite cell proliferation, and improved muscle regeneration (Conboy et al., 2005).

Identification of the factors that have a such critical impact on satellite cell function represents a highly significant avenue of research. Inasmuch as recovery from contraction-induced injury involves activation of satellite cells and regeneration of muscle fibers (see chapter 6), at least for the more severe injuries, factors influencing satellite cell function provide a basis for the mechanism underlying the impairment in recovery from injury in old animals. On the basis of their observation of dramatic enhancement of recovery following severe contraction-induced injury in both adult and old mice by overexpression of heat shock protein (HSP) 70, McArdle and colleagues (2004) speculate that the critical factors involved in the impaired regeneration of muscle in an old host may also be involved in the inability of muscles of old animals to produce HSPs in response to stress.

Increasing Resistance to Contraction-Induced Injury

Despite the high susceptibility of muscles in old animals to injury and their decreased ability to recover, training with lengthening contractions decreased subsequent lengthening contraction–induced injury in both adult and old animals (Brooks et al., 2001; Clarkson and Dedrick, 1988; McBride, 2000; McBride et al., 1995; Ploutz-Snyder et al., 2001). After six weeks of once per week exposure to lengthening contractions, a contraction protocol that initially resulted in a 30% force deficit and morphological evidence of injury in ~10% of fibers in a cross section no longer injured the muscles of either adult or old mice (Brooks et al., 2001). Force deficits one day following an unaccustomed bout of exercise with lengthening contractions of the knee extensors were reduced from 26% to 8% following 12 weeks of twice per week resistance exercise that involved both lengthening and shortening contractions (Ploutz-Snyder et al., 2001). Even a single prior bout of lengthening contractions reduced subsequent lengthening contraction–induced muscle soreness and serum levels of creatine kinase in old human beings (Clarkson and Dedrick, 1988), as well as force deficit and the number of fibers demonstrating morphological evidence of damage in old rats (McBride et al., 1995).

The increased susceptibility to lengthening contraction–induced injury, coupled with impaired recovery for muscles of old animals, suggests potential risks involved in utilizing such conditioning programs in elderly people. In addition, a significant fraction of elderly people may be unable or unwilling to engage in this form of exercise. To address these concerns, Koh and his colleagues (Koh and Brooks, 2001; Koh et al., 2003) tested the hypothesis that overt damage, degeneration, and regeneration were not required to elicit protective adaptations. Muscles were exposed to a protocol of stretches without activation (passive stretches) prior to administration of a damaging lengthening contraction protocol. Passive stretches produced no evidence of overt damage to the muscle, yet exposure to conditioning with passive stretches two weeks before administration of a bout of lengthening contractions resulted in a reduction, compared with values in unconditioned muscles, in the magnitude of the force deficit and the number of damaged fibers at three days (Koh and Brooks, 2001). Damaged fibers were identified in histological sections as those fibers that showed clear morphological evidence of degeneration or regeneration, and their numbers were expressed as a percentage of the total number of fibers in a cross section that exhibited clear signs of degeneration (figure 8.6a).

For EDL muscles of old mice, passive stretch conditioning prior to exposure to lengthening contractions improved force production twofold and reduced the number of damaged fibers by half, compared with values in unconditioned muscles, following lengthening contractions. In addition to reducing force deficits and numbers of damaged fibers following lengthening contractions, conditioning with passive stretches also reduced the accumulation of neutrophils and macrophages in the injured muscles (figure 8.6b). Conditioning with passive stretches provides an exciting alternative to lengthening contractions as a safe and effective method of protecting elderly people from injury (Koh et al., 2003).

While muscles of old animals can be conditioned for increased resistance to injury, the protective adaptations associated with exercise training may be somewhat impaired in old animals. For example, in adult rats, a protocol of 24 lengthening contractions of the tibialis anterior muscles provided complete protection against the force deficit induced by a subsequent identical contraction protocol two weeks later. However, when muscles of old rats were exposed to the same protocol of 24 lengthening contractions, only partial protection was observed following the second bout (McBride et al., 1995). In addition, dorsiflexor muscles of adult mice exposed to weekly bouts of lengthening contractions showed progressive

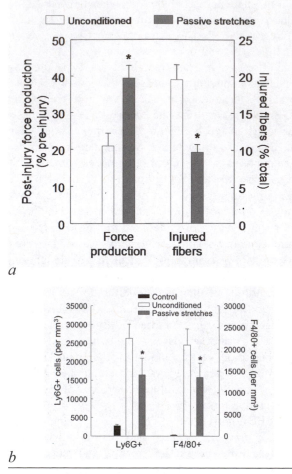

Figure 8.6 Values are presented as means ± SEM for *(a)* force and number of injured fibers and *(b)* concentrations of neutrophils (Ly6G+ cells) and macrophages (F4/80+ cells) in muscles of old mice. Data are from unexercised control muscles (black bars) and from muscles three days after 225 lengthening contractions either without (unconditioned, open bars) or with (conditioned, gray bars) previous exposure to passive stretches. All data for muscles exposed to lengthening contractions are significantly different from control values. Asterisks indicate significant (*p* < 0.05) differences from unconditioned muscles.

JOURNALS OF GERONTOLOGY. SERIES A, BIOLOGICAL SCIENCES AND MEDICAL SCIENCES by TIMOTHY J. KOH, JENNIFER M. PETERSON, FRANCIS X. PIZZA, SUSAN V. BROOKS. Copyright 2003 by GERONTOLOGICAL SOC OF AMERICA. Reproduced with permission of GERONTOLOGICAL SOC OF AMERICA.

reductions in the force deficit until week 4, when the force deficit was eliminated, whereas in old mice a similar level of protection was not demonstrated until week 5 (Brooks et al., 2001). In contrast, exposure to a single conditioning bout of passive stretches resulted in

similar reductions for muscles of both adult and old mice in force deficit and numbers of injured fibers following a severe lengthening contraction protocol administered two weeks later (Koh et al., 2003).

The similarity in the responses of the muscles of adult and old mice in the study by Koh and colleagues (2003) may represent protective adaptations that were truly as effective in old as in young animals. Alternatively, any impairment in adaptation in the old animals may have been masked by the severity of the 225 lengthening contraction protocol used to test for the effectiveness of the conditioning response. The impaired regeneration of muscles in old compared with adult animals suggests that differences in the adaptive responses in adult and old animals may arise only under circumstances in which the muscles must invoke the regeneration process in response to the conditioning protocol (Brooks and Faulkner, 1990; Carlson and Faulkner, 1989; Conboy et al., 2003, 2005; McArdle et al., 2004).

Summary

Muscles can be injured by their own contractions. Activities that require muscles to be stretched during contractions are most likely to injure muscles. The initial injury, which is mechanical, is followed by more severe inflammatory and free radical damage that results in delayed-onset muscle soreness. In young, active animals, muscle fibers are relatively resistant to injury and recover effectively following injuries, whereas fibers in muscles of old animals are more easily injured and recovery may be incomplete. Despite a high susceptibility to injury and an impaired ability to recover, exercise conditioning can protect muscles from injury even in old animals. Thus, the maintenance of "conditioned" fibers in the muscles of old people will decrease the likelihood of injury, which, along with skeletal muscle atrophy, weakness, and fatigability, contributes to physical frailty. Physical frailty impairs performance of the activities of daily living, increases the incidence of falls, and negatively affects the quality of life of old people. Despite the lack of definitive answers to many questions regarding the development of frailty, the longer people can be motivated to maintain a physically active lifestyle, including both endurance and strength training, the higher will be the quality of their life in old age.

ACKNOWLEDGMENT

Work on this chapter was supported by a grant from the National Institute on Aging, AG-20591.

CHAPTER 9

Duchenne Muscular Dystrophy

Tommy G. Gainer, MS; Qiong Wang, MS; Christopher W. Ward, PhD; and Robert W. Grange, PhD

Muscular dystrophy comprises a group of approximately 40 inherited heterogeneous disorders that result in progressive muscle weakness and muscle wasting (Durbeej and Campbell, 2002; Straub and Campbell, 1997). At present, many, but not all, of the genes associated with specific muscular dystrophies have been identified (Durbeej and Campbell, 2002). Duchenne muscular dystrophy (DMD) is the most severe of these diseases and is fatal, affecting 1 in 3,500 boys (Emery, 1993). Although mutations in the gene encoding dystrophin have been recognized since 1987 as the genetic cause of the disease (Hoffman et al., 1987a), the specific pathophysiological mechanisms have not yet been clearly defined. The basic characteristic of DMD is progressive muscle injury that is ineffectively repaired. However, neither the disease mechanisms nor the disrupted repair processes that result in progressive muscle deterioration have yet been clearly elucidated. This chapter focuses primarily on the current hypotheses about mechanisms of muscle degradation in DMD, with very brief comment on repair processes.

Overview

Duchenne muscular dystrophy is a fatal male muscle wasting disease characterized by progressive loss of contractile function due to the absence of the membrane-associated cytoskeletal protein, dystrophin.

Duchenne muscular dystrophy was first described in 1852 by English physician Edward Meryon (Tyler, 2003) but was later named after French neurologist Guillaume Duchenne, who published a series of illustrated articles in 1868 that described key features of the disease (Tyler, 2003). Duchenne examined DMD muscle biopsies and established that the key abnormalities of the disease were hyperplasia of fibrous connective tissue and destruction of the muscle cellular architecture (Byrne et al., 2003).

Genetic and Biochemical Characteristics

The genetic defects associated with DMD include deletions or non-sense mutations in the dystrophin gene (Petrof, 1998). The skeletal muscle isoform of dystrophin is a rodlike cytoskeletal protein localized to the inner surface of the skeletal muscle plasma membrane, the inner layer of the sarcolemma (Sanes, 2004). It forms part of the dystrophin-glycoprotein complex (DGC) linking the cytoskeleton and the extracellular matrix (Straub and Campbell, 1997). The DGC is also called the dystrophin-associated protein complex (DAP), as not all proteins in the complex are glycosylated (Ozawa, 2004). This chapter uses the term dystrophin-glycoprotein complex (DGC) to include all proteins in the complex. The specific function or functions of dystrophin and the DGC are not completely known, although it is believed that they could serve both mechanical and signaling roles (Durbeej and Campbell, 2002; Roberts, 2001). The absence of dystrophin could render the membrane susceptible to mechanical injury, leaky to Ca^{2+}, or susceptible to disrupted signaling. These effects alone or in combination could dramatically disrupt muscle function.

Dystrophin-deficient muscle is characterized by increased permeability to endogenous macromolecules flowing out of and into the muscle cell. A classical marker for DMD is an elevated serum muscle creatine kinase concentration (Blake et al., 2002). In addition, muscle fibers stain positively for endogenous extracellular proteins such as albumin and immunoglobulins such as IgG and IgM (Blake et al., 2002). These clinical features suggest that the membranes of dystrophic muscle cells are fragile and leaky, and that this contributes to the progressive force loss and ultimate death of the muscle cell.

In DMD, the protein dystrophin and the associated proteins of the DGC are absent, and this leads to severe pathophysiology of skeletal and cardiac muscle. The two main features of DMD pathophysiology are (1) progressive degeneration of muscle tissue with replacement by noncontractile fat and connective tissues (Blake et al., 2002; Cozzi et al., 2001) and (2) progressive muscle weakness (Straub and Campbell, 1997). The disease process is characterized by progressive rounds of myofiber necrosis followed by rounds of muscle fiber regeneration. Fiber regeneration is identified histologically based on the presence of centralized nuclei (Schmalbruch, 1984). Grouped degenerating and necrotic fibers are characteristic of DMD muscle biopsies even before muscle weakness is clinically observed (Blake et al., 2002), with evidence of myofiber degeneration in boys at approximately age 3 years. Ultimately, the combination of progressive fibrosis and muscle fiber loss results in muscle wasting, weakness, and failure. Much has yet to be learned about both the pathophysiological mechanisms of DMD, and about why repair processes are ineffective.

Clinical Features

Duchenne muscular dystrophy is characterized by childhood onset, a prevalence in boys, occurrence in children within the same family, progressive muscle weakness yet a gradual increase in size of many affected muscles (pseudohypertrophy), and abundant fibrosis and adipose tissue at later stages of the disease (Emery, 1993). The onset of DMD is evident at approximately age 3 to 5 years. One-third of DMD patients show intellectual impairment, and both speech and motor development are delayed. For example, walking is delayed approximately five months compared to that in a normal child (Emery, 1989). The muscle involvement is bilateral and symmetrical and affects the lower limbs first. A common clinical sign is pseudohypertrophy, in which muscles such as the gastrocnemii, or calf muscles, swell and appear larger than normal. The proximal muscles are more affected than the distal muscles, with contractures evident at the elbows, knees, and hips. Progressive muscle degeneration driven by presently undefined mechanisms compromises the child's ability to support himself. With

disease progression, the child develops thoracolumbar kyphosis (a prominent outward curve) and scoliosis (a lateral curve) of the spine (Karol, 2007). Additional features are a waddling gait due to weakness in hip abductors, lumbar lordosis due to weak gluteal muscles, and the need for Gower's maneuver due to weakened knee and hip extensors. Gower's maneuver involves using the hands to support the body's rise to a standing position from the floor or a chair. Eventually the child loses all ability to support himself. By age 12, 95% of patients need to use a wheelchair. Cardiopulmonary function steadily deteriorates. By age 20, 90% of DMD patients die, most from cardiac failure and respiratory insufficiency (Emery, 1993).

Dystrophin Structure, Glycoprotein Complex, and Gene

Dystrophin is a large protein that links the muscle fiber cytoskeleton with laminin in the extracellular matrix via the dystrophin-glycoprotein complex. The dystrophin gene is very long and prone to mutations that result in the premature truncation of the dystrophin transcript. When this occurs, functional dystrophin protein is not made, and DMD is the outcome.

Protein and Glycoprotein Complex

Dystrophin protein is expressed in both invertebrates and vertebrates. Vertebrate dystrophin is located at the cytoplasmic membrane of skeletal, cardiac, and smooth muscle; at the skeletal myotendinous junction; and at synapses in the central nervous system (Roberts, 2001). The 427 kDa cytoskeletal protein dystrophin is composed of 3,685 amino acids (Kapsa et al., 2003). It is a member of the β-spectrin/α-actinin protein family, which is characterized by an amino-terminal actin-binding domain. Dystrophin has four separate regions: (1) an amino-terminal actin-binding domain, (2) a central rod domain, (3) a cysteine-rich domain, and (4) a carboxyl-terminal domain (Blake et al., 2002). The cysteine-rich domain may be the most critical because it interacts with the intracellular tail of β-dystroglycan, which links dystrophin to the sarcolemma membrane and preserves the entire DGC (Roberts, 2001; Straub et al., 1997); the C-terminal binds additional proteins of the DGC (figure 9.1). Dystrophin represents about 5% of surface-associated sarcolemmal cytoskeletal proteins (Hoffman et al., 1987a; Ohlendieck and Campbell, 1991).

Dystrophin is considered a key structural element in the muscle fiber in that it links cytoskeletal actin to the DGC, which in turn forms a connection through the sarcolemma membrane to the basal lamina (figure 9.2). The DGC along with additional proteins forms riblike lattices on the cytoplasmic face of the sarcolemma, known as costa-

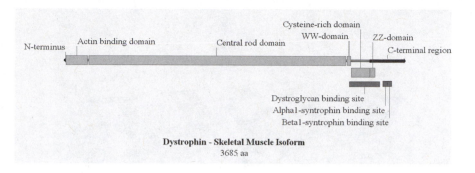

Figure 9.1   The 427 kDa skeletal muscle isoform of dystrophin. The protein has four distinct regions: (1) N-terminal domain, (2) central rod domain, (3) cysteine-rich region, and (4) C-terminal domain. The WW and ZZ domains bind β-dystroglycan (Ozawa, 2004). This figure was created in Vector NTI using the amino acid sequence obtained from NCBI (Accession No. NM_004006, NP_003997).

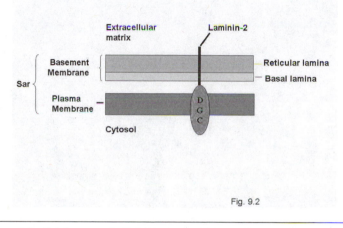

Figure 9.2   Schematic structure of the link between the DGC in the sarcolemma and laminin-2 in the basement membrane (DGC: dystrophin-glycoprotein complex).

Data from Sanes, 2004.

meres, that facilitate force transmission between active and nonactive fibers (figure 9.3) (Campbell, 1995; Rybakova et al., 2000). This complex of proteins also likely acts as a signaling conduit between the outside and inside of the fiber (Lapidos et al., 2004). Dystrophin is also found in abundance in the myotendinous junction and is thought to facilitate the transmission of forces from the muscle fibers to the tendon (Tidball and Law, 1991). The complete absence of expressed dystrophin in DMD leads to the loss of the DGC (Ohlendieck et al., 1993) and thus a disrupted costameric lattice (Williams and Bloch, 1999). Loss of the key structural DGC elements results in the characteristic progressive muscle degradation seen in DMD, the mechanisms of which are still not clearly defined.

The DGC is made up of several subcomplexes, including the dystroglycan complex and the sarcoglycan: sarcospan complex, and the peripheral proteins of the cytoplasmic dystrophin-containing domain (figure 9.4). The amino or N-terminal of dystrophin binds cytoskel-

etal actin, and its carboxy- or C-terminal region binds the dystroglycan complex. The dystroglycan complex is composed of α- and β-dystroglycan. β-Dystroglycan, an integral membrane protein, interacts with dystrophin in the cytosol and with α-dystroglycan, which in turn binds to laminin-2 in the basal lamina. The sarcoglycan:sarcospan complex, composed of α, β, γ, and δ sarcoglycans and sarcospan, is important for stabilizing the dystroglycan complex in the sarcolemma (Blake et al., 2002; Roberts, 2001; Straub and Campbell, 1997). In muscle, the cytoplasmic dystrophin-containing domain contains α-dystrobrevin that binds directly to dystrophin. β-Dystrobrevin is found in the DGC of nonmuscle tissues. Another component of this domain is the syntrophins, adapter proteins that link membrane-associated proteins, dystrophin, and dystrobrevin to the DGC (Blake et al., 2002; Roberts, 2001). Associated with the syntrophins is neuronal nitric oxide synthase (nNOS; Stamler and Meissner, 2001). Several other proteins with less well defined functions are

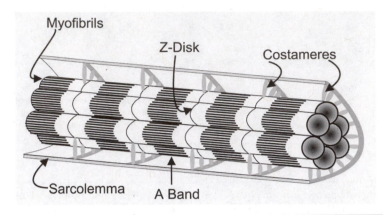

Figure 9.3   Costameric organization in skeletal muscle.

Reprinted, by permission, from J.M. Ervasti, 2003, "Costameres: The Achilles' heel of Herculean muscle," *J Biol. Chem.* 278(16): 13591-13594.

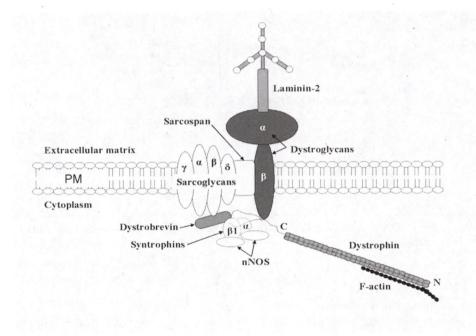

Figure 9.4   The dystrophin-glycoprotein complex. The DGC is composed of distinct integral membrane protein subcomplexes that span the sarcolemma and link the cytoskeleton to the extracellular matrix. PM: plasma membrane.

Reprinted from *Current Opinions in Genetics Development*, Vol. 12, M. Durbeej and K.P. Campbell, Muscular dystrophies involving the dystrophin-glycoprotein complex: An overview of current mouse models, pp. 349-361, Copyright 2002, with permission from Elsevier.

associated with the complex, including syncoilin, biglycan, and filamin-2 (Roberts, 2001). Absence of various DGC components leads to other muscular dystrophies (e.g., limb-girdle) (Straub and Campbell, 1997).

The Dystrophin Gene

The dystrophin gene is located on the short arm of the X chromosome at locus p21 (Koenig et al., 1987). The dystrophin protein was identified in 1987 (Hoffman et al., 1987a). It is one of the largest genes (2.4 Mb) characterized to date, with 79 coding exons and seven tissue-specific promoters for seven protein isoforms. Three promoters produce full-length messenger RNA (mRNA) transcripts (~14 kb) whose 427 kDa proteins differ only in their amino-terminal sequences. These isoforms are brain (B); muscle (M), expressed in both skeletal and cardiac muscle; and Purkinje (P), expressed in both cerebellar Purkinje cells and skeletal muscle (Blake et al., 2002). Besides a full-length gene transcription product, the dystrophin gene also has at least four internal promoters, which give rise to shorter dystrophin transcripts that encode truncated C-terminal isoforms. These isoforms are primarily expressed in nonmuscle tissues, including the brain, central and peripheral nervous system tissues, and lung, and are thought to provide the necessary binding sites for a number of dystrophin-associated proteins (Schofield et al., 1994). However, the molecular and cellular function of these isoforms has not yet been elucidated.

Mutations in the dystrophin gene can be classified into two main categories according to the size of the deletion: large and small. The vast majority of large deletions are clustered around two mutation "hot spots," which result in the loss of part of the rod domain or the actin-binding domain of the protein, or both (Kunkel et al., 1989). The rod domain can have large in-frame deletions without serious clinical consequences, which has facilitated the design of dystrophin "mini-genes" for gene therapy approaches. One-third of DMD cases are caused by very small deletions and point mutations, most of which generate premature stop codons (Lenk et al., 1993; Roberts et al., 1994). These small mutations are distributed evenly throughout the gene (Gardner et al., 1995), but because only a few protein products have been detected so far, it is likely that the corresponding transcripts or the truncated proteins are unstable (Hoffman et al., 1987b).

Scientific Investigation

The study of DMD is very robust as scientists attempt to understand disease mechanisms and develop effective treatments. There are several animal models that are used to study DMD, including the mdx mouse and the golden retriever dystrophic dog. Because there is a protein similar to dystrophin known as utrophin that is thought to compensate when dystrophin is absent, a mouse model deficient in both proteins has also been developed (mdx:utrophin-deficient).

Two Basic Approaches

There are two general scientific approaches to the investigation of treatment strategies for a disease such as DMD. One is to explore and identify potential sources of treatment in the near term that will result in improved quality of life and longevity for those afflicted. Use of such treatments, although they do not provide complete recovery, does offer some relief and much-needed hope that a cure will soon be found. The second approach is to discover the mechanism or mechanisms that represent the disease processes. Understanding these mechanisms should lead to more directed therapeutics. These approaches necessarily occur in parallel.

Completely effective treatments for DMD patients are not yet available, but several encouraging possibilities are being tested, including drugs and gene therapy (see review by Kapsa et al., 2003). While some drugs have been approved for use in human DMD patients, gene therapy has been limited to animal models.

Experimental Models for Duchenne Muscular Dystrophy

As with investigations of many human diseases, suitable animal models are needed to explore and define the disease mechanisms. Fortunately, humans are not the only species that exhibit muscular dystrophy. Mutations in the dystrophin gene also occur in mice, chicken, hamster, cat, and dog and lead to mild to severely debilitating phenotypes. Several dog breeds, especially the golden retriever, are prone to developing canine X-linked muscular dystrophy (CXMD), which is genetically homologous to DMD. As in humans, myofiber necrosis and regeneration begin at an early age, followed by extensive connective tissue proliferation and a premature death from cardiac and respiratory failure (Cozzi et al., 2001). Although the golden retriever muscular dystrophy (GRMD) dog model is the animal model that most closely mimics human DMD, the cost of maintaining a colony has limited its use. The most common experimental models involve dystrophic mice (e.g., mdx).

The mdx mouse has a naturally occurring single point mutation along the 5' end of the dystrophin gene, in exon 23, which causes a premature transcriptional stop and results in the expression of a truncated protein product in skeletal muscle (Sicinski et al., 1989). The mdx mouse exhibits several DMD characteristics, including cardiomyopathy (Khairallah et al., 2007), a prominent characteristic of human DMD (Engel and Ozawa, 2004). The muscles of the mdx mouse undergo degeneration and regeneration cycles similar to those in DMD; but these begin at about age 3 to 4 weeks and end at about age 10 weeks (Rezvani et al., 1995), whereas muscle degeneration in DMD is progressive (Mendell et al., 1995). Muscle stability in mdx is attributed in part to upregulation of utrophin, a protein homolog of dystrophin (Khurana et al., 1991; Wilson et al., 1994). Except for the mdx diaphragm (Stedman et al., 1991), the limb muscles appear to stabilize sufficiently to cope with normal movement, and the mice have a normal life span (Torres and Duchen, 1987; Stedman et al., 1991). The pathophysiology and clinical features of dystrophin deficiency in the mdx are not the same and are not as severe as in humans.

Utrophin is expressed in both human and mouse skeletal muscles (Love et al., 1993). It is completely homologous to dystrophin (Tinsley et al., 1992) and, like dystrophin, is transcribed from multiple promoters (Blake et al., 1995). In adult skeletal muscle, utrophin is exclusively localized to the neuromuscular junction (Ohlendieck et al., 1991) and is believed to play a role in the formation and maintenance of acetylcholine receptors and synaptogenesis (Campanelli et al., 1994; Grady et al., 1997a).

Utrophin expression is increased in dystrophin-deficient humans and mdx mice (Khurana et al., 1991; Matsumura et al., 1992). As expression increases, the pathophysiolology becomes less severe, especially in mdx transgenic mice overexpressing utrophin (Matsumura et al., 1992; Tinsley et al., 1998). Tinsley and colleagues (1998) demonstrated that a transgenic mdx mouse line overexpressing full-length utrophin, localized at the sarcolemma, prevented the dystrophic pathophysiology from occurring in hindlimb and diaphragm muscles. The extent of myofiber necrosis depended on the amount of utrophin present in the muscle. Therefore, in the absence of dystrophin, the localization of utrophin to the sarcolemma compensates and rescues the animal from DMD (Tinsley et al., 1998). Unfortunately, although utrophin is expressed in skeletal muscle of human DMD patients, a similar protective effect is not observed (Grady et al, 1997b; Deconinck et al., 1997).

Another DMD model is the mdx:utrophin$^{-/-}$ (mdx:utrn$^{-/-}$) mouse, in which both dystrophin and utrophin are absent. The absence of utrophin alone (i.e., in utrophin-deficient mice) results in only modest changes at the neuromuscular junction where it is primarily localized, but physiologic function is very similar to that in control muscles (Grady et al., 1997a). Because utrophin could compensate for the absence of dystrophin, a mouse was developed that lacks both proteins. The mdx:utrn$^{-/-}$ mice exhibit nearly all symptoms of DMD, including reduced growth, diminished mobility, limb contractures, muscle weakness, spinal deformities (e.g., kyphosis), cardiomyopathy, myofibrosis, and premature death (Grady et al., 1997b; Deconinck et al., 1997). Additionally, skeletal muscle mass and stress-generating capacity are significantly reduced compared to those in controls (Grady et al., 1997b; Deconinck et al., 1998). Thus, the mdx:utrn$^{-/-}$ model appears to be more appropriate than other models for investigating the fundamental mechanism(s) of onset of dystrophic pathology in skeletal muscle.

Issues for Investigation

Although the genetic basis of DMD is known, the mechanistic basis for the onset of progressive muscle wasting is not. This has led to numerous hypotheses, many of which can be generally classified as mechanical hypotheses or signaling hypotheses.

Mechanisms of Muscular Degeneration

As already noted, the absence of dystrophin and the DGC is thought to render the membrane more readily damaged by mechanical stress. Thus, one prominent hypothesis is that absence of dystrophin leads to a more injury-

susceptible sarcolemma (Petrof et al., 1993). Alternatively, or in addition to this suggestion, disrupted cell signaling may lead to the onset of DMD (Crosbie, 2001). Among the many signaling pathways that may be disrupted are those that are regulated by calcium (e.g., Ca^{2+}-dependent proteolytic pathways). Thus, regulation of Ca^{2+} homeostasis appears to be critically important. In addition, a clear inflammatory profile is associated with DMD. At present, it is not evident which of these many possibilities is a key initiator of DMD, or how they may interact with disease progression. In the following discussion we briefly describe the mechanical hypothesis and provide a more detailed description of the various signaling pathways that may be involved in DMD pathophysiology.

The mechanical hypothesis suggests that there is a structural deficiency in the skeletal muscle sarcolemma and at the myotendinous junction when dystrophin and the DGC are absent, and that this contributes to the degeneration of the muscle tissue and decreased force output. Convincing data reveal that intact skeletal muscles of older dystrophic animals (age 6-10 weeks and older) are susceptible to contraction-induced injury (Petrof et al., 1993); however, the skeletal muscles of maturing dystrophic mice do not appear to be as susceptible (age 9-12 days; Grange et al., 2002). In addition, stiffness is similar for dystrophic compared to wild-type skeletal muscle during early maturation (age 14-35 days; Wolff et al., 2006). Stiffness is the force produced for a given muscle length change. Thus, the absence of dystrophin and the DGC in relation to dystrophic membrane stability may change with maturation (i.e., muscle may become more susceptible to injury with age).

There is also evidence to suggest that the interaction between actin and myosin is compromised and that this accounts for some of the force loss characteristic of dystrophic skeletal muscles both during early maturation and as the mice age (Lowe et al., 2006). The effect of the absence of dystrophin in the fiber–tendon linkage is difficult to assess because of the difficulty in obtaining mechanical measures for this isolated component of force transmission in either the presence or absence of dystrophin. There is a pressing need to understand the mechanical (and signaling) characteristics of dystrophic skeletal muscle at all ages, but particularly during early maturation. If dystrophic skeletal muscles could tolerate and positively adapt to physical activity at an early age, then properly prescribed exercise, which is known to increase strength and endurance in normal skeletal muscle, might become a viable treatment. However, as recently recommended by Grange and Call (2007), the details of exercise prescription should be determined first in dystrophic mice and then in dystrophic dogs as a bridge to exercise prescription in boys with DMD.

In addition to playing a role in membrane stability, dystrophin and proteins of the DGC may function to support signaling processes within the myoplasm, across the sarcolemma, and into the extracellular matrix. Therefore, in their absence, any signaling functions supported by the DGC may be disrupted. However, it is not clear how the DGC proteins interact to coordinate cellular processes or whether a single or a coordinated series of events leads to the development of muscle degradation (Ervasti, 2007).

Recent studies on dystrophin-deficient muscle have focused on some of the potential signaling pathways that may be associated with the DGC. These include but are not limited to the roles of

- integrins in the extracellular matrix, particularly the $\alpha7\beta1$ integrin;
- nitric oxide;
- calcium signaling;
- activation of the calcium-dependent calpain proteases; and
- the immune system.

Integrins

Integrins are a family of transmembrane cell surface receptors that function by interacting with extracellular matrix proteins to play important roles in cell migration, cell shape formation, and cell–cell interactions. Integrins are heterodimers formed from α and β chains. Integrins transmit signals from outside the cell to the cytosol and cytoskeleton, or from within the cell to the extracellular matrix, and contribute to the modulation and activation of extracellular matrix ligand binding (Burkin and Kaufman, 1999).

In skeletal muscle, integrin $\alpha7\beta1$ is the predominant integrin that interacts with the DGC by binding laminin. It serves several roles during development. It is expressed on cells going into early limb budding, appears after differentiation during primary fiber formation, directs cells to laminin-rich sites of secondary fibers surrounding primary fibers, and indirectly participates in myoblast fusion. In adult fibers, $\alpha7\beta1$ integrin is enriched at myotendinous and neuromuscular junctions and localizes peripherally around the myofiber. Overall, it is believed to promote the terminal and peripheral muscle fiber cohesion that is important to neuromuscular connectivity, muscle integrity, and force generation (Burkin and Kaufman, 1999).

In patients with DMD, as well as in mdx:utrn$^{-/-}$ mice, $\alpha7\beta1$ integrin is increased due to increased RNA expression of the $\alpha7$ chain. The protein is found in hindlimb and diaphragm muscles and is normally confined to the junctional sites but expands out to extrajunctional regions in DMD. Enhancing the expression of the $\alpha7\beta1$ integrin reduced the symptoms associated with dystrophin deficiency and restored muscle viability in mdx:utrn$^{-/-}$ $\alpha7\beta1$

transgenic mice (Burkin et al., 2001). These findings suggest that integrins can compensate for the absence of dystrophin and the DGC.

Nitric Oxide Signaling

Neuronal nitric oxide synthase is a heme-containing oxidoreductase that catalyzes the conversion of L-arginine to L-citrulline and generates the free radical nitric oxide (NO; Bredt and Snyder, 1994). In skeletal muscle fibers, nNOS is found in close association with syntrophins in the DGC, and evidence suggests a decrease or mislocalization of nNOS in dystrophic models (Chang et al., 1996). In striated muscle, recent work has demonstrated an important role for NO in the normal physiological modulation of contractile function, sarcoplasmic reticulum Ca^{2+} release, and glucose metabolism (Martinez-Moreno et al., 2005; Hare, 2003; Eisner et al., 2003; Anzai et al., 2000; Pessah and Feng, 2000; Stoyanovsky et al., 1997). In addition, NO promotes vasodilation by stimulating guanylate cyclase and increasing cyclic guanosine monophosphate (cGMP) production in smooth muscle cells.

In dystrophin-deficient skeletal muscle, nNOS is reduced (Chang et al., 1996). It is believed that the reduction in nNOS content decreases NO production (Rando, 2002). During normal muscle activity, NO likely contributes to functional hyperemia to ensure adequate blood flow to the working muscle (Lau et al., 1998; Grange et al., 2001; Crosbie, 2001). The vasodilatory role of NO may be compromised in mdx mice, as well as in young DMD patients (Thomas et al., 1998; Sander et al., 2000). Furthermore, overexpression of an nNOS transgene in mdx mice diminished the dystrophic pathophysiology in soleus muscles, likely by providing NO as an anti-inflammatory molecule (Wehling et al., 2001). Thus, the positive effects of NO may be compromised in dystrophin-deficient muscle (Rando, 2002).

Calcium Signaling

Of the several hypotheses put forward to explain the onset and progression of DMD, disrupted calcium homeostasis is the most well studied. Calcium is a primary regulatory ion for skeletal muscle contraction and also represents an important signal for many cellular processes. For example, calcium homeostasis is critical to muscle contraction as well as calcium-dependent proteolysis in muscle fibers; we discuss the latter in a separate section (Blake et al., 2002).

At rest, the normal myofiber cytosolic free calcium concentration (hereafter, calcium concentration implies the free concentration, and the terms "cytosolic" and "intracellular" are used interchangeably) is maintained at ~50 nM (Berchtold et al., 2000). Secondary to membrane activation by action potential depolarization, the cytosolic calcium concentration is transiently increased ~100-fold, which results in activation of the contractile filaments and contraction. Upon cessation of the action potential the membrane is repolarized, and calcium is quickly resequestered by the sarcoplasmic reticulum to restore a low intracellular calcium concentration.

Several observations suggest that calcium homeostasis is perturbed in dystrophin-deficient muscles. Using fluorescently labeled calcium chelators, spectroscopic studies demonstrated that the total calcium content in both DMD patients and DMD fetuses was elevated (Bertorini et al., 1982, 1984). A marked increase in the intracellular calcium concentration was also found in cultured human dystrophic and mdx muscle fibers (Turner et al., 1988) and mdx myotubes (Bakker et al., 1993).

Additional studies revealed that increased intracellular calcium is likely to appear in specific regions of muscle fibers. In mdx myofibers, local calcium increases occurred more often peripherally at the sarcolemma rather than deep inside fibers when challenged by increased external calcium (Turner et al., 1991). A threefold greater calcium level was reported for the subsarcolemmal region of mdx compared to wild-type mouse fibers at ages 3 to 5 weeks (Mallouk et al., 2000). Calcium level at the sarcoplasmic reticulum (SR) is almost 50% greater in mdx than in control myotubes (Robert et al., 2001). Taken together, these data indicate increased intracellular calcium in dystrophic muscle fibers and myotubes, potentially localized near the sarcolemma and SR.

Based on the observations of Ca^{2+} disruption in dystrophic muscle, the membrane hypothesis of DMD (Hutter, 1992) was reconceptualized by many to include the possibility that a primary onset of the dystrophic process is due to mechanical stress that induces local microruptures and resealing of the sarcolemma (Turner et al., 1988). Evidence of this phenomenon is manifest primarily as a "leak" of large endogenous muscle-specific enzymes out of the myofibers as well as the presence of exogenous molecules (e.g., Evans blue dye) in viable myofibers. Furthermore, it has been proposed that the microdamage and repair process results in transient increases in subsarcolemmal Ca^{2+} that promote the insertion of Ca^{2+} leak channels in the dystrophic sarcolemma. The chronic presence of these channels results in further localized increases in Ca^{2+}, which then perpetuate the dystrophic process of protease activation and dysregulation of mechanosensitive Ca^{2+} channels (Franco and Lansman, 1990; Hopf et al., 1996a, 1996b; Turner et al., 1993) in mdx myofibers and myotubes. As the dystrophic process progresses temporally, the alterations in Ca^{2+} signaling are further exacerbated by abnormal Ca^{2+} cycling through the storage organelles, the mitochondria, and the SR.

This belief that damage is an initiating factor has recently been challenged in favor of primary alterations in sarcolemmal Ca^{2+} signaling, especially during mechanical stress, as a precipitating factor in the dystrophic process. These reports have been recently reviewed (see Allen, 2004; Whitehead et al., 2006).

The role of calcium-dependent processes as *initiating events* of muscle wasting in DMD has been challenged, however, by observations that uptake of [45]calcium in 14-day-old mdx muscle was not different from that in age-matched control muscle but was increased significantly over control values in 40-day-old mice (McArdle et al., 1994). Thus, it was concluded that increased intracellular calcium in mdx muscle was secondary to the onset of the degenerative process.

In contrast, on the basis of gene expression profiling of muscle biopsy samples from DMD patients 6 to 9 years old, Chen and colleagues (2000) hypothesized that disruptions in calcium handling in DMD fibers lead to metabolic crisis and altered developmental programming that prevents functional fiber regeneration. These authors speculated that an increased cytosolic calcium concentration could result from leaky dystrophin-deficient membranes.

Despite the argument about initiating factors in the dystrophic process, it is well accepted that a common point of pathophysiologic progression of the process is calcium-dependent proteolysis in dystrophic muscle fibers. Increased calcium influx into muscle fibers and the reduced calcium-handling ability in dystrophic muscle fiber could lead to broad calcium-dependent proteolysis and therefore facilitate dystrophic muscle fiber necrosis.

Calcium-Dependent Proteolysis

The three major proteolytic pathways in muscle are the lysosomal cathepsin, the calcium-dependent calpain, and the adenosine triphosphate (ATP)-dependent ubiquitin proteasome pathways (Cooney et al., 1997). Calcium-dependent proteolysis appears to be largely responsible for the elevated calcium-sensitive proteolysis in mdx myotubes, as evidenced in a study showing that pretreatment of myotubes with ammonia (an inhibitor of lysosomal proteolysis) did not prevent protein degradation (Alderton and Steinhardt, 2000a). In summary, an increased resting intracellular calcium level is evident in dystrophic compared to control myotubes and fibers, and it appears to arise mainly from the extracellular pool via either transient membrane lesions or abnormally opened calcium channels. The increased calcium could lead to the local activation of proteases, which may modify calcium leak channels to cause further calcium ingress (Alderton and Steinhardt, 2000a, 2000b). A deteriorating cycle of increased proteolysis is formed, and this eventually leads to muscle fiber necrosis.

Calpains are calcium-dependent proteases that likely contribute to the pathophysiology of dystrophic muscle. Membrane disruptions in mdx fibers caused by long-term contractions resulted in an influx of intracellular free calcium into the sarcoplasm (Alderton and Steinhardt,

2000a). This influx could lead to the abnormal activation of calcium-specific transient receptor potential channels (TRPC) by calcium-dependent proteases such as calpains (Vandebrouk et al., 2002). Modification of the channels near wound sites could then cause further calcium ingress into the dystrophic cell (Alderton and Steinhardt, 2000a, 2000b), setting up a debilitating degenerative cycle.

Recent studies have shown that calpain concentration or levels of auto-proteolytic modification (or both) vary over the course of the disease in mdx mice, coincident with the particular stage of the disease process (Spencer et al., 1995). For example, total calpain content is increased in peak-necrotic mdx mice at age 4 and 14 weeks, but not in prenecrotic mice at age 2 weeks. In addition, the auto-proteolytic activation of μ-calpain was highest during peak necrosis and returned to levels near control values in regenerated muscle. Calcium-dependent calpain proteolysis could yield modification of TRPC channels, which leads to further calcium entry and exacerbates calcium-dependent proteolysis and degradation of the cytoskeleton, particularly near the sarcolemma. These outcomes could therefore lead to ultimate degradation and necrosis of the muscle fibers.

In addition to their role in direct cell structure degradation, calpains could also play a regulatory role in cellular functions because they cleave proteins between, rather than within, functional domains and therefore leave the functional domains of their substrates intact (Berchtold et al., 2000; Carafoli and Molinari, 1998). Calpains do not just respond to signaling cues but, once activated, may help regulate signaling. Potential regulation could be achieved by the cleavage of specific transcription factors, cytosolic enzymes, signaling molecules, or some combination of these. These effects could influence gene expression, cell function, and signaling pathways. C-Fos and c-Jun transcription factors are highly sensitive to calpains in vitro (Pariat et al., 2000). Calpains are known to downregulate protein kinase C (PKC) in a variety of cell types (Wang and Yuen, 1997). Many of the PKC target substrates are components of signal transduction pathways and include proteins that regulate ion channels, growth factor receptors, and structural and regulatory proteins of the cytoskeleton (Wang and Yuen, 1997).

Calpains may also be involved in the signaling pathway of apoptosis, since cleavage of p53 (a tumor-suppressor protein) leads to a small number of defined fragments, a step considered necessary to trigger apoptosis (Carafoli and Molinari, 1998). In addition to the potential role of calpain in DMD, calpain involvement is directly exemplified in limb-girdle muscular dystrophy type 2A (LMGD2A). In LMGD2A the structural sarcolemmal proteins are normally expressed, but there is a deficiency in the muscle-specific calpain p94 (Richard et al., 1995).

On the basis of the findings just reviewed, it is reasonable to suggest that calpains could also be involved in the pathophysiology of DMD through disruption of signaling pathways, which eventually leads to the degradation and necrosis of the muscle fiber. However, the mechanisms and time course of these pathways are not known at present.

Inflammatory Response in Duchenne Muscular Dystrophy

An inflammatory signature is clearly present in older DMD and mdx muscles (Spencer and Tidball, 2001; Chen et al., 2000; Porter et al., 2002, 2004). Dystrophin-deficient muscle is infiltrated by a variety of immune cells including cytotoxic T-lymphocytes, macrophages, and dendritic cells (Spencer and Tidball, 2001; Chen et al., 2000; Porter et al., 2002, 2004). The advent of the DNA microarray has enabled examination of the global posttranscriptional expression of mRNA to help identify molecules and physiological events in DMD (Chen et al., 2000; Tkatchenko et al., 2000, 2001; Porter et al., 2002, 2004). Such research has also identified inflammation as a predominant profile of the disease (Chen et al., 2000; Porter et al., 2002, 2004). Examination of global gene expression in human and mdx muscle has provided some clues to mechanisms potentially promoting inflammation. Again, however, it remains unclear whether or not inflammation initiates the pathophysiology of DMD.

Spencer and colleagues (1997) showed that cytotoxic T-lymphocytes were present in mdx muscle during the early stages of pathophysiological onset and that they contributed to myofiber necrosis via apoptosis, likely mediated by perforin. Chen and colleagues (2000) examined gene expression in older human DMD muscle and showed that a variety of markers corresponding to inflammation were upregulated. Immunostaining confirmed that dendritic cells, mast cells, and macrophages were present in DMD fibers during the middle stages of the disease and that gene expression changed for markers involved in collecting these cells around the muscle. Finally, Porter and colleagues (2002) showed that a variety of inflammatory markers were upregulated in older mdx mice, aged 8 weeks. These included a variety of cytokines, chemokines, and myeloid and lymphoid markers, suggesting that dystrophin-deficient muscle was affected by macrophages, neutrophils, lymphocytes, mast cells, and T- and B-cells.

Because antigen presentation is necessary for the activation of immune cells, Spencer and colleagues (1997) suggested that dystrophin-deficient muscle must present an antigen that attracts immune cells. Although the specific antigen is unknown, there are several potential sources. Muscle damage resulting from a weakened membrane could cause a cytosolic leak of proteins into the extracellular matrix. The proteins released, which are not normally seen by the immune system, could activate immune cells, that is, T-lymphocytes. Alternatively, dystrophin-deficient muscle may express proteins on the membranes that are recognized as antigens because they are not expressed in normal tissue (Spencer et al., 1997). Targeting the immune system with therapeutic agents may significantly attenuate the progression of DMD.

In summary, the muscle degeneration in DMD is progressive and ultimately fatal. The precise mechanisms by which the pathophysiology is initiated are at present not defined. Possibilities include a weakened membrane, disrupted signaling, activation of calcium-dependent proteases, activation of the inflammatory system, or some combination of these.

Repair Processes

Normal muscle fibers that are injured can be repaired. As described in chapter 6, muscle regeneration following injury is critically dependent on satellite cells, including their activation, proliferation, differentiation, and finally their fusion to form multinucleated myotubes (Renault et al., 2000). Many of the factors regulating these processes have been identified (Martin, 2003), and additional regulatory signals continue to be elucidated (Yan et al., 2003; Stupka et al., 2004). It is likely that the same repair processes are activated following muscle fiber degeneration in DMD. However, the multiple cycles of degeneration and regen-

eration may deplete the satellite cell pool (Jejurikar et al., 2003). In addition, T-cells may be involved in the onset and continued progression of fibrotic processes that also counter regeneration (Morrison et al., 2000). For both possibilities, the specific mechanisms responsible are not clear.

As noted in figure 9.5, there are several potential and not necessarily mutually exclusive mechanisms of onset and progression of DMD. However, we are still left with two key questions: (1) What are the mechanisms that initiate and prolong degeneration, and (2) why are the regenerative processes that are evident in the repair of normal muscle overwhelmed by those of dystrophic degeneration? Answers to these complex questions should ultimately lead to effective therapeutics (for review see Kapsa et al., 2003).

Summary

Dystrophin is an important protein in the cytoskeletal structure of skeletal muscle fibers. In DMD, mutations in the dystrophin gene eliminate the expression of dystrophin, causing severe progressive skeletal muscle wasting that ultimately kills young men by their early 20s. The functions of dystrophin are not yet completely understood, nor are the events that initiate the severe pathogenesis of DMD. Although dystrophin is considered to function as a mechanical stabilizer to provide structural integrity to the sarcolemma, it is not clear whether

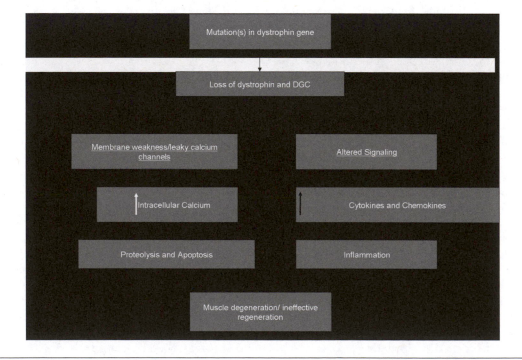

Figure 9.5   Potential scheme of pathogenesis and progression in Duchenne muscular dystrophy. DGC: dystrophin-glycoprotein complex.

a weakened membrane is an onset mechanism of DMD or a consequence of disease progression.

Dystrophin and the proteins of the DGC may coordinate signaling between the myoplasm and extracellular matrix. In the dystrophin-deficient muscle, signaling may be disrupted, and this may cause a variety of pathophysiological events. The $\alpha7\beta1$ integrin and nNOS may be involved in onset of DMD because they associate with the DGC, and in its absence their functions could be compromised. Despite these possibilities, at present it is still not clear how dystrophin and the DGC function as a signaling scaffold.

Alterations in the myofiber calcium homeostasis are thought to underpin the progression of the dystrophic process. It is clear that an increase in intracellular and intraorganellar Ca^{2+} (i.e., SR, mitochondria) is a common factor in the dystrophic progression and that this increase in myoplasmic $[Ca^{2+}]$ acts to increase calcium-dependent proteolytic activity. Whether or not altered Ca^{2+} signaling plays a role in the initiation of the dystrophic process, however, is an unanswered question. Thus, it is important to understand the mechanisms leading to compromised calcium homeostasis, as well as its time course, in dystrophin-deficient muscle.

An inflammatory response dominates dystrophin-deficient human and mouse muscle as evidenced by both DNA microarray and immunohistochemical studies. However, neither the time course nor the mechanisms that initiate the immune response are well understood, particularly during early maturation. Identifying molecules that target immune cells may be a promising method for developing therapeutic agents that counteract inflammation to slow or prevent the progression of DMD.

The mechanisms and their time course in the pathogenesis and progression of DMD are at present not clearly defined, nor are the reasons for ineffective muscle regeneration. Developing completely effective therapeutics depends on resolving the mechanistic details of both the degenerative and regenerative processes in DMD.

CHAPTER 10

Estrogen and Gender Effects

Peter M. Tiidus, PhD

Estrogen and possibly gender can influence the degree of postexercise muscle damage and inflammation. It is also possible that estrogen affects mechanisms associated with muscle repair. This chapter summarizes the evidence from both animal and human studies for an estrogenic influence on skeletal muscle damage and inflammatory responses following various forms of muscle damage. It also addresses the physiological mechanisms by which estrogen may be able to exert such an influence on skeletal muscle. The chapter also covers potential gender differences in response to muscle damage and the implications of menopause-related estrogen loss for muscle damage, repair, and function in older females. The latter is particularly relevant to the continuing health-related controversies surrounding hormone replacement therapy in older women.

Estrogen Effects on Tissues

Estrogen is primarily associated with sexual characteristics and function in females. However, research has also demonstrated that estrogen has a number of other effects and at least some influence in most tissues and organs:

• **Muscle damage and repair.** In addition to its potential to influence muscle damage, inflammation, and repair (Tiidus 2005), which is the primary focus of this chapter, estrogen has been reported to influence skeletal muscle growth, gene expression, metabolism, contraction characteristics, and maintenance of muscle mass. The physiological mechanisms by which estrogen can exert such an influence vary and may involve specific interactions with

estrogen receptor-α or -β or both (Nilsson et al. 2001); other membrane, cytosolic, or nuclear interactions (Nadal et al. 2001); or its antioxidant potential (Moosmann and Behl 1999). Some of these mechanisms have potential relevance to estrogenic influence on skeletal muscle damage, inflammation, and repair and are addressed later in this chapter.

- **Muscle metabolism and growth.** Estrogen has been reported to affect muscle metabolism by enhancing lipid oxidation and consequently downregulating utilization of carbohydrates both at rest and during submaximal exercise (D'eon et al. 2000). Estrogen has also been reported to influence in vitro muscle and myoblast cell growth (Kahlert et al. 1997), as well as to influence in vivo development of muscle size during growth in female mice (Sciote et al. 2001).

- **Other possible effects.** Estrogen may be a factor in slowing or possibly reversing the age-related decline in muscle mass in postmenopausal females (Sipilä et al. 2001; Sorensen et al. 2001) and may also influence muscle fatigue, time to peak tension, relaxation time, and twitch characteristics in rodent muscles (McCormick et al. 2004; St. Pierre-Schneider et al. 2004; Hatae 2001). Kadi and colleagues (2002) have reported that estrogen can alter the expression of myosin heavy chain proteins in fast and slow skeletal muscles of rodents; and

Ogawa and colleagues (2003) have suggested that this hormone can also have indirect effects on skeletal muscle in rodents by influencing their spontaneous self-selected activity levels and patterns as assessed by wheel running and open field movement.

- **Damage and inflammation limitation in several organs and tissues.** It is instructive and relevant to note that estrogen has been shown to influence damage and inflammatory responses following various forms of trauma in a variety of tissues and organs other than skeletal muscle. Specifically, numerous studies have noted that estrogen can significantly attenuate indices of damage and inflammation following trauma in, for example, the brain, neurons, and other neurological tissues (Sribnick et al. 2004; Moosmann and Behl 1999); liver (Harada et al. 2001); skin (Ashcroft et al. 1999); vascular endothelium (Karas et al. 2001); and cardiac muscle (Xu et al. 2004; Squadrito et al. 1997). Thus similar attenuating effects of estrogen as reported in skeletal muscle should not be viewed as unusual. In fact, the question is not *whether* estrogen can influence skeletal muscle damage, inflammation, and repair, but rather to what degree this is possible in humans and to what extent this effect is physiologically and functionally significant (Tiidus 2005).

It is likely that the mechanisms by which estrogen may be attenuating damage in these tissues are ubiquitous to all tissues in which estrogen may be able to bestow a protective effect (Dubey and Jackson 2001; Sribnick et al. 2004). These mechanisms, as well as other possible means by which estrogen may influence damage and inflammation in skeletal muscle (Tiidus 2001; Bär and Amelink 1997), are further discussed later in this chapter.

Estrogen Effects on Skeletal Muscle Damage

A considerable body of evidence demonstrates that estrogen diminishes indices of damage in cardiac muscle following, in particular, ischemia-reperfusion injury in both isolated and whole-body human and animal models (i.e., Dubey and Jackson 2001; Xu et al. 2004; Node et al. 1997; Persky et al. 2000). However, the influence of estrogen on indices of damage in skeletal muscle, particularly following exercise or eccentric muscle contractions and in humans, is less extensively documented.

Creatine Kinase Activity in Animal Models

Among the earliest reports of estrogen's influence on aspects of muscle damage were those from animal studies that documented attenuation of creatine kinase (CK)

activity in plasma of humans or in animal models. Elevation in circulating CK activity has been extensively used as a common clinical measure of exercise-induced muscle disruption (Clarkson and Sayers 1999; Tiidus and Ianuzzo 1983). Creatine kinase is a metabolic enzyme found primarily in muscle that, due to its intracellular location and solubility, can traverse the muscle sarcolemma membrane if the membrane is disrupted or damaged. Circulating CK activity is a relatively inexact indicator of actual structural muscle damage. However, changes in circulating CK level can be used as a sensitive indicator of muscle sarcolemma disruption resulting from exercise or muscular activity (Clarkson and Sayers 1999).

Shumate and colleagues (1979) performed one of the first studies to suggest that estrogen was a factor in diminishing postexercise elevations in circulating CK activity and (by extension) membrane disruption. This study showed that females had significantly attenuated levels of circulating CK elevation following exercise relative to males, which the authors suggested may have been due to gender differences in estrogen (Shumate et al. 1979). Subsequently, a series of rodent-based studies clearly demonstrated a direct attenuating effect of estrogen itself on leakage of CK directly from skeletal muscle (Bär and Amelink 1997):

- Bär and colleagues (1988) reported that normal female rats (which naturally have higher estrogen levels than male and ovariectomized female rats) had significantly lower levels of circulating CK activity than ovariectomized female rats or male rats following running exercise. In addition, supplementing the male and ovariectomized female rats with estrogen prior to exercise resulted in an attenuation of postexercise circulating CK activity similar to that seen in the normal females.

- As further support for the direct connection between estrogen and the loss of muscle CK via sarcolemma disruption, Amelink and colleagues (1990) isolated muscles from male, female, and ovariectomized female rats that had either received or not received estrogen supplementation. They electrically stimulated contractions in these muscles in vitro to simulate exercise and in all cases found that the muscles from animals treated with estrogen supplementation showed significantly diminished CK loss (figure 10.1).

- A number of other more recent studies have also shown that estrogen supplementation in male or ovariectomized female rodents significantly diminishes postexercise or postinjury elevations in serum CK activities (Stupka and Tiidus 2001; Sotiriadou et al. 2003; Feng et al. 2004).

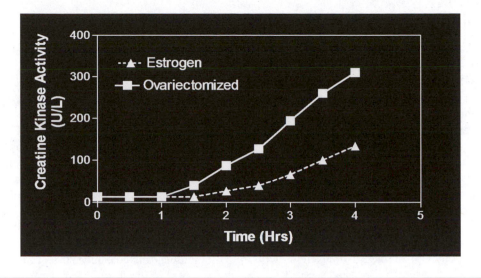

Figure 10.1 Basal creatine kinase release from isolated soleus muscle in vitro in ovariectomized rats and in ovariectomized rats treated with estrogen.

Data modified from Amelink et al., 1990, and Clarkson and Sayers, 1999.

Collectively these studies provide compelling evidence for the protective and stabilizing effect of estrogen on muscle sarcolemma membrane integrity and for its ability to diminish exercise- or damage-induced muscle sarcolemma disruption, particularly in rodent models.

Creatine Kinase Activity in Human Studies

No human studies have yet directly examined the influence of estrogen or estrogen supplementation on indices of postexercise muscle sarcolemma disruption or CK release. However, a number of studies have addressed male–female differences or pre- versus postmenopausal female responses for postexercise circulating CK activities. In females, resting serum CK levels correspond inversely with estrogen level, with pregnant women having lower levels than nonpregnant women and the latter having lower levels than premenarcheal girls (Bundy et al. 1979). However, some conflicting findings on serum CK activity responses in males versus females following exercise have been reported (Clarkson and Hubal 2001). Most studies using running and eccentric quadriceps muscle exercises have shown diminished postexercise serum CK elevations in females versus males (Shumate et al. 1979; Apple et al. 1987; Stupka et al. 2000). These findings on gender differences have typically been interpreted as indicating effects of differing estrogen levels on postexercise muscle membrane damage and stability (Clarkson and Hubal 2001). However, other studies have not consistently shown such gender differences (i.e., Sorichter et al. 2001). Some investigations

(i.e., Buckley-Bleiler et al. 1989) comparing pre- and postmenopausal females are confounded by age-related differences in serum CK clearance; thus these measures may not be valid as indicators of estrogen effects on postexercise membrane stability in comparisons of these cohorts (Clarkson and Hubal 2001).

Heat Shock Proteins

Other indirect evidence of the potential for estrogen to influence postexercise muscle disruption involves its effects on heat shock proteins (HSP). Heat shock proteins are rapidly synthesized in various tissues in response to many forms of stress, including exercise and muscle damage. Their primary physiological function is to act as "chaperones" or guiding agents for the proper mediation of protein folding following synthesis on the ribosome. They are also important in the assembly of proteins and their proper membrane positioning, particularly during times of upregulated protein synthesis following recovery from exercise or muscle damage (Paroo et al. 2002). Two studies have shown that estrogen attenuates specific HSP (HSP70) expression as well as HSP70 messenger RNA (mRNA) synthesis in skeletal muscle of rodents following exercise (Paroo et al. 1999, 2002). These results have been interpreted as indicating a protective effect of estrogen on exercise-induced muscle damage, since HSP70 is induced by exercise stress (Paroo et al. 2002). It has also been recently reported that estrogen attenuates indices of oxidative stress in injured muscles of rats (Feng et al. 2004). However, other studies have not yielded evidence for estrogenic attenuation of exercise-induced oxidative stress in rodents (Tiidus et al. 1998).

Other Indicators of Muscle Damage in Animal Models

Evidence for estrogen as an attenuating influence on muscle structural disruption is limited to a small number of studies that address this question only indirectly through examination of gender-based differences. No studies have focused directly on the effects of estrogen itself on muscle structural disruption following exercise, and the few animal studies of gender differences in postexercise muscle structural damage have not always produced consistent results. The most comprehensive of these studies, by Komulainen and colleagues (1999), showed that for up to 96 h after downhill running, male animals generally experienced greater and earlier histopathological indices of damage to muscle structural proteins (desmin, dystrophin, actin) and a greater degree of muscle fiber swelling than females. The authors also noted much higher levels of β-glucuronidase activity at 48 and 96 h postexercise in muscles of male compared to female rats. β-Glucuronidase is a lysosomal enzyme that is activated in response to muscle damage, and changes in its activity are considered good quantitative indicators of the muscle disruption and repair processes (Salminen and Kihlstrom 1985).

Results from another study showed greater disruption of histochemically determined muscle banding patterns in male versus female rats at 48 h postexercise (Amelink et al. 1991). In contrast to these findings, Van der Meulen and colleagues (1991) saw no histochemically determined differences in degree of muscle damage in female and male rats following exercise of greater duration and intensity than used by Amelink and colleagues (1991). Additionally, Moran and colleagues (2007) recently found no effect of estradiol status on the degree of soleus muscle eccentric and isometric force loss immediately following eccentric contractions in ovariectomized mice; however, they did not measure longer-term force loss or recovery. Interestingly, they did find that estrogen positively influenced maximum isometric muscle forces, possibly by influencing the fraction of strong-binding myosin in muscle.

A recent study addressed the effects of gender on muscle lesions and damage in mice with inherited Duchenne-like muscular dystrophy (mdx mice). The findings suggested that female hormones such as estrogen were at least partially responsible for the lower levels of muscle damage indicators and lesions found in the female versus male mdx mice (Salimena et al. 2004).

Thus, although few investigations have dealt with gender-based differences in susceptibility to muscle damage in rodents, the majority have shown that females present lower levels and fewer indicators of muscle damage than males following exercise or with muscular disease. However, more research is necessary, particularly employing models that directly manipulate estrogen levels, to fully substantiate these findings (Tiidus 2005).

Other Indicators of Muscle Damage in Human Studies

Human studies of male–female differences in susceptibility to muscle damage have generally used indirect measures of muscle damage such as muscle swelling, soreness, stiffness, and loss and regain of muscle force (Clarkson and Hubal 2001). These investigations have generally not shown any significant advantage for females over males in these indirect indicators of degree of muscle damage induced by eccentric exercise or in the rate of recovery from such damage (Sayers and Clarkson 2001; Clarkson and Hubal 2001; Rinard et al. 2000). However, it has recently been reported that males have greater variability and that a greater number of males are "high responders" for circulating CK activity following eccentric exercise–induced muscle disruption (Kearns et al. 2005). Further investigations are under way to determine susceptibility of individuals to more severe muscle damage.

In contrast to these studies, others have shown that chemical indicators of postexercise oxidative stress found in the blood may be affected by estrogen in humans. Blood levels of oxidative stress indicators such as derivatives of lipid peroxidation and other macromolecule peroxidation by-products, as well as changes in antioxidant levels, have been used as markers of muscular and systemic oxidative damage induced by exercise. Since their source is nonspecific, it is hard to definitively conclude that changes in these markers are directly indicative of muscular oxidative stress or disruption. Nevertheless, Dernbach and colleagues (1993) reported that female rowers had lower blood indices of oxidative stress than males during an intense training cycle. Ayers and colleagues (1998) reported higher indices of postexercise lipid peroxidation in amenorrheic female athletes than in normal females. Joo and colleagues (2004) noted that higher estrogen levels in the follicular phase of the menstrual cycle corresponded with attenuated increases in exercise-induced oxidative stress indicators relative to the luteal menstrual phase in normal females. However, Chung and colleagues (1999) failed to see such menstrual cycle–related differences in exercise-induced oxidative stress in normal females.

The only direct measures of gender differences in postexercise muscle damage in humans have been reported by Stupka and colleagues (2000, 2001). The authors examined biopsy samples from quadriceps muscles of males and females following eccentric exercise for histochemical indices of muscle damage. Although Stupka and coworkers (2000, 2001) found no consistent gender differences in damage, the potentially large variability inherent in biopsy assessments of human muscle damage following eccentric exercise may mask any subtle gender-based differences (Beaton et al. 2002).

Studies on human gender differences in susceptibility to exercise-induced muscle damage have been relatively few in number, have been based primarily on indirect indicators of damage, and generally have shown little gender difference. Thus it appears that gender-based differences in muscle damage, at least in humans, may be of minor physiological consequence (Tiidus 2005). Nevertheless, research directly examining the potential for estrogen to diminish exercise-induced structural muscle damage in humans or in animal models has yet to be performed.

The potential effects of estrogen on muscle damage and postexercise recovery may have particular relevance for postmenopausal females in whom estrogen levels are drastically reduced (Tiidus 2003). This cohort should be more closely examined relative to younger females, as pre- and postmenopausal females may be more likely to manifest greater estrogen-based differences in susceptibility to muscle damage than are seen in human gender-based comparisons (Tiidus 2005). For example, Roth and colleagues (2000) found greater evidence of ultrastructural muscle damage in older (65-75 years) than in younger (20-30 years) females following a period of strength training. No such differences existed between younger and older males; hence this study could rule out age as the primary factor in these differences. The authors at least partially attributed the apparently greater susceptibility to weight training–induced muscle damage in older versus younger females to their lower estrogen levels (Roth et al. 2000). This possibility needs to be further examined in research specifically focused on postmenopausal females.

Reduction of Postexercise Inflammatory Response

Probably the most consistent experimentally observed effect of estrogen on skeletal muscle is its attenuating influence on neutrophil infiltration (Tiidus 2003; Kendall and Eston 2002; Tiidus et al. 2001; Stupka and Tiidus 2001). Exercise-induced muscle damage and membrane disruption are typically followed by standard inflammatory-related responses such as leukocyte infiltration. The presence of the leukocyte subgroup, neutrophils, can be detected in muscle less than 1 h after exercise (Belcastro et al. 1998), followed some hours later by macrophage influxes (Tiidus 1998; Malm et al. 2000). Neutrophils and macrophages play important physiological roles

in assisting with the clearance of damaged tissue and in the activation of muscle satellite cells and muscle repair processes (Chalmers and McDermott 1996; Tiidus 1998).

Neutrophil Infiltration

Neutrophils in particular have been reported to increase inflammation-related damage following their influx into injured skeletal muscle because of their ability to produce oxidants and free radicals, which are important in removing damaged tissue (Walden et al. 1990; Tiidus 1998; Sribnick et al. 2004). The attenuation of neutrophil influx into muscular tissue following injury may be associated with a reduction in further inflammation-related damage (Walden et al. 1990; Squadrito et al. 1997; Tiidus et al. 2001). The reduction of neutrophil-related inflammation is also important in reducing inflammation-related injuries in a number of other tissues (i.e., Wise et al. 2001; Harada et al. 2001). Other leukocytes such as macrophages invade skeletal muscle several hours after injury and are important in releasing cytokines and other activators of muscle repair processes and muscle satellite cells (Anderson and Wozniak 2004; Hawke 2005). Hence any attenuation or delay in macrophage infiltration of skeletal muscle following damage could delay or diminish muscle repair (Mishra et al. 1995; St. Pierre-Schneider et al. 1999; Tiidus 2001).

Tiidus and Bombardier (1999) were the first to report that in comparison to male rats, female rats showed significant attenuation of neutrophil infiltration into skeletal muscles at 24 h following running exercise. When the male rats were given daily injections of estrogen for 14 days prior to the acute exercise, they also exhibited the same blunted increase in postexercise muscle neutrophil infiltration as seen in female rats (Tiidus and Bombardier 1999). This study relied on biochemical quantification of neutrophil presence in skeletal muscle by determining activity of myeloperoxidase (an enzyme found exclusively in neutrophils) in muscle homogenates. Subsequent studies using histochemical and immunohistochemical quantification of neutrophils, along with biochemical analysis, confirmed these earlier findings. Ovariectomized female rats, with or without estrogen replacement, were also exposed to running exercise. Females that had received estrogen replacement exhibited significantly attenuated muscle neutrophil infiltration at 1 h postexercise as determined by histochemical quantification of neutrophils (Tiidus et al. 2001). (See figure 10.2.) Estrogen administration to ovariectomized female rats also diminished muscle neutrophil infiltration at 2 h after injury induced by ischemia-reperfusion as quantified by immunohistochemical identification and count of neutrophils in muscle sections (Stupka and Tiidus 2001).

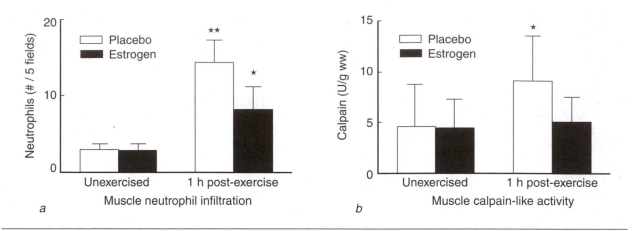

Figure 10.2   Relationship of estrogen administration to 1 h postexercise muscle neutrophil infiltration and calpain activities in ovariectomized rats. Estrogen replacement significantly attenuated the postexercise elevation of both neutrophil infiltration and calpain activity relative to placebo.

Reprinted, by permission, from P.M. Tiidus et al., 2001, "Estrogen effect on post-exercise skeletal muscle neutrophil infiltration and calpain activity," *Can. J. Physiol. Pharmacol.* 79(5): 400-406.

It is possible that differing skeletal muscle fiber types respond with varying degrees of attenuation of postexercise neutrophil infiltration when exposed to estrogen (Tiidus et al. 2005). However, this possibility requires further experimental confirmation. Recently McClung and colleagues (2007) also reported greater neutrophil infiltration in soleus muscles of ovariectomized versus gonadally intact rats following three days of weight bearing subsequent to a period of disuse induced by hindlimb suspension.

Macrophage and Leukocyte Infiltration

St. Pierre-Schneider and colleagues (1999) reported a delay in macrophage infiltration in muscle of female compared to male mice following eccentric contraction–induced injury. They attributed this slower rate of postdamage peak in macrophage infiltration to possible estrogen influences (St. Pierre-Schneider et al. 1999). McClung and colleagues (2007) also found decreased ED1+ and ED2+ macrophage infiltration at seven days after reloading in estrogen-supplemented relative to unsupplemented ovariectomized rats that were recovering from disuse atrophy. Hence it is likely that estrogen may delay and inhibit longer-term macrophage infiltration into damaged or recovering skeletal muscle.

As with measures of muscle damage, no studies have yet directly addressed the effects of estrogen on postexercise leukocyte infiltration into skeletal muscle in humans. Three studies have shown gender-based differences in leukocyte infiltration into the quadriceps muscle following exercise in humans that were at least in part attributed to estrogen (MacIntyre et al. 2000;

Stupka et al. 2000, 2001). Using nonspecific histochemical markers for leukocytes (Bcl-2 and LCA), Stupka and colleagues (2000) reported relatively greater increases in these indices of leukocyte infiltration into eccentrically exercised quadriceps muscles of males versus females at 48 h postexercise. However, a subsequent study by the same group showed no attenuation of either neutrophil or macrophage infiltration into quadriceps muscles of females relative to males following either one or two bouts of eccentric muscle contractions (Stupka et al. 2001). A confounding variable in these studies is the potentially large variability in determination of muscle leukocyte infiltration from human muscle biopsy samples (Beaton et al. 2002).

Using ^{99m}Tc-radiolabeling of neutrophils with subsequent nuclear imaging of the thigh region, MacIntyre and colleagues (2000) also assessed gender differences in muscle neutrophil infiltration, at 2 and 4 h after eccentric quadriceps exercise. They found significantly greater elevations of the neutrophil presence in postexercise muscles of females relative to males (MacIntyre et al. 2000).

There are at present too few studies (and none that have involved direct examination of estrogen), with too many methodological differences and limitations, to enable adequate assessment of the potential influence of estrogen on postdamage muscle leukocyte infiltration in humans.

Possible Mechanisms

A number of mechanisms might be responsible for the ability of estrogen to influence the postdamage skeletal muscle inflammatory response and degree of muscle damage. These mechanisms include antioxidant properties, membrane stabilization, and inhibition of neutrophil and leukocyte infiltration.

Antioxidant Properties

Earlier research demonstrated that estrogen has potent in vitro antioxidant properties (Yagi and Komura 1986; Sugioka et al. 1987). Unlike other steroid hormones such as progesterone and testosterone, which have no antioxidant properties, estrogen possesses a hydroxyl group on its "A" steroid ring, analogous to that found on the antioxidant vitamin E (Sugioka et al. 1987). By donating a hydrogen ion from this hydroxyl group, estrogen may be able to terminate peroxidative chain reactions and thus act as an antioxidant (see figure 10.3) (Sugioka et al. 1987). This has led to suggestions that estrogen's antioxidant properties may protect muscle from exercise-induced damage (Tiidus 1995). However, as noted earlier, the limited data on the ability of estrogen to diminish postexercise muscle oxidative damage in both humans and animal models are inconsistent and inconclusive. Hence it is not yet possible to empirically make a strong case for the primacy of estrogenic antioxidant potential in limiting muscle damage and membrane disruption following exercise, particularly in humans (Tiidus 2005).

Membrane Stabilization

Muscle damage also results in loss of muscle calcium homeostasis due to membrane disruption (Proske and Allen 2005). This leads to muscle damage as a consequence of increased activation of proteolytic enzymes such as calpain and the resultant increase in proteolysis (Belcastro et al. 1998). Estrogen may be able to act as a membrane stabilizer, as evidenced by its ability to diminish muscle CK loss resulting from sarcolemma disruption following exercise-induced muscle damage (Tiidus 1995, 2003). The membrane-related effects of estrogen may result from its potential to directly influence membrane stability, possibly through direct membrane interactions similar to those of cholesterol (Tiidus 2000; Bär and Amelink 1997) or through its interaction with membrane-bound estrogen receptors alpha and beta, both of which are present in skeletal muscle (Lemoine et al. 2003; Wiik et al. 2003). Estrogen may also act to hyperpolarize cell membrane potential by opening calcium-activated potassium channels in cardiac and endothelial cells, resulting in hyperpolarization of membrane potential and thus limiting calcium overload following ischemia-reperfusion injury in cardiac muscle (Node et al. 1997). Although no experimental data have yet been gathered in this regard with skeletal muscle, it is possible that similar mechanisms may be acting in stabilizing skeletal muscle membranes.

It is therefore possible that estrogen may act to diminish activation of proteolytic enzymes such as calpain due to its ability to mitigate postexercise sarcolemma disruption and the resultant loss of calcium homeostasis (Tiidus 2003). We have demonstrated that estrogen supplementation in ovariectomized female rats acts to attenuate calpain activation in skeletal muscle following disruption induced by running exercise (Tiidus et al. 2001) or ischemia-reperfusion injury (Stupka and Tiidus 2001). Nevertheless, further experimental evidence directly relating estrogen supplementation to improved postexercise muscle calcium homeostasis is needed to confirm these suggested mechanisms of estrogen's effect on muscle damage mitigation (Tiidus 2005).

Inhibition of Neutrophil and Leukocyte Infiltration

There are a number of possible mechanisms by which estrogen may act to influence postdamage inflammation, particularly as it relates to neutrophil infiltration of skeletal muscle. One possible mechanism, as noted earlier, has to do with the potential for calpain proteolytic activity to produce neutrophil chemoattractant peptides (Belcastro et al. 1998). By inhibiting calpain activation following muscle damage, as previously described, estrogen may reduce the degree of neutrophil chemoattractant peptide production via calpain activity in skeletal muscle following damage (Tiidus 2003). We have previously noted the correlation of these events in rodent skeletal muscles (Tiidus et al. 2001).

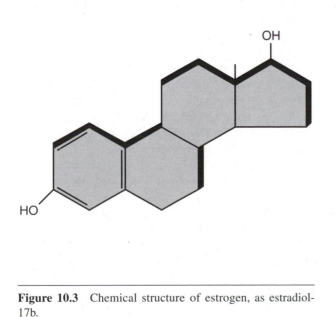

Figure 10.3 Chemical structure of estrogen, as estradiol-17b.

Muscle damage has also been associated with an increase in apoptosis and apoptotic signaling leading to possible increases in muscle cell death and regeneration (Phaneuf and Leeuwenburgh 2001). Estrogen has been reported to decrease calpain and caspase-3 activity in damaged neural tissues, resulting in a reduction in downstream apoptotic signaling in these tissues (Sribnick et al. 2004). Our laboratory has also reported decreased calpain activation in skeletal muscles of estrogen-supplemented rats following running exercise (Tiidus et al. 2001), suggesting that some of the same apoptosis-diminishing effects of estrogen may also be occurring in skeletal muscle. Recently, a preliminary report noted a co-decrease in muscle apoptotic markers and muscle damage in female compared to male rats following eccentric contractions (Sudo et al. 2005). Further follow-up research will be needed to confirm the potential for estrogen to diminish apoptotic signaling in damaged skeletal muscle and the extent to which this may influence the degree of resultant muscle disruption.

Other studies have linked estrogen to the attenuation of leukocyte infiltration and damage following disruption in other tissues. Although no data currently exist to suggest that such mechanisms exist in skeletal muscle, it is instructive to consider these possible mechanisms.

It is well known that estrogen enhances the production of nitric oxide (NO) in various tissues (Van Buren et al. 1992). Several tissues in female animals contain higher amounts of NO synthase (NOS) than in male animals (Laughlin et al. 2003), possibly as a consequence of greater estrogen exposure. Estrogen has also been shown to increase the activity of endothelial NO synthase (eNOS) via receptor-mediated actions (Simoncini et al. 2000). Estrogen may also have a direct stimulatory influence on NOS activity (Harada et al. 2001). Nitric oxide released by eNOS can inhibit leukocyte adhesion following ischemia-reperfusion injury (Prorock et al. 2003). Consequent to ischemia-reperfusion injury in animal models, estrogen has been shown to upregulate NOS in endothelial tissue and hence mitigate leukocyte rolling, adhesion, and infiltration into damaged endothelia (Prorock et al. 2003). Additionally, increases in NO in skeletal muscle can inhibit calpain activation (Koh and Tidball 2000) and thereby also possibly enable estrogen to reduce secondary inflammation and damage via this mechanism. Hence, while no experimental data are yet available regarding this effect in skeletal muscle, the evidence from cardiac and endothelial tissues suggests that estrogen may also be able to mitigate leukocyte infiltration and damage in skeletal muscle via its influence on NO production.

Work involving neural tissues has also suggested that estrogen may be able to limit the binding of the proinflammatory transcription factor NfκB to DNA, thereby limiting a number of inflammatory-related responses (Sribnick et al. 2004). These responses could include inhibition of the transcription of leukocyte chemo-attractants such as matrix metalloproteinase-9 and monocyte chemo-attractant protein 1 in addition to upregulation of NO.

Several studies have presented evidence of tissue protection from damage and leukocyte infiltration via other estrogen receptor–mediated mechanisms (Karas et al. 2001; Pare et al. 2002), including the aforementioned influence on calcium-activated potassium channels (Node et al. 1997).

The mechanisms by which estrogen may be able to influence leukocyte infiltration, inflammation, and damage in skeletal muscle subsequent to injury are not yet known. However, it seems likely that one or more of the mechanisms just discussed and possibly others are directly or indirectly involved in multifactorial ways. Further research is necessary to clarify the potential interactive and inhibitory effects of combinations of estrogen and progesterone in influencing leukocyte infiltration following tissue damage. Recently Xing and colleagues (2004) reported that while estrogen inhibited inflammatory cell infiltration into injured endothelia, progesterone, although it had no independent effect itself, negated this inhibitory effect of estrogen. Understanding these possible mechanisms and interactions may be critical to understanding estrogen's influence on muscle damage responses in both younger and postmenopausal females.

Possible Influence of Estrogen on Skeletal Muscle Repair

Evidence generally agrees that inhibition of neutrophil infiltration into skeletal muscle and other tissues results in reduced injury and thus an enhancement of repair and recovery (Tiidus 2001). However, the infiltration of specific macrophages is important for the activation of muscle satellite cells and their critical participation in subsequent skeletal muscle repair (Chargé and Rudnicki 2004). A potential delay or inhibition of macrophage infiltration into skeletal muscle following injury (St. Pierre-Schneider et al. 1999) may therefore theoretically inhibit skeletal muscle recovery (Tiidus 2001). Nevertheless, as previously noted, the overall effects of estrogen on skeletal muscle and other tissues tend toward reduction of damage and more rapid regeneration following injury. Hence, if estrogen does eventually prove to be inhibitory to macrophage infiltration and subsequent activation of satellite cells in skeletal muscle, estrogen's other inflammation- and damage-inhibiting and regeneration-enhancing effects may be sufficient to outweigh this potential disadvantage.

Several trophic factors either secreted from macrophages or generated within the damaged muscle itself have been demonstrated to stimulate satellite cell activation and muscle regenerative processes following damage. These include insulin-like growth factor-1 (IGF-1), hepatic growth factor (HGF), and interleukin-6 (IL-6), as well as other cytokine-related growth factors (Chargé and Rudnicki 2004). The interactions of these and other regulatory factors in muscle satellite cell activation and muscle repair regulation are complex and not yet fully explained. Nevertheless, estrogen has been shown to influence the concentrations of at least some of these factors, particularly IGF-1 and HGF, in skeletal muscle (Hawke and Garry 2001; Chargé and Rudnicki 2004). This influence of estrogen on IGF-1 and HGF levels in skeletal muscle may occur via regulatory mechanisms associated with estrogen-enhanced NO production in muscle (Prorock et al. 2003). Hence it is theoretically possible that estrogen may be able to positively influence muscle satellite cell activation and subsequent repair via NO-mediated enhancement of satellite cell–activating factors such as IGF-1 and HGF. More research is obviously needed to discern the potential for these possibilities. These potential mechanisms are particularly intriguing since some preliminary research in animals suggests that estrogen may indeed positively influence postdamage repair and satellite cell proliferation in skeletal muscle (Salimena et al. 2004; Tiidus et al. 2005).

Salimena and colleagues (2004) reported that male mice with Duchenne-like dystrophy (mdx) mice exhibited increased skeletal muscle sarcolemma permeability, inflammation, and decomposition at 6 weeks of age. Age-matched mdx females with intact ovaries exhibited significantly diminished signs of necrosis and inflammation and additionally exhibited increased numbers of regenerating muscle fibers and satellite cells. Ovariectomized mdx females had significantly fewer regenerating muscle fibers (Salimena et al. 2004). The authors suggested that these findings point to a role for female hormones in promoting satellite cell proliferation, muscle regeneration, and mitigation of inflammation at least in the mdx model of muscle myonecrosis (Salimena et al. 2004). Additionally, Brown and colleagues (2005) examined the effects of ovariectomy and estrogen replacement on muscle mass recovery following unweighting and reloading in adult rats. They found that ovariectomized rats failed to recover muscle mass while animals supplemented with estrogen recovered muscle mass at a normal rate. Although the mechanisms of estrogen action were not investigated in this study, these findings add further support to suggestions that estrogen can influence muscle recovery and possibly size accrual following damage or atrophy.

Our laboratory has also recently reported preliminary findings suggesting that estrogen may play a role in enhancing muscle satellite cell proliferation following exercise (Tiidus et al. 2005). At three days following a bout of downhill running, male rats with or without estrogen supplementation were examined for muscle satellite cell numbers in cross sections of type I and type II muscle fibers (see figure 10.4). Interestingly, while both groups of animals increased muscle satellite cell numbers (as detected immunohistochemically by Pax7 antibody), there were significantly higher numbers of satellite cells in both type I and type II muscle from the estrogen-supplemented animals (Tiidus et al. 2005). This suggests that estrogen may be able to influence muscle satellite cell proliferation—possibly via one or

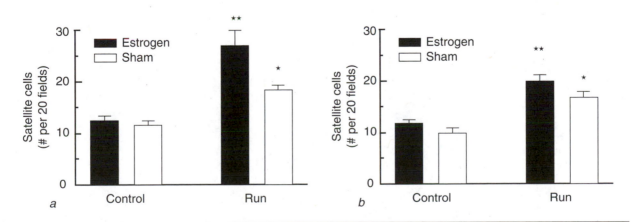

Figure 10.4 Pax7 antibody-detected myogenic satellite cell content in (a) soleus and (b) white vastus muscles of estrogen-supplemented and sham-supplemented control or downhill run male rats at 72 h after downhill running. Estrogen supplementation resulted in significantly greater numbers of satellite cells in both soleus and white vastus muscles following downhill running.

From P.M. Tiidus, M. Deller and X.L. Liu, 2005, "Oestrogen influence on myogenic satellite cells following downhill running in male rats: a preliminary study," *Acta Physiologica Scandinavica* 184(1): 67-72.

more of the mechanisms suggested earlier—following exercise-induced muscle disruption. The implications for estrogen-enhanced muscle repair, particularly in humans, cannot yet be predicted from these preliminary animal studies. Nevertheless, these findings point to the need for more research to determine the potential implications of estrogen in muscle regeneration, particularly as it may relate to postmenopausal females (Tiidus 2005).

Summary

Experimental evidence gleaned primarily from rodents suggests that estrogen may have a mitigating effect on skeletal muscle damage, inflammation, and leukocyte infiltration following disruption induced by exercise or ischemia-reperfusion injury or during recovery from atrophy. Some theoretical and empirical data also suggest that estrogen may play a role in muscle regeneration and recovery after injury or atrophy. However, more research will be needed to confirm the physiological significance in humans of any potential ability of estrogen to mitigate damage and inflammation and enhance regeneration, as well as estrogen's applicability to muscular health in postmenopausal females.

CHAPTER 11

Diabetes

Lisa Stehno-Bittel, PhD, PT; Muhammel Al-Jarrah, PhD, PT; and S. Janette Williams

Diabetes mellitus is a disease that is rapidly increasing in reported new cases each year. It has been termed an epidemic in the United States (Bardsley and Want 2004), due to an aging American population and the explosion of obesity (Jack et al. 2004). Estimates are that 18.2 million people in the United States, or 6.3% of the population, have diabetes. Diabetes mellitus is now the sixth leading cause of death in the United States, with a worldwide prevalence that is steadily increasing (Segal 2004). The mean age at diagnosis for type 2 diabetics decreased from 52 years to 46 years of age within a 12-year time span (Koopman et al. 2005). This means that more people will be living with diabetes for a greater number of years, adding urgency to the need to find ways of controlling or curing this disease.

The cause of diabetes continues to be unclear, although both genetics and environmental factors such as obesity and lack of exercise play a role. Diabetes mellitus is characterized by the body's improper use of insulin or failure to produce insulin, the most important hormone in regulating blood glucose levels. As blood glucose levels rise (typically after a meal), insulin is secreted into the bloodstream by the pancreatic beta cells (Khan and Pessin 2002). The insulin reaches target cells and binds to insulin receptors, and then a chain of events allows the cells to utilize glucose.

The American Diabetes Association's "Standards of Medical Care in Diabetes" defines four types of diabetes that are summarized in "Classifications of Diabetes" (Association 2004). Type 1 diabetes, formerly known as juvenile onset or insulin-dependent diabetes, results from destruction of the cells in the pancreas that produce insulin, the beta cells. It is an autoimmune disorder in which the body creates antibodies against proteins associated with the pancreas (Tisch and Vedevit 1996). Patients develop extremely high blood glucose levels and must be placed on insulin therapy in order to survive. Type 1 accounts for 5% to 10% of all diagnosed cases of diabetes. Type 2 diabetes is the most common form of diabetes, accounting for nearly 90% of the diagnoses. This classification of diabetes was formerly known as adult onset or non-insulin-dependent diabetes. Type 2 diabetes results from the inability of the target cells to utilize insulin, along with a progressive loss of insulin secretion by the pancreas. While type 2 diabetes runs in families, indicating a strong genetic predisposition, the main predictor of type 2 diabetes is obesity. The prevalence of type 2 diabetes is increasing, particularly among youth, in parallel with the continuing rise in obesity.

Gestational diabetes is diagnosed only during pregnancy and typically goes away after the birth of the child (Blayo and Mandelbrot 2004). Gestational diabetes affects about 4% of all pregnant women, which amounts to 135,000 cases in the United States each year. The fourth category of diabetes is called "Other Specific Types of Diabetes." According to the World Health

This chapter is dedicated to our family members for their unlimited love, support, and patient understanding of our dedication to diabetes research: Doug, Aubrey, and Cole Bittel; Jomana, Razan, Hashim, and Layth Al-Jarrah; Pearl Gibson and Wendell Williams.

Classifications of Diabetes

Type 1 Diabetes

- Autoimmune disease destroys the insulin-producing cells of the pancreas.
- The majority of patients have positive auto-antibodies to one of the following:
  - Insulin
  - GAD (glutamic acid decarboxylase)
  - IA-2 and 2B (tyrosine phosphatases)
- Rate of destruction of insulin-producing cells varies.
- Patients are rarely obese and are prone to other autoimmune diseases.
- Type 1 was previously known as insulin-dependent or juvenile onset diabetes.

Type 2 Diabetes

- Patients show insulin resistance with relative insulin deficiency.
- Plasma insulin levels may be normal.
- Autoimmune destruction of the insulin-producing cells does not occur.
- Obesity is common (or high percent body fat).
- Hyperglycemia develops slowly.
- The person has a genetic predisposition for type 2 diabetes.
- Type 2 was previously known as insulin-independent or adult onset diabetes.

Gestational Diabetes

- The woman manifests glucose intolerance with pregnancy.
- If the condition continues after pregnancy (longer than six weeks), it is no longer gestational diabetes.
- Gestational diabetes occurs in 4% of all pregnancies in the United States.
- It increases risk of cesarean section, birth defects, and chronic hypertension.
- It increases risk of diabetes developing in the mother in the future.

Other Types of Diabetes

- Diseases of exocrine pancreas (pancreatitis, neoplasia, trauma)
- Induced by drugs or chemicals (glucocorticoids, thiazides, dilantin)
- Infections (cytomegalovirus)
- Genetic defects (congenital)

Organization and the American Diabetes Association, any condition that results in an increase in plasma glucose levels is defined as diabetes, including genetic defects in pancreatic beta cell function, diseases of the exocrine pancreas, injury to the pancreas after surgery or trauma, malnutrition, or drug-induced or chemical damage to the pancreas (Association 2004). Diagnoses that fall into this category may account for 1% to 2% of all diagnosed cases of diabetes. Finally, prediabetic states such as metabolic syndrome are now being described in the literature. However, the effects of these prediabetic conditions on muscle physiology and function are not well understood and are not covered in this chapter, which focuses on the role of type 1 and type 2 diabetes in muscle physiology.

The complications of diabetes are the systemic effects of long-term hyperglycemia. Chronic hyperglycemia can lead to the development of nephropathies, neuropathies, and retinopathies, along with poor tissue healing (Group 1993). Complications involving the large vessels lead to high blood pressure, stroke, and heart attacks. Discussion of the effects of diabetes on skeletal muscle is complicated by the fact that one of the major causes of hyperglycemia in type 2 diabetics is a defective use of glucose in skeletal muscle. Thus, skeletal muscle not only suffers from the effects of hyperglycemia but also plays a role in causing the high blood glucose.

Alterations in Glucose Transport

Skeletal muscle, which makes up nearly 40% of the body mass of humans, is the primary tissue responsible for the utilization of glucose in response to insulin (Koistinen and Zierath 2002). In fact, skeletal muscle accounts for 90% of the glucose uptake when insulin is released (DeFronzo et al. 1981). Normally, high glucose levels in the blood (for example, after a meal) are sensed, and in response the pancreas releases insulin (figure 11.1a). The insulin molecules travel through the blood to the skeletal muscle where they activate receptors, which allows glucose to enter the muscle cell. In type 1 diabetes, the pancreas beta cells are destroyed by the body's immune system. Thus, when glucose levels in the bloodstream levels increase, the pancreas cannot respond by releasing insulin (figure 11.1b). Without insulin to activate the receptor cascade, glucose cannot be taken up by the skeletal muscle cell.

In type 2 diabetes, insulin can be released from the pancreas, but it has little or no effect when it reaches the skeletal muscle (figure 11.1c). This condition is termed insulin resistance (Kahn 1978). In addition, type 2 diabetics typically have reduced insulin release from the pancreas compared to nondiabetics, which exacerbates the problem (figure 11.1c). Exercise and a balanced diet promote insulin sensitivity, while inactivity and excess caloric intake contribute to insulin resistance (Kirwan and del Aguila 2003).

With impaired regulation of glucose uptake, diabetic patients are more prone to hyperglycemic states and have little or no response mechanism to correct blood glucose levels when they fall too low. In patients with type 1 diabetes experiencing episodes of low blood glucose, defects in skeletal muscle physiology again play a role. Normally, skeletal muscle would halt or slow down glucose uptake when blood glucose levels are low. This compensation constitutes 80% of the response to a hypoglycemic event (Cohen et al. 1995), leaving diabetics at risk for both hyper- and hypoglycemia. Thus, even though the primary defect of type 1 diabetes resides within the pancreas, not skeletal muscle (figure 11.1b), defects in the skeletal muscle glucose-buffering mechanism still have tremendous adverse effects on whole-body glucose metabolism.

Insulin Signaling Pathways

In order to appreciate the interplay between diabetes and skeletal muscle, one must understand the insulin signaling pathway both in normal muscle cells and in

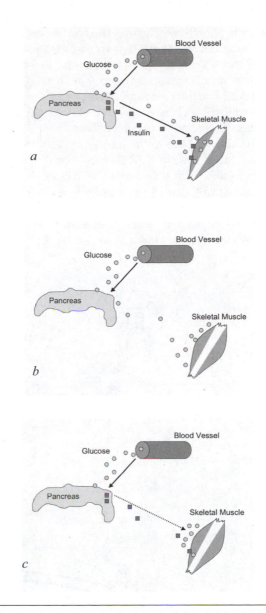

Figure 11.1   (a) In the healthy individual, blood glucose (spheres) levels rise, stimulating release of insulin (squares), which allows glucose transport into the skeletal muscle. (b) In type 1 diabetes the pancreas cannot produce insulin, so glucose cannot enter the skeletal muscle cells. (c) In type 2 diabetes, the pancreas produces smaller amounts of insulin and the insulin signaling pathway is ineffective, allowing only limited amounts of glucose to enter the cell.

diabetes, as well as the role of exercise in stimulating glucose transport.

Basal Activity in Normal Muscle Cells

The normal insulin pathway is activated once insulin binds to its receptor on the plasma membrane. The receptor is a transmembrane glycoprotein with two extracellular subunits that serve as insulin-binding sites and two intracellular subunits that contain phosphorylation sites (Kirwan and del Aguila 2003; Kasuga, Hedo et al. 1982) (figure 11.2). Upon stimulation, the receptor autophosphorylates, leading to activation (phosphorylation) of the insulin receptor substrate-1 (IRS-1) and related proteins (Kasuga, Karlsson, and Kahn 1982). Subsequently IRS-1 acts as a docking site for a number of downstream proteins such as phosphoinositide 3-kinase (PI3K) (Cantley 2002; Kirwan and del Aguila 2003) (figure 11.2). Signaling through PI3K initiates several biological effects, including increased glucose metabolism; glycogen, lipid, and protein synthesis; gene expression; and cell growth and differentiation. Once PI3K is activated, it produces lipid second messengers that stimulate downstream proteins, which include a kinase known as Akt (also called protein

kinase B). Akt, among other things, stimulates translocation of the insulin-responsive glucose transporter (GLUT4) from intracellular storage vesicles to the plasma membrane. Akt belongs to an interesting family of proteins in that they appear to be able to substitute for each other; if Akt1 levels decrease, then Akt2 or Akt3 can step in and fill the role of Akt1 and thus the signaling cascade continues. Redundancy of the signaling pathway ensures its robustness.

Glucose transport is ultimately mediated by GLUT4, a protein normally inserted in membrane of intracellular storage vesicles that fuse with the plasma membrane when the insulin pathway is activated. Once in the membrane, the GLUT4 protein allows for the influx of glucose into the cell (James 2005). GLUT4 is the rate-limiting step in glu-

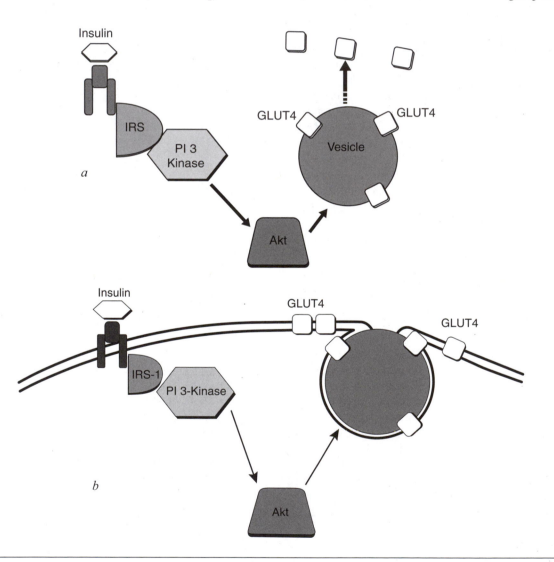

Figure 11.2   *(a)* When insulin reaches the skeletal muscle cell, it binds to the insulin receptor on the plasma membrane. The receptor autophosphorylates and subsequently activates a nearby protein called the insulin receptor substrate-1 (IRS-1). IRS-1 activation starts a cascade of signaling events that include activation of phosphatidylinositol 3-kinase (PI 3-kinase) and Akt, which in turn stimulates the translocation of vesicles containing the glucose transporter, GLUT4. *(b)* When the vesicle reaches the plasma membrane it fuses with the membrane, and GLUT4 transporters are inserted into the membrane. With GLUT4 present in the plasma membrane, glucose on the outside of the cell can now enter the cytosol.

cose metabolism in skeletal muscle (Koistinen and Zierath 2002) and is responsible for most of the insulin-stimulated increase in glucose transport. Production of GLUT4 is carefully regulated in muscle, which is not surprising given the pivotal role that the protein plays in the regulation of insulin-stimulated glucose uptake by tissues.

Basal Activity in Diabetic Cells

Knowing the insulin signaling pathway in muscle and other target tissues allows the identification of the signaling molecules to determine where there is a loss of the signaling cascade associated with diabetes. Most agree that the skeletal muscle insulin receptor itself is not altered in diabetes (Krook et al. 2004), even though early reports concluded that receptor autophosphorylation was decreased with type 2 diabetes (Goodyear et al. 1995). Consensus has been reached on the involvement of IRS-1 and PI3K activation in defective diabetic signaling. Numerous studies conclude that activation of these two proteins is decreased with type 2 diabetes (Bjornholm et al. 1997; Goodyear et al. 1995). Figure 11.3 illustrates the fact that IRS-1 and PI3K are altered with diabetes. The most important difference noted with diabetes is a decrease in IRS-1 activation as measured by phosphorylation. In contrast, total protein levels of IRS-1 in skeletal muscle do not appear to change with diabetes (Krook et al. 2000).

In addition to causing changes in IRS-1 and PI3K, diabetes also alters the next major signaling molecule in

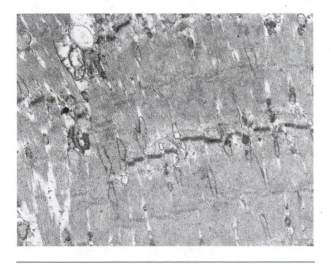

Figure 11.3 In diabetes, a block in the insulin receptor substrate-1 (IRS-1) and phosphatidylinositol 3-kinase (PI 3-kinase) signaling molecules is most likely responsible for insulin resistance. In skeletal muscle the insulin receptor may be activated, but it cannot communicate with IRS-1 or other downstream molecules. The thick frame around each indicates that these molecules are unable to communicate with the neighboring molecules. The end result is a lack of action by insulin.

the pathway, Akt (Krook et al. 1998). The data on this are confusing because, as already noted, there are several isoforms of Akt that appear to be able to substitute for each other so that the signaling activation can continue. In Kahn's laboratory, activation of Akt2 and Akt3 was found to be normal in skeletal muscle from obese type 2 diabetics (Kim et al. 1999). In more recent work, knocking out the Akt2 gene in skeletal muscle of mice resulted in a diabetes-like syndrome, implicating Akt2 as important in diabetes (Cho et al. 2001). Further research is necessary as the Akt story continues to be sorted out.

Whether levels of GLUT4 change with diabetes is not clear. But increased production of muscle GLUT4 in transgenic animals has been shown to reverse insulin resistance associated with obesity or diabetes (Zorzano et al. 2005). How could excess GLUT4 correct insulin resistance when the diabetes defect appears to be upstream? One possibility is that excessive GLUT4 protein causes an increase in the localization of GLUT4 in the plasma membrane independent of insulin signaling, meaning that without insulin stimulation, GLUT4 already exists in the plasma membrane and is transporting glucose into the cell. Thus, it could be that the insulin pathway is bypassed because of GLUT4 overabundance.

Muscle Contraction and Glucose Transport

In addition to the insulin-stimulated uptake of glucose in skeletal muscle, there is a mechanism to transport the GLUT4 to the membrane. This second pathway can be activated in an insulin-independent manner when muscles contract. In fact, a single bout of exercise immediately results in a translocation of GLUT4 from its intracellular site to the plasma membrane in a manner that does not require insulin (Fushiki et al. 1989). The mechanism of GLUT4 translocation associated with muscle contraction is unknown, but it does not involve PI3K or Akt activation (Goodyear et al. 1995). One molecule that has been proposed for contraction-stimulated glucose transport is mitogen-activated protein kinase (MAPK) (Krook et al. 2004). MAPKs are protein kinases involved in a variety of physiological functions including cell survival, proliferation, and programmed cell death. There appears to be a molecular switch in muscle so that during exercise the muscles use an insulin-independent pathway to transport glucose but then, within 3 to 6 h after exercise, revert back to the insulin-sensitive pathway (Kirwan and del Aguila 2003). This insulin-independent pathway may explain how physical exercise can improve glucose tolerance in patients with type 2 diabetes (Albright et al. 2000). Stimulation of glucose uptake by muscle contraction would lower blood glucose levels and improve metabolism even in conditions of insulin resistance. However, this suggestion should be taken with caution, as evidence suggests that both the insulin-sensitive and the contraction-induced glucose transport pathways are attenuated in persons with diabetes (Peltoniemi et al. 2001).

Diabetes and Oxidative Stress

In metabolizing oxygen, the body produces an array of dangerous molecules including superoxide molecules and free radicals. These molecules react with proteins, cell membranes, and DNA to cause irreversible damage known as oxidative damage. The damage can be blocked by antioxidants that are normally produced by the body and that can be obtained through diet. Oxidative stress represents an imbalance between the production of reactive oxygen species and the antioxidant defense system that keeps them in check (Betteridge 2000). By the mid-1990s, growing evidence suggested that hyperglycemia is associated with the production of reactive oxidative intermediates, along with a disturbed antioxidant defense mechanism (Matkovics et al. 1997; Parthiban et al. 1995; Kelley and Simoneau 1994). Since oxidative stress is a direct result of hyperglycemia, the classification of diabetes is not too important when one is considering oxidative damage. Anyone with diabetes, whether type 1 or type 2, who is exposed to chronic hyperglycemia will likely have an imbalance in the oxidative pathway, leading to oxidative damage.

In type 2 diabetes, insulin sensitivity is inversely correlated with the levels of plasma free radicals (Paolisso et al. 1994), and a potential direct link between oxidative stress and insulin-stimulated glucose uptake may be involved (Rudich et al. 1998). In diabetic patients and in animal models of diabetes, the plasma free radical concentration is increased (Paolisso et al. 1994) while the antioxidant defense mechanism is impaired. In adipocytes, prolonged oxidative stress causes a 60% decrease in insulin-stimulated GLUT4 translocation to the plasma membrane by affecting steps distal to the activation of IRS-1 and PI3K (Tirosh et al. 1999).

Increased oxidative stress is present in the skeletal muscle of diabetic animal models. After 3, 6, and 12 weeks of diabetes, the activity levels of a variety of proteins involved in antioxidant activity were reduced in rabbit skeletal muscle (Gumieniczek et al. 2001). Similar results have been obtained in the lattisimus dorsi of other diabetic animal models (De Angelis et al. 2000; Matkovics et al. 1997). Changes at the genetic level also contribute to the imbalance. Specifically, expression of heat shock protein 72 and heme oxygenase messenger RNA (mRNA) is decreased in patients with type 2 diabetes. These decreased mRNA levels correlate with insulin-stimulated glucose disposal and markers of muscle oxidative capacity (Bruce, Carey et al. 2003).

The exact downstream effects of oxidative stress on muscles have only recently been elucidated. Synthesis of muscle creatine kinase and both heavy and light chains of myosin in the gastrocnemius of diabetic rats is impaired, suggesting that oxidative stress triggers a cascade of events that lead to impaired muscle repair. The inability to repair damaged skeletal muscle is a characteristic feature of uncontrolled diabetes. Vitamin E administration to diabetic rats reverses the oxidative imbalance and improves muscle gene transcription (Aragno et al. 2004), reinforcing the suggestion that oxidative stress plays a role in diabetes-related muscle damage. The data provide evidence that the pathogenesis of diabetes involves perturbations to the antioxidant defense mechanism within skeletal muscle.

Fatty Acid Metabolism and Diabetes

The metabolism of free fatty acids by skeletal muscle is extremely important. Plasma free fatty acid oxidation rates account for 50% to 80% of the oxidation utilized during moderate aerobic exercise. Other fat sources utilized include intramyocellular triacylglycerol within muscle (Stellingwerff et al. 2007). Numerous reports have connected altered fatty acid metabolism in skeletal muscle with diabetes.

Fatty Acid Uptake

Along with defects in the glucose uptake mechanism and the free radical scavenging system, diabetes appears to be associated with defects in the ability to metabolize fatty acids. Plasma concentrations of free fatty acids play an important role in determining the rate of fatty acid uptake by skeletal muscle. Elevation of plasma free fatty acid levels during insulin release induces skeletal muscle insulin resistance. Skeletal muscle cells from subjects with type 2 diabetes have defects in free fatty acid metabolism that are retained in vitro (Kelley and Simoneau 1994). These defects include reduced oxidative enzyme activity, increased glycolytic activity, and increased lipid content (Schrauwen and Hesselink 2004).

In type 2 diabetes, an accumulation of fatty acids in nonadipose tissues such as heart, liver, and skeletal muscles has been documented. Studies indicate that diabetics store excess triglyceride within skeletal muscle cells (Bruce, Anderson et al. 2003) and that this storage is linked to insulin resistance. Human studies have shown that muscles of subjects with type 2 diabetes take up fatty acids but that the uptake is not associated with an increase in fatty acid oxidation (Blaak 2004). In giant sarcolemmal vesicles prepared from human skeletal muscle, long chain fatty acid transport rates are upregulated approximately fourfold and are associated with an increased intramuscular triacylglycerol content in individuals with type 2 diabetes (Bonen et al. 2004). Interestingly, the increase in skeletal muscle intracellular lipids is similar in type 1 and type 2 diabetes (Crowther et al. 2003).

Fatty Acid Metabolism

Electron micrographs illustrate that muscles from persons with type 2 diabetes have fewer and smaller mitochondria than those from nondiabetic controls (Kelley et al. 2002). We have noted similar changes in the cardiac muscle of diabetic rats (Searls et al. 2004). Figure 11.4a shows normal muscle structure in a micrograph of a cardiac cell from a nondiabetic animal. Note the tightly packed myofibrils and the abundance of intact mitochondria. With diabetes (figure 11.4b), the number of myofibrils decreases significantly, and the mitochondria are swollen (white arrows) and in many cases ruptured. Skeletal muscle tissue from the diabetic rat

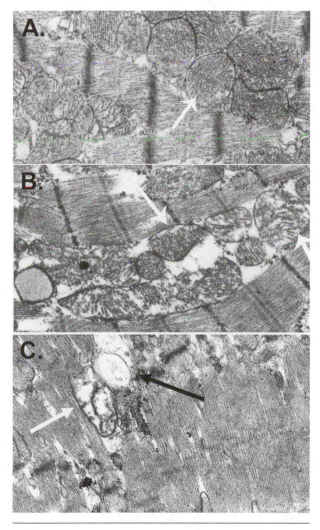

Figure 11.4 Electron micrographs from (a) a nondiabetic cardiac muscle cell, showing healthy myofibrils and an abundance of intact mitochondria (white arrow); (b) cardiac muscle of a diabetic rat, with ruptured mitochondria (white arrow) and fewer myofibrils; (c) skeletal muscle of a diabetic rat with few mitochondria (white arrow) and a large lipid droplet (black arrow).

Micrographs generously provided by Dr. Yvonne Searls.

illustrates similar pathology, but not to the level seen in the cardiac muscle. In skeletal muscle, the myofibril numbers are not affected by diabetes (figure 11.4c). However, the number of mitochondria is drastically reduced, and swollen, ruptured mitochondria are present. In addition, large intracellular lipid droplets are clearly distinguished in the skeletal muscle cells (black arrow).

Mitochondrial Changes

An increased concentration of fatty acids at the mitochondrial membrane, as seen in diabetes, leads to the uptake of neutral fatty acids inside the mitochondrial matrix, which in turn leads to mitochondrial damage (Hamilton and Kamp 1999). Several mitochondrial genes have been associated with type 2 diabetes. One of these genes codes for a mitochondrial intramembrane protease, known as presenilins-associated rhomboid-like protein (PSARL). Expression of the PSARL-coding gene was reduced in the skeletal muscle of diabetic rats, but after exercise training it rose again, successfully mitigating the rats' diabetic symptoms (Walder et al. 2005). PSARL production in human skeletal muscle was correlated with insulin sensitivity as assessed by glucose disposal during a hyperinsulinemic-euglycemic clamp (Walder et al. 2005). Other genes have been shown to be upregulated with diabetes, including skeletal muscle genes that code for acetyl Coenzyme A carboxylase-2 (Debard et al. 2004) and cytochrome oxidase I and III (Antonetti et al. 1995).

Even with the recent identification of mitochondrial gene mutations associated with type 2 diabetes, the exact mechanism of diabetes is unclear and the presentation of the symptoms is variable, ranging from impaired glucose secretion to normal secretion with insulin resistance (Gerbitz et al. 1996). Mitochondrial dysfunction may not be a downstream effect of diabetes but rather a part of the problem. Evidence indicates that a decrease in mitochondrial density and function may lead to an increase in fatty acyl Coenzyme A, which can feed back onto IRS-1, decreasing its ability to be phosphorylated by the insulin receptor (Lowell and Shulman 2005). This could point to a cause of the defective IRS-1 function in diabetes.

Extracellular Matrix Changes

To this point, our discussion of diabetes-induced change in muscle has been limited to intracellular events. However, diabetes causes extracellular changes that lead to muscle dysfunction as well. The extracellular matrix is a complex network of glycoproteins, polysaccharides, and other macromolecules secreted from cells into the extracellular space. The matrix provides a supportive framework, directly influencing various cellular characteristics including shape, motility, strength, flexibility, adhesion, and cellular signaling. The proteins of the extracellular matrix are exposed to the same glucose concentration as is found in the blood. In addition, many of these proteins have an extremely slow turnover and thus experience extended exposure to hyperglycemic conditions. High glucose concentrations increase the synthesis and decrease the degradation of extracellular matrix, leading to a net increase in matrix components.

When matrix proteins are exposed to a hyperglycemic environment for a period of time, cross-links form between the proteins; these correlate with diabetic complications (Ramasamy et al. 2005). The cross-links are most commonly found on collagen (DeGroot 2004) and in skeletal muscle (Jerums et al. 2003; Nagasawa et al. 2003). The rate at which the cross-links form is directly related to the glucose concentration of the environment, making their formation a risk for persons with type 1 or type 2 diabetes. A drug that reduces the formation of glucose-enhanced cross-links has been shown to decrease the rate of progression of diabetes complications in rats. Antioxidants that block cross-link formation in organs also block their formation in skeletal muscle (Nagasawa et al. 2003). However, the extent of cross-link formation with diabetes is not as great in skeletal muscle as in other tissues (Alt et al. 2004).

Although skeletal muscle changes with matrix cross-linking are not well studied, this process is known to be highly important in connective tissue such as tendons. Cross-link formation clearly changes the mechanical properties of tendon tissue. Rabbit Achilles tendons exposed to a glycating environment for four weeks or more showed significant increases in maximum load, energy to yield, and toughness (Reddy et al. 2002). In a time-dependent manner, the stiffness and tensile stress of the tissue increased in the glycated but not the nonglycated tendons. These mechanical changes coincided with a 28% increase in the amount of insoluble collagen in glycated tendons (Reddy et al. 2002). Collectively, these biomechanical and biochemical changes suggest that diabetes-induced cross-link formation of the extracellular matrix alters the biomechanical properties of the tissue, most notably increasing stiffness. One can assume that similar changes are occurring in the extracellular matrix of skeletal muscle, making the muscles less pliable.

Vascular Complications

Some of the most clinically significant complications of diabetes are the structural abnormalities found in blood vessels, resulting in a disruption of normal function of capillary beds (Busija et al. 2004). A hallmark of these pathological processes is a significant and progressive thickening of the blood vessel wall. Vascular complications are common in type 1 diabetic patients, who have a significant reduction in skeletal muscle blood flow even when they tightly control their glycemic levels to within normal ranges (Johansson et al. 2004). While decreased blood flow to the skeletal muscles may result in symptoms including activity-induced pain or cramping, these symptoms appear mild in the context of other complications occurring with diabetes, including kidney disease and blindness. This may explain why the literature in this area is very sparse. However, in rare cases of diabetes, infarction of the skeletal muscle vessels can occur.

Skeletal muscle infarctions were first described in 1965. They usually occur in persons with long-standing type 1 diabetes (Habib et al. 2003) and develop after other complications including nephropathy, retinopathy, or neuropathy (Grigoriadis et al. 2000). Some cases of bilaterally affected calf muscles have been reported (Habib et al. 2003), while the more common symptoms are unilateral and rarely appear in the upper extremities. Little or no evidence of embolic material can be found in review of the literature of skeletal muscle infarcts (Trujillo-Santos 2003). The pathogenesis is likely secondary to diabetes-associated arteriosclerosis and microvascular disease. In support of this concept, some patients with diabetes infarctions have shown severe distal peripheral vascular disease. In addition, alteration in the coagulation cascade due to diabetes may be involved, including impaired tissue plasminogen activator and hypercoagulability (Trujillo-Santos 2003). While chronic infarcts can lead to tissue necrosis and the progressive loss of both strength and muscle bulk (Lee et al. 2004), most cases can be corrected, without complications, with bed rest. In some cases, anticoagulants are recommended (Bjornskov et al. 1995). Recently administration of the proinsulin C peptide increased blood flow in skeletal muscles (Johansson et al. 2004) even in cases without signs of infarct. It is possible that a low level of vascular compromise may be present in skeletal muscle with diabetes, but it rarely rises to the level of being symptomatic.

Exercise

The American Diabetes Association guidelines on exercise in patients with diabetes summarize the value of exercise in the diabetes management plan (Association 2004). The recommendation is a regular physical activity program for all patients with diabetes. The program should be adapted to accommodate the complications that each individual has. Regular exercise has been shown to improve blood glucose control, reduce cardiovascular risk factors, contribute to weight loss, and improve well-being. Furthermore, regular exercise may prevent type 2 diabetes in high-risk individuals (Lazar 2005; Bruce and Hawley 2004). In addition, exercise can have direct effects on diabetes-induced skeletal muscle changes.

The goal of an exercise regimen is very different for a person with type 1 diabetes versus someone with type 2 diabetes. In type 1 diabetes, exercise may be prescribed as a means to prevent or halt the progression of complications, especially cardiovascular disease associated with diabetes. In contrast, exercise for a sedentary person with obesity-induced prediabetes or a recent diagnosis of type 2 diabetes may actually reverse the diabetes so that medications would not be needed to control blood glucose. This is the case because of the robust effect of exercise on insulin resistance in the skeletal muscle. Type 1 diabetics have little or no insulin resistance; thus exercise cannot delay or reverse the onset of diabetes in that population.

Aerobic Exercise Effects

Aerobic exercise has been shown to lead to variety of adaptations in skeletal muscle, one of which involves protection against low-grade inflammation associated with diabetes (Petersen and Pedersen 2005). Additionally, aerobic exercise can have direct effects on the molecules involved in diabetic hyperglycemia. Nonobese elderly men with impaired glucose tolerance performed 12 weeks of endurance exercise training (60-70% of the heart rate reserve) (Kim et al. 2004), and their total GLUT4 protein expression increased significantly. Other findings included a decrease in the intramuscular triglyceride levels, an increase in the fatty acid oxidation capacity, and an increase in the number of capillaries within the vastus lateralis muscle. All of these changes strongly support the conclusion that endurance exercise can limit the development of type 2 diabetes (Kim et al. 2004).

Even a single bout of exercise can be beneficial. For example, transcription of the GLUT4 gene is transiently activated after a single acute session of exercise. GLUT4 protein levels increase as much as two- to threefold after a few days of repeated exercise (Holmes and Dohm 2004). In other studies, eight weeks of exercise training increased insulin-stimulated glucose usage primarily by increasing GLUT4 protein expression, without enhancing insulin-stimulated PI3K signaling (Christ-Roberts et al. 2004). Studies of the transcription regulators that promote expression of the GLUT4 gene have identified two sets of DNA sequences that are important for metabolic regulation and for increased transcription of the gene in response to exercise (Holmes and Dohm 2004). The mechanisms that activate these transcriptional regulators during exercise remain a very important area of research in this field.

As mentioned previously, most data point to proteins upstream of GLUT4 as being defective in diabetes, including IRS-1 and PI3K (see figure 11.3). Yet exercise appears to correct the skeletal muscle problems by increasing the production of GLUT4, the downstream protein. This process is analogous to what occurred in the animal models in which overexpression of GLUT4 resulted in a reversal of diabetic symptoms (Zorzano et al. 2005). Thus, while the molecular problem associated with diabetes may not be in GLUT4 itself, increasing GLUT4 production can overcome the diabetes-induced defect, wherever it is located in the pathway. Thus, a mechanism for the beneficial effects of exercise on skeletal muscle has been elucidated. The increase of glucose transport across the plasma membrane due to enhanced levels of GLUT4 will lower interstitial glucose levels, thereby lowering the person's blood glucose level. In addition, when glucose is within the cell it will be better utilized in an exercise-trained muscle due to upregulation of the metabolic pathways.

In addition to demonstrably improving glucose utilization, exercise induces correction of fatty acid metabolism. Fatty acid transporters increased following seven days of muscle activation (Koonen et al. 2004). In cases of impaired glucose tolerance, 12 weeks of aerobic exercise not only altered GLUT4 gene expression but also decreased the triglyceride levels within the skeletal muscle cells and increased the skeletal muscle fatty acid oxidation capacity (Kim et al. 2004). While fat oxidation rates have been shown to decrease after ingestion of high-fat diets, that trend is reversed with exercise (Achten and Jeukendrup 2004).

Resistance Training

While the positive benefits of aerobic endurance exercise training are widely understood, there is still confusion about the effects of resistance training in diabetes control (Andersen et al. 2003). In a study of strength training in which the participants trained only one leg three times per week for six weeks, leg glucose clearance was increased in the trained legs. Strength training increased the protein content of GLUT4, insulin receptor, and Akt, as well as glycogen synthase levels and activity (Holten et al. 2004). In contrast, another study showed that six months of high-intensity progressive resistance training did not induce any changes in elderly diabetic patients compared to simple weight loss programs (Dunstan et al. 2005). However, when resistance exercise is combined with weight loss, additional benefits can be identified, such as improved glycemic control and increased muscle strength in older patients with type 2 diabetes (Dunstan et al. 2002). With conflicting data on the effects of resistance training for diabetics, such an exercise program should be initiated with caution, especially for persons at risk for high blood pressure.

Exercise Prescription

Aerobic exercise should be prescribed for both type 1 and type 2 diabetes. For type 1 diabetics, the improved effects of antioxidants and the possibility of lowered blood glucose are profound, due to improved glucose utilization. While people with type 1 diabetes who exercise will still be dependent on exogenous insulin, they will likely be able to control their blood glucose levels more accurately with consistent aerobic exercise.

For the type 2 diabetic, the benefits of aerobic exercise are multiple. First, weight reduction is an immediate benefit for diabetes control. Sometimes the loss of 15 to 20 lb (7-9 kg) can transform a person who depended on medications to control blood glucose levels to one who no longer requires any medication for glucose control. In addition to weight loss, exercise induces molecular changes in muscle that reverse the ineffective muscle utilization of glucose and fatty acids. Finally, exercise reverses the molecular deficits associated with diabetes by increasing the production of the glucose transporter, GLUT4.

Prescribing exercise for the diabetic patient should be undertaken with care. The person's risk for cardiovascular complications should be determined, including blood pressure levels, cardiovascular status, and kidney function. Studies indicate that patients with diabetes often lack the warning signs of cardiovascular disease (MiSAD 1997); thus an exercise stress test is warranted before initiation of any vigorous exercise program. Other signs of diabetes complications should be assessed before the start of an exercise program; this should include ruling out neuropathies, loss of vision, or foot impairments. If complications exist, it may not be necessary to prohibit aerobic exercise, but the exercise should be matched to the complication. For example, a person with severe loss of sensation in the lower extremities should not engage in high-impact exercises, including running. Swimming would be a better alternative. Likewise, people with a loss of vision should not exercise outdoors where uneven surfaces are common.

In prescribing aerobic exercise one should follow general guidelines concerning medications, food intake, and the timing of exercise. "Exercise, Medication, and Food Intake With Diabetes" summarizes these guidelines. It is best for type 1 diabetics to exercise 1 to 3 h after a meal, but not when the insulin medication action is peaking. If the blood glucose levels are low before exercise, the person should have a small snack and retest blood levels again before exercising. Type 1 diabetics should not exercise late in the afternoon or in the evening, as they may experience an exercise-induced hypoglycemic event 4 to 6 h after the cessation of the exercise. If they are exercising late in the day, this rebound hypoglycemic event may occur when they are sleeping and could put them at risk for severe hypoglycemia.

Type 2 diabetics should also monitor their blood glucose levels prior to exercising. Patients should be educated so that they understand that exercise can cause hypoglycemic events in some people. When individuals begin an exercise program, it is a good idea for them to maintain blood glucose levels that are at the high end of normal (around 140 mg/dl) to allow an exercise-induced drop that will not lead to a hypoglycemic event.

For type 2 diabetics starting an exercise session with low blood glucose, it is not always wise to suggest that they have a snack. After all, the point of exercise for the type 2 diabetic is weight loss, which is a different goal than for those with type 1 diabetes. "Exercise and Glucose Levels" provides guidelines to assist in deci-

Exercise, Medication, and Food Intake With Diabetes

Type 1 Diabetes

- It is best to exercise 1 to 3 h after a meal, and not when insulin is peaking.
- If blood glucose is low, people should have a snack before beginning a workout.

Type 2 Diabetes

- Blood glucose should be monitored regularly, especially by people with new oral medication prescriptions.
- Exercise alone may reduce blood glucose, causing dangerous hypoglycemia if the person is on medication
- Snacking will undermine exercise effects. Weight reduction is the goal.
- Caffeine, nicotine, and alcohol should be avoided for at least 3 h before exercise.
- Isometric exercises should be avoided especially if high blood pressure is an issue.

Exercise and Glucose Levels

- Avoid exercise if fasting glucose is >250 mg/dl with ketones in the urine.
- Avoid exercise if fasting glucose is >300 mg/dl, regardless of ketone levels.
- Ingest added carbohydrates if glucose levels are <100 mg/dl.
- Avoid exercise if fasting glucose is <60 mg/dl.

sions about eating before the exercise session, as well as clear criteria for when those with either type 1 or type 2 diabetes should avoid exercise due to blood glucose levels or the presence of ketones. Persons with diabetes should be taught how to monitor ketone levels in the urine. Diabetics should not exercise (either aerobically or with resistance training) if their blood glucose levels are above 250 mg/dl with ketones present in the urine. If their blood glucose rises above 300 mg/dl, even without ketones, they should not exercise. If the blood glucose is below 100 mg/dl, they should have a snack and retest their levels later (this applies especially to type 1 diabetics). Finally, no one should be exercising if his or her blood glucose is below 60 mg/dl.

These guidelines are especially helpful to people exercising for the first time. After a pattern of blood glucose changes in response to exercise has developed, it may not be necessary to follow the guidelines as strictly. For example, some diabetics simply cannot control their blood glucose levels in the range suggested for exercise. These people have been termed "brittle diabetics"—no matter how hard they try to control their glucose with diet and medication, their levels bounce from high to low. For these people, an ethical question arises. Should they exercise when their levels are outside the blood glucose guidelines and risk complications, or do clinicians halt their exercise and thereby prohibit any possible benefits? This is a question that a book chapter cannot answer and that can be addressed only on a case-by-case basis by qualified personnel.

Summary

Skeletal muscle has a unique relationship with diabetes. It can exacerbate the disease process or retard progression by playing a critical role in the control of blood glucose. Molecular defects in the insulin pathway responsible for glucose uptake into muscle cells occur in the IRS-1, PI3K, and Akt proteins as well as at other sites. They ultimately reduce the translocation of glucose into the cell through the GLUT4 transporter protein. Aerobic exercise directly improves glucose transport by increasing the production of GLUT4 in the skeletal muscle membrane. Other defects in skeletal muscle associated with diabetes include deficient fatty acid metabolism, oxidative stress, and changes in the extracellular matrix. While exercise may be able to inhibit the effects of altered fatty acid metabolism in skeletal muscle, the most effective mechanism for inhibiting the effects of oxidative stress and extracellular matrix cross-linking is maintenance of blood glucose levels that range below 126 mg/dl. Within the context of this goal, exercise plays an important role in limiting the negative physiological effects of diabetes. This underutilized therapeutic intervention can convert a person in a type 2 diabetic state, necessitating daily medication, to one who is free from medications. The link between this conversion and molecular changes in skeletal muscle has been well documented, and points to the critical role that skeletal muscle plays in the formation and control of diabetes.

CHAPTER 12

Workplace and Other Overuse Injuries

Mary F. Barbe, PhD, and Ann E. Barr, PT, PhD

Musculoskeletal disorders in the workplace have accounted for one-third of occupational illnesses associated with lost work time in U.S. private industry since the mid-1990s (e.g., Bureau of Labor Statistics databases) and cost U.S. health care consumers tens of billions of dollars annually. The prevalence rate of work-related musculoskeletal disorders was 53% in the European Union in the late 1990s (Dupre 2001); and in Canada in 2005, as many as 41% of work-related injuries and disorders resulted from overexertion during pulling, pushing, lifting, or carrying objects (e.g., Alberta Human Resources and Employment 2006 databases).

Epidemiological research has elucidated ergonomic risk factors associated with the development of work-related musculoskeletal disorders (WMSDs). These risk factors include (1) highly repetitive motions; (2) forceful motions; (3) awkward postures, particularly those that are static or at extremes of joint range of motion; (4) prolonged and forceful mechanical coupling between the body and the work environment, as in leaning against or pounding on an object; (5) vibration, either of the whole body (as in standing in a vibrating environment) or of limb segments (as in holding a power tool); (6) cold temperatures; (7) poor lighting; and (8) psychosocial stress involving time pressure, low control over work pace or decisions, or difficult relationships with supervisors or coworkers (National Research Council and Institute of Medicine 2001). Obviously, the etiology of WMSDs is multifactorial, and many workers experience multiple risk factors. Activities outside of work may also contribute to the development of WMSDs through the same

mechanisms, leading to an increase in risk or severity of a disorder or both.

While the ideal management of WMSDs would include prevention through the reduction of known risk factors, it is difficult to protect workers from risk factors in the workplace and elsewhere. Health care providers may not have the opportunity to intervene preventively but must respond to a worker's problems long after the initiating risk factors were encountered or after prolonged risk factor exposure. Therefore, clinicians must understand the pathophysiological injury mechanisms stimulated by workplace risk factors.

Muscle Response to Injuries

Muscle injury is described extensively in chapters 1, 3, and 4. Injury to most cell types is reversible; if the insult is mild and short-lived, the cell may be able to withstand the insult and completely return to normal (Banasik 2000). If the insult is mild and persistent, or both, a structural or functional change may occur within the cell and surrounding tissues to enable these structures to withstand the ongoing stress (adaptation). Sometimes, though, the insult is so severe, prolonged, or repetitive that cell death occurs (irreversible cell injury; degeneration). The extent of cell injury depends on the severity and duration of the insult and also on the prior condition of the cell and surrounding tissues. Well-nourished and -adapted cells and tissues may withstand injury better than aged or only partially repaired cells and tissues. Another key reaction of cells and tissues to injury is inflammation, which leads toward regeneration or repair if it is not too extreme.

Chapter 12 Abbreviations

ATP—adenosine triphosphate
Ca^{2+}—calcium ion
CRP—C-reactive protein
ECM—extracellular matrix
ECRB—extensor carpi radialis brevis
EMG—electromyogram
H_2O_2—hydrogen peroxide
IL-1α—interleukin-1 alpha
IL-1β—interleukin-1 beta

IL-6—interleukin-6
Na^+-K^+—sodium-potassium pump
NSAID—nonsteroidal anti-inflammatory drug
O_2^-—superoxide
OH^-—hydroxyl radical
TNF-α—tumor necrosis factor alpha
UBMA—upper body musculoskeletal assessment
WMSD—work-related musculoskeletal disorder

The common reactions to injury in skeletal muscle may or may not be beneficial to the individual cells within the muscle tissue (figure 12.1). Adaptive, beneficial responses include hypertrophy (increased cell size) and fiber splitting. Nonbeneficial reactions include atrophy (decreased cell size) and necrosis (cell death). Three other reactions—inflammation, repair, and regeneration and remodeling—often overlap and are described in chapters 4 and 6. The major benefit of an inflammatory response, as discussed in chapter 4, is to promote clearance of dead cells and debris as well as repair of damaged myofibers and surrounding tissues. However, as discussed later in this chapter, an inflammatory response can also be harmful to healthy cells and tissues surrounding an injury site if it is not kept in check.

Hypertrophy (enlargement of muscle fibers via addition of newly constructed myofibrils) is a common response to increased physiological demands on muscle. Such adaptive responses, which are discussed in chapter 6, lead to increased skeletal muscle mass and strength. Muscle cells may return to their normal size after the increased demand has ended. However, muscle tissue may not completely return to its original size because of persistent structural changes in the surrounding connective tissues.

Atrophy may result from chronic disuse, denervation, ischemia, nutrient starvation, persistent cell injury, or aging (for review, see Allen et al. 1999). Disuse atrophy occurs in skeletal muscle with prolonged reduction in use, for example after extended bed rest or immobilization or in chronic joint disease. Denervation atrophy occurs following loss of innervation of a muscle. This type of atrophy results in losses in individual muscle fibers, decreasing the overall skeletal muscle size. A sublethal and chronic hypoxia also results in cell atrophy, although full-blown ischemia leads to cell death. Poor blood supply to muscles can lead to chronic nutrient starvation; poor diet and nutrient distribution can also cause cellular starvation. Persistent cell injury from chronic inflammation can lead to muscle atrophy as well.

Hypoxia usually results from inadequate blood delivery. Blood vessel injuries cause blood loss and thus cellular hypoxia. Another cause of hypoxia is a prolonged mechanical deformation of tissues leading to an obstruction of blood flow. Decreased intracellular oxygen delivery to the mitochondria can stall cellular adenosine triphosphate (ATP) production and ATP-dependent pumps, including Na^+-K^+ and Ca^{2+} pumps. Ischemia-reperfusion injury results in cell hypoxia and the abnormal generation of reactive oxygen molecules (oxygen free radicals) in and around the cells (see Gute et al. 1998). Reactive oxygen species, such as superoxide (O_2^-), hydrogen peroxide (H_2O_2), and hydroxyl radicals (OH^-), damage cell membranes, denature proteins, disrupt cell chromosomes, and help initiate inflammatory cascades. White blood cells recruited to the area release lytic enzymes, additional oxygen free radicals, and other chemicals that further damage cells in the area. Irreversible cell injury occurs when the insult is so severe or prolonged that the cells are unable to adapt or to repair themselves.

Cellular and physiologic responses of muscle to injury induced by single bouts of exercise are covered in chapter 4. Other types of muscle injury, such as repeated microtrauma, result in fibrosis (Stauber et al. 1996) and chronic inflammation (Barbe et al. 2003), which further drives the fibrotic response. Stauber and colleagues (1996) found that repeated cycles of muscle strain at fast rates resulted in intercellular-extracellular matrix thickening (evidenced by increased fibronectin deposition) and the formation of intercellular collagen struts. Nikolaou and colleagues (1987) also observed fibrotic repair (fibroblast granulation tissue in repaired zones) after muscle strain injury despite resolution of a previous inflammatory process. Fibrotic repair generally leads to reduced biomechanical strength. In the following section we review experimental findings in animal models of WMSD, as well as findings from humans with these disorders.

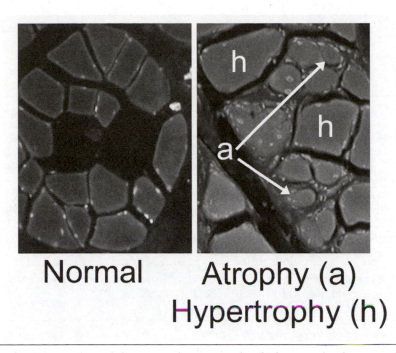

Normal   Atrophy (a)
Hypertrophy (h)

Figure 12.1 Photos showing *(a)* normal muscle in cross section compared to *(b)* denervated and contracted muscle, which contains both atrophied and hypertrophied myofibers.

Workplace- and Overuse-Related Muscular Disorders

To the extent possible, we discuss experimental findings in animals that have been corroborated by recent studies in humans. The use of animal models is particularly important in research on the early development of WMSDs because of lack of access to the tissues of healthy, nondisabled workers with preclinical symptoms. It is the early tissue changes that may provide clinicians and ergonomists with the necessary clues to prevent future, irreversible tissue damage.

Research Using Animal Models

Substantial work over the past two decades has elucidated the mechanisms of muscle injury and their relationship to the type, magnitude, and frequency of contractions. To summarize, eccentric (i.e., lengthening) contractions, particularly of high frequency, produce greater direct and indirect injury, inflammation, and long-term structural tissue changes (refer to chapters 1, 2, and 5) than other types of contractions. While high-magnitude contractions worsen muscle injury and its sequelae, even low-magnitude contractions may induce pathophysiological changes, particularly if they are performed over a period of weeks to months. In this section we discuss key findings on injury, inflammation, and repair and degeneration in muscle tissues in animal models of WMSD or repetitive motion. Table 12.1 lists the findings from various animal models of muscle damage due to repetitive motion.

Nonvoluntary Repetitive Motion Injury Model

Forced lengthening through a physiological range of motion of rat soleus or tibialis anterior muscle during electrical stimulation to tetany leads to injury in as few as five sets of 10 stretches (Fritz and Stauber 1988) and shows a dose-dependent increase in muscle damage between 0 and 300 stretches (Hesselink et al. 1996). Evidence for such injury includes myofiber disruption as well as extracellular matrix (ECM) changes that evolve over a five-day period postexercise. Within 24 h, mononuclear cells are detected, and myofibers take on a swollen appearance with centralization of nuclei (Hesselink et al. 1996). Thereafter, myofibers undergo loss of cross-sectional area (i.e., retraction) with subsequent widening of the interstitial spaces between myofibers, which is evidence of injury-induced edema. By 48 h postexercise, a shift from sulfated to unsulfated proteoglycans and the presence of heparin sulfate in the ECM of muscle tissue indicate that myogenesis has begun. At five days, sulfated proteoglycans again are observed and heparin sulfate undergoes degeneration, suggesting a conclusion to the repair process.

The significance of the findings regarding the ECM molecules, proteoglycans and heparin sulfate, lies in their regulatory function in inducing myogenesis and

Table 12.1 Selected Animal Studies of WMSDs and Repetitive Motion Showing Effects on Muscle Tissues

Authors	Model and protocol	Findings for each study categorized based on tissue and functional changes
Stauber et al. 1994; Stauber et al. 1996	Rat: forced lengthening of soleus muscle at slow (10 mm/s) or fast (25 mm/s) strain rates, 3/week for 4-6 weeks	[3]Hypertrophy, ↑muscle mass, ↑myofiber area (adaptation) after slow stretch; [1 or 3]↑muscle mass, ↓myofiber area after fast stretch; [3 or 4]myofiber splitting and ↑type II A fibers (regeneration) after fast stretch; [3]collagen struts after slow stretch; [4]clear fibrosis after fast stretch
Stauber et al. 2000	Rat: forced lengthening of tetanus toxin–induced hyperactive soleus muscle 50 strains/day, 5/week for 6 weeks, followed by 3 months of cessation of toxin-induced hyperactivity and normal cage activity	[2]Hypervascularity; [3]↓muscle mass, ↓myofiber area; [4 & 5]↑noncontractile tissue, ↑collagen content; [5]incomplete recovery of tissue changes after 3 months
Barr et al. 2002	Rat: reaching and grasping task, 1 reach/30 s, 45 mg of force 2 h/day, 3 days/week for 12 weeks	[2]No increase in serum IL-1α (only serum examined); [6]no motor changes
Barr, Amin, Barbe 2002; Barr et al. 2002; Barbe et al. 2003; Clark et al. 2003	Rat: reaching and grasping task, 1 reach/15 s, 45 mg of force 2 h/day, 3 days/week for 8-12 weeks	[1 & 4]↑hsp72 in distal forelimb and palm by week 3, nerve demyelination beginning week 9; [2]bilateral ↑ in macrophages in nerve and all muscles examined in week 3-6, ↑COX2 and IL-1β in cells of muscles, tendons, CT of distal forelimb and palm, ↑serum IL-1α; [4]bilateral ↑intraneural fibrosis (CTGF and collagen type I) in weeks 8-12; [6]↓NCV of median nerve of reach limb in weeks 9-12, ↓reach rate and task participation, maladaptive movement pattern (raking) beginning in week 4
Clark et al. 2004	Rat: reaching and grasping task, 1 reach/15 s, 180 g of force 2 h/day, 3 days/week for 12 weeks	[2]Bilateral ↑ in macrophages in median nerve; [4]bilateral ↑ intraneural fibrosis (CTGF and collagen type I); [6]bilateral ↓NCV of median nerve, ↓reach rate and task duration, maladaptive movement pattern (raking), bilateral ↑paw withdrawal response threshold to tactile stimulation, bilateral ↓grip strength
LeMay et al. 1990	Rat: exposed to open-field stress (placement in large, bright white pen) 15, 30, or 60 min/day for up to 10 days	[2]↑Serum IL-6 at all exposure times compared to nonexposed controls; [2]↑serum IL-6 positively correlated with stress exposure time; [2]↑body temperature above controls with stress 30 and 60 min exposures; [3]adaptation of IL-6 and body temperature with repeated stress exposures (days 9 and 10)
Hesselink et al. 1996	Rat: isometric or eccentric (forced lengthening) of tibialis anterior 20 times/min for 3-15 min (60-300 contractions)	[1 & 2]↑Fiber swelling, centralization of nuclei, infiltration of mononuclear cells in eccentric exercise of 180 contractions and greater; [1]↑fiber swelling in isometric exercise with 300 contractions; [6]↓peak isometric torque after 60 contractions in eccentric more than isometric
Fritz and Stauber 1988	Rat: forced lengthening of soleus 5 × 10 repetitions with 15 s rest between bouts	[1]↓Chondroitin 6-sulfate around damaged fibers through 72 h postexercise, ↑concanavalin A in interstitial space; [4]↑unsulfated chondroitin proteoglycans around damaged fibers at 48 h postexercise, ↑heparan sulfate around damaged fibers at 72 h postexercise

[1]Injury/degenerative changes; [2]inflammatory/proliferative changes; [3]adaptive tissue changes/tissue reorganization; [4]repair ± regeneration or scarring (fibrosis); [5]pathological remodeling; [6]functional changes (e.g., behavioral or biomechanical).

Abbreviations: COX2 = cyclooxygenase 2; CT = loose areolar and synovial connective tissue; CTGF = connective tissue growth factor; hsp72 = inducible form of heat shock protein 70/72; IL-1α = interleukin-1 alpha; IL-1β = interleukin-1 beta; NCV = nerve conduction velocity; WMSD = work-related musculoskeletal disorder.

providing the scaffolding for proper myofiber orientation during the repair process. Of further significance is the observation that, if given sufficient time without subsequent injurious activity, muscles will undergo effective repair and restoration to their preinjury state or even hypertrophy. However, repetitive motion injury differs substantially from the acute muscle injury model just described; it involves low-magnitude contractions of all types (isometric, concentric, and eccentric), at a high frequency, performed over long periods (i.e., weeks to months). Several investigators have adapted animal models to study the pathophysiology of muscle injury resulting from low-intensity repetitive motion.

In a modified eccentric loading protocol using their rat soleus model, Stauber and colleagues (1994, 1996) studied the effects of both fast (five bouts of 10 repetitions in 5 s) and slow (five bouts of 10 repetitions in 13 s) stretching for three days per week for one month. The slow- and fast-stretch groups showed an increase in soleus muscle mass of 12.8% and 10.4%, respectively. The muscle mass increase in the slow-stretch group was accompanied by muscle hypertrophy and collagen strut formation in the ECM—both indicators of tissue adaptation to increased applied load. On the other hand, the muscle mass increase in the fast-stretch group was accompanied by a widening of interstitial space, increased variability in muscle fiber size with an increase in the number of small and type II A fibers, fiber splitting, infiltration of inflammatory cells (macrophages), and fibrosis. These pathological changes have similarities to those of myopathic diseases and are indicative of a lack of adaptation. The presence of small myofibers, for example, raises the possibility that the fast-stretch muscles are undergoing regeneration or failed healing due to the inability of the normal repair process to keep pace with the repeated muscle injury. This finding represents the essence of repetitive motion injury: The persistent presence of an injury stimulus outpaces tissue repair. Furthermore, recovery from such pathophysiological changes has been shown to be slow, even with complete cessation of the repeated strains for a period of three months (Stauber et al. 2000)—thus highlighting the importance of prevention in the management of such disorders. To cite one example of a preventive measure, Stauber and Willems (2002) compared histopathological evidence of muscle damage in their rat model of forced lengthening using two different interstretch rest breaks, long (180 s) and short (40 s). The longer rest break between stretches prevented such histopathological changes, while the shorter rest break resulted in evidence of muscle damage.

Voluntary Repetitive Motion Injury Model

The finding by Stauber and colleagues of inflammatory cells in injured muscle tissues was not surprising.

Yet the presence of inflammation in repetitive motion injury has not been clearly described with respect to its dose dependence, onset, contribution to signs and symptoms, and role in tissue repair or fibrosis. Work in our laboratory has focused on the inflammatory response to repetitive motion injury of numerous tissues, including muscle, tendon, peripheral nerve, bone, and serum, in a model of repetitive reaching and grasping in the rat (Barbe et al. 2003; Barr and Barbe 2002, 2004; Barr et al. 2002, 2004a, 2000, 2003; Clark et al. 2003, 2004). Unlike previous models of repetitive motion, our model uses a voluntary and functional movement regimen with postural and exposure constraints that permits us to observe behavior as well as to examine tissues in response to different levels of task demands (see Barbe et al. 2003). While we cannot control the loading protocol to the same level of precision as the previously described models, we can use such studies as a frame of reference for interpreting the nature and magnitude of our tissue findings. What we gain by not controlling task exposure precisely is a holistic model that allows observation of animal behavior in response to task exposure.

Using this model we have demonstrated that localized inflammation of injured distal forelimb muscles of the reach limb begins within three weeks of a highly repetitive (8-12 reaches per minute), low-force (<10% maximum voluntary grip force) reaching regimen. Evidence of muscle injury includes the presence of the inducible form of heat shock protein 72 (an indicator of cell distress) and fray at the myotendinous junction of the long digital flexors of the forelimb (an indicator of mechanical stress; figure 12.2) (Barr et al. 2000; Barbe et al. 2003). Inflammatory changes include increased numbers of infiltrating, phagocytic macrophages in the lumbrical muscles, long digital flexors, and surrounding loose connective tissues, which peak at five weeks of task performance. We also observed increased expression of the proinflammatory cytokine interleukin-1 alpha (IL-1α), which peaks between three and five weeks of task performance (figure 12.3), and increased expression of cyclooxygenase-2, a proinflammatory mediator in the arachidonic acid pathway that produces prostaglandins with free radicals as by-products. Free radicals have been implicated in progressive vascular damage and cytotoxicity, thereby worsening pathology in tissues experiencing a prolonged inflammatory response. Prostaglandin E$_2$ has been implicated in muscle pain.

The muscle injury observed in our model is accompanied by a local inflammatory response in the form of increased infiltrating macrophages (Barbe et al. 2003). This same inflammatory response has also been observed in median nerve tissues of rats performing

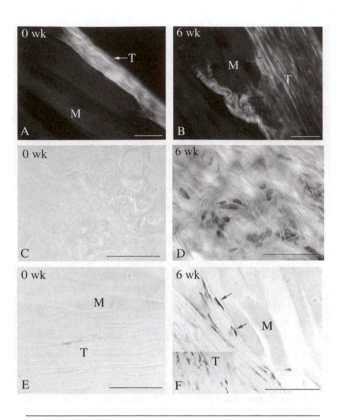

Figure 12.2   Photos showing control rat tissues *(a, c, e)* in comparison to *(b)* microfray at the myofiber-tendinous junction; *(d)* macrophages (dark spots) in connective tissues, *(f)* forelimb flexor muscles, and (inset in *f*) forelimb tendons of rats performing a high-repetition, low-force task for six weeks. M = muscle; T = tendon; arrows indicate macrophages in myofibers at tendon junction. Bar = 50 μm.

From M.F. Barbe et al., 2003, "Chronic repetitive reaching and grasping results in decreased motor performance and widespread tissue responses in a rat model of MSD," *J Orthopedic Res* 21(1): 167-176.

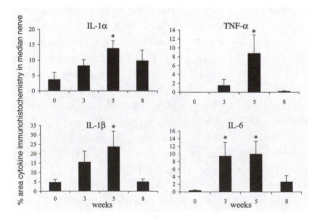

Figure 12.3   The proinflammatory cytokine, IL-1α, increase in median nerve with performance of a high-repetition, low-force task.

Reprinted from *Journal of Neuroimmunology*, Vol. 167, T. Al-Shatti et al., Increase in inflammatory cytokines in median nerves in a rat model of repetitive motion injury, pp. 13-22, Copyright 2005, with permission from Elsevier.

either a highly repetitive low-force or a highly repetitive high-force (50% maximum grip force) task regimen (Clark et al. 2003, 2004). In addition, the median nerve and surrounding tissues show evidence of fibrosis and decreased nerve conduction velocity across the carpal tunnel, with these impairments increasing with increasing task demands (figure 12.4). Results on the dose dependence of the inflammatory responses in muscle tissue are similar to the findings in the median nerve (Barbe et al. submitted).

Behavioral changes consistent with discomfort or movement dysfunction coincide with the tissue changes and also demonstrate dose dependence. These behavioral changes include decreased reach rate (an indicator of animals' inability to maintain task pace), decreased time on the task (an indicator of overall discomfort), and an increase in an abnormal movement pattern known as raking (repeated attempts to reach a target, causing a paradoxical increase in movement repetition) (Barbe et al. 2003; Barr et al. 2000). These behavioral changes begin to emerge as early as three weeks of task performance in a group that performs a highly repetitive and forceful task, and continue to decline through 12 weeks (Barr and Barbe 2004; Clark et al. 2004). With a highly repetitive, low-force task regimen, the behaviors decline at five to six weeks of task performance and then rebound to baseline levels (Barbe et al. 2003; Barr et al. 2000; Barr and Barbe 2004; Clark et al. 2003). When rats perform a task regimen with low repetitions (four to six reaches per minute) and low force, the reach rate and time of task participation do not decline, although the raking behavior does emerge in approximately 50% to 60% of animals by seven weeks of task performance (figure 12.5) (Barr and Barbe 2004; Barr et al. 2002). Taken together with the tissue response, the behavioral changes strongly support findings in the epidemiological literature on WMSDs—that highly repetitive tasks alone are associated with muscle injury and inflammation, with potential long-term tissue scarring and fibrosis, and that the combination of highly repetitive and forceful tasks increases both the tissue response and the ensuing behavioral dysfunction. The finding that task demands can be kept low enough to prevent behavioral changes also suggests that muscle injury may be avoided by reduction in task dose levels.

In addition to the local muscle tissue response induced by our repetitive reaching paradigm in the rat, we have observed evidence of a widespread and even systemic inflammatory response (Barbe et al. 2003; Barr et al. 2002, 2004a). Infiltrating macrophages and proinflammatory cytokines (IL-1α and IL-1β) have been observed in multiple anatomical sites and tissue types (figure 12.6), as well as in serum (figure 12.7)

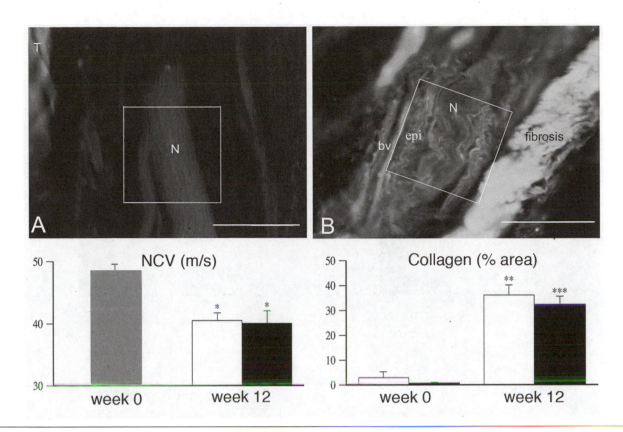

Figure 12.4 Increased collagen type I, indicative of fibrosis, was present in and around the median nerve of *(b)* a 12-week task rat compared to *(a)* a control rat. Nerve conduction velocity (NCV) decreased in median nerves of reach (white bar) and nonreach (black bar) limbs after 12 weeks of performing a high-repetition, high-force task. This decrease coincided with increased collagen around the nerve. *$p < 0.01$; bv = blood vessel; epi = epineurium; N = nerve; bar = 50 µm.

Reprinted, by permission, from B.D. Clark et al., 2003, "Median nerve trauma in a rat model of work-related musculoskeletal disorder," *Journal of Neurotrauma* 20(7): 681-695; reprinted from Clark, Al-Shatti, Barr, Amin, Barbe. Performance of a High-Repetition, High-Force Task Induces Carpal Tunnel Syndrome in Rats. J Orthop Sports Phys Ther. 2004; 34(5): 244-253, with permission from the Orthopaedic and Sports Physical Therapy Sections of the American Physical Therapy Association.

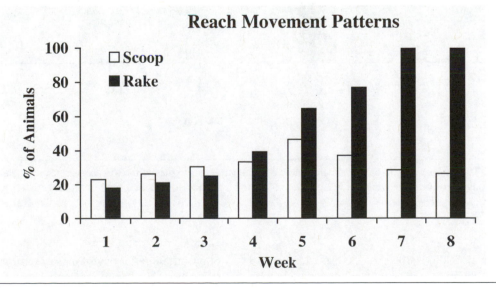

Figure 12.5 A raking-type reaching mode emerged in 100% of the animals by week 7 of performing a high-repetition, low-force task.

From M.F. Barbe et al., 2003, "Chronic repetitive reaching and grasping results in decreased motor performance and widespread tissue responses in a rat model of MSD," *J Orthopedic Res* 21(1): 167-176.

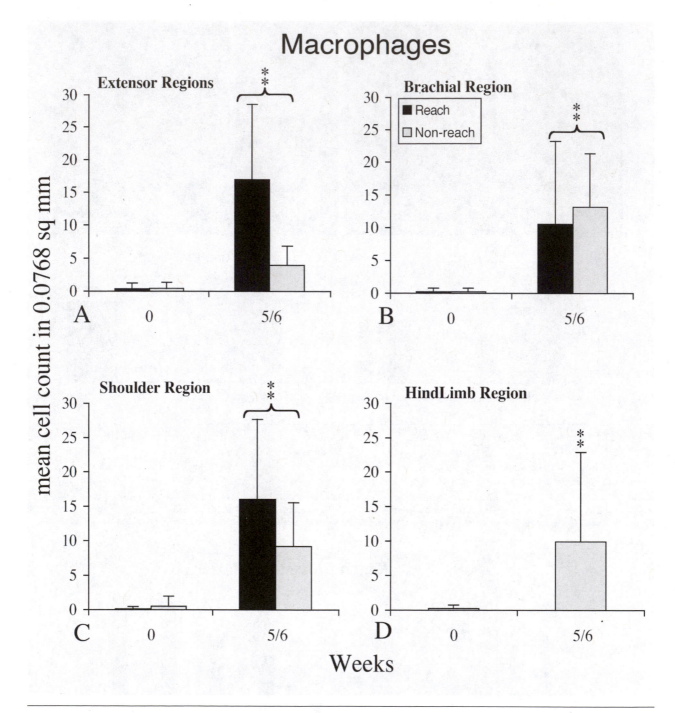

Figure 12.6   The number of macrophages increased in musculotendinous tissues of the forearm extensor, brachial, shoulder, and hindlimb regions after five to six weeks of performance of a high-repetition, low-force task. **$p < 0.001$.

From M.F. Barbe et al., 2003, "Chronic repetitive reaching and grasping results in decreased motor performance and widespread tissue responses in a rat model of MSD," *J Orthopedic Res* 21(1): 167-176.

(Al-Shatti et al. 2005; Barbe et al. 2003). In a quadruped species, the presence of inflammatory mediators in anatomical sites distant from the reach limb may arguably result from localized exposure of these sites due to their participation in the task, but the presence of proinflammatory cytokines in serum is indisputable evidence of a systemic inflammatory response.

We have proposed a mechanism whereby the presence of such a systemic response may lead to amplification of local tissue injury and inflammation through

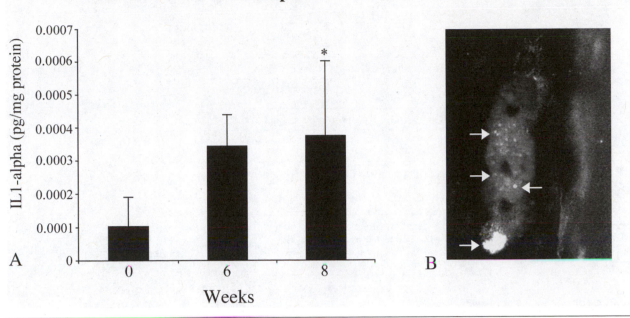

Figure 12.7   *(a)* Serum levels of interleukin-1 alpha (IL-1α) increased significantly in rats during eight weeks of performance of a high-repetition, low-force task compared to controls (*$p < 0.05$). *(b)* Photomicrograph illustrating a muscle macrophage expressing ED-1 immunoreactivity (bright white areas indicated by arrows) that coexpresses IL-1α (gray areas).

From M.F. Barbe et al., 2003, "Chronic repetitive reaching and grasping results in decreased motor performance and widespread tissue responses in a rat model of MSD," *J Orthopedic Res* 21(1): 167-176.

a complex and vicious cycle that renders tissues more susceptible to future injury at constant levels of task demands. In this proposed mechanism, depicted in figure 12.8, a unilateral, repetitive activity induces a localized inflammatory response. If the injury stimulus ceases, an acute inflammatory episode is resolved and healing occurs. If the injury stimulus persists, circulating cytokines will have effects on tissues not directly involved in task performance or global physiological effects. Initiation of the systemic response sensitizes tissues, both local and distant in relation to the injury stimulus, and causes further upregulation of proinflammatory cytokines. Hence the cycle of widespread and chronic effects is propagated.

The "Sickness Response" to Muscle Damage

Also indicated in the proposed mechanism (figure 12.8) is the potential occurrence of global physiological effects due to circulating proinflammatory cytokines. These physiological effects are known as the "sickness response," a constellation of psychoneuroimmunological effects of proinflammatory cytokines, specifically IL-1β, TNF-α, and IL-6, that have been extensively studied in animal models over the past decade. These responses include fever, weakness, listlessness, hyperalgesia,

allodynia, decreased social interaction and exploration, somnolence, decreased sexual activity, and decreased food and water intake (Watkins and Maier 1999; Goehler at al. 1997; Kelley et al. 2003; Dantzer 2004). The sickness response is adaptive; that is, it results in behavior that minimizes energy expenditure in order to allocate metabolic resources to fighting infection or disease (Dantzer 2004). Furthermore, the sickness response was shown to be a motivational state with respect to feeding behavior in rats. Aubert and colleagues (1995) demonstrated that rats injected with IL-1β reduced the frequency of lever presses to receive a food reward but readily ate food that was freely presented. From a motivational standpoint these results suggest that IL-1β produces an aversion to foraging, which is an energy-intensive activity, rather than to feeding per se.

The mechanism of action of the proinflammatory cytokines on such behavioral responses has been partly elucidated but is still a subject of intense research. Recent attention has been given to the possible role of proinflammatory cytokines in the etiology of depression and other mood disorders, particularly among cancer patients treated with proinflammatory cytokine therapy (Capuron and Dantzer 2003). Psychological stress in a rat model has been shown to increase serum levels of IL-6, a cytokine implicated in the acute phase response

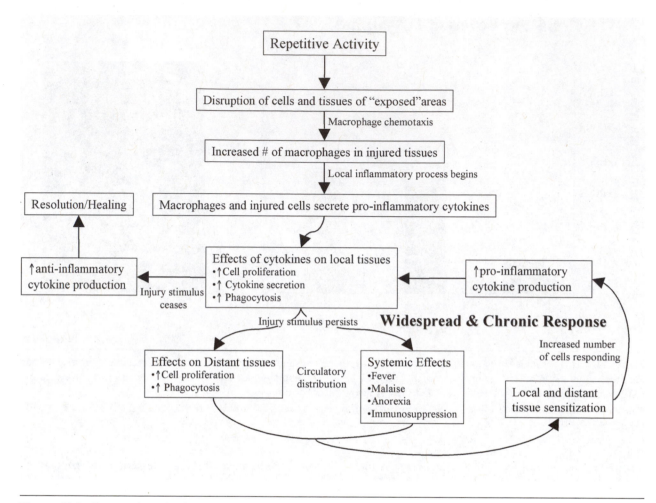

Figure 12.8   Proposed mechanism for the role of systemic distribution of cytokines in widespread WMSD symptoms. ↑ = increase.

Reprinted, by permission, from A.E. Barr, M.F. Barbe and B.D. Clark, 2004, "Systemic inflammatory mediators contribute to widespread effects in work-related musculoskeletal disorders," *Exercise and Sport Sciences Reviews* 32(4): 135-142.

of inflammation (LeMay et al. 1990). The possibility for patients with chronic inflammatory conditions to succumb to the depressive effects of local and systemic proinflammatory cytokines has implications in the management of WMSD. Symptoms of depression and anxiety have been reported in numerous epidemiological and clinical studies of patients with WMSD (e.g., Mathis et al. 1994; Leclerc et al. 2001; Keogh et al. 2000; Weigert et al. 1999). These findings may be attributed to the sickness response, thus suggesting a physiological basis for such symptoms. In our model of repetitive motion injury in the rat, for example, the decreased task participation behavior was accompanied by somnolence, listlessness, and decreased exploration. These behaviors suggest a withdrawal response consistent with sickness behavior. We hope to investigate the relationship between such behaviors and systemic inflammation in future studies in our laboratory.

In summary, animal models of muscle injury, inflammation, and repair in response to repetitive motion tasks have provided rich information regarding the likely mechanisms underlying the development of musculoskeletal disorders in the workplace. These laboratory-based studies have generally supported findings in the epidemiological and clinical literature on workers with WMSD. In the next section, we discuss recent studies in humans that corroborate, to the extent possible, the findings in these animal models.

Research on Humans

Stauber and colleagues (1994) confirmed the mechanisms of injury with maximal eccentric activation as seen in their rat model in a study of biceps brachii in human subjects. Needle biopsies were obtained 48 h after subjects performed 70 maximal isokinetic eccentric contractions of the elbow flexors through a range of 120° at an angular velocity of 120°/s with a 10 s rest between contractions. In addition to significant increases in pain and decreases in relaxed elbow angle seen at 48 h postexercise, histological examination of the biopsied tissues revealed the interstitial widening and increased sulfated proteoglycans in the ECM characteristic of muscle injury. Furthermore, macrophages were observed in the interstitial spaces near blood vessels and in or near necrotic muscle fibers, which confirmed the onset of a delayed inflammatory response to the muscle injury. In a subsequent study, Gibala and colleagues (1995) observed immediate (within 20 min of maximum eccentric exercise) myofibrillar disruption of human biceps brachii myofibers as evidenced by discontinuities in the normal myofibrillar banding pattern seen with electron microscopy. The proportion and severity of disruption were greater with eccentric than with concentric contractions.

While a number of researchers have studied the characteristics of muscle tissues in patients with WMSD (table 12.2), the challenge lies in understanding the mechanisms leading to those observed changes, which are typically examined late in the process of WMSD development (for a perspective on this topic, see Barr and Barbe 2002). For example, Dennett and Fry (1988) performed open biopsies on affected first dorsal interosseous muscles in patients with WMSD and found histological and ultrastructural changes in muscle fibers consistent with denervation or ischemic loss (or both) of type II muscle fibers and hypertrophy of type I fibers. Larsson and colleagues (1990) demonstrated the presence of cellular pathology, in the form of ragged red myofibers, related to mitochondrial dysfunction in trapezius muscle biopsies from symptomatic assembly line workers who were exposed to static loading of the trapezius. The observed changes were consistent with localized hypoxia and were correlated with reduction in muscle blood flow. In biopsies of extensor carpi radialis brevis (ECRB) muscle from patients with work-related epicondylitis, Ljung and colleagues (1999) observed "moth-eaten" muscle fibers, myofiber necrosis, a shift from type II B to the more oxidative type II A myofibers, and increased evidence of myofiber regeneration. These changes are consistent with adaptation to a relative ischemic state and were postulated by Ljung and colleagues to be caused by chronic overactivity in the ECRB muscle. However, some of the patients in this study had a recent history of underuse of the affected upper limb and of steroid injections into the affected muscle, and it is difficult to rule out the influence of these confounding factors on the observed morphological changes.

Trapezius Injury

The majority of studies in humans showing direct structural or physiological muscle changes have involved the trapezius muscle. This is an important site for work-related myalgia resulting from upper limb–intensive tasks given that this muscle has supportive as well as action roles in upper limb function. It is also a frequently affected region in disorders of this type. In studies of male forest-harvesting machine operators or crane operators with work-related trapezius myalgia, Kadi, Hagg, and colleagues (1998) found hypertrophy of type II A fibers, increased vascularization, moth-eaten fibers, centralization of nuclei, presence of developmental myosin, and accumulation of desmin. All of these changes were considered indicative of pathology. An increase above nonoperator control values in the cross-sectional area of type I and II A fibers among these affected workers, also observed in occupational controls without myalgia, was interpreted as an adaptive hypertrophy indicative of chronic exposure to low-level static loading of the trapezius. In females with work-related trapezius myalgia (Kadi, Waling et al. 1998; Lindman et al. 1991), different findings included an increase in cross-sectional area of only type I myofibers, which was interpreted as a hypertrophic adaptive response, and a low capillary to fiber area ratio for both type I and II A fibers that was directly related to reported intensity of pain and hypothesized to cause oxidative stress in these undervascularized tissues. In both males and females with work-related myalgia, the presence of cytochrome c oxidase–negative muscle fibers suggested an oxidative crisis within muscle cells exposed to chronic work-related loading, perhaps through hypoxia (for a review, see Hagg 2000).

An impressive series of studies addressed numerous morphological and physiological muscle changes in female cleaners with trapezius myalgia (Lindman et al. 1991; Larsson, Björk, Elert, and Gerdle 2000; Larsson, Björk, Henriksson et al. 2000; Larsson et al. 2001; Rosendal, Blangsted et al. 2004; Rosendal, Larsson et al. 2004; Larsson et al. 2004). In addition to the morphological evidence for myofiber damage (e.g., moth-eaten fibers; ragged red fibers, characterized by a subsarcolemmal and intermyofibrillary accumulation of mitochondria that appear red with Gomori trichrome stain), the investigators found a decrease in the cross-sectional area of type II fibers (in contrast to the studies discussed in the preceding paragraph) and a correlation between the proportion of ragged red fibers and type I fibers. They also found evidence of cytochrome c oxidase–negative fibers, but these were present in all groups (cleaners with myalgia, cleaners without myalgia, and noncleaner controls) (Larsson, Björk, Henricksson et al. 2000). Regression analysis showed that the prevalence of cytochrome c oxidase–negative ragged red fibers was partly age dependent, irrespective of myalgia.

This last finding is of particular interest given the protective role of estrogen, which stabilizes muscle cell membranes following injury and attenuates leukocyte infiltration in exercise-induced muscle damage (Tiidus 2003; Stupka and Tiidus 2001). The average age of the

Table 12.2 Selected Human Studies of WMSDs and Repetitive Motion Showing Effects on Muscle Tissues

Authors	Subjects and protocol	Findings for each study categorized based on tissue and functional changes
Dennett and Fry 1988	Subjects: 29 pts with painful chronic hand WMSD Measurements: histochemical and morphological examination of first dorsal interosseous biopsies	[2]↑Inflammatory cells; [3]↑type I fibers, ↓number and hypertrophy of type II fibers; [3]mitochondrial changes
Kadi, Hagg, et al. 1998; Kadi, Waling, et al. 1998	Subjects: 21 females with trapezius myalgia (Kadi, Waling, et al.); 10 male machine operators with trapezius myalgia, 9 male machine operators without myalgia, and 6 male healthy controls (Kadi, Hagg, et al.) Measurements: histochemical and morphological examination of trapezius biopsies; pain via VAS	Females: [3]↑CSA and proportion of type I fibers compared to controls from previous study (Lindman et al. 1991); [5]↓capillary: CSA type I fibers in myalgia with high pain scores compared to low pain scores; [5]COX-negative type I fibers > with high pain scores compared to low pain scores Males: [3]↑frequency of type II A fibers; [3]capillarization; [1]centralization of nuclei and developmental myosin in myalgia group compared to controls; [1 & 5]↑mitochondrial disorganization and COX-negative fibers in myalgia group and occupational controls compared to healthy controls
Larsson et al. 1990	Subjects: 17 female assembly workers with work-related trapezius myalgia Measurements: histochemical and morphological examination of trapezius biopsies; local blood flow via laser-Doppler flowmeter	Histological: [1]moth-eaten fibers and ragged red type I fibers worse on more painful side; [1]centralization of fiber nuclei; [1]atrophic fibers; [4]fiber splitting Blood flow: [1 & 5]↓blood flow positively correlated with pain and ragged red fibers
Larsson, Björk, Henriksson, et al. 2000; Larsson et al. 2004	Subjects: 25 female cleaners with work-related trapezius myalgia (CM), 25 female cleaner controls without TM (CC), 21 healthy female teacher controls (TC) Measurements: histochemical and morphological examination of trapezius biopsies	[5]↓Capillary:fiber CSA in CM compared to CC; [1]moth-eaten fibers in CM and CC (4%) > TC (2%); [1]prevalence of ragged red fibers related to working as a cleaner and having tender point in trapezius; [5]COX-negative fibers correlated with age regardless of occupation
Larsson, Bjork, Elert, and Gerdle 2000; Larsson et al. 2001	Subjects: 25 female cleaners with work-related trapezius myalgia (CM), 25 female cleaner controls without TM (CC), 21 healthy female teacher controls (TC) Exercise: 150 maximum isokinetic shoulder flexions through full ROM at 1.05 rad/s Measurements: histochemical and morphological examination of trapezius biopsies, EMG_{RMS}, EMG_{MPF}, RPE, shoulder flexion isokinetic strength and endurance	[6]Functional: strength CM and CC > than TC; [6]endurance CM < CC and TC, RPE CM > TC > CC, ↓relaxation (i.e., ↑EMG_{RMS}) during passive shoulder extension in CM and CC and decreased with increasing age Histological: [3]↓type II fiber CSA in CM and CC compared to TC; [3]difference between CSA of type I and type II fibers greater in CM and CC than in TC; [1]prevalence of ragged red fibers positively correlated with proportion of type I fibers Correlation between functional and histological: [1]prevalence of ragged red fibers correlated with EMG_{MPF}; [1, 3, 6]prevalence of ragged red fibers, age, fiber type proportion and CSA, and occupation were significant regressors of inability to relax during passive shoulder extension
Lindman et al. 1991	Subjects: females with chronic trapezius myalgia and healthy female controls Measurements: histochemical and morphological examination of trapezius biopsies	[3]↑Type I fiber CSA; [5]↓capillary:CSA for type I and II fibers; [1 or 5]↓ATP and ↓phosphocreatine in type I and II fibers in pts compared to controls

(continued)

Table 12.2 *(continued)*

Authors	Subjects and protocol	Findings for each study categorized based on tissue and functional changes
Ljung et al. 1999	Subjects: 26 pts with lateral epicondylitis >7 months Measurements: histochemical and morphological examination of ECRB muscle biopsies	[1 or 5]Abnormal muscle NADH staining, muscle necrosis; [2]no evidence of muscle inflammation; [3]↑type II A fibers and muscle fiber regeneration
Nemet et al. 2002	Subjects: 23 healthy adults (15 male and 8 female) Exercise: 10 min of unilateral wrist flexion at 40% MVC Measurements: serum metabolites, catecholamines, growth factors, PBMCs, adhesion molecules, and cytokines sampled via heparin-lock catheter from exercised and nonexercised forearms	Postexercise: ↑lactate in exercise arm only; [3]↑GH, Epi (> in exercise arm), Norepi (> in nonexercise arm); [3]VEGF; [2 or 3]IL-6; [2]IL-1ra and [2]PBMCs in both arms; [4]↓FGF-2 in both arms, no gender differences Postexercise recovery period: ↓lactate to BL at 60 min; [3]peak GH level at 10 min and ↓ to BL at 120 min; [3]↓Epi to BL at 30 min; [3]peak Norepi at 10 min, remained slightly elevated at 120 min; [2]↑VEGF; [2 or 3]IL-6 and [2]IL-1ra continued through 120 min; [4]↓PBMCs to BL at 30 min; [4]↑FGF-2 at 120 min
Rosendal, Larsson, et al. 2004	Subjects: 20 healthy females, 19 females with trapezius myalgia Exercise: 20 min repetitive, low-force exercise (placement of wooden pegs in a desktop peg board) Measurements: microdialysis measurement of interstitial metabolites, pain and PPT, blood flow	Preexercise: [1 or 5]↑lactate, pyruvate, glutamate, serotonin, and pain in myalgia group; [6]↓PPT in myalgia group; [5 & 6]serotonin correlated to pain and glutamate correlated to pain and PPT in myalgia group Exercise: [1 or 5]↑lactate and pyruvate in myalgia group, but not in control group; [3]↑glutamate in both groups Postexercise: [5]↑blood flow and ↑pyruvate prolonged in myalgia group; [4]lactate returned to BL in myalgia group; [4]glutamate returned to BL in both groups
Rosendal, Blangsted, et al. 2004; Rosendal, Sogaard et al. 2004	Subjects: 6 healthy males Exercise: 20 min repetitive, low-force exercise (placement of wooden pegs in a desktop peg board) Measurements: microdialysis measurement of interstitial metabolites and IL-6, serum metabolites and IL-6, EMG$_{RMS}$, EMG$_{MPF}$, RPE	Exercise: [6]↑EMG$_{RMS}$ and RPE, lactate; [1]potassium and [2 or 3]IL-6; [6]↓EMG$_{MPF}$ Postexercise: ↑lactate and [1]potassium for 10 min, returned to BL by 20 min; [1]↑pyruvate 20 min postexercise; [2 or 3]peak IL-6 levels 30 min postexercise; serum IL-6 nonsignificantly increased 45%
Stauber et al. 1990	Subjects: 3 healthy males and 2 healthy females Exercise: 70 maximal isokinetic, eccentric movements of the elbow flexors through 120° ROM at 120 °/s Measurements: perceived soreness, relaxed elbow joint angle, histochemical and morphological examination of biceps brachii biopsies	[6]Functional: progressive ↑perceived soreness immediately and 48 h postexercise, progressive ↓relaxed elbow joint angle immediately and 48 h postexercise Morphological: [1]↑area of interstitial space; [1]presence of mononuclear cells and [1]degranulating mast cells; [1]necrotic fibers (2%) Histochemical: [1]↑albumin and fibrinogen in widened interstitial space and in damaged fibers; [4]↑chondroitin 6-sulfate and chondroitin 4-sulfate proteoglycans and concanavalin A in widened interstitial space

(continued)

Table 12.2   *(continued)*

Authors	Subjects and protocol	Findings for each study categorized based on tissue and functional changes
Gibala et al. 1995	Subjects: 8 healthy males Exercise: resisted eccentric movements of elbow flexors using dumbbell on one side, resisted concentric movements of elbow flexors on other side Measurements: low-velocity and high-velocity isokinetic strength and isometric strength of biceps brachii; PTT, TPT, HRT, and MRD via percutaneous stimulation of biceps brachii; EMG$_{RMS}$; morphological examination of biceps brachii biopsies	[6]Functional: ↓isometric and low-velocity and high-velocity isokinetic strength concentric arm immediately postexercise, ↓isometric and low-velocity isokinetic strength eccentric arm for 96 h postexercise, ↓high-velocity isokinetic strength eccentric arm for 72 h postexercise, ↓isometric and low-velocity isokinetic strength eccentric arm compared to concentric arm for 96 h postexercise, ↓high-velocity isokinetic strength eccentric arm compared to concentric arm 72 h postexercise, ↓PTT and MRD concentric arm immediately postexercise, ↓PTT and MRD eccentric arm for 96 h postexercise, ↓HRT eccentric arm for 48 h postexercise, PTT and MRD concentric arm > eccentric arm for 96 h postexercise, EMG$_{RMS}$ eccentric arm < concentric arm during exercise Morphological: [1]↑fiber disruption in both concentric and eccentric arm immediately and 48 h postexercise; [1]fiber disruption greater eccentric arm compared to concentric arm immediately and 48 h postexercise Relationship between functional and morphological changes: [6]%↑isokinetic strength in eccentric and concentric arms correlated with [1]% severely disrupted fibers 48 h postexercise

[1]Injury/degenerative changes; [2]inflammatory/proliferative changes; [3]adaptive tissue changes/tissue reorganization; [4]repair ± regeneration or scarring (fibrosis); [5]pathological remodeling; [6]functional changes (e.g., behavioral or biomechanical).

Abbreviations: ATP = adenosine triphosphate; BL = baseline; COX = cytochrome c oxidase; CSA = cross-sectional area; ECRB = extensor carpi radialis brevis; EMG$_{MPF}$ = mean power frequency electromyography; EMG$_{RMS}$ = root mean square electromyography; Epi = epinephrine; FGF-2 = fibroblast growth factor; GH = growth hormone; HRT = one-half relaxation time; IL-1ra = interleukin-1 receptor antagonist; IL-6 = interleukin-6; MRD = maximum rate of torque development; MVC = maximum voluntary contraction; NADH = nicotine adenine dinucleotide reductase; Norepi = norepinephrine; PBMCs = peripheral blood mononuclear cells; PPT = pressure pain threshold; pts = patients; PTT = peak twitch torque; ROM = range of motion; RPE = rating of perceived exertion; TPT = time to peak torque; VAS = visual analog scale for pain perception; VEGF = vascular endothelial growth factor; WMSD = work-related musculoskeletal disorder.

cleaners in the studies by Larsson and Lindman was 46 ± 10 years, placing the older subjects in the menopausal or postmenopausal age range. Therefore, the finding of myofiber distress (as indicated by cytochrome c oxidase–negative ragged red fibers) in older subjects may indicate that they lacked protective levels of estrogen. This study also showed that a significantly larger proportion of cytochrome c oxidase–superpositive fibers were also ragged red fibers in women with work-related myalgia, although whether this is an adaptive or a pathological response is still unclear. Point tenderness of the trapezius on palpation was also associated with increased prevalence of ragged red fibers. Furthermore, the proportion of ragged red fibers was positively correlated with the mean power frequency of the trapezius electromyogram (EMG) during an endurance-type test using an isokinetic dynamometer. Such a frequency shift of the EMG signal is indicative of muscle fatigue. Thus, these investigators were able to relate muscle morphological changes to positive physical examination findings and functional movement deficits.

In a further attempt to relate exercise performance and myalgia in female cleaners with WMSD, Larsson, Björk, Elert, and Gerdle (2000) demonstrated that while cleaners were stronger than age-matched controls, cleaners with trapezius myalgia had a lower level of endurance and a higher degree of perceived fatigue following a fatiguing shoulder flexion isokinetic exercise protocol. Of further note was the decreased ability of cleaners, both with and without myalgia, to relax the trapezius during the passive portion (i.e., shoulder extension phase) of the exercise protocol. While this EMG finding was also partly related to advancing age in the cleaning group, when taken together with the histological and clinical findings it indicates an abnormality in muscular oxidative metabolism among these women with chronic, occupation-related trapezius loading.

One of the limitations of studying workers with myalgia is the difficulty in determining the causality of their tissue and behavioral responses. Presumably, the initiating injury stimulus is long since past, and the condition of the tissues and the motor control system has been substantially altered from the preinjury state. Therefore, it is impossible to conclude whether histochemical changes cause or follow the physiological mechanisms leading to the patient's current clinical presentation. Such uncertainty makes intervention planning, particularly prevention, difficult at best. The animal models discussed earlier have helped enormously to determine the early development of repetitive motion disorders.

Acute Low-Force Repetitive Exercise

Recently, investigators have begun to develop studies in humans that allow confirmation of some of the physiological effects of acute, low-force repetitive exercise. Such studies are possible with the use of microdialysis technique, in which small, minimally invasive catheters of semipermeable membranes are inserted into the muscle ECM during activity. The researcher then samples dialysate at appropriate time intervals in order to follow the metabolic changes in the muscle tissue.

Using microdialysis technique in healthy subjects and in women with chronic work-related trapezius myalgia, Rosendal, Blangsted, and colleagues (2004) and Rosendal, Sogaard, and colleagues (2004) have shown that highly repetitive, low-force arm movements performed for 20 min cause increases in interstitial muscle lactate, potassium, pyruvate, and IL-6. Increased lactate is indicative of an increase in anaerobic metabolism, or glycolysis. The authors concluded that the rise in anaerobic metabolism may have been caused by decreased local blood flow, inhomogeneous muscle activation, or both (Rosendal, Blangsted et al. 2004). Interstitial potassium accumulation plays a role in the regulation of blood flow, development of muscle fatigue, and sensation of pain (Rosendal, Blangsted et al. 2004). The cytokine IL-6 has pleiotropic (i.e., both pro- and anti-inflammatory) effects and has been implicated as one factor that causes exercise-induced enhancement of the immune system in low-grade inflammatory conditions (Starkie et al. 2003). It may also work in a hormonal fashion to inhibit glycogen synthase activity and facilitate glycogen phosphorylase activity, thereby increasing muscle energy supply during exercise (Kanemaki et al. 1998). Interleukin-6 also induces lipolysis (Lyngso et al. 2002), and it is upregulated in persistent pain syndromes (Winkelstein 2004). Therefore, it has immunological, neurological, and metabolic regulatory effects in muscle tissue. Women with trapezius myalgia had higher levels of interstitial glutamate and serotonin than healthy controls before exercise, which correlated with muscle pain intensity and pressure pain threshold; and glutamate increased in both groups with exercise (Rosendal, Larsson et al. 2004). These substances both act to increase the sensation of pain (Babenko et al. 1999; Zimmermann and Herdegen 1996).

In this short-term protocol with low exercise intensity, no changes in serum concentration in these metabolites or cytokines were observed. This is not surprising considering that in our rat model of low-force, repetitive reaching and grasping, tissue levels of inflammatory mediators, including cytokines, increased gradually over a period of weeks and preceded increases in proinflammatory cytokines in serum (Barr et al. 2000; Barbe et al. 2003; Barr and Barbe 2004). In another study using more intensive exercise (40% maximum voluntary contraction) of the wrist extensors for only 10 min, serum levels of IL-6 and IL-1 receptor antagonist increased (Nemet et al. 2002). Hence, an exercise-induced systemic inflammatory response has been demonstrated in humans.

In a study recently completed in our lab, we examined the serum of patients with WMSD of less than three months' duration for the proinflammatory markers TNF-α, IL-1β, IL-6, and C-reactive protein (CRP). In these early-onset cases, severity of symptoms and physical examination findings were positively correlated with TNF-α, IL-1β, and CRP (figure 12.9; Carp et al. 2007). Interleukin-6 showed a more complex relationship, with elevation in mildly and severely affected patients but not in moderately affected patients. This finding reflects the multiple and sometimes competing roles of this cytokine.

In summary, the human studies just discussed provide evidence that the performance of maximal eccentric activation tasks, low-intensity repetitive motion tasks, and other work tasks leads to muscle injury, inflammation, and repair. In the next section we describe some possible approaches to intervention for the management of WMSD as suggested by this research.

Potential Interventions

There is increasing awareness of the need to understand treatment principles for WMSD. Injury-induced release of proinflammatory cytokines can be addressed by a number of clinical interventions. Use of antagonists to specific cytokines, such as inhibitors of TNF-α, might prove useful in severe cases of WMSD. However, this type of drug would be effective only if administered during the inflammatory stage of the disorder. Blood tests for the presence of elevated serum levels of inflammatory cytokines could serve as indicators of a systemic inflammatory process. Unfortunately, a TNF-α inhibitor would not inhibit functionally redundant cytokines such as IL-1 and thus may be of limited use. This treatment regimen has yet to be tested in early-onset WMSDs.

Nonsteroidal anti-inflammatory drugs (NSAIDs) are often taken by workers to treat pain as it arises during a work week and are also often prescribed at the time of treatment if pain is present. However, the available scientific data do not support their use. Recent studies have suggested that although NSAIDs decrease the inflammatory response (number of inflammatory cells), they also may delay the recovery process and rate of muscle fiber regeneration. In animal models

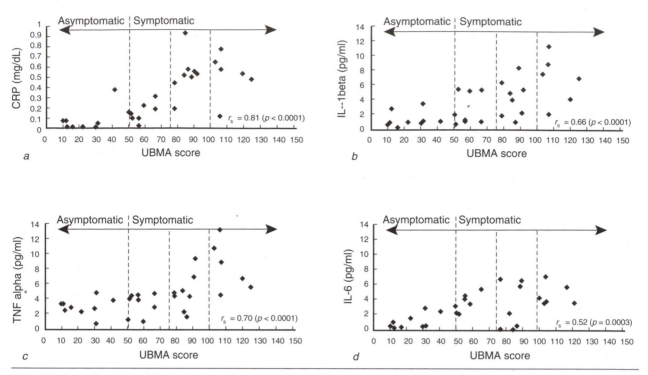

Figure 12.9   Inflammatory biomarker concentrations versus upper body musculoskeletal assessment (UBMA) scores for all symptomatic ($n = 22$) and asymptomatic ($n = 9$) subjects showing Spearman's rank correlation coefficients (r_s). (a) C-reactive protein (CRP); (b) interleukin-1β (IL-1 beta); (c) tumor necrosis factor α (TNF alpha); and (d) interleukin-6 (IL-6).

Reproduced with permission from S.J. Carp et al., 2007, *Clinical Science,* Vol. 112, pp. 305-314. © The Biochemical Society.

of experimental strain, administration of the NSAIDs piroxicam (Feldene), flurbiprofen, or diclofenac beginning immediately postinjury was initially beneficial (Almekinders and Gilbert 1986; Lapointe et al. 2002; Mishra et al. 1995). A few days postinjury, treated rats showed improved contractile functions, significantly less edema, and decreased levels of immune cells and muscle enzymes in the injured areas compared to untreated or placebo-treated injured rats. Unfortunately, continued administration of the NSAIDs did not maintain the improved contractile function, but instead resulted in a decline compared to values in untreated injured rats. Continued administration of the NSAIDs also resulted in delayed resolution of the macrophage response (greater number of macrophages) and delayed myofiber regeneration compared to what occurred in untreated injured rats. Thus, a generalized repression of inflammation appears to delay regenerative responses, apparently by repressing clearance of damaged myofibers by macrophages and perhaps their secretion of proliferative growth factors that would lead to repair and a regain of strength.

Treatments for injury-induced edema such as pulsed ultrasound or ice would be useful only if workers were instructed to self-administer or otherwise access them immediately, which is not typical in the workplace. Workers generally seek medical attention long after the initial swelling phase has passed.

Summary

All of the findings in animal and human experiments on local and systemic increases in metabolites and cytokines lend credence to the hypothesis that localized tissue-level responses to exercise loading lead to the resolution of any injury through normal repair processes as long as the initiating stimulus is removed. The findings also show that low-intensity exercise can cause injuries to muscle through mechanical disruption, hypoxia, or by-products of metabolism. Once injury occurs, an immune response is initiated. In cases in which the injury stimulus persists, this immune response is amplified and perpetuated until its mediators are distributed throughout the circulatory system. Such circulatory distribution may cause tissue effects in distant anatomical sites and may have psychoneuroimmunological effects on mood and behavior. Furthermore, if task intensity is high enough, the normal tissue repair process may be interrupted to the extent that permanent tissue changes occur, thereby rendering tissues more susceptible to future injury even at previously innocuous levels of task intensity. Obviously, this vicious cycle of injury, inflammation, scarring, and behavioral dysfunction should be avoided.

CHAPTER 13

Muscle Function During Human Gait

Michael Pierrynowski, PhD

Voluntarily generated muscle force, acting across mobile body segments, is the motor that drives human activity. Rarely acting in isolation, our 650-plus named muscles allow us to breathe air, chew food, manipulate objects, and most importantly explore our environment. Frequently, environmental exploration requires that we independently move our body from one location to another. For most people, the most economical way to do this is to walk with a cyclical motion pattern. Observations of personal walking patterns within identical environments reveal that they differ between individuals. Potentially more importantly, the walking pattern often varies within an individual over time. These time-varying walking patterns between and within individuals imply that the muscles used in walking must be varying their force patterns. Of interest in this book are the time-varying alterations in muscle force output within an individual due to muscle damage.

This chapter is divided into four sections. The first section presents a schema for characterizing skeletal muscle function during activity. Of importance is the concept that a muscle's contribution is influenced by its cross-sectional area, length, velocity, force output, and placement. A muscle with a small cross-sectional area, or one with a short or long length relative to its rest length, or one that is generating force at high shortening speed has a reduced ability to generate force. Further, if the muscle acts through a small moment arm, it contributes little to the provision of motion. Alterations to a muscle's active cross-sectional area–length–velocity–force envelope, such as those due to muscle damage, modify the role of that muscle during normal function.

The second section outlines techniques for measuring muscle function during gait. To fully understand the impact of skeletal muscle damage on walking, we need to know how the locomotor muscles function. This section presents an overview of how movement scientists characterize muscle function during gait and more specifically discusses the techniques used to quantify muscle force patterns within a clinical gait analysis and their limitations.

The third section presents both graphic and verbal descriptions of normal muscle function during gait, specifically characterizing the action of seven major muscle groups involved in normal locomotion. The functions of these muscles are discussed within the context of the role of the locomotor muscles during normal gait. During gait these muscles act in concert to maintain dynamic joint stability, as well as providing propulsion and shock absorption.

The last section of the chapter describes the impact of muscle damage on gait. Muscle damage is defined as any insult to a locomotor muscle group that reduces its ability to contribute to normal function during gait. As will be seen, an insult to a single muscle group influences gait by shifting its contribution to other muscles. Fortunately, the human locomotor muscular system is highly redundant, with several muscles capable of performing a given function. Although patients with muscle damage can still walk, the normal pattern will be disrupted.

Characterizing Muscle Function

Activated skeletal muscle produces tension that is transferred via elastic structures to the skeleton. These elastic components exist within the contractile machinery and the tendons. Stretching inactivated muscle generates tension due to a passive elastic component parallel to the {Zahalak, 1990 148 /id}force-generating muscle fibers and the tendon. The passive tension–length function of a muscle varies with the percentage of connective tissue (Zahalak 1990){Zahalak, 1990 148 /id}{Zahalak, 1990 148 /id}{Zahalak, 1990 148 /id}. A skeletal muscle without elastic elements can be conceptually modeled as a contractile element (Winters 1990). The contractile element generates a linear force that is transmitted through the elastic components to the skeleton through changing moment arms to generate angular and translational effects at a joint. The action of external and gravitational forces on the skeleton, in concert with the action of a set of the skeletal muscles, causes the given muscle-tendon unit to shorten, to remain at a constant length, or to lengthen.

Muscle force output is thought to arise from the summation of the interactions of cross-bridge sites on specialized actin and myosin contractile filaments (Huxley and Simmons 1971; Pollack 1983). Muscles with a greater number of parallel active filaments, quantified using the muscle's cross-sectional area (CSA), have the potential to generate greater force. Relative motion between these numerous filaments allows a muscle to change length and generate force (Gordon, Huxley, and Julian 1966). One of the earliest predictions from this sliding filament theory of contraction was that as a muscle is lengthened or shortened from its position of maximal overlap, the overlap of the myosin filaments diminishes and the tension developed decreases. Numerous investigations have confirmed this theoretical prediction.

In 1938, Hill determined that the previously well-known inverse relationship between muscle force and speed of contraction near rest length could be represented by a hyperbolic function that relates increasing shortening velocity to decreasing force output. Additionally, the parameters that defined the hyperbolic equation were associated with the muscle's energy production during contraction. However, Hill's equation has a number of deficiencies, one of which is that it is confined to concentric contractions even though a large proportion of muscle activity is eccentric (Eston, Mickleborough, and Baltzopoulos 1995). Notably, eccentric muscle contractions generate greater than isometric force outputs, and eccentric contractions prior to concentric contractions enhance force output, each with reduced metabolic cost (Zahalak 1990).

To fully characterize a skeletal muscle's function during activity, one must measure and then interpret its CSA–length–velocity–force–moment arm profile. Although measuring the CSA–length–velocity–force–moment arm profile of in vitro muscles is technically exacting and requires specialized equipment, it is done routinely (Edman, Elzinga, and Noble 1978). Unfortunately, obtain-

ing similar information noninvasively from an in vivo muscle during activity is problematic. Instead, to estimate these profiles, movement scientists measure segment and joint anthropometrics, joint angular positions, angular velocities, and net joint moments of force; model the location and line of action of the muscles crossing these joints; and apply a theoretical load-partitioning strategy (Zajac and Winters 1990). Frequently, subsets of the CSA–angular position–angular velocity–net joint moment of force profile of planar, single-joint, and constant moment arm muscles are used to describe muscle function during activity (Yamaguchi and Zajac 1989).

An alternative approach is to use a dispersive action muscle function model. Championed by Zajac and Gordon (Zajac and Gordon 1989) and more recently by Kepple and colleagues (1997, 2000), this model estimates the action of multidimensional, multijoint, and changing moment arm muscles on all of the body segments. In essence this model couples and simplifies the Newton-Euler equations to directly relate muscle force to the resultant movement. The resulting equation states that each net joint moment of force accelerates all joints in proportion to its magnitude and segmental configuration, and is independent of joint velocity. Thus each muscle has the ability to control motion at all other joints. This model of dispersive action muscle function explains, in part, potential alternative muscle control strategies when some muscles are damaged. However, although it provides insight into control strategy, the model downplays intrinsic muscle function (i.e., CSA–length–velocity–force–moment arm profiles).

Using either approach, movement scientists are able to characterize a muscle's CSA–length–velocity–force–moment arm profile and can quantify a change due to muscle damage. Damage to a muscle has the potential to reduce its active CSA and to limit both length and velocity, all of which reduce the muscle's force potential.

This chapter uses angular position, angular velocity, and net moment of force at the hip, knee, and ankle during a typical gait cycle to represent the length, velocity, and force output of eight hypothetical uniplanar, single-joint, constant moment arm locomotor muscles. Four of these muscles cross the hip and generate pure sagittal plane flexion and extension and frontal plane abduction and adduction moments of force. The other four muscles act in the sagittal plane and produce knee flexion and extension and ankle plantarflexion and dorsiflexion.

Measuring Muscle Function During Gait

The application of scientific principles to the study of human motion began during the Renaissance with the work of such men as Leonardo da Vinci (1452-1519) and Borelli (1609-1679), who studied human and animal muscle function. It was not until the latter half of the 19th century and into the 20th century, when reliable and accurate methods for recording movement were developed, that the laws of mechanics were applied to the human body (Braune and Fisher 1889; Bresler and Frankel 1950; Elftman 1938, 1939). In the past 30 years, the availability of computers has been instrumental in the growth of this branch of science, now called biomechanics, and in particular clinical gait analysis.

The objective of clinical gait analysis is to quantify the measurable components of gait. However, some investigators apply anthropometric, kinematic, and kinetic data to models of the lower limb to estimate the net joint reaction forces and net moments of force. A smaller subset of the biomechanics literature defines models representing the internal structures and control systems in order to partition the net joint moments of force among muscles and to identify the strategies used to activate the muscles during gait.

Characterizing Gait

The aim of clinical gait analysis is to faithfully and repeatedly capture nuances of an individual's gait that are generally not available from other assessments. This section highlights the most common gait measures: anthropometric, temporal and spatial, kinematic, external kinetic, internal kinetic, and control objectives and electromyographic (EMG) measures.

- **Anthropometric measures.** Clinical gait analysis requires knowledge of the characteristics of the patient's modeled body segments. These include segment lengths, location of the patient's joint centers, and the segments' inertial characteristics (mass, center of mass location, moment of inertia). Estimates of the line of action and the muscle moment arms are also required. These values are often estimated from measurements of a set of anthropometric characteristics (i.e., total body mass and height; a segment's length, width, and circumference) followed by the use of regression equations to predict the values of interest.

- **Temporal and spatial measures.** Stride length, cadence, and progression velocity are examples of the most commonly reported temporal and spatial gait characteristics used to describe the global outcome of a person's gait pattern. They are in close company with right- and left-foot stance, swing and double support times, step length, and stance width, each of which defines a characteristic of a person's intralimb gait pattern. At an even more detailed level, the temporal durations of intralimb gait phases (loading response, midstance, terminal stance, preswing, initial swing, midswing, and terminal swing) are provided within a temporal-spatial clinical gait analysis report. Recently, detailed examination of the temporal sequence of stride intervals has provided insight into how an individual integrates sensory feedback, inherent mechanical properties, and central nervous system drive to control gait (Gates, Su, and Dingwell 2007).

- **Kinematic measures.** Kinematic outcomes describe the linear and angular motions of the modeled segments and joints. Typically the segments include the pelvis, thighs, legs, and feet and the hip, knee, and ankle joints. Since each segment and joint has three linear motion components (forward-backward, outward-inward, upward-downward) and three angular motion components (flexion-extension, abduction-adduction, internal-external rotation), the seven segments and six joints have a total of 78 motion components at each instant of time. These kinematic values are typically presented as figures with the horizontal time axis representing one gait cycle and the motion components individually displayed on the vertical axis. Since human gait displays intrasubject variability, a typical gait session includes a number of "good" trials. The motion components from these trials are then averaged, and a measure of variability is calculated and displayed.

- **External kinetic measures.** External kinetic outcomes encompass the interaction of the body with its external environment. Investigations using these measures primarily examine the forces and pressures exerted by the foot or feet on the ground during independent ambulation, but the forces exerted on the floor by gait aids (crutches, walkers), the forces exerted by the hands on a wheelchair, and the interaction of the air against the moving body have also been examined. Investigators obtain useful clinical information when examining the magnitude, location, and direction of the ground force reaction vector in relation to the hip, knee, and ankle joints during gait. If the ground reaction force vector is located in front of the hip or ankle, or behind the knee, it applies a moment of force to flex that joint. To prevent joint flexion, the appropriate extensor muscles activate.

- **Internal kinetic measures.** Internal kinetic clinical gait analysis focuses on the causes of motion. Calculation of the net forces and net moments of force (torque), as well as the powers across the modeled body joints and the energies and momenta of the modeled body segments, provides insight into how the neurological system is attempting to generate a purposeful integrated pattern of body action. Of interest in this chapter, kinetic analyses identify the net moments of force generated by the locomotor muscles within their length and velocity constraints.

- **Control objectives and EMG measures.** The last step in understanding gait mechanics is the inclusion of internal modeling to calculate the muscle and articular forces. However, the human musculoskeletal system is mechanically redundant, and the classical equations of kinetic analysis permit an infinite number of solutions to satisfy the net joint moment of force equations. In cases in which it is necessary to resolve this dilemma, investigators have used two different approaches. One is to hypothesize that the body activates muscles to satisfy objectives such as minimizing fatigue, maximizing power output, and reducing joint loading (Crowninshield 1978, 1983; Hatze 1981; Herzog 1987). Suitable biomechanical analysis identifies those objectives that the body is attempting to fulfill, which allows estimation of the individual muscle forces. The other approach is to use knowledge of the sequencing of the muscles, obtained through EMG records, to identify the active muscles (White and Winter 1992; Winter et al. 1993).

Methodology and Limitations

Almost every site where clinical gait analysis is performed uses a unique set of hardware, protocols, and software to perform analyses. In fact, the lack of clinical gait analysis standardization has been recognized as a major reason why clinical gait analysis has not received wider acceptance. In addition, the multitude of on–off outcome measures, the absence of a standardized database with which to compare results, and emphasis on methodology and gait descriptors instead of outcomes that explain the reasons for the observed gait pattern have combined to limit acceptance of clinical gait analysis (Andriacchi 1998; Davis 1997; Frigo et al. 1998; Kopf, Pawelka, and Kranzl 1998; Mulder, Nienhuis, and Pauwels 1998; Rozendal 1991; Whittle 1996).

Nevertheless, standards seem to be emerging with respect to the components of a typical clinical gait analysis. This section outlines a generic hardware complement, a typical clinical gait analysis protocol, and a commonly used software approach for generating outcomes, also noting major deviations from this generic clinical gait analysis.

Hardware

A device to capture the four-dimensional time–motion characteristics of selected body points is the core feature of a clinical gait analysis site. These kinematic data acquisition systems are often video based, although electronic devices using optics, magnetism, or sound are available. A variety of targets are affixed to the patient to indicate the selected body points. These targets can range from simple passive markers to electronically sophisticated active sensors.

To simultaneously measure the motion of the patient's right- and left-side body segments, the kinematic data acquisition system is housed in a large room where the patient walks along a predefined linear pathway—the walkway. Embedded within the floor of the walkway are multiple force platforms positioned to record the action of the patient's right and left feet against their surfaces. Often the walkway is raised to incorporate the platforms, or the floor has been cut into to accommodate the platform thickness.

The patient typically also carries, via belt or rucksack, a multichannel analog data relay system that is used to record foot–floor contact events (foot switches) and muscle activation patterns (EMG). This analog system is connected to a central digital data collection computer via a cable or telemetry. A patient-carried data logger is sometimes used.

Lastly, almost all clinical gait analysis sites have several computers to assist with the collection, processing, and presentation of results. These computers can quickly process large amounts of clinical gait analysis data.

Protocol

After informed consent or assent has been obtained, a brief observational gait analysis is performed. This session may or may not involve use of a video camera, although using a camera is preferred for patient documentation and presentation purposes. During the observational gait analysis session, the decision is made whether to proceed with the clinical gait analysis. Patients who would be excluded from proceeding to a clinical gait analysis are those who cannot walk independently, have reduced or minimal stamina, or do not comprehend simple commands.

After an appropriate set of anthropometric measures has been obtained, the foot-switch, EMG, and body landmark targets are attached to the patient. Often these procedures are the most time intensive, since incorrect placement of these devices would provide biased clinical gait analysis outcomes. Placement of the body segment targets is critical. These markers, often placed on anatomical landmarks, define frames of reference that specify the joints' rotational axes; incorrect placement severely offsets the abduction-adduction and internal-external angular kinematics. Correct placement of the EMG electrodes also is crucial for comparison of a particular patient's results across time or to a normative database. Different electrode placements provide data from different parts of the same muscle, from a different muscle altogether, or from a composite of several muscles simultaneously (cross-talk).

The patient is then instructed to walk along the walkway as naturally as possible, such that each foot cleanly contacts the force platform. Several accommodative trials are often required, especially with children and severely affected adults. Sometimes patients are asked to hold their arms near their chest to prevent obscurement of hip markers. After an acceptable number of "good" gait trials, the foot switches, body landmark targets, and electrodes are removed.

Software

The least standardized component of a clinical gait analysis is often the software used to generate the report. Since a unique set of hardware and different protocols are invariably used, specialized software is required for recognizing the data and processing them appropriately.

Most sites use inverse dynamics to calculate the net forces, net moments of force, and powers acting across the modeled joints (hips, knee, ankles). Inverse dynamics starts with knowledge of the system's motion and then calculates the unknown net forces and net moments of force that must exist to produce the observed motion pattern. Unfortunately, inverse dynamics provides only time-varying estimates of one force and one moment of force at each joint during gait. Although this single force and moment of force is resolved into three components, it does not provide detailed information about each muscle crossing each joint. Since a multitude of muscles (and other soft tissues) cross each joint, each with the potential to contribute to the net force and net moment of force estimate, additional internal body modeling and numerical strategies must be used to estimate the action of individual muscle force–time profiles. Because these approaches require additional modeling and computational power, the results from most clinical gait analysis sites terminate with joint kinetics.

Limitations

In an evidence-based approach to clinical practice, clinical outcome measures should be valid (truthful), reliable (producing consistent results), and sensitive to change (detecting important clinical change). Although clinical gait analysis quantifies the anthropometric, temporal and spatial, kinematic, and external and internal kinetic features of gait, it is important to note its limitations. These limitations can be broadly categorized as technical, practical, and philosophical. Technical limitations include invalid frame of reference definitions, poorly validated anthropometrics, unreliable marker placements, and limited comparative databases. Practical limitations have to do with the intrusiveness of attached hardware and force plate targeting, which alters normal gait pattern, and the limited number of gait trials available for averaging. Philosophical limitations consider which outcomes are important to the patient and clinician—aesthetics and movement economy or the standard clinical gait analysis biomechanical outcomes. Collectively, these features limit the ability of clinical gait analysis to truthfully, consistently, and meaningfully capture a patient's gait pattern.

Muscle Function During Normal Gait

The primary objectives of gait are to move the body safely and economically to a desired location. To achieve these objectives, the locomotor muscles apply moments of force across the hip, knee, and ankle joints to propel the body forward while providing trunk stability and shock absorption economically (Perry 1992).

Examining the impact of muscle damage on gait requires knowledge of the normal muscle function profiles to which the damaged muscle function profiles can be compared. The expected function of the locomotor muscles during gait can be presented in verbal, graphical, or mathematical function formats. This section presents graphical depictions and brief verbal descriptions of the primary locomotor muscles' actions. Readers who would like a more detailed presentation of individual muscle action during gait may consult a number of excellent references (Inman, Ralston, and Todd 1981; Perry 1992; Whittle 1996).

Graphic Description

Defining a muscle as a single muscle-tendon unit coursing from origin to insertion, one description of a muscle's action during gait presents its time-varying force output as the muscle changes length either concentrically or eccentrically (or isometrically). Given that the force–length–velocity profile of the locomotor muscles cannot easily be quantified, the angular position, angular velocity, and net moment of force behavior of key locomotor joints are presented instead. These joint-measured outcomes correspond to the length, velocity, and force output of key locomotor muscle groups. Additionally, multiplying the angular velocity by the net moment of force output provides a measure of muscle power generation or absorption across a joint. Joint power generation is necessary for the basic gait requirements of holding the body upright within the gravitational field (stance stability) and propelling the body forward. Joint power absorption selectively controls body segment deceleration (shock absorption) and assists with economical muscle force output. Joint power generation or absorption accelerates and decelerates the body segments of the walking system to enable gait.

The time-varying joint angular positions, angular velocities, net moments of force, and power outputs can describe the function of the locomotor muscles crossing a joint. For example, if at a given time within the gait cycle the hip joint is in flexion (angular position), is extending (angular velocity), and requires an extensor net moment of force to counterbalance externally applied forces (gravity, inertia, ground reaction force), one infers that long-length, eccentrically active hip flexor muscle action is present. Collectively, interpretations of these joint outcomes provide valuable information about the function of the key muscles during normal gait.

It is important to note that the translation of joint to muscle function involves several assumptions. Primarily, one assumes that the locomotor muscles can be clustered into a small number of single-joint muscles acting in a single anatomical plane, each with an unchanging moment arm. Cursory examination of the locomotor musculature system demonstrates that this assumption is incorrect. Many of the muscles are multijoint, and their moment arms vary with joint angulation. Secondly, one assumes that simultaneous agonist–antagonistic muscle activation at a given joint is absent. However, to maintain stance stability and shock absorption during gait, muscle coactivation is required as demonstrated by numerous EMG investigations (Arsenault, Winter, and Marteniuk 1986; Sutherland 2001; Winter and Scott 1991). Nevertheless, in spite of these limitations, this relatively simplistic model of the muscles of the locomotor apparatus does provide a useful window into the role and function of the major locomotor muscle groups.

Figures 13.1 through 13.4 display the joint position, angular velocity, net moment of force, and power curves for the hip (sagittal and frontal planes) and the knee and ankle (sagittal plane) joints during gait. These joints and

planes encapsulate the principal features of the primary action of key locomotor muscle groups.

The gait data for figures 13.1 through 13.4 were obtained from the Normative Gait Database, located at http://guardian.curtin.edu.au/cga/data/index.html. This database, posted by Dr. Chris Kirtley, provides three-dimensional lower extremity temporal-spatial, kinematic, external kinetic, and internal kinetic walking data. Multiple stride data from 20 healthy men and women (ages 15-55 years) were normalized, averaged, and tabulated. Note that this normative database has several acknowledged limitations. The volunteers walked with a slower than expected gait speed and reduced stride length; therefore, compared to subjects who walk faster, they present with reduced plantarflexion at push-off, less

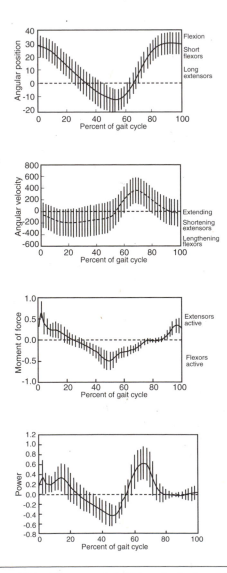

Figure 13.1   The mean (±1 *SD*) sagittal plane hip angular position, angular velocity, net moment of force, and power for 20 asymptomatic adults during a normal gait cycle, with muscle function indicated.

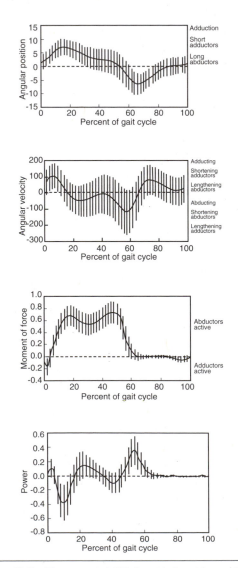

Figure 13.2   The mean (±1 *SD*) frontal plane hip angular position, angular velocity, net moment of force, and power for 20 asymptomatic adults during a normal gait cycle, with muscle function indicated.

hip adduction, reduced hip extension moment of force in terminal stance, and reduced ankle and knee powers. Given these limitations, the magnitude of the joint positions, velocities, moments of force, and powers must be interpreted with caution. However, of importance to this chapter are the direction (polarity) and relative size of the "hills" and "valleys."

Figures 13.1 through 13.4 present the average, and variations on (plus and minus one standard deviation), angular position, angular velocity, net moment of force, and power curves for one gait cycle. For each graph, the polarity of the data is indicated (i.e., flexion versus extension). According to standard convention, each gait cycle starts at right (left) heel contact and ends at the next right (left) heel contact. Collectively for these volunteers, the

first 62% of the gait cycle is the stance (foot on ground) phase, and the latter 38% defines the swing phase. These percentages correspond to expected values (Pathokinesiology Service and Physical Therapy Department 2001). The stance and swing phases are frequently subdivided into a number of subphases: initial contact (0% of the gait cycle), loading response (0-12%), midstance (12-31%), terminal stance (31-50%), preswing (50-61%), initial swing (62-75%), midswing (75-87%), and terminal swing (87-100%) (Perry 1992).

A detailed description of the function of key locomotor muscles can be inferred from the data presented in figures 13.1 through 13.4. For example, in figure 13.1, the hip joint during terminal support and the loading response requires an extensor net moment of force to

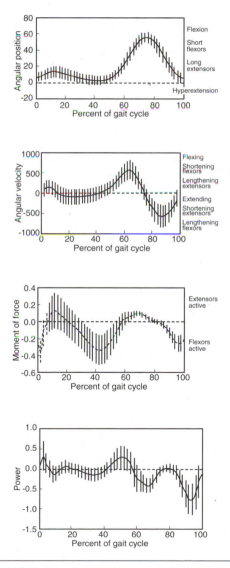

Figure 13.3 The mean (±1 *SD*) sagittal plane knee angular position, angular velocity, net moment of force, and power for 20 asymptomatic adults during a normal gait cycle, with muscle function indicated.

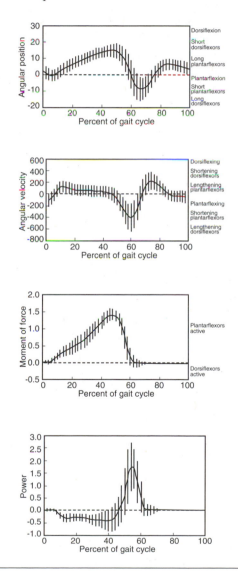

Figure 13.4 The mean (±1 *SD*) sagittal plane ankle angular position, angular velocity, net moment of force, and power for 20 asymptomatic adults during a normal gait cycle, with muscle function indicated.

counterbalance the externally applied flexor net moment of force (third panel). Furthermore, the figure shows that the hip extensors are long (first panel) and slowly lengthening, as well as isometrically active and then slowly concentrically active (second panel). Knowledge of the force–length–velocity relationship of a muscle-tendon unit suggests that in vivo long and eccentrically and then concentrically active muscles are well suited for generating force economically. This extensor force multiplied by its moment arm generates the required extension moment of force for dynamic stabilization at the hip joint.

Verbal Description

Descriptions of the function of each of seven major muscle groups during normal gait were compiled for this section and are presented in table 13.1. The seven muscle groups are the hip flexors, extensors, and abductors; the knee flexors and extensors; and the ankle dorsiflexors and plantarflexors. For each muscle group the gait subphases, with significant moment of force contribution, its length (short, neutral, long), and contraction type (concentric, isometric, eccentric), were catalogued through inspection of figures 13.1 through 13.4. The table also describes the actions of these muscles.

The descriptions in table 13.1 demonstrate that clinical gait analysis is successful in characterizing the ability of the locomotor muscles to achieve three basic locomotion objectives. First, the locomotor muscles must

provide joint stability when the ground reaction force vector attempts to collapse (flex) the hip, knee, and ankle joints. Clinical gait analysis suggests that hip extension during initial contact, knee extension during the loading response, and ankle plantarflexion during terminal stance are all required to prevent lower extremity collapse (Kepple, Siegel, and Stanhope 1997; Perry 1992). In addition, it shows that hip abduction is required during stance. Second, the locomotor muscles must provide gait propulsion. Clinical gait analysis provides evidence of hip flexor activity during preswing and initial swing ("pull-off") and ankle plantarflexion during the loading response and midstance ("push-off") (Kepple, Siegel, and Stanhope 1997; Winter 1980). Lastly, the locomotor muscles must provide shock absorption primarily at the knee through eccentric knee extensor action during the loading response (Eng and Winter 1995).

Impact of Muscle Damage on Gait

Clinicians treating patients with an altered gait pattern frequently use clinical history, observation, and special tests to uncover the underlying primary neuromusculoskeletal impairments contributing to the gait deficit. Impairments such as limited joint range of motion, pain, altered sensation, and changes in muscle activity patterns due to weakness and timing disorders all contribute to

Table 13.1   Summary of the Function of Seven Major Locomotor Muscle Groups During Normal Gait

Muscle group	Gait phase(s) with activity	Length	Contraction type	Description
Hip flexors	Terminal stance to midswing	Long to neutral	Eccentric, then concentric	Active to produce hip flexion; the "pull-off" burst
Hip extensors	Terminal swing to loading response	Long	Isometric	Active to resist hip flexion and keep trunk erect
Hip abductors	Loading response to midstance	Long to neutral	Eccentric, then concentric	Active to prevent pelvis drop to the unsupported side
Knee flexors	Initial and midswing	Neutral	Isometric	Active to decelerate the lower limb prior to initial contact
Knee extensors	Terminal swing to loading response	Long to neutral to long	Concentric, then eccentric	Active to resist knee flexion
Ankle plantarflexors	Loading response to preswing	Neutral to long	Slow eccentric	Active to control forward trunk momentum, then to generate forward propulsion, the "push-off" burst
Ankle dorsiflexors	Loading response	Neutral	Eccentric	Active to control foot slap

altered gait patterns. These primary impairments limit the patient's ability to provide dynamic joint stability, propulsion, and shock absorption (Perry 1992).

Damage to a muscle reduces its ability to generate force. Although the requirements of normal gait typically require a small fraction of its total force–length–velocity capacity, muscle damage may limit its ability to provide adequate force. In this case the patient compensates by presenting with new muscle activation strategies and body postures to circumvent the muscle impairment (Kepple, Siegel, and Stanhope 2000; Perry 1992; Winter 1984; Winter et al. 1990; Winter and Eng 1995).

This section outlines the impact of a single muscle group weakness (often caused by muscle damage) on gait. Specifically, muscle weakness in key locomotor muscle groups alters dynamic joint stability, propulsion, and shock absorption.

This discussion does not include a complete inventory of all possible gait alterations due to muscle weakness. Although multiple muscle group weakness, with its impact on gait, is common in several patient groups, such as patients with muscular dystrophy, myelodysplegia, poliomyelitis, or spinal cord injury, these gait alterations are not discussed here. Readers should consult the literature for information on the impact of multiple muscle weakness on gait in specific pathologies.

Threats to Dynamic Joint Stability

During several gait subphases, the location of the ground reaction force vector, relative to the isolated hip, knee, and ankle joints, would cause lower extremity collapse (Kepple, Siegel, and Stanhope 1997; Winter 1980). These flexion moments of force must be actively counterbalanced by the hip and knee extensors during the loading response and by ankle plantarflexion during terminal stance. In addition, the hip abductors must be active to prevent excessive hip adduction during stance. Damage to any of the joint extensors or hip abductors, preventing their ability to generate appropriate counterbalancing moments of force, requires the locomotor system to alter typical gait presentation to circumvent this deficiency. The following are some common gait alterations.

- **Rearward trunk lean due to weak hip extensors.** When the hip extensor muscles are unable to perform their normal gait function, the patient frequently compensates by assuming a rearward trunk-lean posture for initial contact. This rearward trunk lean shifts the typically anteriorly positioned ground reaction force vector close to the hip joint center, reducing the need for a large hip extensor moment of force. This gait deviation is frequently called a gluteus maximum limp.

- **Forward trunk lean due to weak knee extensors.** This gait alteration occurs when the knee extensors fail to perform their customary role during loading response and the first part of midstance. During loading response, forward trunk lean relocates the ground reaction force vector closer to the knee joint axis, thereby producing a smaller knee flexor moment of force at the knee.

- **Flat foot–floor contact due to weak knee extensors.** At initial foot-to-floor contact, the ground reaction force vector is located posterior to the knee joint's flexion-extension axis and generates an externally applied knee flexion moment of force. This external moment of force is counterbalanced by a muscle-generated knee extension moment of force. Moving the ground reaction force vector close to the knee joint axis reduces the knee extensor moment of force requirement. People can accomplish this by walking with their foot flat at initial contact.

- **Knee hyperextension due to weak knee extensors.** If the knee extensors' moment of force is unable to counterbalance the knee flexion moment of force due to the externally applied ground reaction force at initial foot–floor contact, the patient may present with knee hyperextension during stance. Increased hip extensor and ankle plantarflexor moments of force are needed.

- **Excessive ankle dorsiflexion and knee flexion due to weak ankle plantarflexors.** Prolonged heel contact with excessive knee and ankle flexion in terminal stance is indicative of inadequate ankle plantarflexion moment of force. Compensation for this muscle weakness keeps the knee flexed during the loading response.

- **Lateral trunk lean due to weak hip abductors.** In normal gait during early stance, the ground reaction force position, in the frontal plane, is located medial to the hip joint center that generates an externally applied hip adductor moment of force. Patients with weak hip abductors compensate by leaning the trunk toward the affected side. This trunk shift repositions the ground reaction force vector closer to the hip joint center, producing a smaller externally applied hip adductor moment of force. This gait deviation is also called a gluteus medius limp.

Threats to Adequate Propulsion

During normal gait, the progressive forward motion of the body has been described as one in which the body falls from one limb to the other with small energy injections to account for energy loss (Inman, Ralston, and Todd 1981). Central to this description is a smooth exchange of kinetic and potential energy that improves walking economy (Elftman 1938).

Clinical analysis provides quantitative evidence of the role of the locomotor muscles in propelling the human body during the gait cycle. Specifically, calculation, display, and interpretation of joint power generation and absorption outcomes provide the required information. Through these analyses, persuasive evidence exists that the role of the hip extensors and ankle plantarflexors is to add energy to the walking body. The hip extensors and the ankle plantarflexors add energy (positive power) to move the body forward (Winter 1989; Winter and Eng 1995). It is instructive that the locomotor system uses these two muscle action strategies to generate an economical gait pattern.

An inability of the hip flexors to pull or of the ankle plantarflexors to push the body forward typically presents as a reduced forward progression velocity. However, there is some evidence that one propulsion strategy can dominate to allow a patient to present with a typical progression velocity (Pierrynowski, Tiidus, and Galea 2005).

Threats to Shock Absorption

During the loading response phase of the gait cycle, a large force is applied to the supporting limb and foot. This force, applied within a relatively short time, must be accommodated to prevent potential trauma. During loading response, this impact force is absorbed through eccentric action of the knee extensors and ankle dorsiflexors (Sadeghi et al. 2002). Weakness in either of these two muscle groups potentially presents with the following major gait alterations:

• **Altered limb flexion due to weak knee extensors.** People use a variety of compensations to provide shock absorption during the loading response when their knee extensors are weak. These gait alterations include eccentrically extending the lower leg by activating the hip extensors and ankle plantarflexors and hyperextending

the knee and loading the posterior capsular ligaments. However, neither compensation is ideal, since each can cause muscle fatigue and ligament degeneration.

• **Foot slap due to weak ankle dorsiflexors.** Within the gait cycle, the initial foot–ground contact phase requires ankle dorsiflexor activity to counterbalance the externally applied plantarflexor moment of force created by the rearward and vertically directed ground reaction force applied posterior to a patient's ankle. Typically, the dorsiflexor moment of force is less than the external plantarflexor moment of force, resulting in an eccentric muscular contraction that controls the lowering of the foot to the floor. If the dorsiflexor activity is inadequate, the patient presents with an audible foot-slapping gait. Postural adjustments during the loading response, such as shortened stride length and foot-flat contact, are common gait presentations due to weak ankle dorsiflexors.

Summary

This chapter has provided a schema for characterizing muscle function, described the measurement and results of muscle function during normal gait, and discussed the impact of muscle damage on normal gait. Muscle damage alters the ability of muscles to generate force at different lengths and velocities. Muscle damage was specifically defined as an insult to a single locomotor muscle group that reduces its ability to contribute to dynamic joint stability, propulsion, or shock absorption. Since the locomotor system is redundant, with several muscles capable of performing similar functions, muscle damage may only subtly alter gait. However, severe muscle damage or damage to a particular muscle group may result in significantly altered gait.

CHAPTER 14

Overtraining Injuries in Athletic Populations

Teet Seene, PhD; Maria Umnova, PhD; Priit Kaasik, PhD; Karin Alev, PhD; and Ando Pehme, PhD

Overloading is a natural part of a modern endurance athlete's training process, providing stimuli for adaptation and supercompensation (O'Toole 1998). Accordingly, the training stimulus must be strong enough to induce disturbance of homeostasis so that the body must initiate reactions to adapt. But since repetitive overloading and lack of sufficient recovery lead to overreaching and the overtraining syndrome, the threat of overtraining is a major problem among competitive athletes (Foster, Daniels, and Seiler 1999; Steinacker et al. 1999). The overtraining syndrome is a condition caused by training too hard and for too long, in which the muscles fail to positively adapt to the training and performance decrements. This results in a general breakdown in physiological function. Hooper and colleagues (1995) suggested that volume of training, rather than intensity, may be the major factor contributing to the development of overtraining syndrome. Many top-level endurance athletes train with high volume without showing improvement in performance. Consequently, the main question athletes and their trainers need to answer is how much rest and recovery are necessary in a training program.

Establishing the optimal training regimen is complicated because both the volume of training and the recovery period that is optimal for performance improvement in endurance training are difficult to determine; what is optimal for some athletes may represent undertraining or overtraining for others (O'Toole 1998).

High-volume training protocols both for top athletes and in animal models lead to chronic fatigue and decreased exercise performance and also to changes in skeletal muscle, such as irregular fiber size and distribution and alteration in mitochondria structure (Lambert et al. 1999; Seene, Umnova, and Kaasik 1999). Morphological as well as biochemical findings in our research have shown that during exhaustive endurance training that leads to overtraining syndrome, changes in the skeletal muscle myofibrillar apparatus, especially the destruction of contractile proteins, are the most significant factors in decreasing exercise performance (Seene et al. 2004). Although overtraining syndrome is widely studied, its ultimate causes and pathophysiology are not fully understood (Foster, Synder, and Welsh 1999), particularly at the level of skeletal muscle (Lehmann et al. 1998).

This review deals with the ultra- and molecular structure of overtrained skeletal muscle subjected to too great a volume of work, regeneration of muscle structures, and the effect of these changes on physical work capacity (PWC). Physical work capacity may be defined as the work performed during a single movement, calculated per unit of body weight. In humans, PWC is measured from data on work rate, namely O_2 uptake or O_2 pulse at a given heart rate, such as 170 or 150 beats/min. In small laboratory animals, PWC is most often measured as running until exhaustion at a certain speed (m/min) on a horizontal treadmill and is expressed in kilojoules (kJ).

Insufficient Recovery and Overload

If the time until the next training session is too long, the overcompensation will regress and the original functional state will return, and progressive improvement will not occur. Conversely, if the training stimulus is imposed too frequently, so that it interrupts the overcompensation phase, adaptation will not occur. If the overload has been properly designed, performance will improve progressively (Foster, Daniels, and Seiler 1999). Increased training can take the form of either increased intensity or increased volume. As already mentioned, volume of training rather than intensity may be the major factor contributing to the development of overtraining (Hooper et al. 1995).

At the same time, we know that for endurance athletes, overload is accomplished through a combination of increased volume and intensity. In the training process, volume and intensity are reciprocally related (O'Toole 1998). From this standpoint it is important to stress that alterations in the physiological, metabolic, and biochemical factors during the training process for one type of exercise may limit exercise performance in other types of exercise (Noakes 2000). Endurance training results in increased mitochondrial density, increased capillary supply, changes in key metabolic enzymes, and increased maximal oxygen uptake (Holloszy and Booth 1976) and promotes a transition from type II to type I muscle fibers, which occurs at the expense of the type II fiber population (Thayer et al. 2000).

The nature of the physiological adaptation that occurs in response to endurance training has been extensively studied. In contrast, fewer researchers have evaluated the effect of training volume (particularly an excessive increase in volume) and the recovery state between sessions as measured by the turnover rate of individual myofibrillar proteins and accompanying changes in the myosin heavy and light chain isoform pattern in slow- and fast-twitch muscles (Seene and Alev 1991; Seene et al. 2005).

Animal experiments showed that an exercise volume three to four times greater than that of the recommended training protocol led to overtraining after only four weeks of training (figure 14.1). Rats ran 35 m/min on a horizontal treadmill seven days per week, 2 hrs per day. The weekly training volume for the overtraining group was increased ~3.7 times faster than that for the group exercising according to the recommended protocol. As the rats ran every day, there was not enough time for recovery from volume-induced exercise stress. Training volume was decreased on the sixth day of each week, and PWC was measured on the seventh day. During the fifth and sixth weeks, when daily training volume decreased because the rats were not able to maintain the higher volume, the overtraining syndrome deepened. After four weeks of overtraining, PWC in the overtraining rats decreased by 38%, and after six weeks it decreased by 54% (Seene et al. 2004).

A decrease in performance ability with overtraining appears to be a universal finding. Overtraining results from a rapid increase in training volume and lack of recovery. Problems with recovery seem to occur in animal experiments when the exercise training time reaches 10% of a 24 h period (Seene et al. 2004). A significant decrease in PWC during overtraining as compared to the recommended training protocol suggests that lack of recovery in the training protocol leads to overtraining. If the training stimulus (exercise session) lasts too long and training sessions are so frequent that they interrupt the recovery phase, the necessary adaptation does not occur.

One reason top athletes make a mistake when they use high exercise volumes is that progressive, not sudden, overload is the foundation of successful endurance training. Still, training volume and fast increases in volume are probably only two of the factors that lead to overtraining syndrome. The importance of recovery is evident from the fact that after symptoms of overtraining appear, a much longer recovery time is needed than before (Seene et al. 2004). In the study just discussed, volume was decreased by 60% before symptoms of overtraining were evident and then again after symptoms had become evident. The day after the volume was decreased by 60% the rats were able to tolerate 150% of the prior exercise volume (figure 14.1). After the appearance of overtraining symptoms, on the day after the training volume was decreased by 60% the rats could tolerate only 110% of the prior exercise volume. A further decrease in the volume did not have any effect since the contractile apparatus had been exhausted and the PWC did not recover sufficiently (figure 14.1). From the standpoint of top athletes in endurance events, training protocols are problematic because the boundary between hard training and overtraining is not clear (O'Toole 1998).

Unfortunately, no model exists that one can use to determine the overload training stimulus that will result in optimal improved performance while minimizing the potential to develop the overtraining syndrome (O'Toole 1998). Decreased performance ability seems to be a universal finding in overtrained athletes. In these problematical situations, a balance between training stimulus and recovery is the only guarantee of an increase in performance ability.

Overtraining Effects on Skeletal Muscle

The general principles of musculoskeletal design provide insight into the evolution of locomotor performance and adaptive plasticity, including how a muscle's architecture relates to its functional use and neuromuscular control.

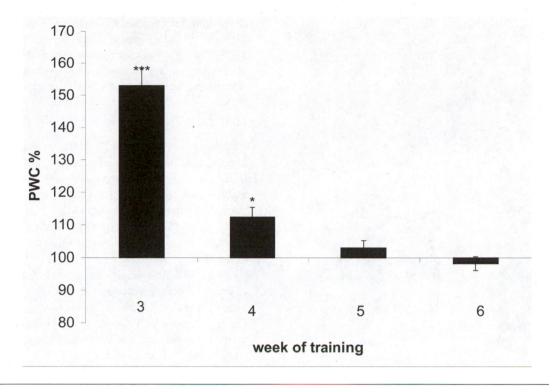

Figure 14.1   Effect on physical work capacity in overtraining rats of a weekly one-day 60% decrease in endurance training volume. Symptoms of overtraining were observed in the fourth week of training. PWC: physical work capacity before training; *p < 0.05; ***p < 0.001 in comparison with the previous day.

Recruitment of different motor units in muscle depends on the character of the contractile activity. As a given motor unit contains only one fiber type, it is important to understand how the ultrastructure of different fiber types is affected by their recruitment during exercise training and overtraining.

Skeletal Muscle Ultrastructure

The overtraining syndrome is accompanied by marked ultrastructural changes in skeletal muscle. As shown in animal experiments, the most typical ultrastructural changes in endurance training occur in slow-twitch (ST) oxidative (O) and fast-twitch (FT) oxidative-glycolytic (OG) muscle fibers (Seene and Umnova 1992). Examples of these changes are lesions in myosin and actin filaments and in the distributed regularity of Z-discs in sarcomeres, destruction of mitochondria, and dilation of the terminal cisternae of the sarcoplasmic reticulum and the tubules of the T-system. During endurance training, ultrastructural changes in FT glycolytic (G) muscle fibers are considerably less evident than in FT OG fibers (Seene and Umnova 1992). Due to the destruction of myofibrils and, mainly, atrophy of FT OG fibers (figure 14.2, *b* through *d*), exercise myopathy may accompany the overtraining syndrome

(Seene, Umnova, and Kaasik 1999; Lehmann et al. 1999). Figure 14.2*a* depicts the condition of myofibrils in endurance-trained muscle. In contrast, in overtrained muscle, myofibrils are thinner because of splitting (figure 14.2*c*). Wide distances between myofibrils (figure 14.2*b*) are due to the disruption of myofibrils (figure 14.2*b*) and disarray of thick myofilaments from myofibrils (figure 14.2*d*).

These changes have been shown to be due to excessive strain in the contracting muscle fiber, not to absolute force developed in the fiber or muscle (Leiber and Friden 1993). Some degree of muscle adaptation occurs in response to each dose of exercise, and various degrees of adaptations are possible (Kibler and Chandler 1998). Overtraining also affects the extracellular matrix. Electron microscopy of affected connective tissue indicates intracytoplasmic calcifications, longitudinal collagen fiber splitting, collagen fiber kinking, and abnormal fiber cross-links (Kibler and Chandler 1998).

Similarities in functional and structural changes in skeletal muscle in glucocorticoid and exercise myopathies have provided a basis for speculation that overtraining-caused myopathy may be a mild form of corticosteroid myopathy (Lehmann et al. 1999; Seene, Umnova, and Kaasik. 1999). Ultrastructural studies have shown disarray of thick myofilaments in FT G fibers in dexamethasone-treated

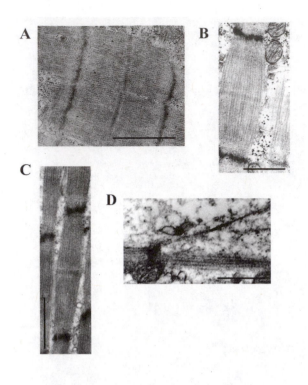

Figure 14.2 Electron micrographs of fragments of FT OG muscle fiber. *(a)* Myofilaments after endurance training (bar: 1 μm). *(b)* Destruction of myofibrils in overtrained skeletal muscle (bar: 1 μm). *(c)* Thin myofibrils in overtrained skeletal muscle (bar: 1 μm). *(d)* Disarray of myofilaments and irregularly located thick myofilaments in overtrained skeletal muscle (1: wide distance between myofibrils; 2: disruption of myofilaments; 3: disarray of thick myofilaments; 4: splitting of myofibrils; bar: 1 μm).

rats. The destructive process in myofilaments begins at the periphery of the myofibrils, spreads to the central part of the sarcomere near the H-zone, and is distributed all over the A-band (Seene et al. 1988). As shown in studies by Hussar and colleagues (1992), the thin filaments and Z-lines are much more resistant to the catabolic action of dexamethasone than the thick filaments in FT G muscle fibers. The myofibrils of G muscle fibers are thinner in dexamethasone-treated rats. This is caused by the splitting of myofibrils. In FT OG fibers, the destruction of thick myofilaments was shown to be remarkably less pronounced. Myofibrils were structurally normal in the dexamethasone-treated soleus (Sol) muscle. There was evidence of increased lysosomal activity in the G muscle fibers and in the satellite cells of dexamethasone-treated rats (Hussar, Seene, and Umnova 1992).

Although there are some similarities between overtrained animals and glucocorticoid-treated experimental animals (decrease in muscle grip strength and PWC, decreased contractile protein synthesis rate, increased degradation rate),

the main differences are that destructive changes occur in different types of FT muscle fibers (Seene 1994; Seene et al. 1995, 2003, 2004). However, differences in myosin heavy chain synthesis, degradation, and turnover rate at the level of different muscles and muscle fibers are still not clear.

Fast-Twitch Muscle Fibers

Studies of overtraining at the ultrastructural level of rat skeletal muscle have revealed that damaging effects depend on the muscle fiber type (Seene et al. 1995; Seene, Umnova, and Kaasik 1999). Due to the destruction of myofibrils and atrophy of muscle fibers (figure 14.2, *b* & *c*), muscle weakness develops (Seene and Viru 1982; Seene et al. 2004). Fast-twitch muscle fibers are the most sensitive to long-lasting exhaustive endurance exercise (figure 14.2).

In FT OG muscle fibers, specific and nonspecific changes have been established, as well as changes that appear in skeletal muscles in other stressful conditions (Seene, Umnova, and Kaasik 1999). Nonspecific changes include the appearance of myelin-containing structures in complexes of mitochondria, absence of mitochondria, and the development of gigantic forms of mitochondria. Specific changes include the destruction of peripheral myofilaments, attenuation of myofibril numbers, and complete destruction of some sarcomeres (Seene, Umnova, and Kaasik 1999).

Due to the destruction of myofibrils, the sarcoplasm-filled spaces between myofibrils increase in size. These spaces contain mitochondria, compact matrix, fragments of T-tubules, fragments of the sarcoplasmic reticulum, and a large number of glycogen granules. Some myofibrils show warped Z-lines (Seene, Umnova, and Kaasik 1999).

The destruction of muscle fiber organelles is accompanied by the activation of lysosomal structures. More extensive and more clearly expressed destructive changes in FT OG muscle fibers lead to the atrophy of these fibers. At the same time, OG muscle fibers manifest morphological signs of the continuation of regeneration processes: a small increase in the number of nucleoli, satellite cells under the basal membrane, activation of the mitochondrial tree, and numerous lipid drops and glycogen granules (Seene, Umnova, and Kaasik 1999).

In overtrained rat muscle, the myofibrils in G fibers are also damaged. Most of the damage is located in the periphery of the myofibrils. This destruction is less evident than that occurring in the OG fibers during overtraining.

Intrafusal Muscle Fibers

Comparison of the effect of exhaustive exercise on the ultrastructure of the muscle spindles in FT and ST muscles shows that, in rat ST muscles, intrafusal muscle fibers exhibit signs of destruction of myofibrils and other cell organelles. Intrafusal muscle fibers located in ST muscle adapt themselves to increased activity with a response reaction similar to that of the extrafusal fibers, although the changes in the intrafusal muscle fibers are considerably smaller.

It has been shown in an animal model that during exhaustive exercise, the actin filaments in the intrafusal muscle fibers are more resistant than other proteins to the action of proteolytic enzymes (Seene and Umnova 1996). In some of the I-discs, the Z-line structure is damaged and becomes undulate. Mitochondria in the intrafusal muscle fibers are lodged in small groups either below the sarcolemma or between the myofibrils in the form of a chain. Regardless of the occurrence of a clearly manifest destructive process in the intrafusal fibers, the regeneration potential during exhaustive exercise is retained (Seene and Umnova 1996). As intrafusal muscle fibers are destined to become slack when the extrafusal fibers shorten; therefore, unless they also shorten to the same degree, due to the gamma motoneurons, ultrastructural changes during exercise training support the idea of an increase in α-γ coactivation during regular physical activity but not during volume-induced overtraining.

Neuromuscular Junctions

Figure 14.3a shows the number and the condition of postsynaptic folds in the normal neuromuscular junction. In contrast, figure 14.3b shows the smeared appearance of the disrupted borders (and the disappearance of entire structures) of postsynaptic folds in overtrained skeletal muscle. The appearance of lysosomes in the FT OG muscle fibers' synaptic terminal, accompanied by the presence of multivesicular formations and large multilayered structures consisting of several parallel membranes and containing a quantity of axoplasm, is typical for overtrained muscle (figure 14.3c).

These multimembrane formations, developed from thickened mitochondria, are located near both small and large synaptic vesicles (Seene, Umnova, and Kaasik 1999). Most of the vesicles are concentrated in the central part of the terminal; only isolated groups of vesicles are located in the active zone of the terminals—in those regions of the presynaptic membrane that correspond to the postsynaptic folds (Seene, Umnova, and Kaasik 1999). Animal experiments have shown that there are large numbers of mitochondria containing very few cristae in these terminals. The axon terminals branch off often, leaving the impression that their range is quite large. There are also regions with a widened synaptic slot and regions that contain membranous structures and vesicles (Seene, Umnova, and Kaasik 1999). It is evident that nerve-muscle transfer in these regions is disrupted.

Around the terminals the number of postsynaptic folds decreases (figure 14.3b), and often they branch off. The numerous mitochondria located in the postsynaptic region are tightly packed with cristae. Lysosome-like formations are located between the mitochondria. Present in the postsynaptic region are some nuclei of the

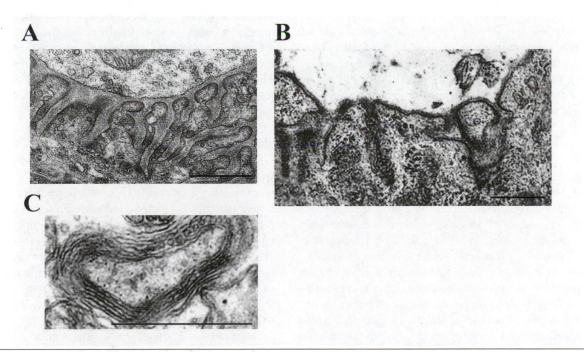

Figure 14.3 Electron micrographs of the neuromuscular junction on an FT OG muscle fiber. (a) Many postsynaptic folds in endurance-trained skeletal muscle (bar: 1 μm). (b) Destruction within the synaptic terminal and decrease of postsynaptic folds in overtrained skeletal muscle (bar: 1 μm). (c) Multilayered membrane structures, consisting of several parallel membranes, in overtrained skeletal muscle (bar: 1 μm).

muscle cell, glycogen granules, ribosomes, and canals of the granular reticulum. The presence of the granular endoplasmic reticulum points to protein synthesis in the region (Seene, Umnova, and Kaasik 1999).

Structural changes in the nerve-muscle synapses of rat OG muscle fibers with endurance training are smaller than those with overtraining, so the structural heterogeneity of the synapse is less clearly manifest. In overtrained muscle the synapse contains synaptic vesicles, and the terminals contain fewer mitochondria. Satellite cells in the postsynaptic region disappear (Seene and Umnova 1992; Seene, Umnova, and Kaasik 1999).

Changes in Skeletal Muscle Biochemistry

It is widely accepted that certain intracellular mechanisms are associated with muscle damage, namely calcium overload, free radical formation, and decrease in energy supply. A fall in cellular adenosine triphosphate (ATP) content is associated with apoptosis, and muscle ATP levels can decrease in response to stress. The release of cellular proteins occurs when cellular ATP falls below a certain critical level, and interference in the energy supply to the muscle membrane is an important factor leading to enzyme efflux.

Neuromuscular Excitability

The muscles of well-trained endurance athletes show increased neuromuscular excitability (NME). On the other hand, reduced NME has been observed in stressed muscles during prolonged exercise. Reduced NME has been interpreted as a reflection of impaired signal transmission to target organs and as a parameter of peripheral fatigue (Lehmann et al. 1998). Neuromuscular excitability has been measured using the minimal current pulse at different impulse durations necessary to induce a single contraction of a given skeletal muscle fiber. Lehmann and colleagues (1999) showed that NME was improved by endurance training but depressed by exhaustive endurance exercise.

Normalization of NME after exhaustive exercise has been observed after rest. Observation of a depressed NME after a single bout of prolonged, heavy exercise showed that this fatigue reaction primarily affects peripheral neuromuscular structures rather than central mechanisms (Lehmann et al. 1999). The mechanism of changes in the excitation–contraction–relaxation cycle during overtraining is multifactorial. One of the factors may relate to the ryanodine receptor, which is not able to elevate sarcoplasmic Ca^{2+} levels sufficiently for muscle contraction after overtraining, resulting in the development of muscle weakness.

Abnormal function of the ryanodine receptor may also lead to an excessive rise in Ca^{2+} levels and thereby to sustained muscle contraction, muscle fiber damage, and apoptosis (MacLennan 2000). Alteration in thick and thin filament proteins can affect Ca^{2+} sensitivity in muscle fibers and cause suboptimal interaction between myosin and actin, which also leads to muscle weakness. Overtraining may cause reduced activity of the sarcoplasmic reticulum Ca^{2+} ATPase (adenosine triphosphatase) pump and prolonged elevation of Ca^{2+} levels, which may lead to stiffness and skeletal muscle relaxation problems.

Growth Factors

Signaling pathways and secondary messenger factors are involved in transcriptional, translational, and posttranslational processes. Insulin-like growth factor-1 (IGF-1) has been shown to affect many steps in the control of gene expression, including cell proliferation, differentiation, and degradation (Adams 1998). It is widely accepted that many of the anabolic effects of growth hormone may result from a growth hormone–stimulated increase in IGF-1 production. Insulin-like growth factor-1 has been shown to stimulate amino acid transportation, which is essential to tissue growth. Recent studies suggest that autocrine and paracrine processes involving muscle-derived IGF-1 may play a pivotal role in linking the mechanical stimulus to the muscle's morphological and biomechanical adaptations (Goldspink 2000). Goldspink (1999) discovered that in response to stretch or increased mechanical activity, the muscle produces a special locally acting isoform of IGF-1 that is directly linked to activation of gene expression necessary for muscle repair, maintenance, and remodeling. The product of this isoform is called mechanogrowth factor (MGF) to differentiate it from the liver IGFs that have more systemic action (Goldspink 1999).

Animal experiments have shown that overtraining decreases the number of satellite cells under the basal lamina of skeletal muscle fibers (Seene, Umnova, and Kaasik 1999). It is well known that satellite cells are a source of muscle regeneration. Unfortunately, muscle damage is unavoidable during overtraining, and the reduced number of satellite cells and the lack of IGF-1 and MGF in overtrained skeletal muscle explain the much lower level of activation of satellite cells (figure 14.4). With appropriate endurance training, the number of satellite cells increases about 3.5 times (Umnova and Seene 1991; Seene and Umnova 1992); satellite cells are activated, fuse with each other, and form myotubes and new muscle fibers (figure 14.4).

Regeneration in overtrained skeletal muscle is slow, as lack of IGF-1 and MGF prevents the activation of satellite cells under the basal lamina of muscle fibers. If, during endurance training, the number of satellite cells in rat skeletal muscle increased about 3.5 times (Seene and

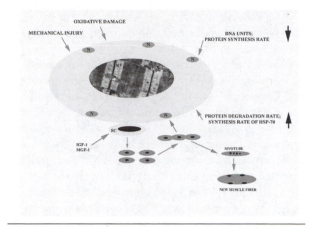

Figure 14.4 Effect of overtraining on the regeneration of skeletal muscle fibers. MF: myofibrils; N: myonuclei; SC: satellite cell; IGF-1: insulin-like growth factor; MGF: mechanogrowth factor; HSP70: heat shock protein; increase; decrease.

Umnova 1992), overtraining led to a decrease in satellite cell number. A decrease in the number of satellite cells means that new fibers are not forming as quickly and damaged fibers do not regenerate appropriately, since satellite cells do not fuse with damaged fibers. Lack of satellite cells also leads to a decrease in differentiation of fibers that are forming, since levels of transcription factors (MyD family), except for myostatin, decrease. Lack of MGF leads to apoptosis. If muscle fibers do not regenerate, muscle atrophy develops, since only myostatin and heat shock protein 70 synthesis increase. Decrease of synthesis rates of muscle proteins, particularly myofibrillar proteins, and increased protein degradation lead to the "wastage" of muscle. Decrease in myonucleus numbers and DNA damage lead to a decrease in DNA units in overtrained muscle. Lack of myonuclei, decreased synthesis, and an increased degradation rate of muscle proteins, particularly myofibrillar proteins (Seene et al. 2004), lead to the development of overtraining myopathy (Seene, Umnova, and Kaasik 1999).

If during overtraining the muscle fibers do not regenerate, muscle atrophy develops since MGF in these muscles is not expressed. Muscle atrophy develops further as myostatin expression in these muscles increases. Overtraining also leads to the decreased differentiation of muscle fibers since transcription factors (MyD family) are not expressed in overtrained muscle (figure 14.4).

Role of Heat Shock Proteins in Overtraining

Overtraining causes heat and metabolic and oxidative stress to skeletal muscle (figure 14.4). Reactive oxygen species (ROS) are involved in the tissue damage (Pan-

sarasa et al. 2002). The reactive species include superoxide anion, hydrogen peroxide, and the hydroxyl radical. Reactive oxygen species may cause cell injuries such as lipid peroxidation, enzyme inactivation, changes in intracellular redox state, and DNA damage. Cells possess enzymatic defense systems to reduce the risk of oxidative injury, that is, superoxide dismutase, glutathione peroxidase, and catalase, which react with superoxide radicals and organic hydrogen peroxides.

There is evidence that an increase in ROS production occurs during endurance exercise and that the resulting oxidative damage arises in muscle, liver, blood, and other tissues (Venditti and Di Meo 1997; Itoh et al. 1998). Exhaustive physical exercise has been associated with enhancement of oxygen consumption in skeletal muscles (Packer 1986), an increase in lipid peroxidation, and inhibition of key mitochondrial enzymes such as citrate synthase and malate dehydrogenase (Packer 1986). In contrast, appropriate endurance training does not lead to functional damage and promotes muscular adaptation.

Heat shock proteins (HSPs) increase the stress tolerance of affected cells and are conducive to the cellular repair process during overtraining (Lehmann et al. 1999). Coincident with slower turnover of muscle contractile proteins, persistently high HSP70 synthesis may indicate a state of inadequate regeneration even after a couple of weeks of recovery from exhaustive exercise (Lehmann et al. 1997).

These molecular chaperones play a universal role in maintaining homeostasis of muscle fibers, and the induction of HSPs in skeletal muscle may be muscle fiber-type specific (Liu and Steinacker 2001) and may occur during overtraining in an effort to counteract the disruption of muscle function (figure 14.4). Cytokines play an important role in the exercise-induced immune reaction and exercise-related metabolic and cellular signal transduction; they are also capable of increasing HSP synthesis (Liu and Steinacker 2001). It is possible, though as yet unproven, that heat shock protein may act as a cytokine in reaction to exhaustive exercise; stimulate tumor necrosis factor-alpha (TNF-α), interleukin (IL)-β, and IL-8 in monocytes; and activate CD14-dependent and Ca^{2+}-dependent pathways (Steinacker and Liu 2002).

Protein Isoforms

Protein isoforms of the skeletal muscle contractile apparatus are interchangeable, and a very large number of phenotypes can be generated by their combination. Only a few stable phenotypes correspond to the main muscle fiber types, and their properties depend on their total protein content. Among these, myosin isoforms are the major determinants of the functional heterogeneity of skeletal muscles (Pette 2001). In skeletal muscle fibers, myosin works within the structure of the sarcomere. Contractile properties of the muscle fibers depend on the myosin heavy chain (MyHC) isoform composition, but the contraction process may be regulated by the presence of isoforms of other myofibrillar proteins such as regulatory and minor proteins (Pette and Staron 2000). Muscle protein content and quality depend on the integrity of the remodeling process of the muscle during endurance training and overtraining. The remodeling basically involves removal of an old protein and replacement with a new one (Booth et al. 1998).

Unfortunately, it is not known how a given MyHC isoform changes in ST and FT muscles during an increase in endurance training volume sufficient to lead to overtraining, or whether MyHC isoforms in different muscles possess differing sensitivity depending on the oxidative potential of the muscle (Seene et al. 2005).

Overtraining increases the degradation rate of skeletal muscle contractile proteins (Seene et al. 2004). Intensive degradation and a decreased muscle protein synthesis rate lead to a decrease in muscle mass (figure 14.4), particularly in FT muscles (Seene et al. 2004).

In rat FT muscles during overtraining, it was shown that the content of DNA per muscle decreased, which might indicate a decreased DNA synthesis rate in skeletal muscle but also a loss of DNA due to a loss of myonuclei consequent to muscle atrophy (Seene et al. 2004). At the same time, the protein:DNA ratio in the plantaris (Pla) and extensor digitorum longus (EDL) muscles increased. The RNA:protein ratio had a tendency to decrease in all muscles studied (Seene et al. 2004).

The synthesis rate of MyHC protein during overtraining decreased in all the rat muscles studied. The synthesis rates of MyHC isoforms differ, however. The synthesis rates of MyHC I and IIa isoforms are faster than those of other isoforms. The IIb isoform has the slowest synthesis rate in both Pla and EDL (Seene et al. 2004). Overtraining did not significantly change the synthesis rate of MyHC isoforms in the EDL muscle; but in the Pla muscle, the synthesis rates of MyHC I and IIb isoforms decreased.

It has been shown that in endurance-trained rats, the turnover rate of the two main muscle contractile proteins, actin and myosin, is faster than in rested muscle. Overtraining caused a decreased turnover rate of MyHC, but not of actin (Seene et al. 2004). The faster turnover rate of MyHC and actin in endurance-trained rats is accompanied by hyperplasia of skeletal muscle; overtraining caused a decrease in turnover rate leading to decreased nuclear domain size (figure 14.4). Thus, changes in the turnover rate of muscle contractile proteins during exercise and recovery reflect the functional conditions of the contractile apparatus in different types of skeletal muscle fibers and have physiological significance (Lehmann et al. 1999). The synthesis rates of actin, tropomyosin, and

myosin light chain (MyLC) are less sensitive to an excessive volume of endurance training than that of MyHC (Seene et al. 1995).

Skeletal muscle is a plastic tissue capable of modifying its contractile and metabolic properties with increased use. For example, endurance training causes significant changes in the relative expression of fast and slow MyHC isoforms, most commonly a shift from the subgroup of fast isoforms to slower oxidative isoforms (MyHC IIa and I) (figure 14.5). The importance of contractile activity in determining the muscle phenotype is widely recognized. An important role in modulating phenotype has been assigned to the motoneuron. It has been demonstrated that the extent and modality of neuronal activity are crucial for regulating contractile properties such as speed, strength, and endurance. World-class marathon runners and ultra-endurance athletes appear to possess remarkably high type I fiber numbers in their trained muscle groups, whereas muscles of sprinters and weightlifters consist predominantly of IIA/IIX fibers (Andersen and Aagaard 2000). It further appears that extreme usage–induced alterations of the muscle MyHC isoform profile result in hybrid fibers that express a combination of MyHC isoforms such as type I/IIA. The extent to which these patterns of MyHC gene expression are conditioned by genetic predisposition or by the specificity of training is unresolved (Baldwin and Haddad 2002).

The effect of endurance exercise on the rat MyHC profile appears to be both muscle specific and dose dependent (Seene et al. 2005). For example, when rodents are trained to run at moderate to high intensity for several weeks, the effects on the MyHC profile of the Sol are manifested only when they run for longer durations (Demirel et al. 1999). In fast muscles, which have a composition bias to type IIX and IIa MyHC isoform expression, both the IIA and IIX isoforms of MyHC are upregulated under these training conditions relative to the sedentary state, whereas the IIb MyHC isoform is significantly down-regulated (Demirel et al. 1999). If running is extended for longer durations, it is possible to induce increased expression of the type I MyHC isoform in fast muscles (Demirel et al. 1999).

Recent findings suggest that the expression of specific myosin isoforms is a relevant mechanism, but not the only one, involved in skeletal muscle heterogeneity and plasticity regulation (Bottinelli 2001).

Myosin Heavy Chain Isoforms

The pattern of MyHC isoforms in skeletal muscle may change for several reasons, including exercise training. Still, we do not know whether other myofibrillar proteins can modulate the functional properties of myosin during exercise training and overtraining.

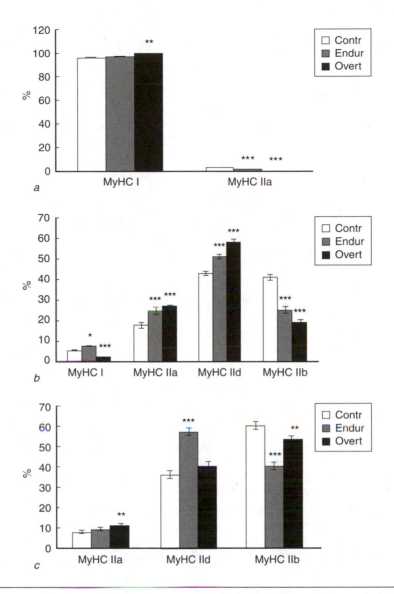

Figure 14.5   Effect of overtraining on the pattern of myosin heavy chain isoforms in fast- and slow-twitch skeletal muscles: *(a)* soleus (Sol), *(b)* plantaris (Pla), *(c)* extensor digitorum longus (EDL). *$p < 0.05$; **$p < 0.01$; ***$p < 0.001$ in comparison with control group; Contr: control group; Endur: endurance training group; Overt: overtraining group.

Slow-twitch muscles are more resistant than FT muscles to high-volume endurance exercise that leads to overtraining; the reason may relate to MyHC composition (Seene et al. 2004). Muscle fibers with predominantly MyHC I and IIa isoforms have relatively high oxidative potential and are recruited more during endurance types of exercise (Seene and Umnova 1992; Seene et al. 2007). These fibers are also more susceptible to oxidative damage by ROS than fibers with predominantly MyHC IIb and IId isoforms. The multidirectional changes in given MyHC isoforms in ST and FT muscles with an increase in training volume not only support the differing resistance of ST and FT muscles to overtraining, but also show the differing sensitivity of a given MyHC isoform to proteolytic enzymes in different muscles (Seene et al. 2005). If in FT Pla muscle the relative content of MyHC I isoforms decreases with an excessive increase in volume leading to overtraining syndrome, and the relative content of MyHC IIa increases, in Sol muscle the content of MyHC I isoform increases and that of MyHC IIa decreases (figure 14.5). The relative contents of MyHC IIa, IId, and IIb in two different FT muscles even change in different directions with an excessive increase in volume (Seene et al. 2005). These changes may be related to differences in oxidative potential between these two FT muscles and their fibers (Seene and Umnova 1992; Seene et al. 2007).

Although it was shown more than decade ago that all endurance training programs lead to a slowing of contraction in skeletal muscle (Guezennec, Gilson, and Serrurier 1990), to our knowledge it is not correct to understand the slowing of the contractile function of a muscle only as a function of the decrease in the relative content of the fastest MyHC isoform (figure 14.5). Therefore, a decrease in the relative content of MyHC IIb isoform in FT muscles during endurance training does not necessarily demonstrate the slowing of muscle contractile properties, since the relative contents of MyHC IId and IIa isoforms are increasing at the same time (figure 14.5). Baldwin and Haddad (2002) have proposed that changes in myosin isoforms during endurance training could be interpreted as quantitative remodeling of muscle in which one isoform is replaced with another that is better suited for adaptation to long-lasting low-level force–generating activity (Baldwin and Haddad 2002). A decrease of MyHC IIb isoform in rat skeletal muscle during endurance training suggests that it becomes energetically more economical to exercise with FT muscles as a result of the changing ratio of the different MyHC isoform (Seene et al. 2005). Changes in the relative content of MyLC isoforms between two FT rat muscles, Pla and EDL, do not show such major differences as changes in the relative content of MyHC isoforms (Seene et al. 2007).

The overtraining and the decrease in PWC that result from excessive increase in endurance training volume are caused primarily by increased degradation and decreased synthesis of MyHC (Seene, Umnova, and Kaasik 1999; Seene et al. 2004). During this process, MyHC composition changes depending on the twitch characteristics of muscle. In both Sol muscle, where the MyHC I isoform dominates, and Pla muscle, where the MyHC IId isoform dominates, the relative content of these predominant MyHC isoforms increases during overtraining. MyHC IIa isoform content decreases in ST muscle and at the same time increases in FT muscle. This may result from the differing sensitivities of a given MyHC isoform in ST and FT muscles to the proteinases, or to differences in expression of MyHC or other myofibrillar protein isoforms, in different muscles (Seene et al. 2003).

Myosin Light Chain Isoforms

Changes in MyLC isoforms and the roles of these isoforms during overtraining have been less extensively studied, but the few existing studies give grounds to believe that MyLC isoforms play an essential role in the process of contraction. As shown in figure 14.6, changes in MyLC isoforms in skeletal muscles during overtraining are much smaller than in MyHC isoforms (figure 14.5).

There is a significant correlation between excessive increase in training volume and decrease in the relative content of MyLC 3f isoform in FT muscles. Some research has shown that expression of MyLC 3f isoform is decreased in most FT muscles with endurance training but increased in Pla muscle (Wahrmann, Winand, and Rieu 2001). Although Hayashibara and Miyanishi (1994) are convinced that the increase in MyLC 1f isoform is an indication of the slow speed of contraction during endurance training, this has not been seen in FT Pla muscle (Wahrmann, Winand, and Rieu 2001). Conflicting data on changes in MyLC isoforms during endurance training indicate that other myofibrillar proteins must be participating in the modulation of contractile machinery during adaptation to endurance training.

Decreases in the synthesis rate of MyHC and myosin-binding C-protein with increased volume leading to overtraining, as well as a significant correlation between the changes in proteins and training volume (Seene et al. 2005), suggest that changes in MyHC isoforms with increased volume are only part of a complex of changes in different individual myofibrillar proteins and their isoforms in overtrained skeletal muscle. There may be several reasons for changes in MyHC composition in skeletal muscles with excessive volume that leads to overtraining syndrome. However, hypoxia (Bigard et al. 2000) and ischemia (Vescovo, Zennaro, and Sandri 1998) have been shown to cause a rise in fast MyHC isoforms in skeletal muscle; and during chronic heart failure, ST fibers tend to be replaced by FT fibers (Gosker et al. 2000). Of course, changes in skeletal muscle with increased training volume are not completely comparable to changes in these pathological situations, but chronic excessive training volume leads to the overtraining syndrome and is accompanied by serious pathological changes in muscle tissue (Seene, Umnova, and Kaasik 1999). In an overtraining situation, then, insult-associated changes in skeletal muscle may be compared with changes that occur in these pathological situations.

Partial correlations showed that the synthesis rate of MyHC, which was negatively correlated with increased training volume, lost its significance in conditions in which the role of myosin-binding C-protein was eliminated (Seene et al. 2005). This proves once again the functional significance of minor proteins in muscle contraction in conditions of overtraining. Myosin-binding C-protein seems to play an essential role in correct thick filament formation during myofibrillogenesis, in modulating muscle contraction, and in increasing the maximum shortening velocity (Hofmann, Greaser, and Moss 1991). C-protein can either bind actin and myosin or affect the mechanical properties of myosin cross-bridges by linking the S2 segment of myosin to the backbone of the thick filament (Hofmann, Greaser, and Moss 1991). Myosin-binding C-protein is very sensitive to high training volume and, together with

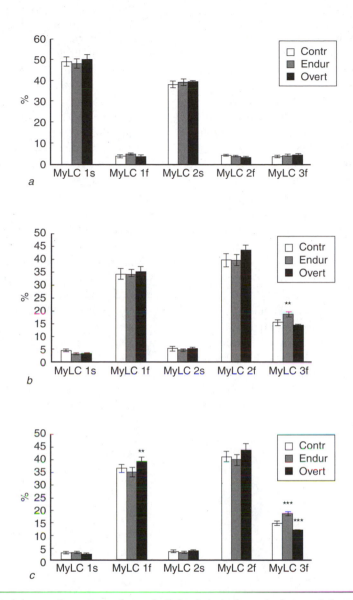

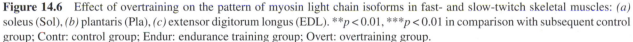

Figure 14.6 Effect of overtraining on the pattern of myosin light chain isoforms in fast- and slow-twitch skeletal muscles: *(a)* soleus (Sol), *(b)* plantaris (Pla), *(c)* extensor digitorum longus (EDL). **$p < 0.01$, ***$p < 0.01$ in comparison with subsequent control group; Contr: control group; Endur: endurance training group; Overt: overtraining group.

MyHC isoforms, may play a key role in changes in the functional properties of contractile machinery during overtraining (Seene et al. 2004).

In addition to the changes in myosin composition, the degradation rate of individual myofibrillar proteins in contractile apparatus increases with increased endurance training volume. The modulation of myosin isoforms during endurance training may also be related to the sensitivity of different MyHC and MyLC isoforms to the action of proteinases (Seene et al. 2003).

Finally, in addition to the changes in MyHC and MyLC isoform patterns, regulatory and minor proteins play an important regulatory role in modulating the function of the contractile apparatus during overtraining.

Summary

In the training process of modern endurance athletes, overload is often improperly designed; frequent training stimulus interrupts the overcompensation phase, adaptation does not occur, and overtraining syndrome develops. Overtraining is accompanied by a decreased rate of synthesis of muscle proteins, particularly myofibrillar proteins, and an increased rate of protein degradation in skeletal muscle. In volume-induced overtraining of skeletal muscles, it is mainly FT OG and ST O muscle fibers that show destruction of myofibrils.

Disarray of thick myofilaments starts from the periphery of myofibrils; wide distances between myofilaments and swelling of mitochondria are signs of the destruction of contractile apparatus on the ultrastructural level.

Intrafusal muscle fibers adapt to overtraining similarly to extrafusal fibers, although the changes in intrafusal fibers are considerably smaller. Overtraining affects neuromuscular junctions as well. Axon terminals branch off, and widened synaptic slots in some regions show that nerve-muscle transfer is disrupted in overtrained muscle. Regeneration in overtrained skeletal muscle is slow, since the number of satellite cells under the basal lamina has decreased. Lack of IGF-1, MGF, and transcription factors leads to apoptosis and atrophy of muscle fibers. A decrease in the number of myonuclei and damage to DNA lead to a decrease in DNA units. Synthesis of HSP70 is greater in overtrained skeletal muscle. Heat shock proteins play a protective role in preserving muscle function and maintaining homeostasis of muscle fibers.

During overtraining, the relative content of MyHC I isoform in ST Sol muscle increases and that of IIa isoform decreases. In overtrained FT Pla and EDL muscles, the relative content of MyHC IIb isoform decreases and that of IIa isoform increases; in Pla muscle, the relative content of MyHC I isoform decreases and that of IId isoform increases. These changes in MyHC isoforms show that the contractile properties of ST and FT muscles are altered in different ways in accordance with muscle oxidative capacity. Changes in MyLC isoforms during overtraining are much smaller than subsequent changes in MyHC isoforms. The most significant changes in MyLC isoforms during overtraining are seen in FT muscles. Regeneration of MyHC 2b and MyLC 3f isoforms, which have high affinity to each other in FT muscle fibers after tissue damage, proceeds at different speeds. Regeneration of MyLC 3f isoform takes about 50% less time than that of MyHC 2b isoform in FT muscle fibers with low oxidative capacity.

The changes in overtrained skeletal muscle described in this chapter reflect the significance of alterations at the cellular and molecular level for diagnosis of overtraining syndrome, and understanding of these mechanisms may be significant in helping to prevent the development of overtraining syndrome among competitive endurance athletes.

ACKNOWLEDGMENTS

This study was supported by the funds of the University of Tartu and of the Estonian Scientific Foundation, grant numbers 4611 and 6501.

CHAPTER 15

Intersubject Variability in Developing Exertional Muscle Damage

Priscilla M. Clarkson, PhD, and Stephen P. Sayers, PhD

Injury or destruction of muscle cells by work or physical activity is known as exertional rhabdomyolysis. This muscle damage can be mild (producing symptoms of delayed-onset muscle soreness) to moderate (producing soreness that is profound and is accompanied by losses of muscle strength and range of motion) to severe, at which point it becomes clinically relevant. Other manifestations of exertional rhabdomyolysis are stiffness, weakness, and an increase in muscle proteins such as creatine kinase (CK) and myoglobin (Mb) in the circulation (Knochel 1982; Schulze 1982; Milne 1988; Knochel 1990, 1992). Severe, or clinically relevant, exertional rhabdomyolysis can lead to renal failure (which can be life threatening), compartment syndrome, or subsequent functional impairment. The reason renal failure may occur is that elevated Mb in the circulation can precipitate in the renal tubules, shutting down kidney function. Because of the ease of assessing blood CK levels, they are often used clinically as a surrogate measure of Mb. No standard level of CK has been established as diagnostic for serious rhabdomyolysis, but five times normal or greater is often used (Brown 2004) in conjunction with other indicators, such as myoglobinuria, which occurs when high levels of Mb in the blood "spill over" into the urine.

Severe exertional rhabdomyolysis is quite rare. For example, of 16,506 candidates who took a firefighter fitness test in New York during 1988-1989, 32 were hospitalized with rhabdomyolysis and renal failure (Centers for Disease Control, 1990). Why selected individuals and not others succumbed to severe exertional rhabdomyolysis is unclear. However, it is striking that, in a group of people exercising with apparently similar intensity and in a similar environment, it may be that only one person develops severe rhabdomyolysis. This chapter examines the intersubject variability in the development of moderate to severe exertional rhabdomyolysis and possible explanations for that variability.

Factors Affecting Development of Rhabdomyolysis

During the first day of police training at a police academy in Agawam, Massachusetts, 50 recruits were asked to perform numerous calisthenics, including sit-ups, push-ups, and jogging (Clarkson 1993). At about 4:00 p.m. that day, one 25-year-old cadet collapsed on the track during a run and was transported to the hospital. He was diagnosed with acute renal failure and placed on dialysis. Complications occurred that caused damage to the liver, and he was transported to another hospital for a liver transplant and subsequently died (Hassanein et al. 1991). Eleven other cadets were hospitalized, but only two were placed on dialysis, and they fully recovered. Thus, 3 of the 50 recruits (6%) were diagnosed with severe exertional rhabdomyolysis after being exposed to overexertion exercise through the course of the day. Factors that may have precipitated these cases were the cadets' limited fluid intake and the air temperature (the day was warm). Most cases of severe exertional rhabdomyolysis, especially those that resulted in renal failure, have occurred in warm or hot environments where dehydration was a factor (Vertel and Knochel 1967; Knochel 1982). However, this is not always so.

When people are asked to overexert themselves beyond a "tolerable limit," some can incur severe rhabdomyolysis (Greenberg and Arneson 1967; Demos et al. 1974; Olerud et al. 1976). Two such cases were of adults who were encouraged by exercise leaders in a local health club to overexertion during their exercise routine (Springer and Clarkson 2003). One case was a fit 22-year-old female college senior without any underlying disease state who typically exercised by running 3 to 5 miles (4.8-8 km) daily and lifted weights five days per week. She performed squats and lunges and worked the calf muscles to exhaustion, causing her legs to shake noticeably and requiring the trainer at the club to assist her in walking from one exercise machine to another. She reported that she was adequately hydrated and that the room temperature was comfortable. She was urged on despite her statements that she wanted to quit. The other case was a 37-year-old relatively unfit man who had been a multiple-sport athlete in college and had participated in Olympic team handball. He had been relatively sedentary since college except for his active job as an emergency physician. He had no underlying disease and took no medications except for the nutritional supplement creatine. The personal trainer at the local health club encouraged him to perform excessive exercise, and he vomited after the session.

Both these people presented with myoglobinuria 48 h after the exercise sessions and were treated for severe rhabdomyolysis. In the first case, the woman came to the hospital two days posttraining with a CK of 234,000 U/L; in the other case, the man sent his samples to a lab and had a peak CK at five days postexercise of 70,158 U/L. These were healthy individuals in whom the primary factor precipitating the event appeared to be overexertion encouraged by a trainer. It is quite likely that these trainers had encouraged and motivated many other clients to work at their maximal capacity—so why is it that these particular individuals developed severe rhabdomyolysis?

Our laboratory has been involved in the study of exertional rhabdomyolysis for about 20 years. We expose subjects to a unilateral elbow flexion eccentric (muscle-lengthening contraction) exercise to induce mild to moderate rhabdomyolysis that is resolved without incident (e.g., is not

clinically relevant). Our laboratory has used this exercise model for about 20 years and found it to be safe, with no alteration in renal function (Clarkson et al. 2005c), likely because only one limb and a relatively small amount of muscle mass are exercised. Performance of exercises that are biased toward eccentric contractions results in muscle damage because of the increased strain these place on muscle tissue (Lieber and Friden 1993). However, we have noticed that some subjects are "high responders" to this exercise. They present with profound strength losses, extreme elevations in CK, exaggerated swelling, intense muscle pain, or some combination of these (Sayers et al. 1999).

Several predisposing conditions have been identified as factors in the development of severe rhabdomyolysis upon exposure to excessive exercise. These include malignant hyperthermia (Hackl et al. 1991; Kojima et al. 1997; Davis et al. 2002), calcium adenosine triphosphate (ATP) deficiency (Poels et al. 1993), carnitine palmityl transferase deficiency (Mantz et al. 1992; Katzir et al. 1996), abnormal dystrophin (Figarella-Branger et al. 1997), alpha-sarcoglycan deficiency (Mongini et al. 2002), myoadenylate deaminase deficiency (Baumeister et al. 1993), and hemangioma steal syndrome (Knochel 2000). However, these conditions are mostly associated with recurrent exercise intolerance or gross muscle function abnormalities and can be ruled out as the underlying cause of the spontaneous severe rhabdomyolysis induced in otherwise healthy individuals exposed to excessive exercise.

Why some people and not others incur severe muscle damage in response to a laboratory test or to "real-life" exercise is not known. Secondary factors associated with the development of clinically relevant exertional rhabdomyolysis include genetic conditions (e.g., sickle cell trait), training status, dehydration, bacterial or viral infections, abrupt alterations in the diet, drugs, medications, alcohol, and use of ergogenic aids (Vertel and Knochel 1967; Knochel 1990; Marinella 1998; Braseth et al. 2001; Wirthwein et al. 2001; Pretzlaff 2002). The nutritional supplement creatine is very popular with athletes, and one of the two people we have referred to used creatine, which was suspected as a possible secondary factor in the development of rhabdomyolysis. However, creatine supplementation has been reported not to alter thermoregulatory responses during exercise (Volek et al. 2001), and we have found that creatine does not exacerbate exercise-induced muscle damage (Rawson et al. 2001). Commercially available nutritional supplements do not need to meet the same quality control standards as pharmaceuticals and could be contaminated with other ingredients, including stimulants such as ephedrine and caffeine (Kamber et al. 2001). While nutritional supplements may serve as secondary factors, these and the other secondary factors just listed have not fully explained many cases of exertional rhabdomyolysis in otherwise healthy people.

Variability in Responses to Eccentric Exercise

In the following sections we describe the variability of several indicators of muscle damage in response to eccentric exercise.

Creatine Kinase and Myoglobin

Several studies have demonstrated a large intersubject variability in CK and Mb responses to eccentric exercise (Newham et al. 1983; Nosaka et al. 1992; Nosaka and Clarkson 1996). The exercise we use in our laboratory to induce temporary, repairable muscle damage involves a series of 24 to 70 maximal eccentric contractions of the elbow flexor muscles. In 1992, we noted a range of 6740 U/L to 24,200 U/L in peak CK activity following eccentric contractions of the elbow flexors in six subjects (figure 15.1), with similar variability in Mb concentration (Nosaka et al. 1992). We later reported that the range of postexercise blood CK activity of 59 college-aged men who performed maximal eccentric contractions was 96 U/L to 30,810 U/L (Nosaka and Clarkson 1996).

Recently we examined changes in CK and Mb in a group of 208 adults who performed the eccentric elbow flexor exercise (Clarkson et al. 2005c). Creatine kinase levels were elevated 6420%, 2100%, and 311% above baseline on days 4, 7, and 10 after the exercise, respectively; Mb was 1137%, 170%, and 28% above baseline on days 4, 7, and 10 days after exercise, respectively. Of the 208 participants, 173 showed an increase in CK of less than 15,000 U/L; 11 had CK levels between 15,000 and 20,000 U/L; 16 had CK levels between 20,001 and 30,000 U/L; seven had CK levels between 30,001 and 50,000 U/L; and one reached a peak CK activity of 80,000 U/L. Creatine kinase and Mb responses were significantly correlated ($r = 0.80$, $p < 0.01$). The average levels of CK and Mb on each of the four days are shown in figure 15.2.

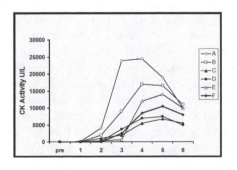

Figure 15.1 Serum creatine kinase activity before (pre) and six days following eccentric exercise of the elbow flexors in six research subjects (A-F).

Modified from Nosaka et al., 1992.

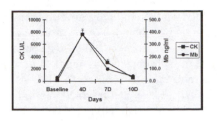

Figure 15.2 Creatine kinase (CK) activity and myoglobin (Mb) concentration in the blood before and 4, 7, and 10 days after eccentric exercise of the elbow flexors (N = 208).

Modified from Clarkson et al., 2005c

While the variability between subjects is high, the variability within a subject may be stable. Chen (2006) separated subjects based on their increase in CK activity after a standard eccentric exercise into high (greater than 10,000 U/L), medium (2000 to 10,000 U/L), and low responders (less than 2000 U/L). These subjects repeated the same exercise one year later and demonstrated very similar CK responses. In other words, those subjects who were high responders on the first exercise were also high responders when repeating the exercise one year later.

Data from animal studies have shown that estrogen appears to protect muscle from contraction-induced injury (Tiidus 2001; Tiidus et al. 2001; Tiidus 2003). Therefore we sought to determine whether women would have less of an increase in CK in response to exercise than men. An earlier study using aerobic cycling as the exercise stress did show women to have a smaller rise in CK (152 U/L ± 111) compared with men (664 ± 546). That study used a small sample size (20 subjects); the increase in CK was small, and the CK responses were quite variable (Shumate et al. 1979). We recently examined CK response after elbow flexion eccentric exercise in 43 men and 58 women and found that men had a mean increase of 12,522 (standard error = 2494) while women had a mean increase of 6617 (standard error = 947) (unpublished observation). Note that the standard error for the men was almost threefold higher than that for the women. Further examination of the data showed that none of the women had an increase in CK over 30,000 U/L but 11% of the men did. These data suggest that more men are high CK responders than women.

Strength Loss and Recovery

The prolonged strength loss after eccentric exercise is considered a hallmark of muscle damage (Warren et al. 1999). Figure 15.3 presents the time course and average strength loss of 208 subjects who performed the eccentric elbow flexion exercise (unpublished data). As can be seen, the average strength loss immediately after the exercise was about 55%, and strength did not fully return by 10 days postexercise. Although not as variable as CK or Mb, there was still a large range in the response. In the larger sample of 208 subjects, strength loss immediately after the exercise ranged from 0% to 91%. At 10 days postexercise, 64 subjects had returned to within 10% of baseline, 53 had returned to within 20%, 40 had returned to within 40%, 34 had returned to within 60% of baseline, and 10 still had greater than 60% loss in strength.

Sayers and colleagues (1999) presented cases of two college-aged male subjects who showed profound strength losses in response to the elbow flexor eccentric exercise (figure 15.4). Case 1 had a 79% loss in strength immediately after the exercise, which increased to 84% at five days postexercise. Strength did not return to within 10% of baseline until 47 days postexercise. Case 2 demonstrated a 74% loss in force immediately after the exercise, and five days postexercise the strength loss was 66%. By 43 days postexercise, there was still a 48% strength loss. In another case report, one male subject took 89 days to fully recover his strength after performing the elbow flexor eccentric exercise (Sayers and Clarkson 2001). During the time of the profound losses in strength, the subjects reported that they were unaware of any impairment and were able to carry out daily activities without difficulty.

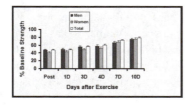

Figure 15.3 Loss in elbow flexor isometric strength in 208 subjects (total) and a subsample of 43 men and 58 women who performed eccentric exercise of the elbow flexors.

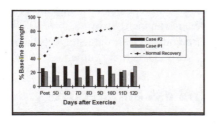

Figure 15.4 Loss in isometric elbow flexor strength immediately (post) and up to 12 days after eccentric exercise of the elbow flexors in two case studies of profound strength loss. Dashed line indicates typical values for normal recovery.

Modified from Sayers et al., 1999.

Sayers and Clarkson (2001) examined force loss and recovery in 98 men and 94 women who performed the elbow flexion eccentric exercise. Thirty-two subjects demonstrated greater than 70% strength loss immediately after exercise; of these, 24 were women and 8 were men. Twenty of these subjects (15 women and 5 men) agreed to have follow-up tests, and the women showed a faster recovery of strength. In our more recent sample of 43 men and 58 women, mentioned earlier, 35% of the women showed greater than 70% loss in strength immediately after exercise while 9% of the men did (figure 15.3). It appears that women lose more strength immediately after exercise but recover more quickly. In fact, the cases we have previously reported that showed a significantly prolonged strength recovery all were men (Sayers et al. 1999; Sayers and Clarkson 2001).

Swelling

The most obvious high response to the elbow flexion eccentric exercise is swelling of the exercised limb. The swelling begins after the exercise, increases up to about five days postexercise, and gradually subsides over the next week (Clarkson et al. 1992; Nosaka and Clarkson 1996). After eccentric exercise of the elbow flexors, the average increase in circumference measured at 4 cm (1.6 in.) above the elbow joint was about 3 cm (1.2 in.) (Nosaka and Clarkson 1996). However, some subjects show profound swelling (i.e., "Popeye" arms; figure 15.5). We have reported that 6 of 204 subjects who performed the elbow flexor eccentric exercise had extreme swelling (Sayers et al. 1999).

Earlier reports suggested that individuals who exhibited symptoms of exertional rhabdomyolysis, including severe swelling, had either a compromise in substrate metabolism or impaired perfusion of the muscle (hypoxia) during exercise (Legros et al. 1992; Trimarchi et al. 2000). On the basis of these reports, Sayers and colleagues (2002) examined whether subjects who had previously demonstrated extreme swelling of the arm in response to elbow flexion eccentric exercise had a metabolic defect. These subjects and a control group (who had previously performed the same eccentric exercise but did not demonstrate swelling) performed a low-intensity anaerobic exercise designed to increase blood lactate concentration. There was no difference in the increase in blood lactate concentration between the "swellers" and "nonswellers" in response to anaerobic exercise, implying that the unusual swelling response was not the result of a dysfunction in energy metabolism. However, the sample size was small (four subjects per group); this was unavoidable since there are so few subjects who demonstrate such profound swelling.

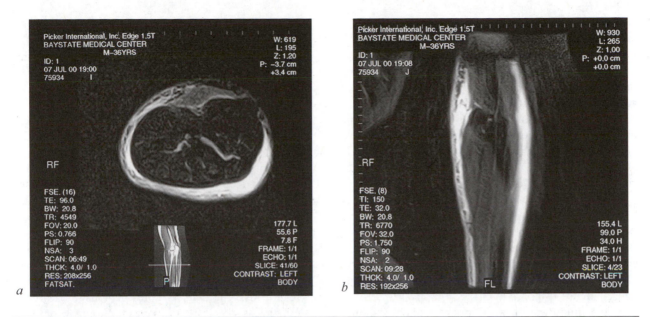

Figure 15.5 Magnetic resonance image of a swollen eccentrically exercised arm obtained several days following eccentric exercise of the elbow flexors. Note the edema (white area) surrounding the upper arm, elbow, and forearm. *(a)* Cross section (inner small photo indicates where the cross section was taken along the forearm) and *(b)* longitudinal section, showing swelling around the elbow area. Note that swelling begins in the upper arm and, due to gravity, tracks down to the forearm where it becomes most noticeable.

Relationships Among Variables

We have described "high responders" with respect to elevations of muscle proteins in the blood, strength loss, and swelling in response to eccentric exercise. Are the subjects who show profound changes in one measure the same ones who show profound changes in another?

Sayers and colleagues (1999) reported two cases in which strength loss was over 76% immediately after exercise and the recovery lasted almost two months. However, the peak CK response for one subject was 18,018 U/L and for the other subject was 8450 U/L—these are relatively low values for "high responders." Case 1 demonstrated severe swelling of the arm, but Case 2 did not. Sayers and colleagues (1999) reported four additional cases of profound swelling. One of these subjects had a peak CK of 38,517 U/L, but muscle soreness was minimal (29 mm out of 100 mm on a visual analog scale [VAS]); the second subject demonstrated 73% strength loss at 48 h postexercise and severe soreness (80.5 mm out of 100 mm on the VAS) (CK was not assessed); the third subject demonstrated 60% strength loss 48 h postexercise and moderate soreness (56 mm out of 100 mm on the VAS); and the fourth subject demonstrated 69% strength loss at 48 h postexercise and severe soreness (87.5 mm out of 100 mm on the VAS). While in general subjects who are high responders for one variable are moderate to high responders for another, this is not always the case.

Genetic Differences in High Responders

Gulbin and Gaffney (2002) examined whether the variability in the markers of exercise-induced muscle damage could be attributed to genetic factors. Sixteen pairs of identical twins performed maximal eccentric contractions using the elbow flexors. The CK, Mb, and strength loss responses differed between the twin siblings. Although the authors concluded that variability following eccentric exercise was not attributable to genetic differences, the relatively small sample size and lack of high CK responders (it does not appear that any subject had a CK value of over 20,000 U/L) make further study of genetic underpinnings warranted.

The recent increase in human genomics research is largely the result of the successful Human Genome Project. The project, started in 1990, was to be completed in 2005, but technological advances made it possible to publish working drafts of the human genome sequence in 2001. Given known nucleotide sequences, it is possible to determine the presence of single nucleotide substitutions; these alterations are termed single nucleotide polymorphisms (SNPs). Single nucleotide polymorphisms are defined as variations within the genetic sequence that occur in at least 1% of a given population. The presence of SNPs can help to explain differences between individuals and has proved useful in identifying an individual's propensity for a disease or disorder. The nomenclature used to describe a gene SNP provides both the change in the nucleotide (substitution) and the nucleotide position where the change occurs. Thus the nomenclature for the SNPs discussed next consists of the abbreviated gene name, the substitution, and the nucleotide position where the substitution occurs. For example, the gene polymorphism for cardiac-restricted ankyrin repeat protein, where there is a substitution of a thymine (T) for a cytosine (C) at nucleotide position 105, would be abbreviated CARP C105T.

We proposed that one or more genetic polymorphisms (SNPs) may confer a propensity to severe exertional rhabdomyolysis. We were able to test whether profound strength losses or elevations in CK and Mb were associated with a particular SNP. Our laboratory was part of a multisite consortium to examine genetic factors underlying the response to changes in physical activity. The consortium was engaged in the National Institutes of Health–funded FAMuSS (Functional Polymorphisms Associated with Human Muscle Size and Strength) study (Thompson et al. 2004). FAMuSS had two objectives:

- To identify all SNPs in the 500 most highly expressed muscle proteins
- To determine, in 1000-plus subjects, which SNPs contribute to baseline elbow flexor muscle size and strength and their response to 12 weeks of resistance exercise training

In the FAMuSS study, Drs. Hoffman and Devaney at Children's National Medical Center (Washington, DC) discovered several novel SNPs, including those in cardiac-restricted ankyrin repeat protein (CARP C105T), myosin light chain kinase (MLCK C37885A and MLCK C49T), Akt1 (-C12273A, -G783T, and -G171T), nicotinamide N-methyltransferase (NNMT G5082T), synaptogyrin (SYNGR2 C886T), and insulin-like growth factor-1 (IGF-1 C1245T). All are proteins that have been found to be involved in muscle structure or function. We also found strong associations of Akt1 genotypes and previously identified alpha actinin 3 (ACTN3 C1747T) genotype with body composition and muscle strength phenotypes (Hoffman and Nader 2004; Clarkson et al. 2005a). We then determined whether these SNPs were associated with increases in CK, Mb, or strength loss after elbow flexion eccentric exercise. In other words, could genetic variations (polymorphisms) in these muscle proteins dictate the susceptibility of individuals to large increases in CK activity and Mb concentration in the blood or profound strength loss and delayed recovery after strenuous eccentric exercise?

The SNPs that were associated with the increase in CK and Mb following elbow flexion eccentric exercise were

those of MLCK (Clarkson et al. 2005b). MLCK (C49T) was strongly associated with the increase in CK activity and Mb concentration at four days postexercise, which was the time point of the peak response (figure 15.6). Table 15.1 presents the genotypes and allele frequencies for the MLCK SNPs. Both the homozygous wild-type and the heterozygous group showed significantly lower CK activity and Mb concentration than the homozygous rare allele. Observe in figure 15.6 that subjects homozygous for the rare allele (TT) had the largest elevations in CK and Mb; and although it was not statistically significant, there appeared to be a dose-response association. Another polymorphism, MLCK C37885A, was associated with CK on days 7 and 10 postexercise, with the heterozygote having higher values than the homozygous wild type (only one subject possessed the homozygous rare allele). MLCK C37885A was also associated with postexercise strength loss; the heterozygotes demonstrated greater strength loss (–57.0% ± 17.9) than the homozygous wild type (–49.7% ± 20.3).

MLCK functions to phosphorylate myosin's regulatory light chain (RLC) (Sweeney et al. 1993), and RLC plays an important modulatory role in force development (Sweeney and Stull 1990; Szczesna et al. 2002). However, RLC is not readily phosphorylated in type I fibers and thus acts predominantly in type II fibers (Grange et al. 1995; Szczesna et al. 2002). We (Clarkson et al. 2005b) suggested that MLCK C49T and C37885A polymorphisms increase the ability to phosphorylate RLC, which would generate more tension during the stretch of the eccentric contraction (strain) and more consequent damage, since damage is related to strain on myofibrils (Lieber and Friden 1993).

The association of MLCK C49T and MLCK C3788A genotypes with increases in CK and Mb is an important clinical finding because subjects with the rare allele may show an exaggerated increase in Mb, which could affect renal function. Individuals possessing rare alleles for MLCK C49T and perhaps MLCK C3788A may be

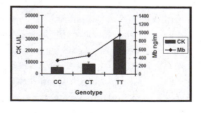

Figure 15.6 The association of myosin light chain kinase (MLCK) C49T genotype with peak creatine kinase (CK) and myoglobin (Mb) increase following eccentric exercise of the elbow flexors.

Modified from Clarkson et al., 2005b.

Table 15.1   Genotype and Allele Frequencies for MLCK SNPs

Gene	SNP	Genotype N (%)		
		Wild-type homozygote (CC)	Heterozygote (CT)	Rare allele homozygote (TT)
MLCK	C37885A	121 (77.1%)	35 (22.3%)	1 (0.6%)
MLCK	C49T	97 (63.4)	50 (32.7%)	6 (3.9%)

at higher risk than others for rhabdomyolysis and acute renal failure, especially in situations of heat stress and dehydration. However, it should be noted that these associations account for only a small percent of the variability in responses. Likely other genes are involved, and these remain to be determined.

Summary

Severe exertional rhabdomyolysis occurs in situations in which people are exposed to intense or extreme exercise. This condition becomes clinically relevant when Mb accumulates in the blood to high levels and precipitates in the kidney, increasing the risk for acute renal failure. However, when groups of people are exposed to the same exercise and environmental conditions, it may be that only certain individuals develop severe rhabdomyolysis. Our laboratory investigations have shown that certain people are high responders to standardized exercise in which one arm is exercised to induce temporary, repairable muscle damage without compromising renal function. These individuals show relatively dramatic increases in CK activity and Mb concentration in blood, profound and prolonged losses in muscle strength, and exaggerated swelling and pain. Generally a high responder in one of these variables is a high responder in the other variables, but this is not always the case. Although secondary factors such as viral infection, heat stress, and dehydration contribute to the development of moderate to severe rhabdomyolysis, these factors cannot fully explain why some people and not others are high responders. We have recently investigated possible genetic underpinnings and have found variations (polymorphisms) in a gene that codes for an important muscle protein, myosin light chain kinase, that may make an individual more susceptible to rhabdomyolysis.

PART III

Treatments and Interventions in Muscle Damage and Repair

CHAPTER 16

Massage Therapy

Peter M. Tiidus, PhD

Although massage is widely used as a therapeutic intervention in muscle injury and damage, information on its effectiveness in influencing muscle repair or recovery is surprisingly scant. In 2004, Galloway and Watt reviewed data recorded by British head team physiotherapists at major national and international athletic events over an 11-year period. They found that on average, close to 50% of the team physiotherapists' time was spent administering therapeutic massage to muscles of team athletes. They noted the general lack of positive data on any therapeutic effects massage may have on muscle repair and called for further research to justify the significant provision of time and resources related to massage delivery to high-caliber athletes.

Massage is also widely practiced in clinical settings involving the general public, often as an "alternative therapy" to treat exercise-, overuse-, or trauma-induced muscle damage (Vickers 1996; Tiidus 1997). It has been suggested that massage therapy may be able to enhance postdamage muscle healing by, among other things, improving muscle blood flow, reducing muscle soreness, reducing muscle tension, removing "toxins" such as lactic acid, diminishing exercise-induced muscle damage, ameliorating strength loss and swelling, and augmenting return of muscle strength and function (Cafarelli and Flint 1992; Tiidus 1997).

Despite its wide use and the beliefs about its efficacy, relatively little empirical research has been done on the potential for massage therapy to influence postdamage muscle damage, inflammation, or recovery or other physiological manifestations of damage. In general, the research available tends to suggest that massage either is ineffective or has only limited influence on manifestations of damage or rate of repair.

Massage Effects on Indirect Indices of Damage and Repair

As described in other chapters in this book, muscle damage and repair are manifested in a number of ways. These include muscle membrane disruption and resultant protein and enzyme leakage out of and into muscle cells; muscle inflammatory and inflammatory mediator responses (including influx of leukocytes into muscle cells and increases in inflammatory cytokines); histochemically and biochemically determined manifestations of muscle ultrastructural disruption; peroxidative damage and activation of lysosomal enzyme activities; alterations in muscle calcium homeostasis; and activation and proliferation of muscle satellite cells and chemical mediators of muscle repair.

As noted in table 16.1, none of the directly quantifiable indexes of damage and repair—and few of the indirect measures of muscle damage, inflammation, or repair mechanisms—have been studied with regard to the potential for massage to influence their progress, reaction, or amelioration consequent to damaging insult. In other words, we have no data on the ability of massage to influence most standard histochemical or physiological-biochemical markers used to determine and quantify muscle damage in experimental models. Thus our understanding of whether and to what extent massage therapy can influence

Table 16.1   Research Data on Massage as an Aid to Muscle Damage Repair

Manifestation of muscle damage and repair or reputed repair mechanism	Research on massage effects on muscle damage and repair manifestation*
Membrane disruption; enzyme/protein leak	No data
Muscle cell leukocyte infiltration	No data
Muscle inflammation and cytokines	No data
Muscle ultrastructure disruption and repair	No data
Muscle satellite cell activation and proliferation	No data
Muscle peroxidative damage	No data
Muscle lysosome and protease enzyme activities	No data
Altered muscle calcium homeostasis	No data
Prolonged loss of muscle force and force recovery	Substantial data—no effect
Delayed-onset muscle soreness (DOMS)	Substantial and mixed data—minor effects
Muscle swelling	Limited indirect data—no effect
Muscle blood flow changes	Substantial data—no effect
Muscle lactate clearance	Substantial data—no effect
Muscle scarring and adhesion	Limited indirect data—no effect
Muscle tension, relaxation, length	Substantial data—minor effects
Myofascial trigger points	Limited data—minor effects

*Substantial data = three or more studies; limited data = one or two studies or indirect inferences from related studies; mixed data = contradictory findings; no data = no studies performed.

the course of damage, inflammation, and repair is limited to relatively few studies focused on rather indirect indices of muscle damage and repair.

Muscle Force Loss or Recovery

The most easily quantified indirect marker of muscle damage and repair that has been used to assess the potential influence of massage on muscle recovery from damaging exercise is the degree of muscle strength loss and its rate of recovery. The loss of muscle strength is commonly seen following eccentric exercise–induced muscle damage, and strength returns to normal within 3 to 14 or more days depending on the severity of the damage (Clarkson et al. 1992; Howell et al. 1993). Warren and colleagues (1999) suggested that the loss and regain of muscle strength following exercise-induced damage were among the best noninvasive indicators of disruption of one or more aspects of the excitation–contraction coupling continuum of muscle function and could be used as a reasonably quantifiable index of muscle damage and repair (see chapters 2 and 3 for

details). In other words, if disruption occurred to any step in muscle contraction initiation and completion (from muscle sarcolemma depolarization through sarcoplasmic reticulum calcium release, troponin and tropopmyosin interactions with actin, myosin ATPase activity, etc.), it would ultimately manifest itself as a loss of muscle force. With use of muscle force as an indicator of recovery, one should also remember that a number of other neuromuscular factors that go beyond muscle structural disruption (e.g., changes in the muscle length–tension relationship and excitation–contraction uncoupling) contribute to the loss and regain of muscle force production following exercise-induced damage (Warren et al. 2001; Proske and Allen 2005).

One of the first studies of the effects of massage on muscle strength recovery following eccentric exercise was that of Tiidus and Shoemaker (1995). In this study a licensed and experienced massage therapist performed four days of daily 10 min deep and superficial massage on the quadriceps muscles of one leg of each male subject following intense eccentric quadriceps contractions involving both legs. The other leg served as unmassaged control. In a questionnaire, subjects had indicated their general expectation that the massaged leg would recover faster than the unmassaged leg. Despite this potential placebo effect, the study showed that for up to four days postexercise, daily massage did not improve the rate of recovery of quadriceps isometric or isokinetic peak torque over that of the unmassaged control. These results are depicted in figure 16.1.

At least five other studies have also demonstrated the failure of massage intervention to influence the recovery of muscle force following various forms of exercise. Weber and colleagues (1994) had female subjects do eccentric biceps contractions and then performed massage immediately and 24 h following the exercise on some of the subjects. No differences were found between those who received massage and control subjects in muscle force loss or regain up to 48 h after the exercise bout. Farr and colleagues (2002) also reported no benefits of one 30 min massage intervention performed 2 h after downhill walking on leg muscle force loss or recovery for up to five days following the exercise. Zainuddin and colleagues (2005) also found no effect of one 10 min massage intervention on post–eccentric exercise biceps muscle force recovery at three to four days postexercise. Similarly, Jonhagen and colleagues (2004) found no effects of three daily massage treatments of one leg relative to the other leg on force recovery following eccentric quadriceps exercise. In addition, Dawson and colleagues (2004) reported no benefit of repeated massage interventions on one leg relative to the untreated leg in force recovery following a half-marathon race.

In addition to performing simple contraction force measurements, Farr and colleagues (2002) and Jonhagen and colleagues (2004) assessed massage effects on the loss and recovery of functional performance of leg muscles. These studies examined the effects of massage on the one-leg long jump (Jonhagen et al. 2004) and one-leg high jump (Farr et al. 2002). In neither study did the leg receiving massage differ from the control leg in rate of performance recovery following eccentric exercise. An earlier study (Drews et al. 1991) also showed no effects of daily 30 min massage on performance or recovery of elite cyclists engaging in a four-day stage race.

It is clear from these studies that a single massage treatment or treatments repeated over several days do not significantly affect muscle force or functional recovery following various types of exercise-induced muscle injury. Hence, if, as Warren and colleagues have suggested (1999), measures of muscle force or function following exercise-induced muscle damage are the most valid and reliable quantitative markers of muscle recovery or repair, then at least by these measures, massage has no positive influence on muscle repair or recovery. Nevertheless, more research still needs to be done on the potential effects of massage on many other biochemical and histochemical indices of muscle damage and repair, as well as in other forms of clinical or overuse muscle injury on which there are currently no data.

Soreness, Swelling, and Pain

The sensation of muscle soreness may be related to muscle inflammatory responses to muscle damage and repair processes (Proske and Allen 2005). If massage were to influence muscle damage, repair, inflammation, or related events such as muscle swelling or edema, or the sensitization of the group III and IV afferent nerves

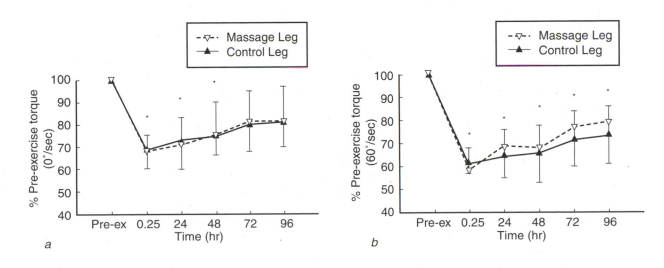

Figure 16.1 Time course of post–eccentric exercise quadriceps peak torques in massaged and control legs at (a) 0° and (b) 60°/s contraction speeds. No significant differences (p > 0.05) between massage and control leg are seen at any point. *Torque < pre-exercise torque (p < 0.05).
Modified from data in Tiidus and Shoemaker, 1995.

that transmit soreness sensation, it might conceivably influence muscle soreness (Tiidus 2002). Few data exist relative to the potential for massage to influence muscle swelling, inflammation, or edema or any of the other factors just mentioned. Smith and colleagues (1994) hypothesized that massage performed within 1 to 2 h of eccentric exercise may act to disrupt the inflammatory and muscle swelling processes. However, their indirect measures of general inflammatory responses following massage could not demonstrate such an influence.

The effects of massage on muscle swelling or lymph flow have also not been extensively reported in humans (Callaghan 1993). Using thigh circumference as a crude measure of muscle swelling, Dawson and colleagues (2004) were unable to demonstrate an effect of massage in hastening the reduction of swelling in subjects following a half-marathon race. Similarly, Hart and colleagues (2005) found no effect of massage on triceps surae muscle swelling up to 72 h following eccentric exercise. Several related studies on the ability of massage to enhance skin lymph flow, or to ameliorate swelling in the fingers or arms of women following surgery for breast cancer, have not provided conclusive evidence for a prolonged effect of massage or similar manual therapies on these factors (Tiidus 2002). For example, a recent study examined the effects of "manual lymph drainage (similar to massage)," compression bandaging, or both in reducing lymphedema in breast cancer–related mastectomies (McNeely et al. 2004). The authors concluded that compression bandaging was more effective than manual drainage in reducing lymphedema in this population, and only minor effects of manual drainage could be noted (McNeely et al. 2004).

An older study using anesthetized dogs with hypoprotein-induced leg edema did report a slight increase in lymph flow as a consequence of kneading and stroking massage of the leg (Elkins et al. 1953). However, the increase was small and not different from that with simple active or passive movements of the legs by the experimenters (Elkins et al. 1953; Callaghan 1993). Hence the effectiveness of massage in influencing muscle swelling and edema, particularly in humans, remains largely untested; and other related studies have not shown strong promise for massage to influence lymph flow or swelling in other tissues or conditions. Hence, despite the lack of any direct experimental evidence, it seems unlikely that massage could influence muscle soreness by significantly reducing muscle swelling following muscle damage.

Application of a second sensation such as massage to a sore muscle could theoretically increase discharge from other low-threshold sensory fibers and thereby temporarily interfere with soreness sensation transmitted by type III or IV afferents and with the transmission of this sensation signal up the spinal cord to the brain (Armstrong 1984). However, the current experimental evidence suggests that the potential influence of massage on delayed-onset muscle soreness is at best minor and transitory. Although a number of groups have attempted to assess the effects of massage treatment on subjective perceptions of muscle soreness, the data have been mixed. Weber and colleagues (1994) reported a lack of effect of two massage treatments on muscle soreness sensation 24 and 48 h after eccentric biceps exercise in healthy females. Jonhagen and colleagues (2004) also reported no significant effect of three daily massage treatments on the level or duration of soreness sensation up to three days following eccentric contractions of the quadriceps muscles of healthy subjects.

On the other hand, Farr and colleagues (2002) reported an attenuation of soreness in leg muscles at 24 h after downhill walking exercise, consequent to a single massage treatment at 2 h postexercise, relative to the untreated leg. This attenuation of soreness relative to the unmassaged limb was no longer evident between 48 and 120 h postexercise. Tiidus and Shoemaker (1995) also found a small but significant reduction in soreness sensation only at 48 h (but not at 24, 72, or 96 h) postexercise consequent to daily massage treatments of quadriceps muscles following eccentric exercise. Hilbert and colleagues (2003) also reported lower soreness sensation at 48 h postexercise (but not at other time points) in hamstring muscles following massage.

In an earlier review of the literature, Ernst (1998) noted the variability in findings on the potential for massage to affect muscle soreness. He also suggested that many early studies on massage and muscle soreness were methodologically flawed. Cheung and colleagues (2003) and Weerapong and colleagues (2005) reviewed more recent literature and noted that massage has been reported to have minor and varying results with respect to muscle soreness sensation. Even in studies showing that massage was modestly effective in temporarily diminishing delayed-onset muscle soreness, the effects were much smaller than those of modest physical activity; physical activity was much more consistent and effective in diminishing muscle soreness sensation (Cheung et al. 2003; Saxon and Donnelly 1995).

These results are similar to conclusions from recent reviews on the effectiveness of massage in treating any type of musculoskeletal pain (Ernst 2004; Moraska 2005; Lewis and Johnson 2006). These reviews concluded that despite some promising findings in the literature, there was not yet convincing evidence for the effectiveness of massage in controlling musculoskeletal or other pain, and that the notion of massage as an effective intervention for controlling pain or reducing muscle soreness has not yet been convincingly demonstrated through rigorous clinical trials (Ernst 2004). Thus if massage has any effect on muscle soreness, it is small, variable, transitory, and of lesser magnitude than the effect that can be brought about by light exercise of the affected muscles (Tiidus 1997).

Blood Flow

A popular suggestion is that massage may be effective in treating muscle damage by increasing muscle blood flow (Goats 1994b). Some early studies (i.e., Wakim et al. 1949; Hovind and Nielson 1974) showed small increases in muscle blood flow consequent to massage, while others (e.g., Ebel and Wisham 1952) showed no such effect. All such studies prior to the 1990s used versions of venous occlusion plethysmography or xenon (Xe) clearance techniques to assess muscle blood flow consequent to massage. Problems with these techniques included possible motion artifact from massage, the inability to measure blood flow during massage itself, and possible confounding influences of hyperemia induced by injection of the substance to be cleared (Tiidus and Shoemaker 1995; Hinds et al., 2004).

Several more recent studies have employed Doppler ultrasound to measure arterial and venous blood flow during massage. Doppler ultrasound is a much more reliable measure of blood flow than earlier techniques and can be used during the performance of massage (Tiidus and Shoemaker 1995). These studies have consistently demonstrated no effects of massage on muscle arterial or venous blood flow. The first of these investigations (Tiidus and Shoemaker 1995) measured both femoral artery and femoral venous blood flow continuously during 10 min of effleurage massage performed on the quadriceps muscles by an experienced massage therapist. These measures were performed on muscles experiencing muscle soreness following eccentric exercise, as well as on muscles that had largely recovered from such exercise. At no time during the massage treatment did arterial or venous leg blood flow change from resting values. In fact, the trend was for muscle arterial and venous blood flow to move toward reduction during the massage (Tiidus and Shoemaker 1995). On the other hand, light voluntary contraction of the quadriceps muscles had an immediate and significant effect in increasing arterial blood flow to the leg.

In a subsequent study, Shoemaker and colleagues (1997) examined the effects of 10 min of three types of massage (effleurage, petrissage, and tapotement) on arterial blood flow to the leg (large muscle group) and forearm (small muscle group) using Doppler ultrasound. Effleurage massage involves rhythmic pressure strokes along the longitudinal axis of the muscle group; petrissage consists of kneading and squeezing motions over the muscle mass; and tapotement involves percussive motions of the hands or fingers along the muscle length (Goats 1994; Shoemaker et al. 1997). Tapotement in particular has been attempted as a form of muscle stimulus prior to competition and has been touted by massage practitioners as potentially the most effective massage technique in enhancing muscle blood flow (Callaghan 1993). Despite these claims, neither femoral (leg) nor brachial (forearm) arterial blood flow was significantly

different from resting values during the performance of any of these techniques (figure 16.2). These results conclusively demonstrate that typical manual massage techniques do not enhance arterial blood flow into either smaller or larger muscle groups while they are being performed. Hence, if muscle blood flow were actually a factor in enhancing muscle healing, mild muscle contractions would be far more effective in inducing such responses (Tiidus and Shoemaker 1995; Tiidus 1997).

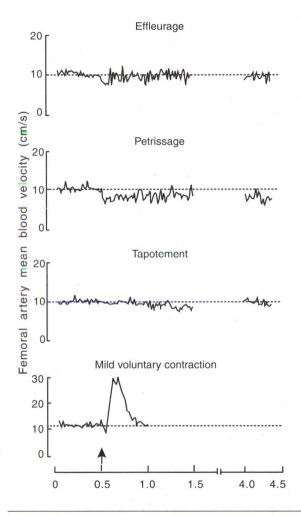

Figure 16.2   Femoral artery mean blood velocity at rest and during 5 min of effleurage, petrissage, and tapotement forms of massage and following a brief voluntary quadriceps contraction. The dotted line represents average resting blood velocity. Effleurage and petrissage massage altered the variability of blood velocity response, but overall blood velocity during effleurage, petrissage, or tapotement massage was not statistically different from resting values ($p > 0.05$). The voluntary contraction significantly elevated blood velocity over rest ($p < 0.05$). Tracings are the mean of 10 subjects with second-by-second resolution. Arrow indicates onset of massage or exercise.

Reprinted, by permission, from J.K. Shoemaker, P.M. Tiidus and R. Mader, 1997, "Failure of manual massage to alter limb blood flow: measures by Doppler ultrasound, *Medicine and Science in Sports and Exercise* 29(5): 610-614.

One potential weakness of these studies is that while able to measure arterial blood flow, they were not able to distinguish between potentially different effects of massage on muscle and skin blood flow downstream from the measurement point. A recent study used compressed air "massage" on rabbit muscle. The findings suggested that such artificial "massage" did result in muscle capillary dilation as determined via light microscopy from muscle biopsy samples (Gregory and Mars 2005). How applicable these findings are to manual massage in intact human muscle and to subsequent changes in microcirculation is unknown. Another recent study addressed this issue in humans by measuring both skin and muscle blood flow using laser Doppler and Doppler ultrasound flowmetry, respectively (Hinds et al. 2004). Effleurage and petrissage of the quadriceps raised both skin temperature and skin blood flow during the massage; however, as in the previous studies, this investigation also showed no effects of massage on femoral artery blood flow to the leg.

Mori and colleagues (2004) also reported increased skin blood flow with massage to the lumbar region. Supporting these findings, another recent study showed that up to 15 min of massage induced significant increases in skin temperature but produced only minor alterations in deep vastus lateralis muscle temperature (Durst et al. 2003).

These findings raise the possibility that muscle blood flow may actually decrease during massage; since femoral artery flow was unchanged from rest, some of that finite blood flow might actually have been diverted away from the muscle and to the skin during massage (Hinds et al., 2004). Hence, as previously suggested, if muscle blood flow is indeed a factor in muscle healing (which is by no means certain), then massage may actually do more to hinder than to help this process.

Muscle Lactate

Naive claims have been made that massage may be able to remove muscle or blood "toxins," that is, blood lactate, and thereby influence muscle healing and muscle soreness (Hemmings et al. 2000; Tiidus 1999). It is of course well established that lactic acid does not cause muscle soreness, nor (as indicated in the other chapters of this book) is it a factor in exercise-induced muscle damage (Armstrong 1984; Clarkson and Sayers 1999; Proske and Allen 2005). Hence, any suggestion that removal of lactic acid or other unspecified "toxins" from muscle can somehow benefit muscle soreness sensation or aid in muscle repair can be dismissed as having no physiological basis.

It has been repeatedly demonstrated that massage has no influence on the clearance of blood lactate following exercise (Cafarelli and Flint 1992; Gupta et al. 1996, Hemmings et al. 2000; Monedero and Donne 2000; Robertson et al. 2004). No studies have addressed the ability of massage to affect actual muscle lactate clearance. However, since massage does not influence muscle blood flow or lymph flow as discussed in the previous section, there is no basis for suggesting that it could affect muscle lactate clearance. On the other hand, mild exercise, which does enhance muscle blood flow, can significantly speed up blood lactate removal following exercise (Gupta et al. 1996; Monedero and Donne 2000). It should also be noted that blood lactate clearance may not actually be the prime factor associated with enhanced performance induced by active recovery from exercise.

Massage Effects on Fibrosis, Muscle Tension, and Mood

Other factors associated with muscle damage include scar tissue formation or fibrosis and feelings of muscle "tightness" or tension. The latter may also have effects on or be influenced by mood states. Although many claims are made regarding the ability of massage to influence scar tissue formation or muscle "relaxation," there is surprisingly little empirical evidence for these effects, particularly in human muscle. In contrast, there appears to be significant evidence that massage can elevate mood states in both healthy and clinical populations.

Fibrosis "Scarring" or "Adhesion"

Fibrosis or scarring can commonly occur during the healing process in muscle that is recovering from contusion-induced injury (Sato et al. 2003). Formation of fibrosis can prevent complete healing of the muscle and limit subsequent muscle function (Huard et al. 2002). It has been alleged that massage, particularly friction massage, can break up such muscle scarring and reduce muscle "adhesions" (Goats 1994). Despite these claims, almost no evidence exists as to the potential for massage to actually affect postdamage muscle scarring or to break up unspecified "adhesions." No studies have directly examined the ability of massage to influence muscle scar, fibrosis, or adhesion formation or removal; and the limited studies on the ability of massage to influence scar formation or reduction following injury to skin or other tissues have not been promising (Ramey and Tiidus 2002). For example, several studies have shown no effect of massage in reducing hypertrophic scars or their formation or in the skin of burn victims (i.e., Blaha and Pondelicek 1997; Patino et al. 1999). Hence claims that massage can remove scar tissue or adhesions from muscle or any other tissue are not currently supported by scientific evidence.

Muscle Tension, Length, and "Trigger Points"

Some have suggested that massage can reduce muscle stiffness, tension, knots, or spasms. While anecdotal evidence for these benefits exists (Goats 1994), empirical evidence for such potential benefits is scarce and difficult to acquire.

There is some limited evidence for reduction in muscle sympathetic activity following massage, as well as a reduced H-reflex response during but not following massage (Callaghan 1993; Morelli et al. 1990; Sullivan et al. 1991). A further study showed reduced muscle electromyographic (EMG) activity in healthy volunteers following massage (Longworth 1982). Two recent studies also demonstrated improved hip flexibility and hamstring length in athletes following various forms of massage (Hopper et al. 2005a, 2005b). The physiological and functional importance of these improvements and their potential benefits, relative to improvements seen with a general warm-up, have not yet been determined.

In addition, six weeks of massage had only minor effects on myofascial "trigger points" (tender or tight points in the muscle that may be associated with pain and muscle spasm) in patients experiencing chronic muscle pain (Gam et al. 1998). In fact, other treatments may be more effective than massage in treating trigger points (Huguenin 2004). The reader is referred to chapter 20 of this book for a more thorough review of trigger points and their treatment.

General Relaxation

Massage does seem to have the ability to enhance general relaxation and improve mood state in athletes, sedentary individuals, and clinical patients (i.e., Hemmings et al. 2000; Moyer et al. 2004; Mok and Woo 2004). However, any effects of improved relaxation specifically on muscle recovery and regeneration are undocumented and unknown (Weerapong et al. 2005). Hence, preliminary data seem to suggest some limited influence of massage on muscle tension and relaxation. Additional well-controlled research is needed, particularly with subjects experiencing muscle pain, spasm, and stiffness, to provide a full understanding of the potential and limitations of massage in treating these symptoms of muscle dysfunction and damage.

Summary

Much more well-controlled research needs to be done before the effectiveness and limitations of massage therapy with respect to muscle repair and regeneration can be fully assessed. Current evidence does not support massage therapy as a useful or effective treatment in enhancing most aspects of recovery from muscle damage, suppression of inflammation, or enhancement of muscle repair. It appears that the use of massage therapy to treat muscular injuries is not generally justified by current research and that its effectiveness as a treatment for muscle damage cannot be upheld.

CHAPTER 17

Ultrasound

Dawn T. Gulick, PhD, PT

Despite its discovery in the mid-1800s, ultrasound did not become a treatment modality until the 20th century. It was realized that deformation of a crystal could produce a measurable current that became known as the piezoelectric effect. The reverse piezoelectric effect involves the application of a high-frequency alternating electrical current across a crystal. The result is deformation and resonation of the crystal. Studies in the early 20th century demonstrated the influence of ultrasound (US) on living tissue (i.e., Chilowsky and Langevin, 1916, in Drez, 1990).

In 1980, Stewart and colleagues reported that 15 million US treatments were rendered in hospitals in the United States. Other studies have also supported the popularity of US as a therapeutic modality in England, Australia, the Netherlands, and Canada (Hecox, 2006). However, after several decades of therapeutic use, there is still significant controversy over the efficacy of this modality, due primarily to a lack of sufficient rigorous research. For example, Robertson and Baker (2001) performed a systematic review of the literature on the effectiveness of US treatment with regard to clinical outcomes from 1975 to 1999. Upon completion of a six-step filtering process, they ended up with only 10 randomized controlled trials to analyze.

Many of the studies reviewed did not provide sufficient information on the parameters utilized. Thus, it is difficult to draw meaningful conclusions from the results. In addition, there were concerns about the assumptions drawn from in vitro experiments, since some of the biophysical effects may not correspond to those under in vivo

conditions (Baker et al., 2001). Fortunately, randomized controlled trials are not the only method of obtaining information about an intervention, and several other studies have been published since 1999. This chapter is an attempt to sift through the relevant research with respect to clinical application of US to the treatment of muscle damage and injury, as well as to provide the reader with at least some scientific information and data on which clinical decisions can be based. First, it is important to present the vocabulary used in the study of US. These terms are defined in "Essential Ultrasound Terminology."

Mechanisms of Ultrasound

The human ear can hear sounds at frequencies of 20 to 20,000 cycles per second (Hertz). Therapeutic US is, by definition, any sound with a frequency greater than 20,000 cycles per second (Belanger, 2002). Yet audible sound and US have similar characteristics. Both types of sound demonstrate attenuation as the sound waves propagate through the conducting medium.

Interaction Between Ultrasound Waves and Mammalian Tissue

Sound waves mechanically oscillate the molecules of a given tissue, setting up a chain reaction with the adjacent molecules. Molecular vibration is the result of cycles of high- and low-pressure waves (Belanger, 2002). These waves take the form of a sinusoidal pattern alternating between positive and negative phases. During these

Essential Ultrasound Terminology

- **BNR = beam nonuniformity ratio:** The ratio of the maximum intensity to the spatial average intensity; the magnitude of the lack of homogeneity of the US beam intensity. The greater the piezoelectric quality of the transducer, the lower the BNR. Generally, the lower the BNR, the less likely a person is to experience "hot spots" or spikes in the intensity that may be uncomfortable. For example, given a treatment intensity of 1.5 w/cm^2, using an US unit with a BNR of 5:1 would result in intensity spikes (SPI) of 7.5 w/cm^2

- **Cavitation:** The oscillation of small gas bubbles in the blood or tissue fluids.

- **Coupling medium:** A substance used to transmit the US waves from the transducer to the treatment area of the body.

- **Duty cycle:** The relationship of "on" time to "total time" of the treatment. Duty cycle can be expressed as a ratio or as a percentage. For example, if the "on" time is 5 ms and the "off" time is 5 ms, the duty cycle is 1:2 (5 ms on:10 ms total time) or 50%.

- **ERA = effective radiating area:** The area of the US transducer that is capable of transmitting US energy. Because the crystal is secured to the periphery by a metal end plate, there are fewer oscillations from the perimeter of the transducer than the central region. This is why the ERA is always less than the total surface area of the transducer head. The ERA is expressed in square centimeters (cm^2).

- **Frequency:** The number of cycles of sound oscillations (compression and rarefaction) in a second. Frequency is expressed in cycles per second or hertz (Hz).

- **Intensity:** The energy emitted per unit area from the US head. Intensity is routinely expressed in watts per square centimeter.

- **Microstreaming:** The unidirectional flow of fluids in an US field that changes the permeability of the cell membrane. This process has been linked to increased blood flow, tissue repair, and reduction of edema and pain.

- **Mode:** Continuous mode is an uninterrupted stream of US waves. Pulsed mode involves consistent interruptions in the stream of US waves at established intervals.

- **SAI = spatial average intensity:** The average of the US output over the entire conducting surface of the transducer.

- **SPI = spatial peak intensity:** The greatest output over the surface of the US transducer.

phases, the molecules in the tissue undergo compression and rarefaction (dispersion). The high-pressure waves result in molecular compression, and the low-pressure waves cause the molecules to move apart (rarefaction). The higher the ultrasonic intensity, the more vigorous the molecular vibration and the frictional heat generated within the tissues (Belanger, 2002). The vibration continues to propagate within the tissue until the sound energy is dissipated (Sparrow, 2005). Although molecules of dense tissue oscillate more rapidly than those of tissues that are less dense, they also dissipate the energy more readily (Sparrow, 2005).

Furthermore, molecular oscillations of US waves can travel in both longitudinal (along tissue) and transverse (penetrating into tissue) waves. In the convergent region (near field), the US intensity varies, whereas in the divergent region (far field) it is more uniform. The US frequency and effective radiating area (ERA) influence the length of the near field. At a 1 MHz frequency, the near field is one-third the length of the 3 MHz near field. Thus there is an inverse relationship between US resonant frequency and depth of penetration (Draper, Castel, and Castel, 1995). It has been reported that a 1 MHz frequency penetrates to a depth of 5 cm while a 3 MHz frequency can penetrate to a depth of 2.0 to 2.5 cm (Gulick et al., 2004). However, it is the types of tissue through which the sound waves are traveling that determine the direction of the energy flow (Sparrow, 2005). Because of its strong intermolecular bonds, compact bone is the only biological tissue in which US travels in both directions.

In all other tissue, US waves travel longitudinally only (Sparrow, 2005).

When US waves travel through a tissue, they can be conducted, absorbed, or reflected (Drez, 1990). How the US wave behaves is determined by the properties of the tissue through which it is traveling. The law of Grotthus-Draper (table 17.1) describes these properties of US (Draper and Sunderland, 1993). Some tissues provide a significant level of impedance to US energy, thus absorbing the energy and transmitting very little of it. When waves are absorbed, a significant heating effect can be achieved. In contrast, when the tissue impedance is low, the waves are conducted and very little energy is lost (attenuated) to the tissues. The protein content of the tissue, especially the collagen content, is one of the primary determinants of attenuation. Blood and fat are examples of tissues with low protein content that do not absorb the US energy. Muscle, blood vessels, and skin have modest levels of attenuation. Tendon, cartilage, and bone have high levels of attenuation (Sparrow, 2005). As a result of absorption and transmission of US waves, US can have both thermal and nonthermal effects on tissue. These effects and their potential to influence muscle damage and repair are discussed in more detail later in this chapter.

Thus several factors may alter what has been accepted as a simple inverse relationship between the depth of penetration of US waves and frequency. When waves are reflected, the tissue may absorb some of the energy, but most of it will be diverted to the surrounding tissue. Ultrasound waves travel best through homogeneous material. Since subcutaneous adipose tissue is very homogeneous, US waves are only minimally attenuated in this type of tissue and can therefore effectively pass through it to influence deeper tissue. However, as US waves are conducted, they must pass through several layers of different tissues. Maximal temperature increases often occur at soft tissue–bone interfaces. These are locations where the changes in tissue properties result in vast differences in the magnitude of the impedance. When an US wave strikes this interface, it is reflected back toward the surface and can potentially overheat the tissue. Slow and controlled movement of the sound head can mitigate a portion of this problem.

Another important factor related to the effectiveness of US treatment is the quality of the US crystal (Draper, 2002). Crystals are usually made of lead zirconate titanate, plumbium zirconium titanate, or barium titanate. Figures 17.1 and 17.2 display the outputs of two different crystals, both set to deliver 1.5 w/cm^2 (Gatto et al., 1999). Beam nonuniformity ratio (BNR) is the relationship of the maximal intensity of the US beam to the spatial average intensity (SAI). Ideally, a low BNR should relate to treatment uniformity and patient comfort. Yet BNR is a very small "snapshot" of the homogeneity of a crystal. The SAI may be achieved via a consistent output across the crystal or as a result of great variability. For example, an SAI of 1.5 w/cm^2 may result from the crystal's having an output range of 1.0 w/cm^2 to 2.0 w/cm^2 across its entire ERA. On the other hand, an SAI of 1.5 w/cm^2 could result from an output range of 50% of the beam conducting at 0.5 w/cm^2, 25% of the beam conducting at 1.5 w/cm^2, and 25% of the beam conducting at 3.5 w/cm^2. Clearly these are very different US outputs with very different clinical results. This is displayed in figure 17.3 (Young et al., 1999).

It is this variability that can produce "hot" and "cold" spots in the tissue. Moving the US head in a circular pattern at a velocity of 4 cm/s may mitigate a portion of the beam nonuniformity, but even consistent transducer movement may not be enough for some crystal inconsistencies. Furthermore, having 1 and 3 MHz frequencies available within the same unit is convenient, but consideration may need to be given to using different transducers for the different frequencies. Given potentially inconsistent output across a crystal, the clinician may best serve a patient by selecting optimal parameters and then placing the primary emphasis on patient feedback. In other words, select the frequency, duty cycle, and size of the treatment area and then increase the intensity based on patient feedback. When thermal effects are desired, the patient should be instructed to inform the clinician when the maximal heating effect is achieved without discomfort. If the goal is to utilize the mechanical effects of US, this process may not be as simple but is still possible.

Coupling Media

Ultrasound waves cannot be effectively transmitted through air (Balmaseda et al., 1986). Thus, a coupling medium must be used to transmit the waves to the

Table 17.1   Law of Grotthus–Draper

Tissue type	Ultrasound behavior
Tissue high in water content	Ultrasound penetrates through the tissue
Tissue high in protein	Ultrasound is absorbed in the tissue
Bone	Ultrasound is reflected off of bone
Joints	Ultrasound is refracted by the joint
Fat	Ultrasound penetrates through fat

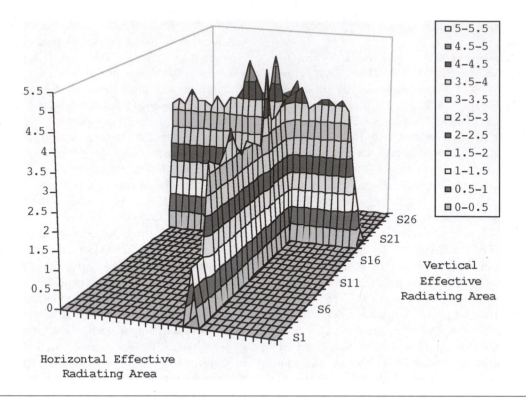

Figure 17.1   Ultrasound output of unit X as measured with a hydrophone at an intensity of 1.5 w/cm².

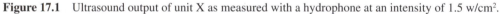

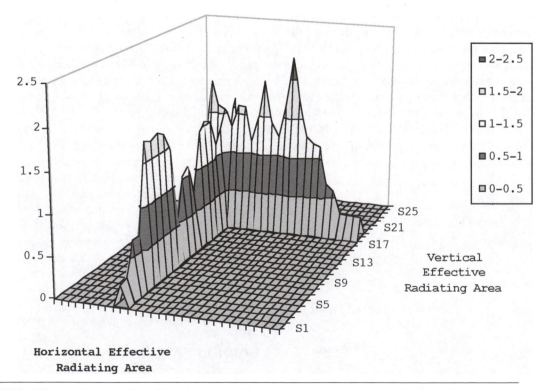

Figure 17.2   Ultrasound output of unit Y as measured with a hydrophone at an intensity of 1.5 w/cm².

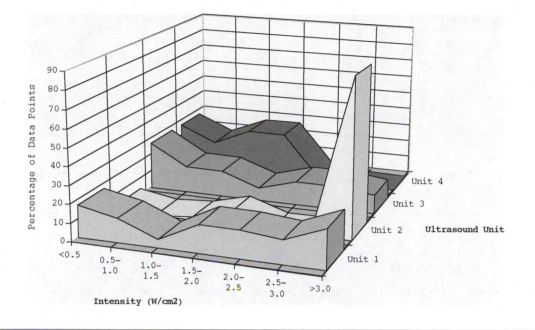

Figure 17.3   Comparison of US outputs of four different calibrated units at an intensity of 1.5 w/cm².

target tissue. The most common coupling medium is a water-soluble gel. Cameron (2003) reported that gel transmits 96% of the US waves. This coupling medium serves clinicians well when they are treating smoothly contoured areas such as the quadriceps muscle, low back, or cervical regions. However, under some clinical conditions, US gel may not be the most appropriate medium. For example, when the treatment surfaces are highly irregular with numerous bony prominences, it can be challenging to maintain contact between the sound head and the skin. Failure to do so results in a loss of US energy and potential damage to the US transducer (due to reflection of the energy).

This was demonstrated by Kimura and colleagues (1998) in a study that compared tissue heating across four US application angles: 90°, 80°, 70°, and 60°. Four trials were performed at each test angle on two different US units. A continuous 1 MHz frequency at 2.0 w/cm² was administered for 5 min. The findings indicated that US treatments requiring angulation of the transducer, like that needed to accommodate irregular surfaces, can fail to produce temperature increases large enough to accomplish therapeutic goals. Greater surface contact between the transducer and the skin results in transmission of a larger portion of the US beam to the treatment area to produce a greater temperature elevation. Figure 17.4 demonstrates that as the angle of application decreases, more US energy may be lost through scattering and refraction of the US beam. Smaller increases in temperature at the 60° angle of application could be attributed to poor coupling.

Given a treatment area with an irregular surface, an alternate approach should be considered. Klucinec and colleagues (2000) examined the transmissivity of a 3 MHz US wave through pig tissue at five intensities (0.2, 0.5, 1.0, 1.5, and 2.0 w/cm²). The US wave transmission through the various interfaces was as follows:

- Gel pad = 109%
- Gel = 100%
- Degassed water–filled latex glove = 79%
- Tap water–filled latex glove = 66%
- Gel-filled latex glove = 50%
- Degassed water bath = 33%
- Tap water = 31%
- Gel-filled condom = 23%

Likewise, Draper and colleagues (1993) compared underwater (tap water) conductivity with gel using a 1 MHz frequency. At an intensity of 1.5 w/cm² for 10 min, the muscle temperature increased by 13.9% with gel and only 6% under water. The lower temperature yield was related to the absorption of the sound waves by the water. Draper and colleagues (1991) demonstrated that the distance of the immersed sound head from the skin was inversely related to the tissue temperature increase. Furthermore, Reid and Cummings (1977) found that 72.60% of US power transmitted through gel while only 59.38% transmitted through water.

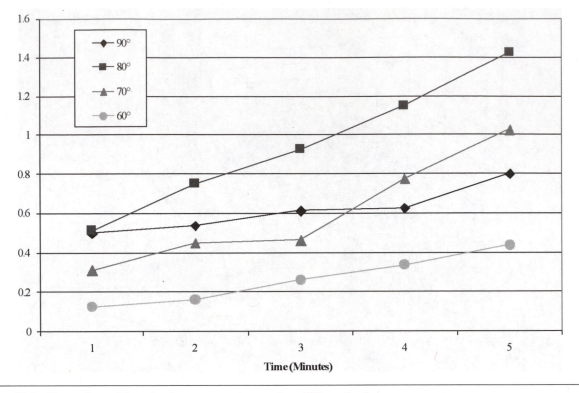

Figure 17.4   Comparison of tissue heating across various angles of US application.

Thus, proper technique is important when one is performing US under water. The transducer should be kept approximately 1 cm from the surface of the skin and moved at a velocity of 3 to 4 cm/s. Care should be taken to minimize the presence of air bubbles or turbulence in the water that could diminish transmission of the sound waves. It may be necessary to significantly increase the intensity of the treatment to achieve the desired goal.

In addition, the temperature of the water for underwater US should be consistent with the desired treatment effect. If thermal effects are the goal, warm water is appropriate; if nonthermal effects are desired, colder water is appropriate. It is also important to use a nonmetal container for treatment. This will minimize the number of reflected US waves in the container. The reflected sound waves may be transmitted not only to tissues other than the desired treatment area, but also to the clinician's hand that is holding the transducer. Besides an irregular surface contour, another circumstance that might call for underwater US is when the patient has difficulty tolerating the pressure of the US transducer on the skin.

If the area being treated does not lend itself to underwater treatment, an alternative method of conducting US waves to target tissue is via gel pads (Merrick et al., 2002). Disposable gel pads are clear, malleable discs that were originally designed for diagnostic US. The disc conforms to an irregular surface to ensure adequate

coupling (figure 17.5). The disc can also be used in the treatment of wounds.

Effects of Ultrasound on Tissue

It is generally accepted that ultrasound has two primary biophysical effects: thermal and nonthermal (mechanical). It may be unreasonable to attempt to completely isolate the components of these two effects. In reality, both occur in varying degrees with the use of different treatment parameters. Ultrasound energy that results in heating will always be accompanied by mechanical effects (Baker et al., 2001). Likewise, one can reduce the amount of US energy transmitted to emphasize the mechanical effects, but this does not totally eliminate tissue heating (Baker et al., 2001).

Thermal Effects of Ultrasound

The thermal effects of US are displayed in figure 17.6 (Rennie and Michlovitz, 1996). Besides heating tissue, US has been reported to increase blood flow, tissue metabolism, enzymatic activity, and oxygen uptake (Rennie and Michlovitz, 1996). Moreover, tissue heating results in an increase in cutaneous thermoreceptor activity and relaxation of the smooth muscles of the blood vessels (Rennie and Michlovitz, 1996). All of these potential

Figure 17.5   Ultrasound technique with a gel pad for irregular surface contours.

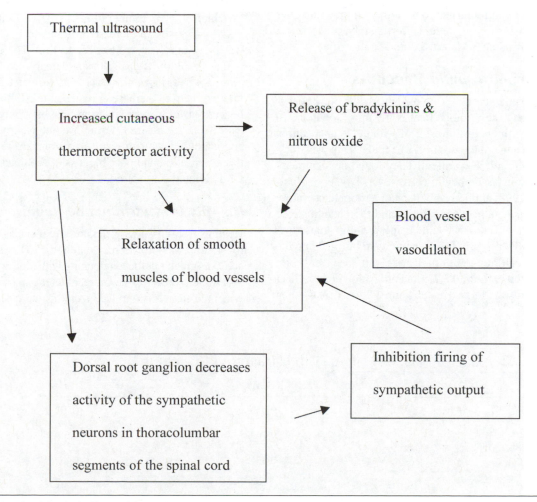

Figure 17.6   Physiological effects of thermal US.

effects of US may theoretically facilitate the healing of tissue. However, US may also enhance the inflammation-induced destruction of tissue when inappropriately used with acute inflammatory conditions (Cameron, 2003).

Thermal Effects

The magnitude of soft tissue heating is influenced by the frequency, intensity, and duration of the US application. Draper, Castel, and Castel (1995) investigated the rate of tissue heating for 1 and 3 MHz US frequencies. Briefly, tissue heating occurred at a rate of 0.2° C/min and 0.6° C/min in 1 and 3 MHz frequency treatments, respectively. Table 17.2 presents a summary of the results.

It is important to bear in mind that tissue heating is accompanied by increased blood flow. Although this may be desirable for bringing nutrients to the damaged tissue, the increased blood flow will also cool the heated tissue as it passes through the treated area. Fabrizio and colleagues (1996) demonstrated a significant increase in blood flow with 1 MHz continuous US at both 1.0 and 1.5 w/cm^2 when treating an area that was two times the ERA. The increase in blood flow plateaued after 5 min of treatment and returned to baseline within 1 min after the conclusion of treatment. Unfortunately, no studies on blood flow using 3 MHz US could be identified.

With Nonultrasound Modalities

In clinical practice, US is often coupled with other modalities; among these, moist heat is the most common. To study the influence of moist heat prior to US application, Lehmann and colleagues (1978) applied a hot pack for 8 min prior to a US treatment. They found no significant difference in tissue heating. However, Draper and colleagues (1998) applied a hot pack or a room temperature pack to each subject for 15 min and then administered 10 min of US at 1.5 w/cm^2 using a 5 cm^2 transducer with a BNR of 1.8:1. Thermistors were inserted into the posterior calf at a depth of 1 and 3 cm. The results are summarized in figure 17.7. The application of moist heat with US resulted in a greater temperature increase than

US without moist heat at both the 1 and 3 cm tissue depth. Furthermore, the hot pack made a more profound impact on the tissue temperature at 1 cm, while US accounted for a more significant portion of the tissue temperature increase at 3 cm. This is consistent with the expectations of each modality. Moist heat is a superficial modality, while US is utilized for deeper heating. So although the application of moist heat influences only the superficial tissue (1-2 cm in depth), there may be slight advantages to the generalized heating effect and the establishment of a temperature gradient to facilitate conduction of the thermal effects of US.

The application of ice prior to the administration of US has been shown to detract from the heating of the deep tissue. This was demonstrated by Draper and colleagues (1995). Subjects were randomly assigned to receive an ice bag to the posterior calf for 5 min followed by US or to receive US alone (without ice). The US parameters were 10 min of treatment using a 1 MHz frequency (BNR 2.2:1) over an area that was two times the ERA, at an intensity of 1.5 w/cm^2. The results are summarized in figure 17.8. These data are consistent with findings from a previous study by Rimington and colleagues (1994) in which a 2° C temperature increase occurred with US alone versus an increase of less than 1° C when US was preceded by an ice bag for 15 min. The conclusion is that despite increasing tissue density, ice before US diminishes the thermal effects. Therefore, ice prior to US may not help to increase collagen extensibility but may be warranted for acute situations in which only mechanical effects are desired. Caution is advised because the application of ice can result in a transient reduction in tissue sensation and possible loss of feedback when the US treatment is rendered.

On Muscle Stretching and Repair

Heating of soft tissue is well known to increase extensibility (Lehmann et al., 1970; Lentell et al., 1992). Heating achieves greater tissue lengthening with the application of less stretch force. Furthermore, greater gains are retained when the stretch is removed from heated tissue as com-

Table 17.2   Rate of Tissue Heating With Ultrasound (°C/minute)

	1 MHz	3 MHz
0.5 w/cm^2	0.04°	0.30°
1.0 w/cm^2	0.16°	0.58°
1.5 w/cm^2	0.33°	0.89°
2.0 w/cm^2	0.38°	1.40°

Adapted from Draper, Castel and Castel, 1995.

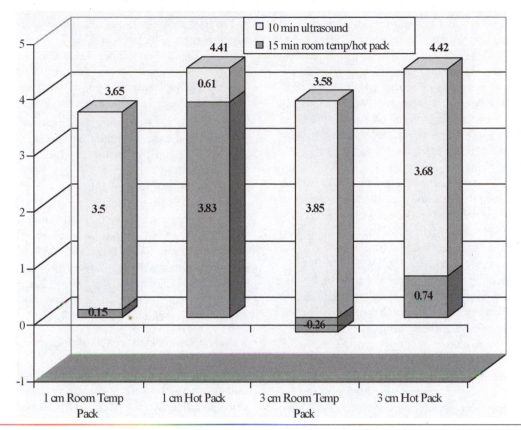

Figure 17.7 A comparison of tissue heating with and without a hot pack before US.

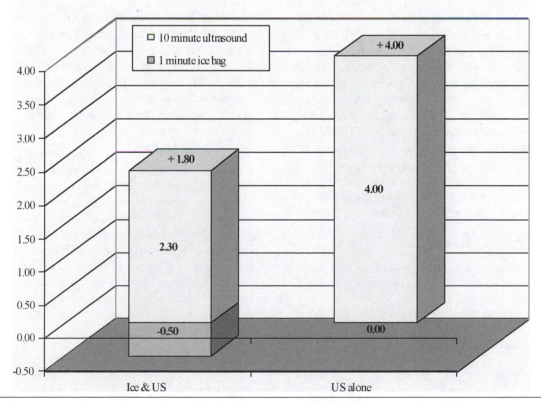

Figure 17.8 A comparison of tissue heating with and without an ice bag before US.

pared to unheated tissue. Maximum results are achieved when the tissue temperature is elevated to 40° to 45° C (104-113° F) and maintained for 5 to 10 min (Lehmann et al., 1970). Griffin and Karselis (1982) attributed the relaxation of polypeptide bonds to the absorption of ultrasonic energy. The benefits of increased tissue extensibility were demonstrated by Wessling and colleagues (1987) in a study showing that static dorsiflexion stretching with US treatment (1 MHz at 1.5 w/cm²) was more effective than stretching alone.

The timing of stretching with respect to US is critical. This was demonstrated in a study by Rose and colleagues (1996) with 3 MHz US and by Draper and Ricard (1995) with 1 MHz US. In the 3 MHz study, the tissue temperature was recorded at a depth of 1.2 cm. After the tissue temperature was increased by 5.3° C (9° F), it fell below the vigorous heating range in less than 4 min. In the case of the 1 MHz US, the temperature was monitored at a depth of 4 cm and declined below the vigorous heating range in 5.5 min. Thus the clinician has a limited "stretching window" in which to capitalize on the enhanced viscoelastic properties of the heated tissue (Draper and Prentice, 1998). Stretching of soft tissue should be performed within 4 to 5 min after the completion of the US treatment. Under some conditions, stretching can even begin during the US treatment and then continue after the US to maximize the outcome.

While stretching can be important in improving joint mobility after injury and restoring muscle and tendon length, its effects on actual muscle damage and healing processes are limited. Data are also lacking on the potential (or lack thereof) specifically of thermal effects of US on muscle damage and repair. As seen from an overview of the limited literature on US effects on muscle damage and repair (see next section), such US studies have not been aimed at partitioning out thermal and nonthermal influences on the very limited outcomes of US interventions on muscle healing. Several recent reviews have pointed to the need to develop specific guidelines for US dose response in treating various conditions and the need to validate the use of US in treating muscle injury (Robertson, 2002; Robertson and Baker, 2001).

Nonthermal Effects

The nonthermal or mechanical effects of US are the result of two physical processes, cavitation and streaming:

- **Acoustic cavitation** involves the formation of gas pockets in the tissue fluids that develop into microscopic bubbles (the Latin root for "cavitation" means "hollow"). These bubbles cause "cavities" in the soft tissue. With the high- and low-pressure waves of the ultrasonication, these microscopic bubbles expand and contract. During therapeutic US (frequencies of 0.85 to 3 MHz; intensities

< 3 w/cm²), the magnitudes of expansion and contraction are equivalent, which results in stable or acoustic cavitation (i.e., the bubbles neither shrink nor grow) (Belanger, 2002). Despite evidence that acoustic cavitation occurs in vitro, Holland and colleagues (1996) have challenged the existence of cavitation in vivo.

- **Acoustical streaming** is the movement of fluids as a result of acoustic radiation. Acoustical streaming can be distinguished further as "bulk streaming" and "microstreaming" (Baker al., 2001). Bulk streaming is the unidirectional flow of fluids as a result of propagation of the US beam (Baker et al., 2001; Drez, 1990); microstreaming is associated with cavitation and forms as eddies of flow adjacent to the oscillating tissue (Baker et al., 2001; Drez, 1990; Williams, 1983). The ultrasonic pressures cause the fluids to stream away from the sources of the acoustic radiation (Drez, 1990). The phenomena of streaming and cavitation are believed to be responsible for enhanced membrane permeability.

When the intensity of US application is kept low or the US is delivered in a pulsed mode, heating of the tissue can be avoided. It has been suggested that under these circumstances, the mechanical effects include increases in skin and cell membrane permeability (Benson and McElnay, 1987), histamine release (Fyfe and Chahl, 1982), macrophage and fibroblast activity (Young and Dyson, 1990), intracellular calcium uptake (Mortimer and Dyson, 1988), and protein synthesis (Young and Dyson, 1990) in various tissues. These processes are essential to tissue healing particularly during the inflammatory phase. Thus any potential influence on these parameters by US could theoretically influence muscle repair. However, as with the thermal effect of US, there are currently no studies specifically addressing the potential of nonthermal effects of US on actual indices of muscle repair.

On Muscle Tissue

Empirical evidence for the effectiveness of US application (whether through thermal or nonthermal mechanisms) in influencing postdamage muscle healing is quite limited. Very few studies have examined the effects of US on actual muscle damage (i.e., histological changes) or on indirect indicators of damage and repair such as muscle soreness, force loss and recovery, or swelling (Tiidus, 1999). As previously noted, recent reviews have concluded that there is currently not enough evidence to establish appropriate dose-response guidelines for US use in treating not just muscle damage, but any sort of soft tissue injury (Robertson, 2002; Baker et al., 2001). However, since these reviews were published, at least three well-controlled studies have indicated that US may be effective in reducing pain when applied to myofascial trigger points (Draper et al., 2007; Gulick et al., 2001; Majlesi and Unalan, 2004). Draper et al.

(2007) reported a significant improvement in the stiffness of myofascial trigger points in the upper trapezius muscle with 3 MHz US delivered at 1.0 W/cm^2 for 5 minutes. Gulick et al. (2001) compared continuous US with gel, continuous US with methyl nicotinate, 20% pulsed US with gel, and sham US rendered for 7 min to myofascial trigger points. The authors reported that the thermal effects of continuous 1 MHz US delivered at 1.5 w/cm^2 with gel and methyl nicotinate were significantly more effective at reducing trigger point pain than nonthermal treatments. It was hypothesized that the thermal effects may have influenced the ischemic cycle of the trigger points to flush the local nociceptors. Furthermore, Majlesi and Unalan (2004) compared "conventional" US with "high-powered" US for the treatment of myofascial trigger points. "Conventional" US was administered for 5 min at 1.5 W/cm^2 prior to cervical muscle stretching. "High-powered" US was delivered through a stationary technique at a maximal tolerable intensity for 4-5 s followed by 50% of maximal intensity for 15 s. This procedure was repeated three times prior to stretching. The "high-powered" US treatment resulted in a significant reduction in pain and a significant improvement in range of motion in less treatments than the "conventional" US. The authors likened the physiologic response of the high intensity US intervention to that of a trigger point injection.

Another common indicator of muscular damage is delayed-onset muscle soreness (DOMS). Known to be associated with eccentric muscle activity, DOMS occurs 24 to 48 h postactivity. Craig and colleagues (1999) and Stay and colleagues (1998) examined the effect of pulsed US on DOMS. Despite varying pulse rates (20-25%), intensities (0.8 to 1.5 w/cm^2), and durations of treatment (7 to 14 min), there was no evidence of a treatment effect in either study. Ciccone and coworkers (1991) compared 1 MHz US at 1.5 w/cm^2 for 5 min with gel versus 10% trolamine salicylate. They found that US with trolamine salicylate controlled pain better than US alone. In fact, continuous US alone increased the symptoms of DOMS. The authors attributed the increased symptoms to the thermal effects of the US and suggested that the trolamine might have offset this increase. This is consistent with the results of Hasson and colleagues (1990), who reported that when 1 MHz US delivered at 0.8 w/cm^2 was pulsed at 25% for 20 min, muscle soreness was reduced and muscular function improved. The authors theorized that the mechanical oscillations of the pulsed US altered vascular permeability and decreased intramuscular pressure. However, two other studies, in which daily US treatments were administered following eccentric muscle activity for several days, showed no effects of US on reduction of soreness sensation (Plaskett et al., 1999; Tiidus et al., 2002).

Plaskett et al. (1999) and Tiidus et al. (2002) also examined the potential of US to influence muscle force

recovery following eccentric exercise–induced muscle damage. Neither study showed any significant influence of daily application of 8 min of pulsed US on muscle force recovery (as an indicator of recovery from muscle damage) up to four days postdamage.

Because of its invasive nature, measurement of more direct indicators of muscle damage and repair in humans is difficult. However, several studies have addressed the effects of US treatment on various indicators of muscle damage and recovery in animal models. Rantanen and colleagues (1999) examined the effects of repeated pulsed US application on muscle regeneration rates in rats up to nine days following a standard contusion injury to the gastrocnemius. They concluded from histological examinations that there were no significant differences in muscle regeneration rates between US-treated and control groups. Two other studies examined the influence of pulsed and continuous US treatments on regain of muscle mass, protein, and fiber cross-sectional area following standard blunt contusion injury to hindlimb muscles of rats (Wilkin et al., 2004; Fischer et al., 2003). Neither of these showed any significant enhancement by US administration of any indicators of muscle recovery or regeneration. One cannot be certain, though, how well these studies were able to mimic actual US application procedures typically used in humans.

Several investigations have addressed the effect of pulsed US on clinical muscle pathology in humans. Binder and colleagues (1985) studied 76 patients with lateral epicondylitis. Patients were treated 12 times with 1 MHz US pulsed at 25% at 1 to 2 w/cm^2 for 5 to 10 min. The authors reported that US enhanced recovery (decreased pain and increased grip) in 63% of the cases. In addition, the US-treated group had a lower incidence of recurrence. Likewise, Harvey and colleagues (1975) reported that US treatment administered three times per week for four weeks to individuals with chronic low back pain resulted in significantly less pain, decreased muscle guarding, and a more rapid improvement in motion than in those who received placebo US or no US treatment.

Hence, on the basis of the very limited and sometimes conflicting evidence available, any effectiveness of US in enhancing postdamage muscle repair is difficult to determine.

On Nonmuscle Tissues

While evidence for the impact of US on muscular tissue healing is limited, a number of studies have suggested that US has the potential to influence healing of tendon, nerve, and connective tissue. Although these potential effects of US may indirectly influence muscle recovery following injury, a discussion of any such effects in actual clinical settings is beyond the scope of this chapter. Any relationship to postinjury muscle recovery of a

US influence on tendon, connective tissue, nerve, and other tissues, while theoretically possible, would involve unwarranted and inappropriate extrapolation of data and render their clinical meaning irrelevant. Much more direct clinical research is needed to establish the effectiveness of US treatment (if any) on all factors related to muscle damage and repair.

Phonophoresis (or Sonophoresis)

Phonophoresis uses US for the transdermal delivery (TDD) of a topically applied drug. The concept of using US in this fashion was first suggested by Fellinger and Schmid in 1954. In the medical model, TDD is used for pain relief (fentanyl), angina (nitroglycerin), elevated blood pressure (clonidine), hormonal therapy (estradiol), motion sickness (scopolamine), and birth control (Byl, 1995; Henley, 1990). In the medical model, the goal is to introduce these medications into systemic circulation, whereas in rehabilitation the goal is to deliver the medication for a local influence. Phonophoresis is often used to treat pain and chronic inflammation associated with severe muscle damage and can be an effective method of enhancing pain control, particularly in chronic muscle injury. Phonophoresis has limited usefulness for treating typical muscle soreness and injury induced by exercise and overuse, as this typically normalizes within days and is typically present only upon palpation or movement.

Transdermal delivery has several advantages over injections or oral medications. Drug concentration at the delivery site is initially higher with phonophoresis (McNeill et al., 1992), and the first-pass effect is avoided (passage of the drug through the gastrointestinal system and metabolism by the liver). Phonophoresis can deliver medication to a larger area than an injection but avoids the need to puncture the skin (Byl, 1995). The most common substances used for phonophoresis are anti-inflammatories, anesthetics (to block pain), and analgesics (to relieve pain).

Mechanisms

Several processes appear to contribute to TDD. Griffin and Karselis (1982) stated that the pressure of the US beam drives the medication through the skin and into the tissue. They reported depths of penetration of 10 cm. According to Byl (1995), on the other hand, evidence is insufficient to warrant the conclusion that US waves "push" medication across the skin. Other factors believed to play a role in phonophoresis include acoustical streaming and tissue heating to dilate the blood vessels and enhance diffusion and kinetic energy (Byl et al., 1993).

The effects of TDD depend on several factors. The anatomic structure of the treated skin is influenced by blood flow, sweat glands, thickness, and hydration. All of these characteristics have been shown to influence the effectiveness of TDD. Likewise, the amount and concentration, rate of delivery, and depth of penetration of the medication determine the success of the process. Ultimately, it is the stratum corneum (SC) that is the protective layer of the skin and that serves as the rate-limiting barrier to prevent penetration of noxious substances into the skin. The mechanical effects of US are capable of changing the permeability of the SC and altering the porous pathways to permit drugs to diffuse into the skin. Once across the skin barrier, the elevated concentration of the medication distributes the medication to adjacent tissue via diffusion (movement from an area of high concentration to an area of low concentration via a concentration gradient). The molecules permeate the dermis and may be absorbed into the capillaries to transfer to the bloodstream (Byl, 1995; Byl et al., 1993).

Menon and colleagues (1994) used mice to demonstrate that phonophoresis induces lipid phase separation. Permeation of hydrophilic tracers was identified in pockets (lacunae) within the SC bilayers. When the tissue was sectioned, the lacunae appeared to have formed interconnected channels that expanded into a continuous reticulum when exposed to US. The investigators did not find similar channels in untreated tissue. Thus, they hypothesized that these channels did not exist before treatment and collapsed soon after treatment, in other words, that the SC is transiently permeable. This was supported by the research of Bommannan, Menon, and colleagues (1992), who found that pretreatment of a tissue with US did not enhance penetration of a subsequently applied medication.

The primary points of drug entry for TDD are the hair follicles, sweat glands, erector pili muscles, and sebaceous glands (Byl, 1995). Hydration of the SC is critical for effective drug transmission through these structures (Byl, 1995). Increased hydration has been reported to increase permeability to hydrophilic drugs almost 10-fold (Byl, 1995). Therefore, it may be appropriate to heat the tissue before phonophoresis to enhance TDD. It has been found that this technique dilates the hair follicles (the primary port of entry), increases kinetic energy, and increases circulation to facilitate drug absorption (Byl, 1995). A temperature increase of 5° C is needed to cause a measurable change in cell membrane permeability (Bommannan, Menon et al., 1992; Bommannan, Okuyama et al., 1992). Thus, both the thermal and mechanical characteristics of US are important components in pretreatment of the tissue to increase cell permeability. Heating *after* TDD increases drug absorption, but at the risk of losing the local medication to the systemic blood flow.

Efficacy

As in the US literature, numerous papers on phonophoresis have failed to include details about the research parameters. This makes it difficult to draw conclusions or make comparisons. In addition, many studies could not confirm that the medication in the formulations used in the research were capable of transmitting US waves (Benson and McElnay, 1987; Benson et al., 1989; Cameron, 2003; Penderghest et al., 1998). Nonetheless, a significant number of studies support the idea that TDD can be achieved. However, the optimal technique appears to be contingent on the status of the SC. Moist heat, tissue hydration, and a preliminary US treatment prior to the attempt to achieve TDD may all prepare the SC to yield enhanced drug penetration.

Given the question of the transmission properties of many US media for phonophoresis, Cameron and Monroe (1992) surveyed clinicians to determine the types of media used. They then studied the transmissivity of several of these media; the results are summarized in table 17.3. Recognizing that clinicians may not have the resources to study every possible medium used in phonophoresis, Cameron and Monroe recommended the following process to test a substance for US transmission.

Table 17.3 Ultrasound Wave Transmission of Coupling Media for Phonophoresis

Medium	Transmissibility
Lidex gel (0.05% fluocinonide)	97%
Thera-gesic cream (15% methyl salicylate)	97%
Mineral oil	97%
US gel	96%
US lotion	90%
Betamethasone (0.06%) in US gel	88%
1% hydrocortisone powder in gel	29%
1% hydrocortisone	0%
10% hydrocortisone	0%
Myoflex cream (10% trolamine salicylate)	0%

- Create a well on the face of the US transducer by wrapping a 2 to 3 cm wide piece of tape around the transducer (figure 17.9).
- Place a layer of the conducting medium to be tested on the surface of the transducer and then fill the remainder of the well with water.
- Turn the US machine intensity up to 1 to 2 w/cm^2.
- If the medium is capable of conducting the US waves, the water will begin to percolate; if the medium fails to conduct the US waves, the water will remain still.

Establishing conductivity is an important first step. Benson and McElnay (1987) examined the conductivity of benzydamine hydrochloride (a nonsteroidal anti-inflammatory) and found it to be a good coupling agent (87-139% transmission as compared to that with degassed water). However, in a later study, Benson and colleagues (1989) compared 0.75, 1.5, and 3.0 MHz continuous treatments at 1.5 w/cm^2 with a 3 MHz pulsed (50%) treatment at 1.0w/cm^2 and sham and control treatments. The authors acknowledged that there could be some concerns about the method used to assess penetration; but when there was no difference in drug absorption or in the amount of benzydamine hydrochloride remaining in the gel after a variety of US treatments, they concluded that benzydamine might not have the physical or chemical properties suitable for transdermal penetration. However, the US treatment time was not included in the explanation of the methodology. It is possible that the duration

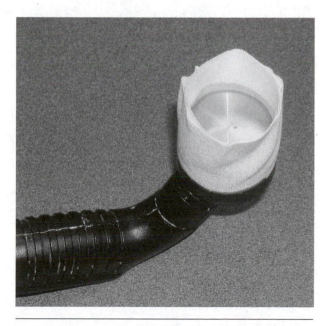

Figure 17.9 Procedure for testing ultrasound transmission.

of treatment was not adequate for sufficient penetration. Similarly, Davick and colleagues (1988) studied radiolabeled cortisol (5% and 10%) in dogs, employing an 870 kHz frequency at a continuous intensity of 0.5 w/cm² for 8 min. Although cortisol was identified in the SC, it was not found in the knee joint. The authors concluded that since the animals were sacrificed immediately, the full effect of phonophoresis may not have been realized.

Kleinkort and Wood (1975) reported that 10% hydrocortisone (HCC) was more effective than 1% in the treatment of tendinitis and bursitis. However, no treatment parameters were specified. In a recent study, Vore and colleagues (2006) examined the TDD of 10% HCC using 0.5, 1.0, 1.5, 2.0, and 2.2 w/cm² for 5 min at a 1 MHz frequency. They reported that there was no indication of HCC in the subdermal unembalmed human tissue via chromatography analysis.

Several studies have used phonophoresis for the treatment of pain. Smith and colleagues (1986) reported a successful reduction in pain with 1 MHz US rendered at 1.5 w/cm² for 5 min with Decadron and lidocaine. However, they were unable to confirm TDD. Moll (1979) rendered six US treatments with a continuous 870 kHz frequency at 1.5 w/cm² for 5 min. Moll phoresed lidocaine and Decadron to treat muscle trigger point–related pain. He reported 88% trigger point pain relief with lidocaine and Decadron, as compared to 56% with US only and 23% with sham US. Gulick and colleagues (2001) also conducted a study using US on myofascial trigger point pain. They used a 7 min treatment with 3 MHz frequency at 1.0 w/cm² to compare continuous US with gel to continuous US with methyl nicotinate (T-Prep), US pulsed at 20% with gel, and sham US. The results were a significant reduction in trigger point discomfort for both the continuous US treatments with gel and methyl nicotinate (MN) in comparison to pulsed and sham US. The authors attributed the enhanced response to the thermal effects of the US and commented on the influence of the MN on the transmission of the US beam. It is hypothesized that MN decreases SC density and hydrates the tissue for enhanced transmission of the US beam. With tissue heating and increased blood flow, the removal of painful local metabolites may have contributed to the reduction in trigger point pain (Dellagatto and Thompson, 1999).

Klaiman and colleagues (1998) matched 49 patients for a variety of musculoskeletal injuries and used 1 MHz US at 1.5 w/cm² for 8 min. They compared the effects of US alone and US treatment with 0.05% fluocinonide gel and examined pain using a visual analog scale and an algometer. They reported that US improved all conditions but that the addition of the fluocinonide gel did not enhance recovery. In contrast, McElnay and colleagues (1987), in a double-blind crossover study, applied fluocinolone acetonide to the skin 5 min before the US treatment. An 870 kHz frequency was used to deliver a 5 min US treatment at 2.0 w/cm². Following 3 h of airtight occlusion, TDD was significantly enhanced.

Table 17.4   Contraindications and Precautions for Ultrasound Treatment*

Contraindications for US treatment	Rationale
Acute trauma	Bleeding may still be prominent.
Cardiac pacemaker	US waves interfere with the electrical conductivity if administered in the area of the pacemaker. US may be safely administered to distal areas of the body with the patient positioned so that the US waves are not directed toward the heart.
Malignant tissue	10 US treatments at 1.0 w/cm² (3 MHz) over 2 weeks resulted in significantly greater tumor growth in mice as compared to those that did not receive US (Sicard-Rosenbaum et al., 1995).
Growth plates	Results unknown.
Eyes, genitals, heart	Thermal effects may be deleterious.
Pregnancy	Limited research (McLeod & Fowlow, 1989) and ethical issues related to studying this population; clinicians opt to avoid the use of US in any manner on a pregnant individual.

Precautions for US treatment	Rationale
Areas of diminished sensation	Patient may not be aware of excessive heating or tissue damage.
Limited ability of patient to communicate	Patient may not be able to provide appropriate feedback.

*As evidenced in a study of tissue heating (Gerston, 1958), metal is not a contraindication for US treatment; as evidenced in a study by Gulick et al. (2006), a joint arthroplasty is not a contraindication for US treatment.

Proper Ultrasound Technique

- Determine the rationale for the application of therapeutic ultrasound.
- Rule out any contraindications.
- Confirm the patient's ability to provide accurate feedback throughout treatment.
- Confirm intact sensation and circulation in the area of treatment.
- Inspect the skin for infections, wounds, or rashes.
- Remove jewelry as needed.
- Position the patient comfortably and so as to provide easy access to the treatment area.
- Select an appropriate frequency to target the desired tissue (1 vs. 3 MHz); depth of ultrasound penetration is inversely related to frequency.
- Select the appropriate treatment area and the corresponding transducer size; treatment area should not exceed two to three times the ERA of the sound transducer.
- Select mode of sound wave transmission—continuous or pulsed.
- Apply and distribute the coupling agent or prepare the vat of water; remember to use water of an appropriate temperature to complement the desired treatment effect (i.e., cooler for non-thermal treatments aimed at edema control vs. warm water when the goal is to increase tissue extensibility).
- Select treatment duration based on duration therapeutic effect.
- Begin treatment by turning up intensity to desired treatment level and continue to move transducer head in a circular pattern at a rate of 3 to 4 cm/s (Kramer, 1984; Michlovitz, 1996). The circular movement (overlapping by about half of the diameter of the sound head) can compensate for the uneven distribution of the US energy (BNR). It is important to keep the sound head surface parallel to the skin surface (Kimura et al., 1998).
- Solicit patient feedback to confirm comfort and desired effect (thermal vs. nonthermal).
- Adjust intensity as needed.
- Terminate treatment, clean up residual gel, and evaluate efficiency of treatment.
- Record treatment parameters and outcome measures.

Note: The new hands-free US applicators, although stationary on the patient, have a built-in scanning mechanism to move the sound beam in a sequential pattern across the applicator head.

Frequency and number of treatments rendered: Treatments should be rendered with the therapeutic goal in mind. It is not uncommon for US treatments to be given daily. However, the number of treatments rendered should be based on the efficiency of the treatment. Ultrasound should not be different from any other modality. If measurable changes are not detected within three treatments, the parameters of the treatment need to be reevaluated and modified or discontinued.

Numerous studies have challenged the effectiveness of US and phonophoresis. It is easy to get confused by the literature regarding what works and what doesn't work. There should be no surprise when patients report not feeling a heating effect with US when the transducer is being moved quickly across the skin, being tilted side-to-side, and treating an area more than three times the ERA. Likewise, the perpetual use of 1 MHz US at 1.5 w/cm^2 for 5 min without regard to the target tissue or the goal of treatment is totally inappropriate.

The steps for rendering an effective US intervention are identified in "Proper Ultrasound Technique." The challenge to the clinician is to apply US with proper technique to enhance the likelihood of an effective intervention. It is critical both to use proper technique and to rule out any contraindications that could make this modality unsafe.

Summary

At present we lack clear evidence to substantiate a positive effect of US treatment in promoting postinjury muscle healing and regeneration. There is a need for more well-controlled research to establish a role (if any) of US treatment in promoting postinjury muscle repair and, if a role is established, to determine optimal treatment dosages and methods. Evidence is stronger for the ability of US treatment to increase muscle heating and for the use of US in phonophoresis to deliver analgesic and anti-inflammatory drugs. However, practitioners are cautioned not to overprescribe US treatment to enhance postinjury muscle repair, as the effectiveness, and indeed the optimal US treatment modalities for such intervention, have yet to be firmly established.

CHAPTER 18

Physical Therapy and Related Interventions

Richard M. Lovering, PhD, PT

The primary goal of rehabilitation in persons with a muscle injury is to maximize function and prevent reinjury. To deal with the impairment, prevent secondary conditions, and provide guidance toward a return to activities, a multidisciplinary approach is required. Current recommendations for treatment vary considerably. Tendinitis may occur with or without associated muscle injury, but the focus of this chapter is on the muscle tissue itself (parenchyma) rather than damage to the surrounding tendons and extracellular matrix (stromal tissue). The goal of the chapter is to present an overview of current treatments for muscle damage and to provide a rationale for evidence-based treatment.

Muscle Injury and Recovery

Muscle injuries are quite common, but the clinical presentation can vary greatly depending on the mechnism of injury, the extent of injury, the muscle(s) involved, and the patient's ability to tolerate pain. Fortunately, most muscle injuries are self-limiting and will heal spontaneously, but the associated morbidity can be minimized with appropriate treatment. An understanding of the physiological mechanisms that underlie injury and recovery of skeletal muscle helps to shape clinical decisions of how to best treat muscle damage.

Muscle Damage

Muscle damage has been defined and quantified in a variety of ways. As discussed in part I, structural damage is evident in histological findings such as disruption of the muscle fiber membrane and its associated cytoskeleton (sarcolemmal damage), fiber splitting, and alterations in sarcomere alignment ("Z-disc streaming"). In addition to histological changes observed after injury, structural damage is further implicated by findings that include a shift in the length–tension curve, loss of force in single isolated muscle fibers, and indications of increased expression or degradation of different structural proteins as seen in biochemical studies. In an animal model of an acute muscle strain injury, we found that immunolabeling for the membrane-associated protein dystrophin was lost in fibers that had sarcolemmal damage (Lovering and De Deyne, 2004) (figure 18.1), while others have found changes in different cytoskeletal proteins, for example desmin (Lieber et al., 1996), using different models of injury. In addition to disruption of structure, other factors likely play a role in the loss of force after injury as discussed in previous chapters.

One problem with many of the biological markers used to assess muscle injury, including those used in animal studies, is that they usually do not correlate with the loss of force. "Muscle damage" is often defined within the context of the assay used to examine it, and no single finding can account for the totality of force loss after injury (figure 18.2). Since full contractile function can persist despite the presence of injury markers, loss of force may be the most valid measure of injury (Brooks et al., 1995) and is probably the most relevant to the clinician.

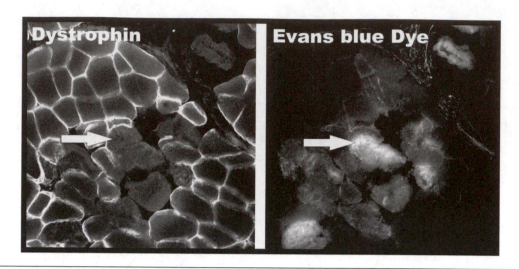

Figure 18.1   Damage to the muscle fiber membrane and associated cytoskeleton occurring after an acute muscle strain injury. Representative tissue section of animal muscle fibers after an acute injury, *(left)* showing loss of dystrophin immunolabeling in fibers with sarcolemmal damage, *(right)* as indicated by the presence of intracellular Evans blue dye. Arrow indicates same fiber in each image.

Adapted, by permission, from R.M. Lovering and P.G. De Deyne, 2004, "Contractile function, sarcolemma integrity, and the loss of dystrophin after skeletal muscle eccentric contraction-induced injury," *Am J Physiol Cell Physiol.* 286(2): C230-238.

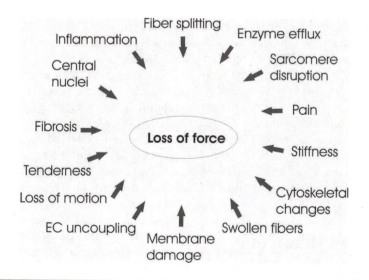

Figure 18.2   Measures of injury. Muscle injury is characterized by many different "markers" of damage. Since no one finding can account for the associated force deficit, loss of force may be the best measure of the totality of injury.

Interventions must be designed to target specific physiological processes so that rational treatments can be developed for a particular injury. Although there is a growing understanding of the issues relating to mechanisms of muscle injury, few controlled studies have examined the role of rehabilitation interventions after injury. Treatment of muscle injuries remains largely dependent upon symptoms, and the self-limiting nature of an injury often determines when to return to activity.

It is difficult to study muscle injuries in humans, as the occurrence is a random event that is difficult to predict and the clinical presentation varies greatly. Therefore many of the data regarding muscle injuries have been obtained from studies on animals, which allow control over many variables and the ability to study mechanisms of injury and recovery. However, the variety of models used to induce injury makes it difficult to compare results. Different determinants of injury are also used in animal studies, which means that injury may be present according to one criterion but not another. Lastly, various markers of injury are present with common training protocols that result in beneficial adaptations, so some findings indicating an "injury" may simply reflect a necessary stimulus for muscle hypertrophy, regeneration, and adaptation.

Recovery

Genesis of skeletal muscle is not restricted to prenatal development, but also occurs in damaged muscle that is "rebuilding." This is a process that recapitulates embryonic myogenesis (Grounds, 1991), and the key player for regeneration in adult skeletal muscle is likely to be the satellite cell (Bischoff, 1989; Schiaffino et al., 1976) situated outside the plasmalemma of each muscle fiber. These mononuclear cells are normally quiescent; however, they are thought to become active with stimulation (e.g., injury). Under appropriate conditions, satellite cells develop into myoblasts, which fuse to form myotubes. Myotubes can then repair, or even replace, damaged muscle fibers. It is generally hypothesized that satellite cells, after several rounds of proliferation, are a determinant factor in the functional recovery of muscle (Allen et al., 1997; Irintchev et al., 1997; Kauhanen et al., 1998; Schiaffino et al., 1976). The importance of satellite cells is underscored in certain pathologies (Maier and Bornemann, 1999) and in atrophy linked to either disuse (Allen et al., 1997) or denervation (Yoshimura and Harii, 1999), as well as in their reduced number in older individuals (Hasselman et al., 1995). Myogenesis and its role after injury are reviewed extensively in chapter 6.

Other mechanisms also contribute to the recovery of force after injury (chapters 12 and 14). Cytoskeletal reorganization as well as formation of new membrane may contribute to the process of recovery (McNeil and Terasaki, 2001; Papadimitriou et al., 1990). Other factors that may affect recovery from injury include changes in gene expression (Barash et al., 2004) and protein content (Chopard et al., 2001) and inflammation (Tidball, 1995; Almekinders and Gilbert, 1986; Mishra et al., 1995; Thorsson et al., 1998). As we will see, it is not always clear which therapeutic interventions hasten regeneration and repair.

Types of Muscle Injury

Muscle damage can result from contusions, ischemia, lacerations, mechanical strain, drug toxicity, and idiopathic causes, as in fibromyalgia or compartment syndromes. Iatrogenic muscle damage (secondary to surgical procedures) is commonly encountered in physical therapy,

as it is often necessary for surgeons to dissect muscles and fascia to reach the underlying structures. Treatment is determined by the type of injury and is directed at the suspected underlying problems. The most common causes of muscle injuries are (1) "muscle strains" induced by lengthening contractions and (2) contusions.

Strains

Muscle strains are among the most common complaints treated by physicians (Garrett, Jr., 1996). Muscle strains account for the majority of sport-related injuries (Zarins and Ciullo, 1983; Brockett et al., 2001), as well as a significant proportion of low back pain (Bartleson, 2001; Glass, 1979). Therefore, the symptoms associated with these injuries have a significant economic impact on both the individual and society as a whole. When an activated muscle lengthens because the external load exceeds the tension generated by the muscle contraction, its action is termed a lengthening ("eccentric") contraction. Although lengthening contractions require less energy, the force generated during a maximal lengthening contraction is approximately twofold the force developed during a maximal isometric contraction; therefore lengthening contractions are more likely to produce damage than either than isometric or concentric contractions (Hunter and Faulkner, 1997). The basic mechanism of higher force generation during a lengthening contraction is not understood. The difficulty of explaining the enhanced force from a lengthening contraction lies in the fact that the force produced is greater than the sum of the measured active force (from an isometric contraction) and the passive force at that given muscle length.

Lengthening contractions are physiologically relevant (Cavagna, 1977; LaStayo et al., 2003) and often occur without causing damage. They produce high forces, which is a goal of strength training (the overload principle). This is evident in strengthening protocols that use lengthening contractions, or "negatives," to increase strength. Not only can lengthening contractions produce more force than other types of contractions, but they can do so at a reduced oxygen requirement (Lindstedt et al., 2001; LaStayo et al., 1999). Thus, the application of exercises using lengthening contractions in elderly people is appealing, as high metabolic demand is sometimes not wanted in this population. Although greater strength gains are achieved using lengthening contractions in training, many studies have shown that the high force produced by eccentric muscle contractions can cause subsequent pain and damage (Armstrong et al., 1983; Brooks et al., 1995; Friden and Lieber, 2001; MacIntyre et al., 1996).

For clinical purposes, a muscle strain can be categorized into three different levels of severity, or grades (table 18.1).

Table 18.1   Grades of Muscle Strain

	Grade I	Grade II	Grade III
Tenderness	+	++	+++
Loss of motion	–	+	++
Pain	+	++	+++
Palpable defect	–	+	+++
Detectable swelling	–	++	+++

- A grade I strain is a mild strain that is symptomatic (tenderness and mild pain) but results in no impairment (full range of motion and no loss of strength).

- A grade II strain results in moderate impairment, marked tenderness, decreased range of motion of the associated joint, and a noticeable loss of strength.

- A grade III muscle strain results in immediate pain, as well as a palpable defect in the muscle surrounded by swelling (edema), indicating rupture of fibers or the whole muscle.

Although the underlying mechanisms of the different grades of muscle strains may be different, the grades are often reported to represent "the amount of torn fibers." Acute muscle strains can be detected with computed tomography or magnetic resonance imaging (MRI) methods (Speer et al., 1993), but diagnosis is typically made based on physical exam and patient history.

It is important to identify symptoms derived from structures above and below the injury. For example, a condition such as sciatica that causes referred pain down the posterior thigh and leg could be misdiagnosed as a chronic grade I hamstring or calf strain. Tenderness of the muscle-tendon unit—not typically present with referred pain—is a hallmark of muscle injury, and the amount of tenderness increases with the grade of strain. Increased pain in the muscle with isometric resistance applied to oppose the muscle action is another indication of injury.

Muscle injuries are typically self-limiting and do not require imaging, but as this noninvasive technology continues to improve and imaging (e.g., MRI) becomes more commonplace, it may play a role in rehabilitation planning and prognosis (Blankenbaker and De Smet, 2004). Plain films, or X rays, are not very useful for imaging muscle pathology unless heterotopic bone formation has occurred within the muscle (see later discussion of myositis ossificans). Unlike X rays, MRI has high sensitivity to the hemorrhage and edema that follow muscle injuries.

This, together with the capability to evaluate multiple anatomic planes, makes it the most suitable technique to evaluate muscle injuries. Muscle strains are revealed best by T2-weighted images, which optimize contrast between injured muscles with edema (increased signal intensity) and normal uninjured muscles (the T1 and T2 relaxation times define the way the protons in tissues revert back to their resting states after an initial radiofrequency pulse) (see figure 18.3). Imaging of muscle injuries does not, however, provide information regarding the underlying mechanisms of injury or the cellular processes that are taking place.

Contusions

Muscle contusions are second only to muscle strains as a leading cause of morbidity from sport-related injuries (Beiner and Jokl, 2001). As opposed to muscle strains, which are contraction induced and are typically "noncontact" injuries, muscle contusions are due to a direct blow to a muscle. Muscle contusions can also result in injury to the underlying bone. Although a fracture is usually evident by clinical exam and confirmed with an X ray, there is often a less obvious injury to the relatively thin periosteum that surrounds the underlying bone.

Muscle closest to the underlying bone is usually the most extensively damaged (Berg, 2000). The accumulation of blood in the muscle tissue from the resulting hemorrhage is termed a *hematoma*, with discoloration (ecchymosis) and marked tenderness. As with muscle strains, a classification scheme has been proposed based on severity (mild, moderate, or severe) of symptoms (Jackson and Feagin, 1973). Although the demarcation between different grades can be vague, the classification schemes of strains and contusions are necessary to help clinicians communicate with one another and to interpret outcome studies.

Myositis Ossificans

Myositis ossificans is characterized by ossification (bone formation) within muscle tissue and is a complication

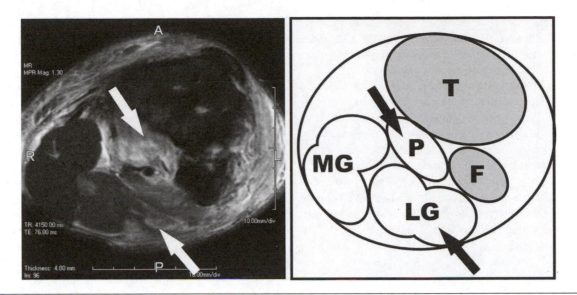

Figure 18.3   Imaging of injured muscles posterior to the knee. Left: A T2-weighted axial magnetic resonance image (MRI) of a 22-year-old male with injury in the lateral gastrocnemius and popliteus muscles. Arrows indicate edema and inflammation. Right: Schematic depiction of the anatomy shown in the MRI. T = tibia; F = fibula; MG = medial gastrocnemius muscle; LG = lateral gastrocnemius muscle; P = popliteus muscle.

of recovery from a muscle contusion. The incidence is likely related to the magnitude of the impact (Jackson and Feagin, 1973) and has been estimated to be as high as 17% of contusion injures (Beiner and Jokl, 2002). Even if ectopic (abnormally positioned) bone is not palpable within the belly of the injured muscle, one should suspect myositis ossificans if the symptoms do not improve or if they worsen with time. In particular, lack of a progressive return to normal range of motion is an indicator sufficient to raise suspicion of myositis ossificans (Ryan et al., 1991). Although a direct blow is the initial cause, there seems to be a consensus that overaggressive stretching of a muscle after a contusion can contribute to the development of myositis ossificans (King, 1998; Lipscomb et al., 1976; Hierton, 1983).

Intramuscular calcification can be visualized with X rays as early as three weeks after injury (Norman and Dorfman, 1970), but a longer time period may be needed (Cushner and Morwessel, 1992). Regardless of when the bony mass can be detected, the radiographic findings and the ability to palpate the growth clearly lag behind the presentation of symptoms. The muscles most commonly affected are proximal limb muscles, particularly those in the anterior-lateral thigh. As with many orthopedic conditions, there are further classifications of myositis ossificans (Arrington and Miller, 1995), but these are usually based on the location of the ectopic bone (e.g., new bone surrounded by muscle vs. attachment to the underlying skeleton) in addition to the severity of symptoms. The source of the cells that form new bone is not clear. Although it seems logical to assume that osteoblasts from

the underlying bone contribute to the ossification within the muscle, bone marrow–derived progenitor cells cannot be ruled out (Charge and Rudnicki, 2004).

The natural history of myositis ossificans closely follows that of a contusion, with a gradual return in range of motion and typically no long-term impairment or disability. For patients who do have persistent symptoms, surgical excision of the ectopic growth can be performed after the bone mass has matured (evidenced by smooth margination on X ray), which usually occurs months after the initial insult (King, 1998).

Concern is warranted when patients' findings are consistent with myositis ossificans but there has been no precipitating traumatic event. Myositis ossificans is the result of a direct injury to muscle and must be differentially diagnosed from a neoplastic growth, such as an osteosarcoma (Dudkiewicz et al., 2001). Unlike myositis ossificans, a tumor is not necessarily associated with an injury, and symptoms, as well as size of the mass, will continue to worsen over time.

Overuse Injuries

Muscle overuse injuries often involve the associated tendons, as in rotator cuff tendinitis, Achilles' tendinitis, patella tendinitis, and carpal tunnel syndrome. It is not known whether the underlying mechanisms differ from those of acute strains, in which the muscle-tendon junction is a common location of inflammation and tenderness. As their name implies, overuse injuries can be due simply to repetitive use of a particular muscle. Faulty biomechanics are also thought to play a role in many work-related overuse injuries (see chapter 12); and a sudden change in the mode, intensity, or duration of training is associated with overuse injuries in athletes.

Several factors may play a role in muscle overuse injuries. For example, in exertional compartment syndrome of the anterior leg muscles (a possible cause of "shinsplints"), the cause of symptoms may be mechanical pressure on the muscles, ischemia, tendinitis of the muscle attachments, or conditions not associated with muscle such as periostitis or stress fractures. Compartment syndrome is a good example of a muscle pathology that may be due to factors other than the muscle itself, such as pes planus ("flat feet"), tight heel cords, and improper footwear.

Current Therapies

Except for complete ruptures of a muscle, displaced avulsions, and recalcitrant symptoms from myositis ossificans, almost all muscle injuries are uniformly treated with nonoperative therapy (Carrino et al., 2000). Standard conservative therapy for either acute muscle strains or acute contusions usually involves rest, ice, compression, and elevation (RICE). Beyond the principle of short-term rest and ice, there is no clear consensus on treatment of muscle injuries. The few studies that focus on treatment concentrate on outcomes, providing few recommendations of therapies to target specific biological processes. Short-term goals include return of strength, flexibility, and endurance, with the presumption that the uninvolved extremity can serve as a "control" for the preinjury status. Many patients return to activities or sports after gaining sufficient joint motion despite lingering strength or endurance deficits, which they may not perceive. Protocols can be developed that are grossly designed to address the different grades of injury, but each rehabilitation plan must be tailored to the individual and the findings from the person's physical exam.

Modalities

Many physical therapeutic agents, or modalities, are available after an injury occurs. One of the most common questions asked by patients after any type of injury is whether to use ice, heat, or some other modality. Since most muscle injuries are self-limiting and will recover spontaneously, the goal of applying modalities is to hasten recovery.

Ice and Heat

There is ample anecdotal evidence, as well as some experimental evidence, that cryotherapy (the application of cold medium) is effective in limiting inflammation after muscle injury (Dolan et al., 2003; Merrick et al., 1999); but these conclusions are not shared by all (Curl et al., 1997; Fu et al., 1997). Cryotherapy has been used to minimize delayed-onset muscle soreness (DOMS) and the enzyme efflux from muscle fibers after injury (Howatson and Van Someren, 2003), but these findings are inconsistent (Paddon-Jones and Quigley, 1997; Yackzan et al., 1984). Most studies of the effects of ice have involved tissues other than muscles (Hubbard and Denegar, 2004).

The physiological basis for cryotherapy is assumed to be a cold-mediated constriction of the microvasculature resulting in decreased edema, but the amount of subcutaneous fat can certainly affect whether or not underlying tissues respond to ice (Myrer et al., 2001; Otte et al., 2002). If used, cryotherapy should be applied as soon as possible after a suspected muscle injury. The microcirculatory effect occurs primarily in the venules, not the arterioles (Smith et al., 1993); and the vasoconstriction and decreased permeability secondary to application of cryotherapy are short-term (Menth-Chiari et al., 1999; Curl et al., 1997). Others have suggested that the diameter of venules increases after cryotherapy, which then allows greater absorption of the edematous fluid (Smith et al., 1993).

Topical moist heat is not recommended in the acute phase immediately after muscle injury. Moist heat has been tried as a means to treat soreness and swelling following injury, but without much success (Jayaraman et al., 2004). It is generally assumed that moist heat applied over a muscle will cause an increase in tissue temperature and therefore increased tissue extensibility. It is further assumed that these changes can decrease susceptibility to injury. One study showed that almost a half-hour of moist heat application was required to obtain a 0.4° C increase in hamstring muscle temperature, and the treatment did not significantly affect hamstring flexibility (Sawyer et al., 2003). Others have used animal experiments to show that the biomechanical properties of muscles are altered after warming, such that failure of the muscle-tendon unit occurred at longer lengths (Strickler et al., 1990; Noonan et al., 1993). However, the muscles were directly exposed to temperature-controlled saline solution rather than topical moist heat. Such studies provide insight into the effects of heat on the mechanical properties of muscle; but information about load to failure may not be directly applicable to a physiological injury, and such findings are more important for injury prevention than for treatment.

Mobilization

Muscle quickly remodels in response to environmental and physiological cues. It should be made clear to patients that the "R" in RICE is for rest, not immobilization. All too often patients, particularly nonathletes, will guard the joint associated with the muscle injury, potentially creating secondary problems. Although a brief course of immobilization after a muscle injury may have some benefit (Jarvinen and Lehto, 1993; Sayers et al., 2000b, 2000a), prolonged immobilization can lead to many secondary problems, such as muscle fibrosis (Huard et al., 2002; Best and Hunter, 2000), joint contractures (Williams, 1997), atrophy (Jarvinen et al., 1992), inhibition of myogenesis (Vierck et al., 2000), decreased tensile strength (Jarvinen, 1977), and increased susceptibility to reinjury (Jarvinen et al., 1992). Animal studies define "early" mobilization as approximately three days after a grade II injury, which was shown to optimize muscle growth and result in better remodeling of the surrounding connective tissue (see "Benefits of Early Mobilization After Muscle Injury") (Jarvinen and Lehto, 1993).

Electric Stimulation

Electric stimulation has long been used in patients with neurological impairments and is used routinely in physical therapy to treat musculoskeletal injuries. Electric stimulation is purported to increase strength, mitigate pain, and reduce edema.

Electric stimulation of muscle with surface electrodes to elicit a contraction is often referred to by the acronym NMES, for neuromuscular electric stimulation (Hultman et al., 1983). Nerves have a much lower capacitance (ability to store charge) than muscle fibers do, so this lower threshold to activation causes nerves to depolarize sooner than the muscle fibers. Therefore, even though a visible contraction occurs when surface electrodes are placed onto the muscle of a patient, it is the motor nerve that is firing, then the muscle fibers it innervates contracting secondarily.

The claim that NMES can improve strength appears valid, especially for muscles that are inhibited and atrophied following a surgical procedure—for example, weak quadriceps following anterior cruciate ligament surgery (Snyder-Mackler et al., 1995). On the other hand, once the individual regains muscle control and is able to develop significant tension in the muscle, NMES appears to be no more effective than using volitional contractions against resistance. The highest level of NMES that a healthy muscle can tolerate is one that produces about 25% of the force of a maximal voluntary contraction (MVC) (Lieber and Kelly, 1991). The claim that NMES can be used to elicit muscle forces greater than 100% of an MVC may sound appealing to athletes; however, the intensity of electric stimulation needed to do this is usually intolerable, and this use of electric stimulation to healthy muscle falls outside the realm of rehabilitation. NMES should be prescribed much like a strength training protocol. The key is the amount of tension developed in the muscle, so it behooves the patient to perform an MVC during the stimulation. Just as in strength training, there seems to be a threshold for obtaining an effect with NMES after which further sessions are not beneficial (Kernell et al., 1987).

Electric stimulation has been employed as an analgesic treatment. This typically involves a current that is below the threshold to induce a muscle contraction and is commonly known as transcutaneous electric nerve stimulation, or "TENS," despite the fact that this acronym could be applied to any electric stimulus applied through the skin. The concept behind TENS is that the smaller sensory neurons depolarize before the larger motor neurons, so using intensities below the threshold of contraction allows constant stimulation of sensory nerves while avoiding the obvious undesirable side effect of constant muscle contractions. TENS is often used with people who have chronic pain but is also used frequently for acute pain such as acute low back pain. The mechanism may involve excitation of sensory fibers, which interferes with transmission of pain from smaller fibers in the spinal cord (the "gate control theory") (Melzack and Wall, 1965), or release of endogenous chemicals that alleviate pain. Regardless of the mechanism, TENS has been used as an analgesic modality since its conception (Wall and Sweet, 1967), and the majority of studies support its effectiveness in ameliorating pain. High-frequency, low-intensity, and continuous stimulation are considered "conventional TENS," but settings on these portable devices can be adjusted in many ways to obtain a protocol that works best for the patient.

Benefits of Early Mobilization After Muscle Injury

- Enhanced muscle regeneration
- Improved vascularization
- Improved tissue remodeling
- Stronger tensile properties

The claim for electric stimulation as a modality to reduce edema is unfounded. There are virtually no controlled randomized trials in humans that address this, and most animal studies do not provide a basis for the use of electric stimulation to reduce edema (Cook et al., 1994; Cosgrove et al., 1992; Dolan et al., 2003; Fish et al., 1991). The muscle "pumping action" provided by NMES can help to reduce chronic edema but is likely no more effective than actively contracting the muscles when possible.

Other Modalities

Many modalities not commonly used in acute injuries are utilized to treat chronic muscle injury. Many of these are employed to reduce pain, but the mechanisms by which they work have not been elucidated. Massage and ultrasound are conventional tools used to treat persistent muscle pain (Tiidus, 1999); these modalities are discussed in detail in other chapters. Acupuncture can elicit a physiological response in muscle, such as an increased blood flow (Sandberg et al., 2003), but some studies report no differences in symptomatic relief between treated and placebo groups (sham acupuncture) (Goddard et al., 2002). Several studies have examined the use of hyperbaric oxygen therapy (HBO) as a novel treatment for muscle injuries. HBO therapy is the inhalation of 100% oxygen inside a hyberbaric chamber that is pressurized to greater than 1 atmosphere (atm). Although some studies suggest a benefit from this type of therapy after muscle injury (Best et al., 1998; Staples et al., 1999; Gregorevic et al., 2000), many do not (Harrison et al., 2001; Germain et al., 2003; Webster et al., 2002; Nelson et al., 1994). It is difficult to identify the mechanisms associated with specific effects of HBO. Some have suggested that HBO compensates for a reduction in peripheral oxygen partial pressure, but there may be a more specific effect on muscle that has not yet been observed. Aside from the mixed results, potential problems in using this therapy include access, cost, and time (one session is typically 90 to 120 min).

Medication

Muscle injury is often defined by the loss of force that occurs immediately after the injury. However, a second decline in force usually occurs days afterward; and this bimodal pattern of force change, as well as the accompanying DOMS, is thought to be due to inflammation (MacIntyre et al., 1996; Warren et al., 2002), which is reviewed in chapter 4.

Anti-Inflammatory Medication

Nonsteroidal anti-inflammatory drugs (NSAIDs) are medications commonly used to treat pain and inflammation following muscle injury. Some studies, however, suggest that NSAIDs can actually delay the recovery process and rate of regeneration (Almekinders and Gilbert, 1986; Mishra et al., 1995; Obremsky et al., 1994). Anabolic steroids and corticosteroids reduce pain and inflammation after muscle injury in the short term but may cause irreversible damage to healing muscle in the long term (Beiner et al., 1999; Levine et al., 2000). As with steroids, this effect of a benefit from short-term use yet harm from long-term use has been observed with some NSAIDs (Mishra et al., 1995). A recent study raised the question of NSAIDs' general effectiveness (Rahusen et al., 2004). The authors concluded that use of NSAIDs after contusion injury was no more effective than treating the muscle with an analgesic that had no anti-inflammatory effects.

Muscle is a highly vascularized tissue, and there is an obvious need for inflammatory cells (neutrophils and macrophages) from the blood to infiltrate the injured area, assuming the presence of necrotic tissue that needs to be phagocytized and removed. However, whether inflammatory cells promote further muscle damage is still not clear. Neutrophils are capable of membrane lysis (Tidball, 2005) and can release cytokines that impede myogenesis (Pizza et al., 2005). However, neutrophils can also release cytokines that enhance myogenesis, and recently identified cytokines from macrophages (Chazaud et al., 2003) can also facilitate myogenesis (Tidball, 2005). Although most clinicians have typically considered inflammation the enemy, recent findings indicate that the inflammation process has a beneficial effect on myogenesis (e.g., Bondesen et al., 2004; Mendias et al., 2004). Such confounding findings may leave the clinician at a loss about how to interpret the literature and unsure about whether the aggregate effect of inflammation is beneficial or detrimental. Clinicians need to take into account other tissues in addition to muscle, however, when deciding whether or not to recommend NSAIDs for any amount of time. Since inflammation can inhibit joint motion, promote fibrosis in the extracellular matrix, and cause pain, short-term use of NSAIDs still seems prudent (Gierer et al., 2005). Studies on anti-inflammatory medications are needed to determine what the best therapeutic agents are and their most effective dosage. The ideal result is likely a prescription leading to a controlled inflammatory response that diminishes the negative consequences to surrounding healthy tissues while maintaining the apparent benefits to the damaged area.

Transdermal Drug Delivery

With iontophoresis and phonophoresis, therapeutic agents are administered through the skin by the application of a low-level electrical current or sound waves, respectively. Ultrasound is a modality commonly used by physical therapists and is the method used to implement phonophoresis (see chapter 17). Iontophoresis uses an electrode of the same polarity as the charge on the drug to drive ionic (charged) drugs into the body (Singh and Bhatia, 1996). A small low-voltage (10 V or less) continuous current (0.5 mA/cm^2 or less) is used to push a drug into the skin. Recently developed electrodes do not require an external power source, but instead have a power supply self-contained within the patch worn by the patient (Bolin and Goforth, 2004). The drugs most commonly used in physical therapy clinics are probably lidocaine and dexamethasone (Costello and Jeske, 1995); but their effectiveness in injured muscle has not been established (Hasson et al., 1992), and injections may be more effective (Gokoglu et al., 2005). It should also be noted that the drugs used are often controlled substances and need to be prescribed by a physician.

The most common complication encountered with iontophoresis is skin irritation and burns. Another potential problem is that the drug enters systemic circulation soon after it passes through the skin. This may diminish the desired "local effect," although the concentration of a given medication at a given site is unknown. Iontophoretic delivery of medications such as NSAIDs does have the benefit of avoiding potential gastrointestinal side effects, which occur in about 10% to 30% of patients receiving oral NSAIDs (Banga et al., 1999). Iontophoretic delivery also avoids drug absorption problems that occur with oral medications, such as first-pass metabolism (chemical and enzymatic breakdown) by the liver, which can significantly reduce the amount of drug that ends up in systemic circulation. Regardless of the delivery method, the issue of how much the inflammatory process should be suppressed after muscle injuries still needs to be studied more fully.

The topical agents that can be successfully delivered through the skin by iontophoresis are limited by their properties, such as size and charge. Electroporation is an alternative method to iontophoresis for transdermal drug delivery. In contrast to iontophoresis, electroporation applies a high-voltage (>100 V) pulse for a very brief duration (1 μs-1 ms). While iontophoresis acts on the drug, electroporation is used to permeabilize the skin (Prausnitz et al., 1993) and provides the potential to include larger molecules in transdermal delivery (Banga et al., 1999). However, unlike the situation with iontophoresis, few studies on electroporation have been done in humans.

Physical Training

The mechanical stress resulting from muscle contractions can cause an injury or a beneficial adaptation. Injury is more likely to occur in muscles that are unaccustomed to the mechanical stress placed upon them; but it has been well documented that subsequent bouts of the same activity, performed days or even weeks after an initial unaccustomed bout, result in significantly less damage (Nosaka et al., 2001a, 2001b; Koh and Brooks, 2001). This phenomenon has been termed the "repeated-bout effect," and the mechanism by which the adaptation occurs is still controversial (see chapter 5). Possible explanations include the elimination of weaker fibers via necrosis (Friden and Lieber, 2001), changes in motor unit recruitment (Warren et al., 2000), changes in the cytoskeleton (McHugh et al., 1999; Trappe et al., 2002; Lehti et al., 2007), and changes in the extracellular matrix (Stauber et al., 1990). Another theory is that the addition of sarcomeres resulting from myogenesis increases the optimal length of a muscle (Brockett et al., 2001), which would result in a net effect of less strain to individual sarcomeres at any given length (Talbot and Morgan, 1998). Obviously the mechanisms of the repeated-bout effect are still being studied, and it is not clear whether this protective effect with regard to overuse injury extends to acute muscle injury. Nonetheless, the implication for the clinician is that resistance protocols—or "strength training" programs—that involve lengthening contractions are important in preventing future contraction-induced muscle injuries (Kirkendall and Garrett, Jr., 2002).

Eccentric Contractions in Training

Ironically, although lengthening contractions are associated with injury, they also provide the most protection against future injury (Koh and Brooks, 2001). Therefore, in an apparent paradox, this type of contraction is also employed during rehabilitation after injury to exploit the benefits of the repeated-bout effect (Croisier et al., 2002; Brockett et al., 2001), even with chronic tendinitis (Svernlov and Adolfsson, 2001; Mafi et al., 2001; Jensen and Di Fabio, 1989). Many patients have a persistent but undetected strength deficit after an acute muscle strain, especially in their "eccentric strength" (Croisier et al., 2002). In the early return to strength training, however, only isometric or concentric contractions should be performed within a pain-free range of motion. Submaximal lengthening contractions early after injury do not provide a benefit over other types of contractions (Donnelly et al., 1992), and maximal lengthening contractions can cause further damage. Therefore lengthening contractions should be utilized as a progression from the other types of resistance training (Sorichter et al., 1995). As with restoring normal range of motion, a *progressive* increase in tension is necessary to avoid further injury (LaStayo et al., 2003). A muscle that has been strained is susceptible to reinjury if the rehabilitation is too aggressive or if the patient returns to activities too soon (Taylor et al., 1993).

Biomechanical studies on lengthening contractions indicate that damage is a function of strain (Lieber and

Friden, 1993), prior activation (Lovering et al., 2005), and initial muscle length (Hunter and Faulkner, 1997) but is not dependent on velocity of lengthening (Willems and Stauber, 2000). Ultimately, it is likely that the total work (force generated by the muscle multiplied by the distance of lengthening) is a primary determinant of contraction-induced injuries (Brooks et al., 1995). These biomechanical factors need to be considered during the eccentric strength training component of therapy. Rehabilitation should not only address regaining strength and flexibility in the injured muscles, but should also include endurance training (Mair et al., 1996) and retraining in agility and sport-specific tasks, which can help to minimize the likelihood of reinjury (Sherry and Best, 2004).

Isokinetic Training

Isokinetic equipment can be an extremely effective tool for evaluation and improvement of strength. This equipment usually provides feedback to the patient in various forms, which often helps to motivate them and allows them to observe their progress. Another benefit is that the therapist can control parameters such as minimum or maximum force, velocity of movement, and range of motion allowed. Furthermore, isokinetic resistance ceases immediately when the patient stops, so the patient will never meet more resistance than he or she can handle. A muscle strain that occurs during squatting with a substantial amount of free weights could prove disastrous, whereas during isokinetic exercise the load is removed as the subject stops resisting. In the later stages of rehabilitation, isokinetic equipment provides a means to load a dynamically contracting muscle to its maximum capacity at all points throughout the range of motion; this is difficult to achieve with other forms of exercise due to a constantly changing muscle length and moment arm.

When used improperly, however, isokinetic equipment can also exacerbate a condition, and for this reason some surgeons may be concerned when a patient with a newly reconstructed ligament or a reattached muscle is placed on this equipment. Other potential disadvantages include equipment cost, the need for trained personnel, and the fact that isokinetic movement is not physiological or functional (no one normally moves a limb at constant velocity or is typically limited to one plane of motion). There is also some experimental data to suggest that strength gains using isokinetic equipment are specific to the velocity and range of motion used during training (Moffroid and Whipple, 1970; Lindh, 1979). Although these findings have been contested, they are consistent with the notion that training needs to be "sport specific." That is, rehabilitation should be tailored to the needs of the specific individual, taking into account the type and speed of contractions.

Prevention

As discussed previously, strength training appears to be beneficial in preventing contraction-induced muscle injury. Many athletes also want to know whether stretching is beneficial or if, instead, "warming up" is sufficient. Passive stretching can activate stretch-activated ion channels (Sachs, 1997), alter gene expression (De Deyne, 2001), change the pattern of firing of intrafusal muscle fibers ("spindles") (Poppele and Quick, 1981), and induce release of growth factors (Yang et al., 1996). Stretching is recommended prior to exercise to avoid muscle injury. If it were effective, then it would have a role during rehabilitation, as one of the goals is to avoid reinjury. So, does stretching "work"?

A passive stretch needs to be transmitted via the connective tissue (tendons, perimysium, and endomysium) in order to reach the muscle fibers. Since muscle is much more compliant than tendon, it will undergo the most elongation (De Deyne, 2001). Because viscoelastic tissues elongate in a time-dependent manner in response to a low-level maintained force (this property is known as "creep"), most clinicians recommend that a muscle be held in a stretched position (Bandy and Irion, 1994; Bonnar et al., 2004). Viscoelastic tissues also exhibit the property of relaxation, however, which means that the tissue will return to its original length over time after the force is removed; and it is debatable whether the gain from stretching is due simply to an improved tolerance to stretch (Halbertsma et al., 1996) or to a maintained adaptive response (Bonutti et al., 1994; Decoster et al., 2004; Taylor et al., 1990).

So, despite its widespread use, it is not clear whether or not standard stretching has an effect on muscle length. But does it prevent injury? Some limited evidence suggests that stretching may contribute to injury prevention (Rodenburg et al., 1994; Hartig and Henderson, 1999; Koh and Brooks, 2001), but most studies conclude that stretching is not advantageous in preventing injury (Lund, 2003; Black et al., 2002; Black and Stevens, 2001; Herbert and Gabriel, 2002). Yet, since there is no evidence to suggest a deleterious effect, stretching before athletic activity is still commonly recommended.

The evidence for actively "warming up" to prevent injury is a little more convincing. Biomechanical studies that employ isometric contractions as a model for warm-up indicate that "cold" muscles are stiffer, so that load to failure occurs at shorter lengths (Safran et al., 1988). As discussed earlier, other studies using passive warming of muscles have yielded similar findings (Noonan et al., 1993; Strickler et al., 1990). Warming up has also been associated with improved muscle performance (Stewart et al., 2003; Mohr et al., 2004), but the mechanisms for this have not been thoroughly examined. Although biomechanical studies suggest some benefit from warming up and performance seems to be improved, there is a dearth of clinical evidence showing a protective effect from warm-up (High et al., 1989; Rodenburg et al., 1994). These studies are limited to overuse-type injuries; no clinical study has yet addressed whether warming up protects against an acute muscle strain.

A contracted muscle can absorb more energy than a passive muscle (Noonan et al., 1993; Crisco et al., 1996) and therefore is able to sustain more force before failure. This may spare structures associated with the muscle from injury, but an activated muscle is more susceptible to damage. So, should one "tighten up" during an accident? For example, should skiers be told to relax their muscles during a fall, or should they fight the fall? As far as the muscle is concerned, it is probably better to relax (Lovering et al., 2005) (and see figure 18.4). However, the debate about whether to relax or tighten up during injury needs to consider other tissues, and this debate has not been resolved.

The best prevention against contusion injuries is making sure that athletes play by the rules of the game and that they wear proper protective equipment during play. Shinsplints and other overuse injuries can be difficult to predict; but once these injuries occur they can become a long-term nuisance. The clinician needs to ensure that the muscles involved have adequate rest and rehabilitation, but environmental factors play a large role in causing overuse injuries. Improper equipment (e.g., forearm tendinitis from a tennis racket not fitted to grip

size or shinsplints from running shoes with too much mileage), improper technique (e.g., inflammation of the supraspinatus muscle in the shoulder from lack of sufficient torso rotation during swimming), improper surfaces (e.g., running on concrete or uneven surfaces like curbs), and improper training (e.g., overaggressive weightlifting) can all contribute to muscle overuse injuries.

Intrinsic anatomic factors can also affect muscles. Many people have a leg-length discrepancy (LLD) and flat feet ("pes planus"), but not every flat-footed person with an LLD needs expensive orthotics. Nonetheless, these factors should be considered when standard care does not resolve long-standing muscle complaints. Poor posture has been alleged to cause many muscle symptoms. For example, with a forward head and rounded shoulders ("slumped posture"), the posterior cervical muscles and anterior shoulder muscles are shortened over time while their antagonists are stretched over time. A slumped posture also alters the position of the shoulder blade (scapula), which can contribute to impingement on the supraspinatus muscle tendon and alter the entire mechanics of the shoulder (Lewis et al., 2005; Bullock et al., 2005; Julius et al., 2004; Greenfield et al., 1995; Schuldt et al., 1986).

Summary

Muscle injury is a clinical diagnosis that can usually be established from the history and physical exam alone. Mild transitory pain from unaccustomed exercise (e.g., lifting weights) that is bilateral and does not impair function is likely more of a "stimulus" than an injury, and should be treated symptomatically. A rupture of a muscle (grade III muscle strain) is usually clinically obvious and may require surgery to repair, but is much less common than grade I and II muscle strains. Moderate persistent pain that prevents one from participating in activities can be classified as a grade I or II muscle injury. Fortunately, muscle has a remarkable ability to regenerate and repair itself. Therefore, most muscle injuries are self-limiting and recovery occurs spontaneously. The only lasting impairment from most muscle injuries is often undetected weakness, although secondary issues such as joint stiffness or tendinitis may persist. Appropriate rehabilitation can affect the rate of recovery and is needed to assess overall flexibility, strength, and endurance before return to athletic activity or work. RICE is still the first aid treatment of choice immediately after injury; but the benefits derived from anti-inflammatory medication, stretching, and several common modalities may depend on when they are used and the type of injury involved.

There is a multitude of data on muscle injury, but much of this information relates more to mechanisms

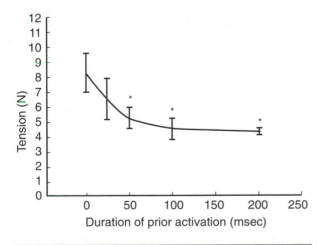

Figure 18.4 Timing of activation affects injury. With use of an animal model, a significant loss of force is seen after a maximal lengthening injury only when the muscle is activated at least 50 ms prior lengthening, suggesting that tightening up before an injurious event is detrimental.

Reprinted from *Journal of Biomechanics*, Vol. 38, R.M. Lovering et al., The contribution of contractile pre-activation to loss of function after a single lengthening contraction, pp. 1501-1507, Copyright 2005, with permission from Elsevier.

than to treatment. The amount of information and the different methodologies used to obtain the data can be overwhelming. Therefore, we frequently see findings that seem to conflict, leaving us to wonder if the information is unintelligible. One of the biggest problems in interpreting the literature on treatment of muscle injuries is the variety of models used to induce injury. Other problems include assorted measures of injury (using "soreness" in one study, a biological marker in another) and the variety of prescriptions for a given therapy. It is entirely possible for a given therapy to be effective in one model of injury but not in another. One cannot discard a therapeutic modality simply because the treatment is ineffective on a given type of injury at a specified frequency, duration, and intensity. Despite the obvious need for treatment to be evidence based, empirical data and anecdotal evidence are still valuable to anyone who is practicing physical therapy or clinical medicine.

ACKNOWLEDGMENTS

Funded, in part, by National Institutes of Health grant 5K01AR053235 and a grant from the NFL Charities. The author would like to thank Dr. Peter Leininger for his comments during preparation of this manuscript, Dr. Michael Mulligan for the MRI, and the laboratories of Dr. Robert Bloch and Dr. Patrick De Deyne for support and mentorship. Support was also provided by the Interdisciplinary Training Program in Muscle Biology at the University of Maryland, Baltimore.

CHAPTER 19

Antioxidant Supplementation

Allan H. Goldfarb, PhD

Numerous substances are purported to be potentially beneficial in the prevention or treatment of damage to skeletal muscle. Unfortunately, research on the ability of these substances to protect skeletal muscle from damage is inconsistent, and the findings on enhanced repair of damage are minimal and inconclusive. Before any consideration of studies in this area it is important to understand that not all exercise-induced muscle damage is the same.

Some exercise-induced muscle damage may result in serious tears and disruptions of muscle and connective tissues. These often require surgical intervention and other therapies and are not considered in this chapter, which instead focuses on the type of exercise that induces transient micro muscle and connective tissue damage. This type of micro damage is often associated with delayed-onset muscle soreness (DOMS) from unaccustomed exercise, although it can also occur with repetitive exercise depending on the forces. This chapter deals with nutritional and nutraceutical interventions that might influence the extent of muscle damage and repair in relation to DOMS. Therefore, it focuses on those substances that might influence inflammatory processes and the production of reactive oxygen and nitrogen species (RONS). The chapter does not cover nutritional and nutraceutical interventions that may influence DOMS but do not act on these two processes; for example, we do not consider nutritional needs for proper structural precursors for muscle and connective tissue.

Potential Mechanisms of Exercise–Induced Muscle Damage

The mechanism of exercise-induced muscle damage during unaccustomed exercise is thought to involve, at least in part, the greater forces imposed on the contractile and connective components within the muscle. Numerous processes occur within the muscle and connective tissue as a result of this micro damage. Apart from the mechanical forces, chemical changes associated with inflammatory and oxidative stress processes (Pizza et al., 2005) may exacerbate the damage. These processes appear to be important not only for degradation of the damaged components but also for repair of the area. However, they may also cause additional damage. In this chapter we focus on nutritional antioxidants that are purported to either prevent or attenuate the damage under these limited circumstances or that may influence repair. Often these substances are given as prophylactic pretreatment agents to influence inflammatory or oxidative stress reactions or both.

A number of processes associated with the formation and production of reactive species (RONS) may contribute to secondary damage associated with DOMS. In addition, inflammatory-mediated processes are activated with lengthening contractions that induce damage and certainly contribute to the restructuring of the muscle (Best et al. 1999; Pizza et al. 2002).

Several inflammation-related processes involve RONS (Cannon and Blumberg 2000), including the inflammatory-mediated processes in which phagocytic leukocytes infiltrate damaged tissue (Klebanoff 1982; Tsivitse et al. 2003). Neutrophils and monocytes can produce RONS during their "oxidative burst" activity in response to stimuli (Cannon and Blumberg). Nicotinamide Adenine Dinucleotide Phosphate (NADPH) oxidase within the lyosome catalyzes the reaction to produce a superoxide anion (O_2^-). Subsequent to the formation of O_2^-, other RONS can also be formed. Reactive species can be produced from the actions of prostanoids or their by-products (Halliwell and Gutteridge 1985; McArdle et al. 1991) and may alter calcium-activated proteases (McArdle et al. 1992). Circulating prostanoids have been reported to be elevated (Cannon et al. 1990; Smith et al. 1993) in response to eccentric exercise. Reactive species have been implicated in the loss of normal calcium handling within the muscle (Essig and Nosek 1997).

Inflammatory cells can infiltrate the muscle and release proteolytic enzymes that work in conjunction with the production of cytokines to remodel the affected tissue. Neutrophils could influence the damage to myotubes (Pizza et al. 2001; McLoughlin et al. 2003) by releasing certain cytokines (e.g., tumor necrosis factor-alpha [TNF-α]). Some of these cytokines impair myoblast differentiation, myotube formation, and protein synthesis (Cassatella 1999; Langen et al. 2002). It is important to note that other cytokines released by inflammatory cells, for example interleukin-6 (IL-6), can enhance myogenesis (Cassatella 1999) or promote macrophage activity that results in myoblast proliferation as has been shown in vitro (Cantini et al. 2002). Therefore, inhibition of the inflammatory process may not only reduce the damage but also, depending on the type of cytokine, influence subsequent repair and remodeling of the muscle. Chapters 1, 4, and 5 discuss the mechanisms of muscle damage in more detail.

Reactive species and the inflammatory processes have been implicated in the secondary damage associated with DOMS (Jackson 2000; Mc Hugh et al. 1999). Several pathways that involve RONS appear to be activated in response to eccentric exercise, including NADPH oxidase production, prostanoid metabolism, disruption of iron-containing proteins, and calcium proteases (McHugh et al. 1999). See figure 19.1 for some of the interaction sites within the cell that could be affected by RONS and inflammatory processes and antioxidants. In addition, numerous studies have shown alterations in the inflammatory process with eccentric contractions (Pizza et al. 2005). Therefore, an intervention that influenced either the RONS or the inflammatory processes might also influence the other. As already noted, the response may differ depending on the factors activated with the inflammatory process.

Several studies have shown that eccentric exercise can increase blood (Close et al. 2004; Goldfarb et al. 2005a;

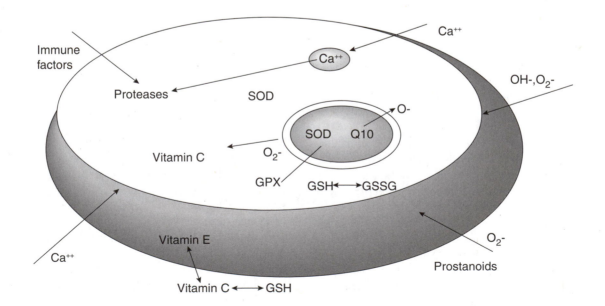

Figure 19.1 Interactions of Oxidants and Antioxidants. CAT= Catalase; GSH= Glutathione reduced form; GSSG = Glutathione oxidized form; GPX= Glutathione Peroxidase; GR = Glutathione reductase; O_2^- = superoxide radical; OH$^-$= hydroxyradical; O$^-$= singlet oxygen; NO$^-$= nitric oxide radical; Ca^{++}= calcium; Q10= Ubiquinone; SR= Sarcoplasmic reticulum.

Lee et al. 2002; Sacheck 2003) and muscle markers (Saxton et al. 1994; You et al. 2005) of oxidative stress; others have not. Differences in findings may relate to the marker examined, the type of exercise, or the condition of the subject. For example, downhill running is not only an eccentric exercise but is also aerobic and therefore could increase oxidative stress through at least two potential avenues in comparison to eccentric resistance exercise.

Table 19.1 lists well-known antioxidants and their potential modes of action.

Intervention studies utilizing a single antioxidant prior to eccentric exercise have yielded mixed results (Goldfarb 1999). The antioxidants have typically been vitamin E or vitamin C but have included ubiquinone, alpha-lipoic acid, and also isoflavonoids. Not every study dealing with these compounds is reviewed here. Some studies have utilized animal models whereas others have involved humans. Several studies have utilized a combination of antioxidants in an effort to ascertain the possible benefits in reducing oxidative stress, in some cases in relation to the damage associated with DOMS.

Since downhill running is an aerobic as well as an eccentric activity, it may differ from an eccentric resistance activity with respect to RONS formation. Theoretically,

Table 19.1   Well-Known Antioxidants and Their Potential Mode of Action

Substance	Mode of action
Lipid phase	
α-Tocopherol (vitamin E)	Quenches singlet oxygen Prevents or nhibits lipid peroxidation Stabilizes membranes
β-Carotene	Quenches singlet oxygen Quenches superoxide radical Precursor for Vitamin A (Provitamin A)
Ubiquinone and ubiquinol	Prevents lipid peroxidation Can spare vitamin E Two-electron reduction of ubiquinone-10
Flavonoids (phenolic plant antioxidants)	Inhibit lipid peroxidation (in vitro) Inhibit lipoxygenase, cyclooxygenase Anti-inflammatory
Genistein	Shows highest antioxidant ability (in vitro)
α-Lipoic acid	Works with glutathione Influences cellular metabolism, glucose homeostasis Influences diabetic neuropathy Can influence insulin signaling
Aqueous phase	
L-ascorbate Vitamin C	Quenches aqueous soluble radicals Quenches singlet oxygen Regenerates vitamin E to reduced form Possibly increases glutathione peroxidase activity in red blood cells Essential for certain hydroxylase enzymes
Glutathione	Scavenges singlet oxygen Regenerates vitamin E (and vitamin C) Scavenges hydroxyl radicals Removes hydrogen peroxide (by peroxidase activity)
N-acetyl cysteine	Helps to increase glutathione levels

the downhill activity should produce a greater RONS response, whereas eccentric resistance activity may result in greater reactive hyperemia and mimic ischemia-reperfusion injury.

Vitamin Supplementation

Vitamins are organic substances that are essential for proper body function. Since the body does not manufacture or synthesize vitamins, they must be provided in the diet. Vitamins are essential for proper actions within the cells including antioxidant protection (vitamin C and E, alpha-lipoic acid), stabilization of membranes (vitamin E), synthesis of connective tissue (vitamin C), and mitochondrial oxidative electron transport (ubiquinone). Often the amount in the diet is less than the recommended daily requirement (vitamin E), or the need for additional intake may increase as a result of an active lifestyle (vitamins C and E).

Vitamin E

Vitamin E is a major lipid-soluble chain-breaking antioxidant that also helps to stabilize membranes. It is usually present in low concentration within the membranes of all cells. Vitamin E can be oxidized or reduced and can interact with vitamin C and glutathione. It has eight isoforms and based on the isoform has different biological activities. The isoform that has the greatest biological activity is the d-α-tocopherol form, also known as α-tocopherol. The biological activity is also affected by linkage to an ester to help with its stability in capsule form. Therefore, its biological activity can vary depending on the isoform and the ester linkage. It has been reported that vitamin E deficiency can result in membrane fragility. Studies showed that vitamin E content decreased after acute exercise in male rats (Bowles et al. 1991) and with endurance training in male rats (Packer et al. 1989), but not after acute exercise in female rats (Tiidus et al. 1993). Cannon and colleagues (1990) reported increased vitamin E within human skeletal muscle in response to supplementation (Cannon et al. 1990). The normal vitamin E requirement for adult men and women in the United States is 8 to 10 mg.

Table 19.2 outlines studies on vitamin E supplementation and its influence on exercise-induced muscle damage in both animals and humans.

Animal Studies

Jackson and coworkers (1983) reported that vitamin E deficiency in the diets of rats (males) and mice (females) increased muscle susceptibility to contractile damage induced by electrical stimulation. An increase in malonaldehyde, a marker of lipid peroxidation, in muscle homogenates from the anterior tibialis from vitamin E–deficient animals occurred only when they were incubated with 10 μM ferric chloride and 100 μM ascorbic acid. This suggests that when the muscle was vitamin E deficient and there was a stress to the cell by the iron load and electrical stimulation, there was an increase in lipid peroxidation and muscle damage. These authors noted that they could not determine that RONS was the mediating factor. However, they did conclude that vitamin E supplementation was able to reduce skeletal muscle damage in these animals.

Oxidative stress and muscle damage were examined in rats that walked downhill. The rats were on either a normal diet or a vitamin E–enriched diet (Warren et al. 1992). Muscle damage was assessed by examination of the muscle cells and measurement of plasma creatine kinase (CK). Oxidative stress was reduced in the vitamin E supplemented animal homogenates when they were exposed to hydrogen peroxide. Typically, CK does not leak out of the cell, and if CK is found to increase in plasma, it suggests that the membrane is more open or leaky. Vitamin E supplementation (10,000 IU/kg diet for five weeks) did not influence muscle fiber damage (30-35%). Plasma CK increased about twofold after exercise independently of vitamin E treatment. Since the vitamin E supplementation did not influence muscle fiber damage and did not alter the amount of CK in the plasma, the authors concluded that vitamin E supplementation did not alter muscle injury but may be helpful in reducing free radical damage from eccentric exercise.

Another rat study using the same dose of vitamin E supplementation (10,000 IU/kg diet) showed reduced protein oxidation in muscle at rest and after exercise (Reznick et al. 1992). Exercise increased protein oxidation within the muscle, and vitamin E supplementation significantly reduced the amount of protein oxidation.

Vitamin E supplementation (250 IU/kg diet for five weeks) attenuated exercise-induced oxidative stress in plasma and skeletal muscle from rats (Goldfarb et al. 1994). The exercise increased thiobarbituric acid reactive substances (TBARS) in plasma as well as lipid hydroperoxides and TBARS in skeletal muscle, with the vitamin E supplementation attenuating the exercise increase. Vitamin E supplementation in rats (750 IU/kg body weight for eight weeks) reduced liver and red gastrocnemius muscle TBARS and protein carbonyls in response to exercise (Sen et al. 1997).

In another investigation, vitamin E treatment (five to eight days of intravenous injections) increased muscle vitamin E content threefold but did not alter lengthening contraction–induced injury to rat muscle in situ (Van der Meulen et al. 1997). Vitamin E did not alter force deficit or change the percentage of muscle fibers damaged. However, serum CK, which increased twofold after three days in the vehicle-treated muscles, did not change with vitamin E treatment. The authors noted that vitamin E helped with membrane integrity to prevent enzyme loss after muscle damage but did not reduce the extent of damage.

Table 19.2 Vitamin E Supplementation and Exercise-Induced Muscle Damage

Supplementation	Model	Exercise	Effect	Reference
Animal studies				
240 mg/kg diet, 42-45 days	Rats and mice	Stimulated electrically	↓ CK and ↓ LDH leakage	Jackson et al. 1983
10,000 IU/kg diet, 5 weeks	Rats	Ran	↓ Protein oxidation (muscle)	Reznick et al. 1992
10,000 IU/kg diet, 5 weeks	60 rats	Downhill walking	= Force = CK = Muscle damage	Warren et al. 1992
250 IU/kg diet	64 male rats	60 min run	↓ TBARS (blood and muscle)	Goldfarb et al. 1994
750 IU/kg body weight, 8 weeks	80 male rats	Run to exhaustion	↓ Liver TBARS ↓ Protein carbonyls (muscle)	Sen et al. 1997
500 mg/kg diet, 3 weeks	36 male rats	60 min run	↓ p65 ↓ MCP-1 ↓ Myeloperoxidase ↓ TBARS	Aoi et al. 2004
Human studies				
300 mg, 28 days	20 college-age males	Incremental to fatigue	↓ MDA	Sumida et al. 1989
400 IU, 48 days	Males 22-24 Males 54-74	Downhill running	↓ CK loss = Recovery	Cannon et al. 1990
800 IU, 48 days	21 males	Downhill running	= IL-1 beta ↓ IL-6 over 12 days ~ TNF-alpha	Cannon et al. 1991
400 mg, 21 days	Males	Box stepping	= Force ↓ 25%	Jakeman and Maxwell 1993
800 IU, 48 days	Males 22-24 Males 54-74	Downhill running	↓ Urinary TBARS ↓ Conjugated dienes (muscle)	Meydani et al. 1993
400 IU, 5 months	30 cyclists	Aerobic training	↓ CK ↓ MDA (serum) = Performance = Lactate	Rokitski et al. 1994a
800 IU, 28 days	12 males	Eccentric actions to failure	= Plasma CK ↑ = Plasma TBARS ↑ = DOMS	Boyer et al. 1996
1200 IU, 30 days	16 males	240 isokinetic eccentric contractions	= Z-band streaming = Torque ↓ = Macrophage ↑	Beaton et al. 2002
1200 IU, 3 weeks	18 males	Whole-body resistance	= CK ↑ = DOMS = Performance = MDA ↑	Avery et al. 2003
1000 IU, 12 weeks	16 males, 71 years 16 males, 26 years	Downhill running	↓ CK ↓ $F_{-\alpha}$-isoprostanes	Sacheck et al. 2003

Equals sign means there was no significant difference; ↓ = significantly decreased response; ↑ = significantly increased response in vitamin E–supplemented compared to placebo or no treatment; ~ = about the same. CK = creatine kinase; LDH = lactate dehydrogenase; DOMS = delayed-

In a recent study, vitamin E supplemented in the diets of rats prevented the increase in myeloperoxidase and TBARS without influencing neutrophil activity in response to moderate exercise (Aoi et al. 2004). Muscle cells from rats without vitamin E supplementation demonstrated increased chemokine expression for both neutrophils and monocytes. Additionally, a nuclear translocation component, p65, was elevated, as were neutrophil chemoattractant-1 and monocyte chemoattractant protein-1 in these isolated muscle cells. In contrast, cells preincubated with high vitamin E demonstrated limited increases in these responses. The authors suggested that moderate-intensity exercise increases these factors and that vitamin E supplementation can reduce the activation of cytokines and adhesion molecules to decrease damage.

These animal studies have yielded mixed results on reduction of muscle damage with various doses of vitamin E. It is clear that vitamin E may attenuate oxidative stress in some cases and may stabilize the membrane so that less leakage of molecules occurs. Vitamin E can influence the inflammatory cascade and redox-sensitive signaling, but in certain instances the muscle damage may be too severe and cannot be overcome by the antioxidant. Differences in the amount of vitamin E used and the severity of the damage have contributed to the inconsistent results in these studies.

Human Studies

The human studies have varied in length of time for treatment (three weeks up to five months), vitamin E dose (about 200 IU to 1200 IU per day), type of exercise, and age of subjects. Some researchers report vitamin E in milligrams. To convert this to international units if the vitamin E is the RRR-alpha-tocopherol form, one divides by 1.49. I have done this for the values in this chapter and rounded to the nearest whole number. The RRR-alpha tocopherol isoform of vitamin E has the highest biological activity of the eight isoforms of vitamin E.

Sumida et al. (1989) reported that 201 IU for four weeks resulted in less lipid peroxidation as indicated by malonaldehyde in plasma in response to incremental exercise to fatigue in healthy young men. Vitamin E for three weeks (268 IU) did not prevent a force decrement in box stepping (Jakeman and Maxwell 1993). In addition to these two studies, others using low doses of vitamin E have often reported no alterations in muscle function when vitamin E was supplemented.

Cannon and colleagues (1990) reported reduced CK loss in young males in response to running downhill and improved recovery in older men when vitamin E was supplemented. The authors also noted that vitamin E supplementation had no effect on circulating neutrophils in response to downhill running but did decrease

IL-6 levels over the 12 days after the exercise (Cannon et al. 1990). This group also found that oxidative stress markers were lower after vitamin E supplementation (800 IU) in both young and older men (Meydani et al. 1993). Decreased CK and malondialdehyde (MDA) were also reported in response to aerobic training (Rokitski et al. 1994a). A recent study demonstrated modest reductions in oxidative stress markers with vitamin E supplementation for two age groups (Sacheck et al. 2003). The young subjects showed less CK after exercise whereas older men showed reduced $F_2\alpha$-isoprostanes both at rest and after the exercise with vitamin E treatment.

In contrast to these studies, two investigations suggest that whole-body resistance exercise with eccentric components can increase DOMS, CK release, and markers of oxidative stress but that vitamin E supplementation does not alter these responses (Avery et al. 2003; Boyer et al. 1996). It is unclear whether there was protection within the muscles themselves, as both studies measured only blood factors. Vitamin E supplementation appeared to lead to a slightly higher CK response than the placebo after an initial exercise, but the CK responses to the subsequent exercise sessions were similar for the two groups of subjects (Avery et al. 2003). This may have been a repeated-bout effect rather than related to the vitamin E. Beaton and colleagues (2002) showed that vitamin E did not protect against effects of 240 eccentric isokinetic contractions. Subjects took either a placebo or vitamin E, and there was no benefit in taking vitamin E with respect to preventing force decrement, macrophage infiltration, or muscle damage as indicated by Z-band streaming.

In summary, vitamin E supplementation has shown mixed results in human studies, and this may be related to the amount of damage produced by the exercise. It appears that vitamin E can reduce leakage of membranes to result in less loss of CK and can reduce several markers of oxidative stress but does not always do so. Vitamin E has manifested little influence on muscle soreness and recovery from DOMS. Additionally, limited information is available on the effect of vitamin E supplementation on inflammatory processes that may influence muscle damage and recovery. In part the research is controversial because the extent of damage has not been controlled and the vitamin E concentration has varied. Another factor that is not controlled is the form of vitamin E. Most papers do not indicate the form (isomer) or the ester linkage, which could have influenced the availability of the vitamin E.

Vitamin C

Vitamin C, the major aqueous antioxidant, has important functions in connective tissue formation and immune function. Vitamin C is in the reduced form and can be oxidized to help regenerate oxidized vitamin E or glutathione. Vitamin C is transported across membranes and is stored throughout all tissues and within the blood. The tissue level is typically 3 to 10 times higher than the plasma level due to a transport shuttle into the cells. The recommended dietary intake for vitamin C for adults is 90 mg. Keith (1997) reported that the vitamin C plasma level was decreased in athletes whose vitamin C intake was 100 mg per day and that 300 mg was needed to maintain their plasma level. Another study, however, showed that vitamin C supplementation for seven to eight months did not alter plasma vitamin C level in athletes (Telford et al. 1992). In contrast, supplementation with vitamin C (1000 mg) in combination with vitamin E (400 IU) increased plasma vitamin C level in young adults (Bloomer et al. 2006). The vitamin C level within the blood and the amount of the supplement have been shown to alter the absorption rate (vitamin C >500 mg decreases absorption) (Levine et al. 1996).

Frei and colleagues (1989) showed that vitamin C is a potent aqueous antioxidant in vitro and especially in blood plasma. Ascorbate can help to regenerate oxidized vitamin E (Niki et al. 1984). A limited number of studies have examined vitamin C supplementation and muscle damage with oxidative stress. Several have focused on supplementation and the exercise-induced oxidative stress response. Table 19.3 outlines studies on vitamin C supplementation and its influence on exercise-induced muscle damage in humans.

Vitamin C supplementation (1 g/day) for either one day or two weeks minimally altered plasma TBARS concentration after 30 min of running at 80% $\dot{V}O_2$max as compared to placebo treatment (Alessio et al. 1997). The total antioxidant capacity (optical radical absorbance capacity, ORAC) in blood at rest and after exercise was unaffected by this dose of vitamin C. Vitamin C supplementation (2 g/day) did not alter the serum conjugated diene (one of the initial steps in the process of lipid peroxidation) concentration immediately after a 10.5 km (6.5-mile) run compared to placebo (Vasankari et al. 1998) but produced an 11% decline in conjugated dienes during the recovery.

Vitamin C supplementation for two weeks (400 mg/day) led to a modest reduction in plasma MDA, an end product marker of lipid peroxidation, compared to placebo in response to an intermittent shuttle run (Thompson 2001a). In addition, there was a very modest decline in IL-6 in the postexercise period with vitamin C as well as a slight decline in soreness.

In another study, designed to show whether the dose of vitamin C would influence oxidative stress markers in the blood, 500 mg given for two weeks attenuated protein carbonyls (a marker of protein oxidation) increase following a 30 min run at 75% $\dot{V}O_2$max compared to placebo (Goldfarb 2005b). In the same study,

1000 mg for two weeks further attenuated the increase in protein carbonyl level following this exercise compared to the placebo or the 500 mg dose. No influence on glutathione ratio changes or lipid peroxidation as reflected by the indirect marker TBARS was seen in this investigation.

Thompson and colleagues (2003) looked at the possible influence of the timing of vitamin C supplementation on its effect with muscle damage. Vitamin C (200 mg) given immediately after a shuttle run had no beneficial effect and resulted in no difference in serum CK, muscle soreness, or recovery of muscle function. Furthermore, IL-6 and MDA did not differ after the exercise in the vitamin C treated groupcompared to the placebo.

To ascertain whether metabolic demand influenced the results from this study, Thompson and colleagues (2004) had subjects run for 30 min at 60% $\dot{V}O_2$max and then 30 min downhill (same speed) at –18% grade. Subjects took either a placebo or vitamin C (400 mg/day) for two weeks prior to the run and for three days after the run. Despite higher plasma vitamin C levels in the vitamin C–treated group, there was no difference in blood markers of muscle damage (CK and myoglobin). Furthermore, IL-6 levels and soreness increased similarly after the exercise in the two groups. The authors concluded that vitamin C pretreatment did not influence the IL-6 response following eccentric exercise with low metabolic demand.

High-dose vitamin C supplementation (3 g/day) was compared to placebo treatment to assess soreness in subjects who performed calf plantar- and dorsiflexion (Kaminsky and Boal 1992). In a counterbalanced design, half the subjects were randomly assigned to vitamin C and half to placebo, and subjects did one-leg exercise. Then they had a three-week washout, received the other treatment, and repeated the activity with the opposite leg. Subjects took the supplement for three days prior to the exercise and for the next four days. Subjects reported significantly less DOMS with the vitamin C treatment. In a similar study performed to confirm these results, 18 young males were randomly assigned to either vitamin C (3 g/day) or placebo treatment (Bryer and Goldfarb 2006). They took pills for two weeks prior to the eccentric exercise and for four days afterward. Delayed-onset muscle soreness was significantly reduced in the vitamin C–treated group, with no difference in changes in blood glutathione status and CK.

Vitamin C (400 mg/day) prior to and during recovery was reported to enhance recovery in muscle function in response to 60 min of box stepping (Jakeman and Maxwell 1993). Vitamin C (2000 mg) with carbohydrate slightly decreased conjugated dienes 90 min after a run but not immediately following the run (Vasankari et. al. 1998).

Table 19.3   Vitamin C Supplementation and Exercise-Induced Stress and Muscle Damage

Supplementation	Model	Exercise	Effect	Reference
1000 mg, 1 day; 1000 mg, 2 weeks	12 humans	80% $\dot{V}O_2$max	= ORAC ≈ TBARS = ORAC ≈ TBARS	Alessio et al. 1997
2000 mg	17 endurance athletes	10.5 km run	↓ CD 11% during recovery	Vasankari et al. 1998
1000 mg 2 h before exercise	9 trained males	Shuttle run, 90 min	= Soreness = MDA = CK	Thompson 2001a
400 mg, 2 weeks	16 males	Shuttle run, 90 min	= CK ≈ Soreness ≈ IL-6 ≈ MDA	Thompson 2001b
1000 mg, 3 weeks	7 trained males	Cycling, 90 min	↑ MDA exercise ↑ MDA rest	Bryant et al. 2003
500 mg, 2 weeks or 1000 mg, 2 weeks	12 males	75% $\dot{V}O_2$max	= TBARS = Glutathione ratio ↓ Protein carbonyls	Goldfarb et al. 2005b
200 mg after exercise	8 subjects	Shuttle run, 90 min	↑ Plasma vitamin C = Soreness = Recovery = IL-6	Thompson et al. 2003
400 mg, 2 weeks	12 subjects	Downhill run, –18% (30 min)	= CK ↑ = Soreness ↑ = IL-6 ↑	Thompson et al. 2004
3000 mg 7 days before and 3 days after exercise	19 adults, mean 35 years	Calf eccentric exercise	↓ DOMS	Kaminsky and Boal 1992
3000 mg 2 weeks before and 4 days after exercise	16 males, young	70 eccentric contractions	↓ DOMS ≈ CK = Glutathione ratio = ↓ MIF	Bryer et al. 2001
400 mg 21 days before and 7 days after exercise	8 subjects, young	Box stepping, 60 min	↑ Recovery 24 h	Jakeman and Maxwell 1993

Equals sign means there was no significant difference; ↓ = significantly decreased response; ↑ = significantly increased response with vitamin C supplementation compared to placebo or no treatment. CK = creatine kinase; LDH = lactate dehydrogenase; DOMS = delayed-onset muscle soreness; TBARS = thiobarbituric acid reactive substances; ORAC = optical radical absorbance capacity; CD = conjugated dienes; MDA = malondialdehyde; IL-6 = interleukin-6; MIF = maximal isometric force. ≈: The authors suggested a possible modest effect that may or may not have reached significance.

In summary, the research has shown that vitamin C may have some beneficial effects on oxidative stress markers in response to exercise, depending on the marker used. In addition, there appears to be some benefit to vitamin C supplementation in relation to DOMS, but the supplement must be taken prior to the activity and may need to be continued during the recovery period. The amount of vitamin C appears to be a factor and may contribute to the discrepancies in the literature. The extent of muscle damage may be a factor as well, as also indicated with the vitamin E studies. More studies are needed to clarify the effect of vitamin C supplementation on the inflammatory process, as the information currently available is limited.

Supplementation With Other Antioxidants

Nutritional components besides vitamins C and E have antioxidant capabilities, including ubiquinone, an important component of the electron transport chain. Ubiquinone supplementation studies have used various doses and time periods. Ubiquinone bioavailability was shown to be slowly enhanced to a limited extent in plasma and tissues in humans (Bhagavan and Chopra 2006). Animal studies have shown that ubiquinone in large doses can be taken up by all tissues. Endurance training increases ubiquinone concentration but does not alter the amount of ubiquinone in the oxidized form (Gohil et al. 1987). Controversy exists about the use of ubiquinone (CoQ) to prevent muscle damage.

Ubiquinone

Ubiquinone, also known as CoQ, is an electron transport intermediate that is an important component of the electron transport chain and also has antioxidant properties. It has been suggested that CoQ is beneficial in the prevention of oxidative stress and may influence muscle damage. Laaksonen and colleagues (1995) reported that CoQ supplementation for six weeks (120 mg/day) did not alter performance in young or older subjects and had no effect on serum MDA levels. This supported previous research indicating that CoQ had no effect on performance or lipid peroxidation (Braun et al. 1991; Malm et al. 1997a). In fact, Malm and colleagues (1997b) reported that giving ubiquinone resulted in greater cellular damage during intense exercise.

In contrast, CoQ administration in rats reduced serum CK and lactate dehydrogenase levels after downhill running compared to control treatments (Shimomura et al. 1991). It has been suggested that ubiquinone supplementation may affect aspects besides oxidative stress and may help with performance (Karlsson 1997). CoQ supplementation in rodents for four weeks reduced lipid peroxidation in liver and heart as indicated by TBARS (Faff and Frankiewicz-Jozko 1997). The CoQ supplementation had no effect on postexercise increases in CK in the serum of these rats after running to exhaustion. In a review, Rosenfeldt and colleagues (2003) identified 11 studies on CoQ supplementation and exercise capacity; about half of these studies showed some positive effect and five showed no effect.

Whether CoQ can enter muscle cells even with an increase in the blood is controversial. CoQ supplementation (120 mg/day) for 20 days in young healthy men increased plasma CoQ levels at 11 and 20 days but did not influence muscle CoQ levels (Svensson 1999). In a study with mice, CoQ supplementation (for 11 weeks at either 148 or 654 mg CoQ10/kg body weight was compared to a standard diet) increased plasma CoQ levels (Kamzalov et al. 2003). This study also showed a dose-dependent increase in CoQ in heart, skeletal muscle, and brain, with most of the CoQ within the mitochondria of these tissues. The results also demonstrated increases in α-tocopherol within the tissues. In an earlier study, however, CoQ supplementation did not increase tissue levels in rodents (Lonnrot et al.1998). It is still unclear whether CoQ levels can increase in all tissues in humans, although levels have been reported to increase in all tissues in animals with high doses (Bhagavan and Chopra 2006). Further investigation in the area of CoQ supplementation is needed to determine whether or not ubiquinone can help to reduce oxidative stress in response to exercise. Current research does not support taking CoQ10 to prevent oxidative stress in order to decrease muscle damage.

Alpha-Lipoic Acid

Alpha-lipoic acid (ALA) or thioctic acid is readily absorbed and converted to dihydrolipoic acid (DHLA) within the cell. Therefore, both ALA and DHLA should be determined in people who are taking supplemental ALA. Dihydrolipoic acid has been shown to help regenerate other antioxidants such as vitamin E and vitamin C when they are in the oxidized form. In addition, ALA can increase the intracellular level of glutathione (Han et al. 1995) and can act as an antioxidant (Packer et al. 1995) to help reduce oxidative stress. When ALA is taken up by various cells it is rapidly reduced by nicotinamide adenine dinucleotide reduced (NADH) or NADPH-dependent enzymes to DHLA (Packer et al. 1995). Studies have suggested that ALA supplementation can reduce oxidative stress in the aging rat heart (Suh et al. 2001), in diabetic rats (Cakatay et al. 2000; Dincer 2002), and in rat brain and liver (Hagen et al. 1999). However, a recent study indicated that ALA supplementation in aged rats increased protein oxidation in heart muscle (Cakatay et al. 2005). The authors suggested that ALA may also have a prooxidant effect.

Few studies have addressed the influence of ALA supplementation on prevention of exercise-induced oxidative stress. Khanna and colleagues (1999) reported that ALA supplementation can increase ALA levels in red gastrocnemius muscle and increase total glutathione levels in liver and blood. In addition, the exercise-induced increase in muscle TBARS and heart damage was attenuated with ALA. Alpha-lipoic acid supplementation has also been reported to improve insulin action and glucose tolerance in obese endurance-trained rats (Saengsirisuwan et al. 2001). Therefore, some of the beneficial effects associated with ALA may involve enhanced insulin signaling (Saengsirisuwan et al. 2004). If ALA does help with prevention of oxidative stress and insulin signaling, it may also help to prevent exercise-induced muscle damage. Unfortunately, there are no studies dealing with ALA and eccentric exercise–induced muscle damage.

Isoflavonoids

Isoflavonoids are a group of plant chemicals (phytochemicals) that have some antioxidant activity and may also be involved in other health-related outcomes. Several isoflavonoids have been examined for their role in preventing a number of disease processes linked to RONS (Wedworth and Lynch 1995; Hollman and Katan 1998). Currently there is no consensus on the effects of these isoflavonoids in preventing disease processes.

Rats fed soy protein were reported to have lower plasma TBARS compared to rats on a casein diet (Madani et al. 2000). Genistein, one of many substances that have been extracted from soy protein, is an isoflavonoid. Genistein and daidzein are isoflavonoids that have antioxidant characteristics due to their ability to donate hydrogen ions from their hydroxyl groups on their ring structures. Genistein has greater antioxidant capabilities because it has two potential sites for interaction whereas daidzein has only one. Genistein was reported to reduce low-density lipoprotein (LDL) oxidation in vitro (Tikkanen et al. 1998), protect against iron-induced microsomal lipid peroxidation (Jha et al. 1985), and significantly increase a number of antioxidant enzymes in mice (Cai and Wei 1996), as well as inhibit DNA damage induced by the Fenton reaction to activate oxidative stress (Wei et al. 1996). In addition, genistein and soy extracts may work through inhibition of a tyrosine kinase. Since genistein has an influence on all these processes, it may have a beneficial effect independent of its antioxidant properties. However, whether genistein can help prevent muscle damage remains unclear.

Few studies have dealt with soy isoflavonoids and exercise. One animal study addressed the effects of genistein supplementation on exercise-induced oxidative stress in rats (Chen and Bakhiet 1998). Rats (1 year old) were supplemented with 500 mg of genistein per kilogram body weight for four weeks or placed on a normal diet. The rats either rested or ran acutely at 22 m/min up a 12% grade until exhaustion. Acute exercise increased plasma genistein levels compared to resting. Genistein had no effect on the circulating antioxidant

enzymes and did not alter exercise-induced increases in liver and muscle TBARS. It appears that genistein at this dose had no effect on exercise-induced oxidative stress.

In contrast, a soy protein in diets of rats, as compared to casein, helped to attenuate the calpain activation in muscle in response to running to exhaustion (Nikawa et al. 2002). Plasma CK was significantly reduced in the soy-treated group in response to exercise, but TBARS and activation of antioxidant enzymes were similar. The authors suggested the possibility that the reduction of protein degradation by the soy protein was through inhibition of the calpain proteolytic process.

In a recent study, soy isoflavones given to humans did not alter exercise-induced oxidative stress (Chen et al. 2005). Subjects received soy isoflavones in pill form (genistein isoflavone extract 150 mg/day) or placebo for four weeks. There was no difference in exercise-induced increases in MDA or decreases in reduced glutathione in the blood between the supplementation and placebo treatment. Isoflavones (120 mg/day 1.3:1 genistein:diadzein) for 30 days were compared to fish oil and placebo treatments in subjects who performed an eccentric protocol of 50 maximal actions at 90°/s (Lenn et al. 2002). There were no differences in DOMS across treatments. The authors observed no significant treatment or time effects for serum MDA, CK, TNF-α, or IL-6 despite a loss of range of motion and strength in the exercised arm. They concluded that their protocol of supplementation was ineffective in reducing or preventing DOMS.

One should note that there were few subjects in each group in this study and that large variability in some of the biochemical parameters was seen over the 72 h after the eccentric exercise. Others have reported that a soy-based supplementation (44 mg/day genistein) for three weeks could attenuate muscle soreness in humans, and also that CK and myeloperoxidase levels were reduced by soy treatment compared to whey treatment (Rossi et al. 2000).

The results with soy are mixed and do not suggest that isoflavones can consistently inhibit oxidative stress induced by exercise. They may, however, work on other mechanisms that might reduce proteolytic breakdown of muscle, and this possibility should be examined in future research.

Combined Antioxidant Supplementation

Several groups have tested combined supplementation treatments in relation to exercise-induced muscular damage. Combinations have included vitamins E and C, vitamins and other antioxidants, and various antioxidants administered together.

The combination of vitamin E (400 IU) and vitamin C (200 mg) for 4.5 weeks in trained subjects reduced CK loss but did not influence TBARS in the circulation after a marathon (Rokitski et al. 1994b). The combination of vitamin E (400 IU) and C (1000 mg) and beta-carotene (30 mg) given for six weeks reduced resting MDA and expired pentane (a marker of lipid peroxidation) levels. However, no difference in the extent of the exercise-induced increase in these oxidative stress markers was observed with the vitamin supplementation compared to no supplementation (Kanter et al. 1993). An antioxidant mixture of 10 mg beta-carotene, 1000 mg vitamin C, and 800 IU vitamin E given for eight weeks helped maintain the glutathione system and attenuated CK loss into the serum in response to running downhill (Viguie et al. 1989). A combination of 9 IU vitamin E and 90 mg ubiquinone for three weeks had little effect on the CK response to a marathon (Kaikkonen et al. 1998). The supplementation neither attenuated oxidation of lipoproteins nor altered muscular damage induced by the exercise despite a higher plasma level of both vitamin E and ubiquinone. It did reduce the amount of LDL oxidized in vitro by copper, but this suggested that the LDL particle was better protected under this extreme condition. This small dose of the two antioxidants was able to enhance the plasma antioxidant capacity but did not influence muscular damage.

Two different doses of vitamin C + E were reported to have no significant influence on blood or tissue markers of muscle damage compared to placebo in 15 trained runners (Dawson et al. 2002). The subjects took either a placebo or vitamin pills (C + E, two pills per day) for four weeks and then performed the initial 21 km (13-mile) run. After a four-week washout, subjects received the other treatment. This time, however, the vitamin dosage was doubled to 1000 mg vitamin C + 1000 IU vitamin E. Blood samples before and after the exercise were examined for CK, MDA, and myoglobin. Six of the subjects had muscle biopsies taken from the gastrocnemius 24 h after the run. The muscles were examined for damage using staining procedures and electron microscopy. The results for the two doses of vitamin supplementation were pooled, as there were no differences between them. Serum vitamin C and E increased significantly after four weeks of vitamin supplementation. Malondialdehyde increased 28% in the vitamin group and 40% in the placebo group after the exercise, but the increases were not statistically different. Serum CK increased 24 h postexercise in both groups, with no differences between treatments. All subjects had some degree of muscle damage prior to the 21 km run as indicated by liopfuscin granules, paracrystalline inclusions, or subsarcolemmal accu-

mulations. Vitamin supplementation did not appear to affect the indices of muscle damage after the exercise. Two of the six subjects showed some evidence of plasma membrane damage. However, there was no marked increase in frequency or severity of muscular damage after the run regardless of treatment.

In another study, the combination of vitamin E and C supplementation for 37 days prior to 300 eccentric contractions of the knee extensors was reported to attenuate declines in muscle contractile force without influencing the amount of DOMS (Shafat et al. 2004). Subjects were assigned to either placebo ($N = 6$) or vitamin treatment ($N = 6$) and were assessed before and up to seven days after the eccentric protocol. The eccentric protocol reduced peak concentric torque 31.6% in the placebo but only 17% in the vitamin group. The vitamin pretreatment attenuated the decline in work in response to the eccentric protocol and also the extent of decline in isometric force. Thigh volumes increased similarly in the two groups after the exercise, as did DOMS. The authors concluded that the vitamin pretreatment reduced muscle function loss after the eccentric exercise.

A study in untrained females showed that combined vitamin E and C plus selenium, given for 14 days prior and for two days after 48 eccentric contractions, attenuated the increase in CK and DOMS (Bloomer et al. 2004). However, there was no protection from the force deficit as indicated by decline in maximal isometric force, and there was little influence on range of motion. It appears that continuing this antioxidant treatment for two days after the exercise could reduce membrane leakage and muscle soreness but had little influence on muscle function. Measurement of the oxidative stress response was another aspect of this study (Goldfarb et al. 2005a). The amount of protein carbonyls in the plasma was significantly attenuated by pretreatment with antioxidants compared to placebo. In addition, MDA at 48 h postexercise was significantly lower; however, the two groups showed similar transient increases in glutathione status immediately and 2 h after the exercise. The authors suggested that the pretreatment offered only partial protection and that further study within the muscle is necessary to determine if indeed this combined antioxidant treatment would be effective in humans.

Rats running downhill that had received a combined vitamin E and C diet had lower amounts of protein carbonyls in their muscles compared to rats given a standard diet (You et al. 2005). Markers of oxidative stress in the blood were less than in the muscle but showed similar changes in the two groups in response to the downhill run. There were no significant changes in MDA and glutathione in either group over time. The authors suggested that the antioxidant diet pretreatment only partially attenuated protein breakdown within the muscle as evidenced by the protein carbonyl data.

A treatment combining vitamin C (12.5 mg/kg body weight) and N-acetyl-cysteine (NAC) (10 mg/kg body weight)—the latter of which can help increase tissue glutathione levels—was given immediately after eccentric exercise (Childs et al. 2001). After 30 eccentric arm curls at 80% of the subjects' 1-repetition maximum, the supplements were given based on body weight. The antioxidant treatment was compared to placebo, and the results suggested that giving vitamin C and NAC after eccentric exercise exacerbated the muscle damage and inflammatory response. There was an increase in most of the markers in the antioxidant group despite an increase in serum total antioxidant status. The authors noted that the supplementation given in this dose after the eccentric exercise probably added to tissue damage and oxidative stress.

Combined mixed tocopherols, flavonoids, and docosahexaenoate (omega-3 fatty acid) were given seven days prior to and for seven days after an eccentric arm curl protocol and compared to placebo (Phillips et al. 2003). Subjects performed three sets of 10 eccentric actions at 80% of their 1-repetition maximum with their nondominant arm. Serum lactate dehydrogenase and CK as well as pain increased, and range of motion declined to a similar extent in the treatment and placebo groups after the exercise. In contrast, IL-6 and C-reactive protein responses after the eccentric exercise were significantly reduced by the antioxidant supplementation compared to the placebo. It appears that this combination of supplements was able to inhibit some of the inflammatory processes but did not seem to influence the leakage of muscle proteins or damage associated with the eccentric protocol. The authors noted that it was unclear whether or not the repair processes were influenced; this is a question that needs to be pursued further.

The effect of 400 IU α-tocopherol acetate, 3 mg beta-carotene, and 20 mg lutein given orally for one month was examined in Alaskan sled dogs (Baskin et al. 2000). Measurements were taken after the first and third days of exercise and then after three days of rest. DNA adducts were decreased with supplementation. Additionally, supplementation increased the lag time of lipoprotein oxidation in vitro, suggesting that the supplement attenuated exercise-induced oxidative damage.

It is worth noting that mice given intraperitoneal injections of superoxide (SOD) dismutase 24 h prior to eccentric contractions showed less muscle damage and retained greater force from isolated muscle contractions compared to intraperitoneal injections of a placebo (Zerba et al. 1990). Therefore, it appears that radical scavengers (antioxidants) given as pretreatment must be able to enter the cell to be effective and must be in close proximity to the location of the injury. This might help to explain why combined antioxidant treatments may be potentially more effective in reducing the amount of damage and perhaps aid in the return of normal function to damaged muscle than using a single antioxidant treatment.

Summary

The use of antioxidants to prevent exercise-induced muscle damage and oxidative stress has shown mixed success. Antioxidants may potentially reduce certain types of exercise-induced muscle damage and oxidative stress, but supplementation probably cannot protect against other types of damage (the initial muscle force damage). Infiltration of macrophages and immune and ROS changes with remodeling of the muscle are probably the processes affected by antioxidants. The exact amounts needed for protection against oxidative damage in vivo are still in question because studies have not adequately controlled for the extent of damage. The timing of the treatment is also important. If the inflammatory process has already begun or if metals (Fenton reaction) have already been released from tissue, the antioxidants could act as prooxidants and instead of being protective could actually exacerbate the situation.

The combination of different types of antioxidants in preventing muscle damage and oxidative stress in response to exercise appears to provide some benefits, but the research is still limited. Comparisons are difficult because of the various doses and agents used in the research that has been done. More research using proper controls is needed to elucidate the effects of phytochemicals as antioxidants in vivo. This is a promising area, but few studies have compared phytonutrients to placebo and to the antioxidant vitamins.

Future research should not only address the release of markers of muscle damage but also measure damage within the muscle, since these appear to be two separate processes. Vitamin E can stabilize the membrane and prevent CK leakage but may not prevent muscle damage. Additionally, studies should use not just one but several markers of oxidative stress; often muscle protein damage is not the same as lipid peroxidation or oxidative stress. Finally, if the extent of damage is too great, pretreatment with antioxidants and neutraceuticals will not be effective.

Table 19.4 outlines research on supplementation with two or more antioxidants and its effects on exercise-induced muscle damage.

Table 19.4   Combination Antioxidant Treatment Effects on Exercise–Induced Muscle Damage

Supplementation	Model	Exercise	Effect	Reference
Humans				
800 IU vitamin E 10 mg β-carotene 1000 mg vitamin C	23 males, trained	65% $\dot{V}O_2$max, −5% grade	↓ CK increase ↓ LDH ↑ GSH	Viguie et al. 1989
592 mg vitamin E 1000 mg vitamin C 30 mg β-carotene 6 weeks	20 males	30 min 60% $\dot{V}O_2$max + 5 min 90% $\dot{V}O_2$max	= Rest MDA ≈ Rest pentane = MDA ↑ = Pentane ↑ with exercise	Kanter et al. 1993
400 IU vitamin E 200 mg vitamin C 4.5 weeks	24 males, trained	Marathon	↓ CK increase = TBARS	Rokitzki et al., 1994b
13.5 mg vitamin E 90 mg Q10 3 weeks	37 males, trained	Marathon	= CK ↑ = LDL oxidation = LDL Copper-induced oxidation\	Kaikkonen et al. 1998
10 mg/kg NAC 12.5 mg/kg vitamin C 7 days after exercise	14 males, untrained	3 sets of 10 eccentric arm curls	↑ LDH increase ↑ CK increase ↑ 8-Iso-PGF$_{2\alpha}$ ↑ ↑ Lipid HOOH ↑ ↓ MPase increase = IL-6 increase	Childs et al. 2001
500 or IU vitamin E 500 mg vitamin C 4 weeks Or twice dose	15 males, endurance trained	21 km run, flat	= CK increase = MDA increase = Muscle damage ($n = 6$)	Dawson et al. 2002
300 mg tocopherols 300 mg flavonoids 800 mg docosahexaenoate (hesperetin + quercetin) 14 days	40 males, untrained	3 sets of 10 eccentric arm curls	= CK increase = DOMS ↑ ↓ CRP increase ↓ IL-6 increase	Phillips et al. 2003
400 IU vitamin E 1000 mg vitamin C 90 μg selenium 14 days before + 2 days after exercise	18 females, untrained	4 sets of 12 eccentric arm curls	↓ CK increase ↓ DOMS ↑ = ROM decrease = MIF decrease	Bloomer et al., 2004
1200 IU vitamin E 500 mg vitamin C 37 days	12 males, moderately active	30 sets of 10 eccentric knee extensions	↓ Peak torque ↓ ↓ Loss of work = DOMS ↑ ↓ MIF decrease	Shafat et al. 2004
400 IU vitamin E 1000 mg vitamin C 90 μg selenium 14 days before + 2 days after exercise	18 females, untrained	4 sets of 12 eccentric arm curls	↓ PC increase ↓ MDA at 48 h = GSSG ↑	Goldfarb et al. 2005a

(continued)

Table 19.4 *(continued)*

Supplementation	Model	Exercise	Effect	Reference
Animals				
400 IU vitamin E 3 mg β-carotene 20 mg lutein 1 month	62 Alaskan dogs	Sled exercise	↓ LDL copper-induced oxidation	Baskin et al. 2000
2000 mg vitamin C + 1000 IU vitamin E/kg diet	56 male rats	90 min treadmill (−16°)	↓ PC increase = MDA = Glutathione status in soleus and vastus muscles	You et al. 2005

Equals sign means there was no significant difference; ↓ means significantly decreased response; ↑ means significantly increased response in supplemented compared to placebo or no treatment; ≈ means about the same. ROM = range of motion; CK = creatine kinase; LDH = lactate dehydrogenase; GSH = glutathione reduced form; MDA = malondialdehyde; DOMS = delayed-onset muscle soreness; TBARS = thiobarbituric acid reactive substances; lipid HOOH = lipid hydroperoxides; MPase = myloperoxidase; PC = protein carbonyls; LDL = low-density lipoprotein; Q10 = ubiquinone; MIF = maximum isometric force; CRP = C-reactive protein; 8-Iso-PGF$_{2\alpha}$ = 8-isoprostane; IL-6 = interleukin-6; NAC = N-acetyl-cysteine

CHAPTER 20

Therapies for Myofascial Trigger Points

Leesa K. Huguenin, MBBS, MS

Myofascial trigger points, as initially described by Travell and Simons (Simons et al., 1998), are a widespread and well-recognized phenomenon in clinical practice. They have most frequently been described as a component of generalized pain syndromes (Harden et al., 2000) but also occur in regional pain syndromes, in conjunction with acute pathologies such as lumbar disc protrusion, and in the athletic setting.

Unfortunately, despite widespread clinical recognition and multiple attempts to further define this muscle pain syndrome, many controversies surrounding trigger points remain. Much of the early literature is based on theories generated from anecdotal reports and clinical experience. While these theories have been challenged, they have yet to be conclusively disproved and remain the cornerstone of our understanding of trigger points and the basis on which most research is performed. Despite the more recent push for a better scientific understanding, it is difficult to design experiments to truly test the theories. We are yet to conclusively "see" a trigger point, and clinical examination is intrinsically difficult due to poor interrater reliability. Consequently, we are left with loose ends when trying to prove theories of pathogenesis. Further, as trigger points are such a widespread phenomenon, it will be quite some time until the data from the many treatment studies on specific populations can be analyzed as a whole so that the outcomes can be applied to trigger points in general.

Although a number of treatments for trigger points are known to be effective, knowledge on the detail of their mechanisms is lacking, and once again we are left with theory extrapolated from multiple studies.

The aim of this chapter is to present the current theories and research to allow readers to appreciate the evidence base and therefore make informed choices in their approach to trigger points, whether these are an isolated phenomenon or a component of a specific pathology.

Defining and Explaining Trigger Points

According to the most common description of trigger points, based on the traditional teachings of Travell and Simons (Simons et al., 1998, pg. 35), trigger points are the "presence of exquisite tenderness at a nodule in a palpable taut band (of muscle)." Trigger points are able to produce referred pain, either spontaneously or on digital compression. Clinically, this translates to localized areas of deep tenderness within a taut band of muscle. Trigger points exhibit a local twitch response (muscle fasciculation) or jump sign (whole-body movement) in response to digital pressure or dry needling.

Simons and colleagues (1998) also produced diagrams of common trigger point locations and their referral zones. It has been noted that these locations have significant similarities to acupuncture points used in traditional Chinese medicine for the relief of pain (Melzack et al., 1977).

Two main types of trigger points have been described (Simons et al., 1998). Active trigger points are symptomatic and are responsible for the patient's presenting with a pain complaint. They may also be associated with less readily definable symptoms such as weakness, paresthesia, or temperature changes and may have associated referred pain. Latent trigger points present with muscle shortening, and, while painful on palpation, are not spontaneously symptomatic. These trigger points may become activated in response to postural issues, overuse, strength imbalance, local injury, or various other stimuli.

Clinical Recognition

Clinical history taking may lead to a suspicion of trigger points as a cocontributor to the pain complaint, but this is usually based on recognition of a syndrome or clinical pattern seen previously rather than a classical clinical presentation that can be applied to all patients with trigger points. Common clinical examples include a generalized pain syndrome with widespread trigger point formation, hyperalgesia, and psychological disturbance such as that seen in fibromyalgia (Wolfe et al., 1992; Harden et al., 2000).

Alternatively, athletes with osteitis pubis often have numerous trigger points through the gluteals and adductors that may be contributing to at least a percentage of their pain complaint. These trigger points may also be partially responsible for maintaining the imbalance of muscle tightness and activity around the pelvis. In patients with hamstring pain of no identifiable cause on magnetic resonance imaging, gluteal trigger points have been shown to reproduce recognizable pain in a large proportion of cases (Huguenin et al., 2005). Likewise, clinically, trigger points often exist in the muscle belly when a tendinopathy is diagnosed at the Achilles or common extensor origin at the elbow and may be contributing to the presenting pain complaint.

On palpatory examination, a trigger point is recognized as a locally tender spot within a taut band of muscle that produces recognizable pain if it is active. Latent points, while painful on palpation, do not produce recognizable symptoms. Snapping palpation (transverse flicking of a muscle band, similar to plucking a guitar string) produces what is described as a local twitch response in the affected band of muscle, and this is also seen with needle entry into a trigger point (Simons et al., 1998).

Unfortunately, most examination findings have poor interrater reliability. The most common examination finding in widespread pain syndromes is local tenderness in taut muscle bands (Wolfe et al., 1992). However, clinical location of these taut bands, visualization of local twitch responses, and identification of active trigger points show poor reliability (Hsieh et al., 2000; Njoo and Van der Does, 1994). Reliability is even poorer when asymptomatic patients are used and the clinician is trying to identify latent trigger points (Lew et al., 1997). Untrained examiners perform significantly worse (Hsieh et al., 2000), but these results can be improved somewhat through extensive group training of observers (Gerwin et al., 1997).

Therefore, local tenderness and the production of recognizable pain are the most reliable clinical findings, whereas the presence of local twitch responses, production of referred pain, assessment of the number of trigger points, and the presence of taut bands are far less reliable.

Theories of Pathogenesis

The most widespread belief regarding the etiology of trigger points is based on the combination of two theories and provides a plausible explanation. There is a third, again plausible, theory that still lacks experimental verification. The first two theories (Simons et al., 1998) are the energy crisis theory and the motor end plate hypothesis; these assume that the site of primary abnormality is the muscle cell, or more specifically the motor end plate. The third theory (Gunn, 1997) suggests that the spinal nerve is the primary site of pathology and that the muscle changes seen clinically are secondary to this.

Much of the emphasis to date has been on the earlier theories; however, as will be seen later in the chapter, the effect of many treatments of trigger points does not seem to depend on direct stimulation of muscle but does require application of a noxious stimulus. It is possible that all three theories contain part of the truth, but how much and which part remain to be verified experimentally.

The Energy Crisis Theory

The energy crisis theory is the earliest explanation of trigger point formation and is extensively discussed in *The Trigger Point Manual* (Simons et al., 1998). The postulate is that under increased demand (recruitment) of a muscle, recurrent micro- or macrotrauma leads to increased calcium release from the sarcoplasmic reticulum and prolonged sarcomere shortening that compromises the local circulation, reducing oxygen supply. In the locally ischemic state, the cell is unable to produce enough adenosine triphosphate (ATP) to initiate the active reuptake of calcium and therefore end the contraction. The accumulation of ischemic by-products (Simons, 1996) is said to sensitize sensory nerves.

Unfortunately, the muscle injury just described as the initiating factor has not been seen experimentally to date. However, the concept of altered metabolism at trigger point sites has been further investigated (Bengtsson et al., 1986). In a biopsy study of fibromyalgia patients, researchers found that high-energy phosphate levels were reduced and low-energy phosphate levels increased at trigger point sites compared to nontender muscle points in both fibromyalgia patients and controls. High-energy phosphates include ATP and phosphocreatine, both of which can donate phosphate for energy conversion. This seems to support the idea of a metabolic derangement at trigger point sites; but the researchers did not find increased levels of anaerobic metabolites at the sites, which would suggest that the derangement is not purely ischemic (see figure 20.1).

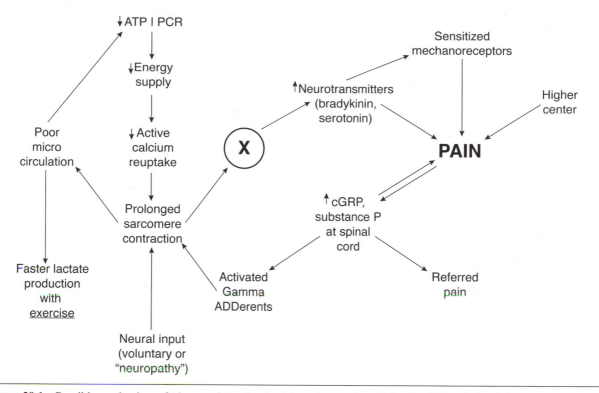

Figure 20.1   Possible mechanism of trigger point pathophysiology, integrating all theories. X = missing link.

Motor End Plate Hypothesis

Although separate theories, the energy crisis theory could well coexist with the motor end plate hypothesis in explaining more about the possible aetiology of trigger points. The motor nerve synapses with a muscle cell at the motor end plate. Needle electromyography (EMG) studies have shown that each trigger point contains minute loci that produce characteristic electrical activity called spontaneous electrical activity (SEA) (Hubbard and Berkhoff, 1993). Spontaneous electrical activity is seen as low-amplitude background noise (50 μV), with super-imposed high-amplitude spike activity (100-700 \μ\V), in a resting muscle (McNulty et al., 1994; Hong et al., 1995; Simons et al., 1995). Initially, it was postulated that this pattern of activity represents stimulation of intrafusal muscle spindle fibers (Hubbard and Berkhoff, 1993), but others have since claimed that it represents motor end plate noise (Simons, 2001; Simons et al., 2002).

The motor end plate theory states that the background noise (motor end plate noise) represents excessive release of packets of acetylcholine (ACh) by the motor nerve terminal next to the muscle cell. Acetylcholine is a neurotransmitter that causes mini-depolarizations of the postsynaptic muscle cell membrane (Simons, 2001). The origin of the spikes is not clear. One suggestion is that they may represent propagated single muscle fiber action potentials occurring as a result of summation of background noise with extra ACh released in response to the needle contact (Simons, 1996). This small amount of propagation may be enough to cause activation of a few contractile elements and may be responsible for some degree of muscle shortening.

Controversy in this area remains, however, as the EMG changes have also been identified at nontrigger point sites in humans (Simons et al., 2002) and have not been seen with surface EMG, which raises the question whether the needle penetration itself is responsible for some of the recordings obtained. That having been said, there is no evidence to date that the needle is responsible for *all* of the signal; and in animal studies of dissected muscle, the signal *is* confined to trigger point sites (Simons et al., 1995). Of further interest, the onset of SEA with needle advancement has been correlated with reported onset of pain (Hubbard and Berkhoff, 1993), and SEA in rabbit muscle has been shown to decrease after dry needling therapy (Chen et al., 2001).

While not providing a reason for the increased motor nerve activity, this theory sits nicely with the energy crisis theory, which relies on increased muscle activity as the instigating factor.

Radiculopathic Model

Proponents of a primary neurological cause for trigger point formation do not agree with the energy crisis or the motor end plate hypothesis (Gunn, 1997; Quintner and Cohen, 1994). Gunn (1997) states that "myofascial pain describes neuropathic pain that presents primarily in the musculoskeletal system" (p. 121). Gunn's proposed radiculopathic model for muscle pain is based on the fundamental premise that denervated structures exhibit supersensitivity. Therefore, partial neural injury should result in a relative increase in sensitivity of the structures supplied by a damaged nerve.

On the basis of clinical observations, Gunn (1997) states that neuropathic nerves are most commonly found at the nerve root rami at the spinal cord, where they are compressed or angulated due to subclinical disc disease and resultant paraspinal muscle spasm. This "neuropathy" then sensitizes structures in the distribution of the nerve root, causes distal muscle spasm, and contributes to other degenerative changes in tendons and ligaments within its distribution that are then perpetuated by the ongoing muscle shortening. If neural injury is the origin of the pathology, it is said to explain the lack of pathology seen in muscle and the sensory, motor, and autonomic changes seen in myofascial pain syndromes.

On the basis of his theories, Gunn (1997) proposes that long-lasting pain relief requires needle treatment to the shortened paraspinal muscles in order to reduce nerve root compression, as well as to trigger points more local to the site of perceived pain.

Quintner and Cohen (1994) argued that the reasoning behind traditional trigger point teaching is circular and fails to allow for the possibility of a nonmuscular cause for the clinical picture observed. They suggest that descriptions of trigger point pain mirror those of neural pain and that phenomena like referred pain and altered local sensations are more easily explained by a neurological cause. To date, no neurophysiological studies have confirmed or refuted these claims. Routine nerve conduction testing has not identified any abnormalities but may lack the sensitivity to do so.

Clinical Precipitants of Formation

Trigger points are thought to form in response to increased or altered muscle demands, which may include prolonged muscle contraction as in workplace postural errors, proximal nerve compression and resultant muscle spasm, and trauma (Simons et al., 1998). Latent trigger points are thought to become activated in response to the same conditions that cause trigger point formation, that is, muscle overload, prolonged muscle contraction, or nerve compression.

Trigger points can also be influenced by central factors such as stress or constitutional illnesses. The sympathetic "flight or fight" response to stress is related to increases in the amount of circulating catecholamines. It has been shown that the EMG activity in trigger points can be reduced by the use of sympathetic antagonists (Chen et

al., 1998) and that it increases at times of stress (McNulty et al., 1994).

Trigger points should be considered in many different clinical pain scenarios. For example, sedentary workers presenting with head or neck pain are likely to have multiple contributors to their pain condition, including thoracic spine tightness, cervical or thoracic disc degeneration, and ongoing postural errors. Resultant trigger points in their trapezius and scalenes may be responsible for part of their presenting complaint. Likewise, a lumbar disc injury is often responsible for significant pain, but elements of this pain may be reproduced with pressure on trigger points in the paraspinal, quadratus lumborum, or gluteal muscles. Osteitis pubis may be accompanied by trigger points in the adductor muscles or the gluteals. Importantly, trigger points can form in either of these regions purely in response to an increase in load, without any specific underlying injury.

The clinical syndromes seen may reflect the referral zones of the active trigger points rather than the location of the points themselves. Examples include posterior thigh pain reproduced from gluteal trigger points or Achilles tendon pain reproduced from calf trigger points. Sole causality should not be assigned to trigger points without a thorough assessment to eliminate any underlying pathological, strength, or postural factors. Likewise, when the pain complaint is out of proportion to the evident local pathology, or when treatment of a local area fails to completely relieve symptoms, trigger points may be considered as contributors and assessed and treated. It is important not to assign significance to trigger points if they do not produce reproduction of a *recognizable* pain. Nonspecific pain does not identify a contributing trigger point according to the classical definition (Simons, 1997). Recognizable pain is one of the essential criteria for diagnosis of an active trigger point.

Investigating Trigger Points

The fact that the exact pathogenesis of trigger points is yet to be proven has made targeted investigation for diagnostic and monitoring purposes difficult, as many traditional modalities lack the sensitivity to detect them. Studies of muscle metabolism are perhaps the most likely to detect true differences in trigger point regions. Unfortunately, the use of magnetic resonance imaging is limited by cost and physical availability.

Pathogenesis

Currently, there is no gold standard pathological test for the identification of trigger points, and much of the research on the pathophysiology of trigger points is aimed at indirectly verifying the common theories for their formation (table 20.1).

Table 20.1   Theories of Trigger Point Pathophysiology

	Instigating factor	Site of primary pathology
Energy crisis theory	Muscle damage	Muscle cell
Motor end plate hypothesis	Overactive motor nerves	Motor end plate
"Radiculopathic" model	Spinal disc disease	Spinal nerve root rami

To date, there is no evidence of inflammation or increased levels of nociceptive neurotransmitters at trigger point sites; this is the major stumbling block for theories assuming that muscle injury is the instigating factor in trigger point formation. The theories implicating a primary neurogenic cause are stronger in this respect, but lack any confirmatory data.

Investigations of the histopathology so far have been inconclusive (Yunus et al., 1986; Yunus and Kalyan-Raman, 1986; Russell, 1997; Bartels and Dannieskold-Samsoe, 1986). Researchers have reported either negative findings or findings of nonspecific fibrosis in the region of trigger points. Although one study noted the presence of rubber band–like transverse adhesions, this finding was obtained at a random site in the quadriceps muscle of chronic pain patients, has not been replicated, and is unlikely to represent local changes at trigger points (Bartels and Dannieskold-Samsoe, 1986). It is noteworthy that no evidence of inflammation or increased numbers of inflammatory cells has been seen in any investigation.

Based on some work that suggested reduced oxygen pressures in muscles at trigger point sites (Lund et al., 1986), one biopsy study indicated that there may be altered levels of high-energy phosphates (ATP and phosphocreatine) in painful muscles of patients with fibromyalgia (Bengtsson et al., 1986); this led to some promising pilot trials on the use of magnetic resonance spectroscopy (31PNMR) to investigate the presence and possible etiology of trigger points. Muscle uses phosphate moieties as a source of energy, and the donor compounds therefore contain less energy as each phosphate is released. 31PNMR detects the amount of each compound present in muscle and can therefore give an idea of its current metabolic status.

A reduction in levels of high-energy phosphate compounds is thought to represent less available energy for the muscle cells and would implicate an ischemic mechanism in trigger point causality. However, biopsy assessment of the affected regions yields no evidence of higher levels of lactate or pyruvate, both of which are metabolic products observed under ischemic conditions (Bengtsson et al., 1986).

Kushmerick (1989) concluded 31PNMR is potentially valuable if normal resting values can be determined accu-rately and muscles can be studied under various states of contraction for documentation of normal responses. An abstract of a pilot study has pointed to some potential changes in patients with myofascial pain compared to controls (Mathur et al., 1988). Lowered phosphocreatine: ATP ratios were found in five of six cases when compared to 22 controls.

One recent spectroscopy exercise study indicated an earlier onset of fatigue, less efficient use of ATP, and a faster local drop in pH (indirect indicator of lactate formation) in patients with fibromyalgia compared to controls (Lund et al., 2003). Again, the researchers postulate a poor microcirculation in the region of trigger points as important in this pain complaint.

Other attempts to further clarify trigger point pathology through imaging have not provided additional information. Conventional ultrasound scanning does not visualize trigger points (Lewis and Tehan, 1999). Early case reports and a retrospective analysis indicated that thermography could identify trigger point sites (Diakow, 1988, 1992). However, other research showed that hot spots were common in fibromyalgia patients and controls and that their locations did not correspond to the clinical locations of trigger points; in addition they were not more tender than control sites (Swerdlow and Dieter, 1992). Therefore, hot spots are likely to represent more superficial alterations in blood flow, such as small arteriovenous malformations, rather than muscle pathology.

One study has suggested that there may be thermographic changes in the referral zones of trigger points (Kruse and Christiansen, 1992), but no convincing confirmatory data have been obtained to date.

Pain Production and Referral

Muscle pain and distinct referral patterns in response to the injection of noxious stimuli were described as early as 1938 (Kellgren, 1938a, 1938b). Both muscle and cutaneous pain are transmitted by group III (A delta, thin myelinated) and group IV (c, nonmyelinated) afferent nerves (Franz and Mense, 1975; Simone et al., 1994). Bradykinin (Franz and Mense, 1975) and prostaglandins (Hedenberg-Magnusson et al., 2001) are well accepted as neurotransmitters involved in this pain response.

Serotonin has also been implicated; but while it increases muscle pain in healthy volunteers, results are not the same in myofascial pain subjects (Ernberg et al., 2000), and use of a serotonin antagonist has not been shown to influence muscle pain in these subjects (Ernberg et al., 2003). Substance P and calcitonin gene-related peptide (CGRP) are likely to be important in altering pain transmission at the spinal cord but are not directly algesic compounds in muscle (Graven-Nielsen and Mense, 2001).

In comparison to intradermal injection, injection of capsaicin into muscle produces low-intensity muscle pain that is highly likely to refer to cutaneous or deeper structures (Witting et al., 2000). The mechanism of this high rate of referral remains to be further investigated but is likely to relate to spinal cord mechanisms.

Of further interest is that mechanoreceptors in muscle have a lowered stimulation threshold in the presence of bradykinin (Graven-Nielsen and Mense, 2001). Thus, although the compounds involved initially have some algesic properties, it is possible that their primary role is to reset neural thresholds so that stimuli previously seen as innocuous (stretch, etc.) are now perceived as painful.

Traditionally, the convergence projection theory is used to explain pain referral from one structure to another (Mense, 1993; Gerwin, 1994). This theory states that when the innervation of two structures converges on the same dorsal horn neuron, the dorsal horn neuron expects the source of the pain to be the structure that most commonly reports painful stimuli. Input from the other structure is misinterpreted by the dorsal horn neuron as coming from the recognized zone.

A modification of this theory is used to explain the widely accepted idea that muscle pain referral must have a central basis. Mense (1996) and Hoheisal and colleagues (1993) have applied stimuli to known receptive fields of specific dorsal horn neurons in rats and found that new receptive fields open in response. Thus, the theory is that not all convergent projections are open all the time, but that dormant spinal cord connections can be unmasked in response to a painful stimulus. Hong (1996) has shown that noxious input can affect multiple levels of the spinal cord by increasing local levels of substance P and CGRP and thereby increasing their sensitivity to noxious input. It is not known whether these two transmitters are locally produced, or released from higher centers, or both (Vaeroy et al., 1989). This neural plasticity may explain how a previously innocuous stimulus such as pressure or tightness becomes painful, as well as how referred pain can be felt at seemingly unrelated body sites, and may be important in the transition from acute to chronic pain (Bendtsen et al., 1996).

Muscle spasm, as seen in many conditions of muscle pain, may relate to a connection between dorsal horn neurons in the spinal cord and gamma afferent neurons. Gamma afferent neurons supply muscle spindles and are responsible for reflex muscle shortening, such as that seen with tendon reflexes. Inhibition of dorsal horn neurons indirectly inhibits discharge from gamma afferents (Xian-Min et al., 1992).

Central Modification of Pain

The central awareness of pain is modulated by various factors. Melzack (1982) further defined his original gate control theory of pain, according to which synaptic connections in the dorsal columns modulate the progression of a stimulus along a pathway *before* it reaches awareness. These gates can be affected by other peripheral stimuli and from higher centers in the brain. Therefore emotional and other stimuli can have an effect on the degree of pain perception by an individual.

Endogenous opioids such as enkephalin and dynorphin groups and β-endorphin are released segmentally in the spinal cord in response to a peripheral noxious stimulus; they depress nociceptive neurons in the dorsal horn (Henry, 1989) and brain and are thought to be responsible for the phenomenon of diffuse noxious inhibitory control (DNIC). In the presence of pain, application of a new noxious stimulus increases central opioid release, the pain gates are closed, and awareness of the preexisting stimulus is reduced (Melzack, 1982).

Higher centers, such as the periaqueductal gray matter, the thalamus, the hypothalamus, and the amygdala, can reduce chronic pain when stimulated. Stimulation of the raphe nuclei suppresses nociceptive nuclei in the dorsal horn. Many of these centers have been shown to have increased activity on functional brain imaging in response to acupuncture treatment (Bialla et al., 2001; Wu et al., 1999). There is also a descending noradrenergic pathway whose precise role is yet to be determined (Henry, 1989).

Proposed Therapies

It is easiest to divide trigger point therapy into invasive and noninvasive options. Manual therapists have traditionally employed noninvasive techniques, but recently there has been a marked increase in the use of invasive therapies, dry needling in particular, to manage trigger points. Anecdotally, all therapies have their supporters. Scientifically, however, few stand up to scrutiny. Even when there is clear evidence of a pain-relieving effect, a clear mechanism for the effect has not been identified. Successful therapies tend to be those that involve application of a second noxious stimulus. Table 20.2 lists proposed therapies for trigger points, their side effects, and recommendations.

Table 20.2   Proposed Therapies for Trigger Points

Treatment	Evidence	Side effects	Recommended for use?	Caveats
Massage/soft tissue therapies	Benefit above stretching alone	No—some posttreatment soreness	Yes	May cause irritable generalized pain to flare up
Stretch	More benefit with counterstimulatory techniques as well	None	Yes—routinely after any trigger point therapy	None
TENS	Pain relief likely	None	Yes	No evidence on duration of benefit
Ultrasound	No effect	None	No	–
Laser	Conflicting results	No	No	May work in some populations
Local anesthetic	Effective, no posttreatment soreness	Muscle necrosis with wrong dose or vasoconstrictors	Yes	Medical practitioners only; loss of twitch responses once local acting
Botox A	Probably effective to some degree	None	No—cost prohibitive	–
Dry needling	Effective, but depth of penetration not important	Posttreatment soreness	Yes	Special training required

Noninvasive Therapies

A large number of potential therapies for trigger points have been suggested. Many rely on the application of a second noxious stimulus, and these have been found to have the best effect.

Stretching

Travell and Simons (Simons, 1997) described application of a vapocoolant spray followed by stretching as the "single most effective treatment" for trigger point pain. A study that was aimed at clarifying this in chronic neck pain patients used pain scales and pressure threshold as outcome measures for response to spray and stretch techniques (Jaeger and Reeves, 1986). This study suffered from the use of contralateral sides as controls in symptomatic patients and was not blinded, but it did show a significant reduction in both reported pain and pain pressure threshold after the intervention. It was postulated that the coolant spray acted as a counterirritant, but from the results provided it is impossible to say whether stretch or spray alone would have produced similar effects.

Investigators in another study tried to extract the counterirritant effect in a home stretch program in which stretching was either preceded or not preceded by ischemic pressure to trigger points (Hanten and Chandler, 1994). The counterirritant of ischemic pressure combined with stretching resulted in a greater improvement in pain scores and pain pressure threshold. Importantly, this confirms a benefit of massage techniques over and above the effect of stretching alone.

Transcutaneous Electrical Nerve Stimulation

Transcutaneous electrical nerve stimulation (TENS), often used in chronic pain, is a stimulus that can act at pain "gates" to reduce transmission of preexisting painful stimuli. Two studies have addressed its use in trigger point pain. Graff-Redford and colleagues (1989) found that high-frequency TENS afforded significant pain relief whereas low-frequency TENS and placebo did not. The pain pressure threshold did not change in any group. Hsueh and colleagues (1997) also found a reduction in pain with the use of 60 Hz TENS compared to placebo, but interestingly found a significantly greater improvement in pain threshold in the active TENS group as well. A third arm of their trial consisted of the use of electrical muscle stimulation, or interferential. This group did not have improvement in pain, but did exhibit a marked improvement in range of cervical lateral flexion. The authors of both of these studies concluded that a central mechanism of pain relief was responsible. There are no data on the expected duration of improvement from these therapies..

Ultrasound

Therapeutic ultrasound was used in a three-armed trial (Gam et al., 1998) in patients with head and neck myofascial pain. Subjects were randomized into three groups: ultrasound with home exercise and massage, sham ultrasound with home exercise and massage, or nonintervention control. Both intervention groups showed significant improvements in the number and sensitivity of trigger points, but the ultrasound conferred no additional advantage. Despite the use of massage and stretching in this trial, there was no reduction in pain scores or analgesic use in the subjects of any group. This contradicted the findings of the studies on stretching discussed earlier, but no explanation was offered for this contradiction.

Laser

Laser is one of the most controversial areas in the treatment of trigger points, and there is no consistent body of evidence on which to base its use in myofascial pain. Despite the use of double-blind trials, results are conflicting. Using infrared laser, Olavi and colleagues (1989) found an immediate improvement in pain threshold, whereas Ceccherelli and colleagues (1989) found improvement only after 12 treatments (24 days) and three months. Snyder-Mackler and colleagues (1986) used helium-neon laser with routine physiotherapy and observed increased skin resistance over trigger points. They postulated that this observation accompanied the resolution of the pathology but provided no data corroborating the association. Ceylan and colleagues (2004) found an improvement in pain scores and an increased urinary excretion of serotonin metabolites with infrared laser. In contrast, Waylonis and colleagues (1988) reported no benefit of five treatments of helium-neon laser over placebo in a double-blinded crossover trial. Thorsen and colleagues (1992) used low-level laser and found that subjective reports of improvement were higher in the placebo group. Altun and colleagues (2005) reported no difference between placebo and treatment groups using low-level laser therapy for cervical myofascial pain.

Invasive Therapies

Injection of local anesthetic or irritant substances into muscle for pain relief has been described in case reports and letters since the early 20th century (Button, 1940; Howard, 1941; Ray, 1941; Souttar, 1923; Steinbrocker,

1944). Most of the substances were irritants to muscle and are not considered for use in modern practice. Local anesthetics, however, are still widely used. Interestingly, very early in the discussion, Steinbrocker (1944) suggested that the mere insertion of a needle somewhere in the region of the pain, without the introduction of analgesic solutions, had been frequently reported to give lasting relief. This is the basis on which investigation of dry needling began, and although he didn't back this statement up with specific trials, to date none of the scientific evidence has convincingly refuted Steinbrocker's statement.

Local Anesthetic

Local anesthetic is certainly the substance most often investigated for injection into trigger points. Many different agents have been used, and most give results equivalent to those with injection of normal saline (Frost et al., 1980; Garvey et al., 1989; Hameroff et al., 1981). One consistent finding is that the pain relief, when seen, well outlasts the expected half-life of the injected solution, suggesting mechanisms of pain relief above the pure pharmacological mechanism of the anesthetic. It is important to note that in these studies it is extremely difficult to have a true placebo, particularly if Steinbrocker's early comments are taken into account. Any needle penetration in the area may result in pain relief, therefore placebo design is very problematic.

Overall, local anesthetic injection is most likely to improve subjective outcome measures (e.g., pain scores) (Garvey et al., 1989; Frost, 1986; Hameroff et al., 1981; McMillan et al., 1997), although improvements in range of motion and pressure threshold have been reported (Hong, 1994; Kamanli et al., 2004). This may well be explained by central modulation of pain as the dominant factor in relief, which may also account for the long duration of pain relief that is seen.

Local anesthetic use can result in reversible myotoxicity, seen as ischemia and necrosis of muscle fibers in the injected region (Benoit, 1978; Foster and Carlson, 1980). These changes are much worse in the presence of vasoconstrictors (adrenaline); but then again, perhaps this is simply the optimum form of counterirritation, whereby a true inflammatory response produces mediators sufficient to reset the spinal transmission of pain and hence result in prolonged pain relief.

Injection of the skin over the trigger point with sterile water has been postulated as an appropriate treatment if counterstimulation is the active mechanism (Byrn et al., 1993). Although Byrn and colleagues (1993) observed prolonged pain relief in their subjects, a large number of injections were used over three treatments. Painful subcutaneous injection is not likely to be tolerated by many patients; this will negatively influence compliance despite the evidence that suggests benefit.

Of interest to those treating mainly acute or athletic pain, injection therapy has been shown to lead to a more rapid onset of pain relief in those with isolated regional pain than in those with pain in multiple regions and longer-standing pain (Hong and Hsueh, 1996).

Botulinum Toxin

Botulinum toxin is produced by the bacterium *Clostridium botulinum* and acts by blocking the release of ACh at the neuromuscular junction. It has been used in trigger point therapy on the basis of the assumption that trigger points involve excessive ACh release from motor nerve terminals.

The research to date is conflicting. One early pilot trial suggested a potential benefit of botulinum toxin over saline at two to three weeks (Cheshire et al., 1994), and a retrospective trial claimed longer pain relief with Botox than local anesthetic or steroid injection for lumbar pain (Carrasco et al., 2003). Recently, Kamanli and colleagues (2004) tested Botox in a three-armed trial against local anesthetic and dry needling; they found that it gave results equivalent to those of local anesthetic in some parameters and was slightly less effective in others and did not recommend it as a first-line treatment due to cost.

In contrast, Wheeler and colleagues (1998), in a well-designed trial, found that saline, 50 units, or 100 units of botulinum toxin produced equivalent pain scores and pain threshold improvements. An interesting but unexplained finding was that subsequent treatment of the nonresponders from all groups with 100 units of Botox resulted in a significantly better response in those who had received 100 units of Botox as their initial treatment. Thus, the evidence for Botox is conflicting, and its cost prohibits its use as a first-line therapy in trigger point pain.

Dry Needling

Dry needling involves multiple advances of an acupuncture-type needle into a trigger point with the aim of reproducing the patient's symptoms, visualizing local twitch responses, and achieving relief of muscle tension and pain (figure 20.2). In an early study, dry needling was found to be equivalent to local anesthetic, corticosteroid, and coolant spray in the treatment of lower back pain (Garvey et al., 1989). In a more recent investigation, Karakurum and colleagues (2001) studied subjects with tension headache and found that muscle dry needling resulted in improvements in pain and neck range of motion equivalent to those with placebo (subcutaneous) dry needling. The methods of measuring range of motion in this study, however, were not clearly stated.

A four-armed blinded trial in patients with myofascial jaw pain showed no improvement in pain pressure threshold but demonstrated equivalent improvements in

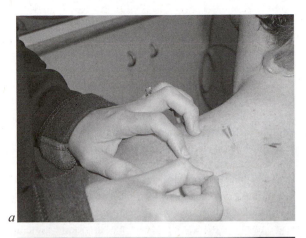

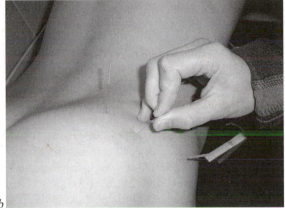

Figure 20.2   Use of dry needling to treat trigger points of the *(a)* trapezius and *(b)* gluteals.

pain intensity and unpleasantness with different combinations of dry needling, local anesthetic, placebo needling (subcutaneous), and placebo local anesthetic (McMillan et al., 1997).

A three-armed comparison of dry needling with local anesthetic injection and Botox showed that dry needling caused more posttreatment pain; and although it resulted in improvements in pain scores and pressure thresholds, these were less than for the other two groups (Kamanli et al., 2004). Quality of life scores also failed to improve with dry needling. The dry needle used in this study, however, was a 25-gauge hypodermic attached to a syringe, and at least 15 advances were used per trigger point. The posttreatment soreness in the dry needle group, as well as the lack of improvement on emotional scales, is hardly surprising. A less traumatic and painful method of dry needling is to use an acupuncture-type needle (which is usually 0.15 – 0.3 mm in diameter, and is more flexible); this results in less muscle trauma and less posttherapy pain, although posttreatment muscle soreness (similar to a bruised feeling) is well recognized with all dry needling.

Another trial involved the use of acupuncture at distal points, local dry needling, and sham laser acupuncture on patients with chronic neck pain (Irnich et al., 2002). Although none of the groups exhibited a large reduction in pain, it is interesting that the acupuncture group improved by 30% whereas the other two groups did not improve at all. In the dry needling protocol, the needle was moved within trigger points until a twitch response was seen. Twitch responses are painful and correlate with posttreatment soreness, and this may have clouded the results seen. The most important point to be gleaned from this trial is the lack of response when a nonnoxious placebo (sham laser) was used.

In a double-blind randomized controlled trial, dry needling versus nonpenetrating placebo needling of the gluteal muscles for trigger point–related referred hamstring pain in athletes resulted in no change in the range of motion measures chosen but produced equivalent improvements in activity-related pain scores in the two groups on a visual analog scale (VAS) (Huguenin et al., 2005). Median running tightness before the intervention was 4 out of 10 in the true treatment group and 5 out of 10 in the placebo group. Median running hamstring pain was 3 out of 10 in both groups. Running hamstring pain improved by approximately 2 points on the VAS immediately, and the relief lasted for 72 h. Running hamstring tightness improved by approximately 3 points over the same time frame in both groups.

A recent Cochrane review (Furlan et al., 2005) presented recommendations regarding the use of dry needling or acupuncture in the treatment of lower back pain, concluding that most of the studies were methodologically weak and that no firm evidence for the usefulness of these therapies in acute back pain could be seen on meta-analysis. However, there is clear evidence for the benefit of these therapies over placebo in the treatment of chronic lower back pain. The Cochrane review concludes by calling for more well-designed trials to clarify this area.

The major drawback to dry needling over local anesthetic injection is the higher incidence of posttreatment soreness (Hong, 1994). This would appear to be maximal in the 24 h after therapy and is usually manageable with heat packs and stretching, but may be intolerable to some patients. Therefore care in patient selection is important. The advantage of dry needling is the ability to reposition the needle until all local twitch responses have been seen, whereas the application of local anesthetic reduces these responses and the clinician is unsure whether or not all the pain-producing points in the trigger point have been treated.

Suggested Mechanisms of Proposed Therapies

Given that dry needling, local anesthetic, and placebo treatment appear to be equally effective, the mechanism of action of all of these therapies needs to be clarified.

Researchers have used various placebo treatments, but all have involved the application of a penetrative or nonpenetrative (but nonetheless noxious) stimulus to the skin that did not reach the muscle and that were much less noxious than the other treatments—hence it was felt that these stimuli were not significant enough to have an effect on pain. Central opioid release is thought to produce a global reduction in pain perception by gating spinal cord pain impulse transmission. Reversal of local anesthetic–induced analgesia has been observed with the administration of an opioid antagonist (Fine et al., 1988). This implicates the endogenous opioid system, which acts to produce hypoalgesia at a spinal cord level to at least a partial extent, in the reduction of pain seen with this therapy. This is the system implicated in the production of the "runner's high" (Koltyn, 2002) and has been suggested to be important in production of the placebo effect (Grevert et al., 1983).

Beyond these suppositions, there is little hard evidence to date on the mechanisms of action of any of the therapies discussed. This is a notoriously hard area to research due to the interactions of so many systems on both the regional and the whole-body level. Stress and the sympathetic nervous system have been shown to increase pain perception, but the effect of these treatments on this system has not been conclusively determined.

Summary

Although trigger point–related pain is widely recognized by health professionals, reliable clinical evaluation and imaging for diagnosis still elude us. Many treatments in widespread use are poorly validated and not necessarily more effective than placebo. The application of a noxious stimulus may be the key to obtaining improvements in pain perception. Less stimulatory interventions, such as laser and ultrasound, have not been convincingly shown to be beneficial. Most stimulatory interventions are able to induce subjective improvements in pain scores if not objectively measurable improvement. Stretch, TENS, injection therapies, and dry needling have all shown benefits. Unfortunately, we have extremely limited data comparing results between different therapeutic approaches—in particular, invasive versus noninvasive—from which to draw clinical conclusions.

Studies of invasive treatment utilizing a placebo intervention have not shown the active treatments to be any more effective. Importantly, the placebo interventions are in themselves stimulatory. The amount of stimulation required to induce analgesia is currently unknown. Despite EMG evidence of changes in the regions of trigger points, muscle penetration does not seem to be necessary to produce an analgesic effect. The evidence is trending toward the idea that the magnitude of the effect is consistent regardless of the therapy chosen, or the depth of needle penetration, as long as some counterstimulation is involved. The relative contributions of local tissue effects and central pain modulation to these clinical improvements require further investigation.

The choice of therapy can, therefore, be guided by patient-specific criteria, the therapist's experience and qualifications, and patient preference. The discomfort induced by the therapy, the likelihood of posttreatment soreness, and the current functional level of the patient are important considerations. Dry needling may not be appropriate for irritable, long-standing chronic pain but may be the treatment of choice for an athlete with regional pain that has not responded to previous soft tissue work. Needle phobias or other known adverse reactions will limit therapeutic choices.

Once one has identified an appropriate treatment, it is imperative to remember that trigger points rarely occur in isolation and that the key to long-term success lies in addressing the precipitating and perpetuating factors in the patient.

CHAPTER 21

Hyperbaric Oxygen and Drug Therapies

Peter M. Tiidus, PhD

The last chapter of this book is a brief overview of two other therapeutic modality categories that have been examined for their ability to influence muscle injury and repair. As the previous chapters have illustrated, not all therapeutic interventions have proved successful in altering or enhancing the course of muscle repair following various types of muscle damage or injury. Many common therapies, such as massage, ultrasound, anti-inflammatory medications, and stretching, have not proved to have significant impact on muscle healing. Other therapeutic interventions, such as cryotherapy, dry needling, nutritional antioxidants, and others discussed in this book, have had only limited success. Many less well-known therapies and interventions in muscle damage and repair remain untested or have been examined in only a few empirical studies. Often these more "esoteric" therapies have also proved to be unsuccessful in influencing muscle repair. Just one example of many such potential treatments is "low-intensity monochromatic infrared therapy" (LIMIRT), which had no effect on muscle soreness or functional recovery over five days following eccentric biceps contractions (Glasgow et al. 2001). The intent of this chapter is not to review all other potential therapeutic interventions that have been suggested to influence muscle repair. Instead the focus is on two areas not yet covered in this book that have gained some public attention.

This chapter deals with the research on the potential of hyperbaric oxygen therapy and several newer drug treatment interventions to influence muscle healing processes. These two areas have generated a small but significant amount of research that is not directly addressed in other chapters and—unlike many other purported therapeutic interventions—at least some positive results.

Hyperbaric Oxygen Therapy

Hyperbaric oxygen therapy can be defined as "the inhalation of 100% oxygen while the treatment chamber is pressurized at more than one atmosphere" (Leach 1998, pg. 489). While over 97% of hemoglobin is typically saturated with oxygen at standard atmospheric oxygen concentration and pressure, increasing oxygen concentration and atmospheric pressure will result in the dissolution of significantly more oxygen in blood plasma (Tabrah et al. 1994). Theoretically, this should result in significantly greater oxygen delivery to muscle.

In addition to the potentially enhanced oxygen delivery, a number of other mechanisms may be affected by the inhalation of hyperbaric oxygen, such that soft tissue and muscle healing may be enhanced. As cited by Best and colleagues (1998) a number of studies have suggested that hyperbaric oxygen therapy as employed in treating burns, carbon monoxide poisoning, or various soft tissue injuries may, among other effects, reduce inflammatory response to injury, enhance collagen deposition, and limit edema formation. This therapy also decreases neutrophil adhesion to injured cells and thereby limits oxygen radical production and oxidative damage consequent to trauma. Many of these factors are also associated with muscle injury, inflammation, and repair processes; and their attenuation by hyperbaric oxygen therapy could potentially diminish muscle damage and aid in muscle regeneration.

In muscle crush injury or ischemia-reperfusion injury, hyperbaric oxygen therapy may also act by helping to normalize muscle oxygen delivery and thus preserve muscle adenosine triphosphate (ATP) and energy homeostasis in situations in which the injury has compromised muscle blood flow or microcirculation (Haapaniemi et al. 1996; Gregorevic et al. 2000). Hyperbaric oxygen treatment has been successfully employed in treating and enhancing healing in a number of acute trauma conditions such as burns, crush injuries, and muscle compartment syndromes (Bouachour et al. 1996; Cianci and Sato 1994; Best et al. 1998; Harrison et al. 2001). Hence it seems logical that hyperbaric oxygen treatment might be successfully employed in other forms of acute or sport- or overuse-related muscle injury.

On the basis of these theoretical possibilities, the use of hyperbaric oxygen in the treatment of sport-related muscular injuries has been advocated in the scientific literature (Babul et al. 2004). Hyperbaric oxygen therapy as a potential means to optimize healing following muscular trauma or overuse injury, particularly among elite and professional athletes, has grown in popularity (Leach 1998). However, research on its effectiveness in treating sport-related muscular injuries in both animal and human models has been mixed. In particular, the results with human subjects have tended to be disappointing.

Animal Studies

Two animal-based studies, using rabbits (Best et al. 1998) and rats (Gregorevic et al. 2000), showed positive results with the use of hyperbaric oxygen therapy in muscle repair and recovery following injury. In the first of these studies, rabbit tibialis anterior muscles were subjected to acute stretch-induced injury (Best et al. 1998). The experimental group animals were subsequently exposed to hyperbaric oxygen (>95% oxygen at 2.5 atmospheres) daily for 60 min for five days. The control animals had no hyperbaric oxygen exposure. At seven days postinjury the muscles were evaluated for force deficit and histochemically analyzed for damage indices. Force deficit is a standard quantifiable marker of muscle injury and recovery and is often used as one of the best overall indicators of the extent of muscle damage and recovery (Warren et al. 1999). The results of this study indicated that animals experiencing five daily exposures to hyperbaric oxygen had significantly greater recovery of muscle force at seven days postinjury than the control animals and also showed significantly less histologically determined fiber disruption and damage.

The second study used a rat model in which extensor digitorum longus muscles were injected with bupivacaine hydrochloride (Marcaine) and their regeneration was evaluated at 14 days postinjection (Gregorevic et al. 2000). Bupivacaine hydrochoride is a myotoxic agent that increases cytosolic calcium and causes, within two days, a degeneration of muscle fibers with a reduction of functional force development of up to 90% (Gregorevic et al. 2000; Carlson and Faulkner 1996). Beginning at 1 h postinjection, rats were exposed to 60 min of 100% oxygen at either 2 or 3 atmospheres pressure or to room air for 14 days. As with the previous study, those animals exposed to hyperbaric oxygen had significantly greater muscle regeneration and functional force return than the control animals. In addition, animals exposed to 3 atmospheres pressure had significantly greater recovery and repair than those exposed to 2 atmospheres (Gregorevic et al. 2000). Hence, both of these animal studies showed positive effects of repeated exposure to hyperbaric oxygen therapy on postinjury muscle regeneration and repair.

Human Studies

Human studies, on the other hand, have not consistently demonstrated such positive results. An early study, which was limited by the lack of a control group, did seem to suggest that hyperbaric oxygen therapy could accelerate recovery from sport injuries in athletes (James et al. 1993). The first comprehensive and well-controlled human study also yielded some positive results (Staples et al. 1999). Staples and colleagues examined the effects of five days of hyperbaric oxygen therapy (1 h a day of 100% oxygen at 2 atmospheres) beginning either immediately or 48 h after eccentric exercise–induced quadriceps injury. Both of the groups receiving hyperbaric oxygen therapy tended to have a slightly greater recovery of muscle force at five days post–eccentric exercise than the placebo (room air) group. Despite this, there were no differences in perception of muscle soreness between groups. Subsequent to these first positive reports, three well-controlled, double-blind studies did not show any significant effects of hyperbaric oxygen therapy on postinjury recovery of muscle force or muscle soreness (Mekjavic et al. 2000; Harrison et al. 2001; Babul et al. 2004).

Mekjavic and colleagues (2000) had subjects perform eccentric exercise with the biceps flexor muscles and then exposed them to either room air or 100% oxygen at 2.5 atmospheres pressure for 1 h beginning immediately postexercise and continuing daily for seven days. The investigators measured muscle force loss and recovery, muscle soreness, and arm circumference (as an indicator of swelling) daily over the seven-day period. None of these indices of muscle damage and recovery differed between the hyperbaric oxygen therapy and the control group during any of the measurement days. Figure 21.1 depicts these results.

Harrison and colleagues (2001) also had subjects per-

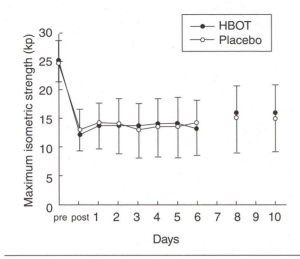

Figure 21.1   Average maximal isometric strength of the elbow flexor muscles before and immediately after a high-force eccentric workout and after each daily hyperbaric exposure (HBOT). There were no significant differences between the hyperbaric oxygen–exposed and placebo group at any time point.

Reprinted, by permission, from I.B. Mekjavic et al., 2000, "Hyperbaric oxygen therapy does not affect recovery from delayed onset muscle soreness," *Medicine and Science in Sports and Exercise* 32(3): 558-563.

form eccentric elbow flexion exercise to induce biceps muscle injury. Hyperbaric oxygen exposure consisted of 100 min at 2.5 atmospheres pressure daily for four days. Isometric strength loss and recovery and muscle soreness measures did not differ between hyperbaric therapy and placebo therapy groups at any of the postexercise time points. This study also measured serum creatine kinase activities, muscle relaxation time, and biceps circumference as determined by magnetic resonance imaging (as indicators of muscle membrane disruption, muscle damage, and muscle swelling, respectively) daily. Again, these indicators failed to demonstrate any effect of hyperbaric oxygen therapy on muscle healing or recovery from eccentric work over the course of the experiment.

The most recent study, by Babul and colleagues (2004), further supports the lack of significant enhancement of postinjury recovery of skeletal muscle with hyperbaric oxygen therapy. Subjects performed eccentric quadriceps muscle contractions and were exposed to either 100% oxygen at 2 atmospheres pressure or normal air for 60 min each day for four days. Muscle soreness, muscle force decline and return, quadriceps circumference, and blood levels of creatine kinase and malondialdehyde were measured every day for four days following the exercise protocol. Malondialdehyde levels in the blood can be used as a marker of oxidative damage to muscles and muscle membranes caused by oxygen radicals. No significant differences between control and hyperbaric oxygen treatment groups were seen on any day in any

of the measures performed. This study concurred with the previous two studies in concluding that hyperbaric oxygen therapy did not appear to accelerate muscle repair following eccentric contraction–induced muscle damage (Babul et al. 2004). It also tended to contradict the earlier study from the same laboratory (Staples et al. 1999), which had presented some limited positive results on at least muscle force recovery.

Interpreting the Studies

The primary difference between the animal studies demonstrating a positive effect of hyperbaric oxygen therapy and the human studies showing no effect appears to be in the severity of the muscle damage induced. Injuries induced by stretch and bupivacaine hydrochloride injection both lead to much more severe and extensive muscle disruption than is typically seen with eccentric contraction–induced muscle injury in humans. As shown with muscle crush injury and ischemia-reperfusion injury, hyperbaric oxygen treatment may be helpful in speeding healing by helping to increase oxygen delivery in vascularly compromised muscles (Best et al., 1998). This may have been a factor in its relative effectiveness in treating the stretch-injured rabbit muscle (Best 1998). On the other hand, the bupivacaine hydrochloride–injured rodent muscles also healed faster when the animals were exposed to hyperbaric oxygen therapy despite no compromise in their vasculature (Gregorevic et al. 2000). Hence hyperbaric oxygen therapy may have been positively influencing one or more other factors involved in muscle damage, inflammation, or repair in this model. These mechanisms are not yet identified but may be related to some of the factors potentially affected by exposure to hyperbaric oxygen that were mentioned earlier in this chapter.

What emerges from these studies is the understanding that hyperbaric oxygen therapy, if it is at all effective in accelerating muscle repair, may be of benefit only if the muscle injury or disruption is quite severe. Its use in the treatment of modest muscle injury in humans appears to have little or no effect and as such is usually not warranted. As well, its effectiveness in speeding healing of more severe muscle injuries has been demonstrated only in animal models and needs to be experimentally verified in humans. Nevertheless, any potential benefits need to be weighed against the known potential deleterious and tissue damaging effects of prolonged exposure to hyperbaric oxygen and the high cost of these procedures (Leach 1998; Mekjavic et al. 2000). When these factors are taken into account, it seems that hyperbaric oxygen therapy for typical muscle damage or injury in humans is not cost- or time-effective and thus should not be used in most circumstances.

Drug Therapies

The most commonly used drug treatments for muscle damage and soreness involve the nonsteroidal anti-inflammatories (NSAIDs). Various forms of NSAIDs have been found to have little or no effectiveness in enhancing strength recovery, muscle performance, or muscle satellite cell activation or in diminishing soreness sensation after exercise-induced muscle damage (Thorsson et al. 1998; Van Heest et al. 2002). However, it appears also that NSAIDs do not impair recovery of skeletal muscle following severe injury in animal models (Vignaud et al. 2005). The effectiveness of NSAIDs is covered in chapter 18 in this book, and these drugs are not discussed further here.

Reports on several newer drug interventions related to muscle repair following trauma have recently appeared in the literature. These studies used animal models to assess the effectiveness of drug treatments on muscle repair following experimentally induced injury; no human-based studies have yet been conducted. Several of these preliminary trials, as outlined next, have yielded positive findings with respect to enhancement of muscle healing.

MS-818

Sugiyama and colleagues (2002) investigated the effects of the synthesized pyrimidine compound 2-piperadino-6-methyl-5-oxo-5, 6-dihydro(7H)-pyrrolo[3,4-d] pyrimidine maleate (MS-818) in treating standard surgical laceration injury in muscles of male rats. Previous studies had shown that MS-818 exhibited neurotrophic effects in both human and animal nerve cells (Fukyama et al. 2000) and that it accelerated in vitro myoblast cell growth in the presence of growth factors such as insulin-like growth factor-1 (IGF-1).

In their study, Sugiyama and colleagues (2002) surgically lacerated the semitendinosus muscles of rats and treated some of the recovering animals with MS-818. They measured the rates of muscle healing by quantifying muscle isometric twitch tension, myogenin messenger RNA (mRNA), and desmin. They also quantified a number of immunohistochemical and histochemical indices of muscle ultrastructural damage and recovery for up to 12 weeks following the surgical muscle damage. The animals treated with MS-818 exhibited significantly greater postinjury regeneration of muscle fiber diameter as well as restoration of muscle isometric twitch force, the latter representing approximately 60% of preinjury force in the treated animals compared to 40% in the untreated animals at 12 weeks postinjury (see figure 21.2).

Histochemical analysis also indicated that the injured muscles from the MS-818-treated animals as compared to the untreated animals benefited from accelerated proliferation and differentiation of muscle satellite cells

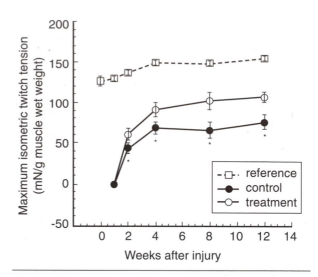

Figure 21.2 Recovery of maximum isometric twitch tension of injured muscle groups with (open circles) or without (closed circles) MS-818 for up to 12 weeks after injury. Open squares show mean values of maximal isometric twitch tension for the reference group at weeks 0 through 12. *$p < 0.001$.

From N. Sugiyama et al, "Acceleration by MS-818 of early muscle regeneration and enhanced muscle recovery after surgical transection," Copyright © 2002 *Muscle and Nerve*. Reprinted with permission of John Wiley & Sons, Inc.

and enhanced transformation of myotubes into myofibers during the muscle healing process. During the early phase of regeneration, the muscles from MS-818-treated animals expressed significantly more myogenin mRNA and desmin, which also suggested a greater activation of muscle satellite cells and greater muscle regeneration (Sugiyama et al. 2002). On the basis of previous studies, it was suggested that MS-818 accelerated postinjury muscle healing by enhancing the influence of muscle growth factors such as IGF-1 (among others) that are known to be generated at the site of muscle wounds (Sugiyama et al. 2002).

Fenoterol

A subsequent study examined the potential of fenoterol, a β_2-adrenoceptor agonist, to enhance muscle repair following bupivacaine hydrochloride–induced degeneration in rat skeletal muscle (Beitzel et al. 2004). Fenotorol has been shown to increase muscle mass due to β_2-adrenoceptor-mediated protein accretion consequent to receptor-promoted increases in muscle cyclic adenosine monophosphate (cAMP). It has been demonstrated that β_2-adrenoceptor-mediated increases in muscle cAMP both promote increased muscle protein synthesis and lead to decreased protein degradation (Navegantes et al. 2002). Fenoterol greatly enhanced muscle mass and force-producing capacity in normal rat skeletal muscles (Ryall et al. 2002).

Following injury to the extensor digitorum longus muscle (EDL), the animals treated with fenoterol exhibited significantly greater recovery of muscle mass and muscle function than placebo-treated controls. At 21 days postinjury, the fenoterol-treated animals had 26% greater muscle cross-section size and muscle mass than the untreated animals. Similarly greater improvements in postinjury muscle function, protein content, and muscle morphology were also seen in the fenoterol-treated animals relative to controls (Beitzel et al. 2004). These findings clearly demonstrated the benefits of fenoterol for hastening the restoration of muscle following myotoxic injury in the rat model (Beitzel et al. 2004). It was suggested on the basis of previous research that fenoterol administration may enhance early stages of muscle fiber regeneration through β-agonist-mediated increases in vascularization, satellite cell proliferation, increased local muscle expression of growth factors such as IGF-1, inhibition of proteolysis and acceleration of protein synthesis, or some combination of these (Beitzel et al. 2004).

Insulin-Like Growth Factor-1 and Decorin

A third study examined the potential of the combination of the growth factor IGF-1 and the antifibrotic agent decorin (for reduction of postinjury fibrosis) to enhance postinjury muscle recovery (Sato et al. 2003). In cases of more severe muscle injury, scar formation and fibrosis can begin about two weeks into recovery and continues over time (Huard et al. 2002). The process of muscle scar formation can hinder full muscle regeneration and strength recovery (Huard et al. 2002; Sato et al. 2003). High levels of transforming growth factor-β1 (TGF-β1) are present in injured skeletal muscle, and TGF-β1 stimulates the deposition of collagens, the overgrowth of the extracellular matrix, and the formation of muscle scar tissue (Li and Huard 2002). Hence, it was hypothesized that inhibiting TGF-β1 expression in injured muscle with the use of decorin, an anti-TGF-β1 human polyglycan, could improve muscle healing following surgically induced damage (Sato et al. 2003).

Sato and colleagues (2003) induced a standard surgical laceration in the gastrocnemius muscles of mice and treated the animals with regular intramuscular injections of decorin and IGF-1, separately or in combination, or placebo while following the recovery of the muscle over four weeks. Administration of IGF-1 and decorin separately or in combination resulted in the development of significantly greater numbers of new regenerating myofibrils and greater myofibril diameter in the injured muscles compared to those in the placebo-treated animals. The

combination of IGF-1 and decorin was additive and was more effective than either treatment alone in increasing regenerating myofibril numbers (Sato et al. 2003). In addition, the extent of muscle fibrosis and scarring during healing was significantly reduced by the use of decorin alone. Administration of IGF-1 alone or in combination with decorin did not significantly contribute further to the reduction of muscle scar formation during healing (Sato et al. 2003).

The restoration of physiological muscle strength following laceration injury was also significantly enhanced by the use of decorin alone in this model (Sato et al. 2003). Hence it appears that limiting muscle fibrosis and scar formation with use of the TGF-β1 inhibitor drug decorin could significantly enhance physiological and histochemical indices of muscle healing following laceration injury in mouse muscles. These findings are supported by other studies from the same group demonstrating improved healing of injured skeletal muscle when scar formation is diminished by other antifibrotic agents, such as suramin (Chan et al. 2003).

It appears from these studies that the use of various drug interventions can accentuate the influence of naturally produced muscle growth factors, satellite cell activation and proliferation, and muscle protein accretion and also reduce muscle scar formation. The use of such drug interventions in relatively severe experimentally induced muscle injury in animal models can significantly enhance muscle repair and recovery. The effectiveness and safety of such interventions in treating severe muscle injury in humans have yet to be demonstrated. It is also not yet known whether interventions like these would be effective or worthwhile in treating less severe muscle overuse or athletic injuries. Nevertheless, these animal studies are promising and suggest the need for continuing research in these areas.

Summary

From studies using animal models with relatively severe experimentally induced muscle injury, it appears that both hyperbaric oxygen therapy and drugs that enhance the action of local muscle growth factors, stimulate muscle satellite cells, or increase muscle protein accretion can successfully improve muscle healing and repair. The usefulness of these therapies with more modest muscle overuse- or exercise-induced injuries may be limited and in the case of drug therapies has not yet been demonstrated in humans. On the basis of the successful enhancement of muscle repair seen in animal studies, further research, particularly with selective drug therapies in human models, is warranted.

REFERENCES

Chapter 1

Balnave, C.D. and D.G. Allen. 1995. Intracellular calcium and force in single mouse muscle fibres following repeated contractions with stretch. *J Physiol* 488 (Pt 1): 25-36.

Brooks, S.V., E. Zerba and J.A. Faulkner. 1995. Injury to muscle fibres after single stretches of passive and maximally stimulated muscles in mice. *J Physiol* 488 (Pt 2): 459-469.

Brown, L.M. and L. Hill. 1991. Some observations on variations in filament overlap in tetanized muscle fibres and fibres stretched during a tetanus, detected in the electron microscope after rapid fixation. *J Muscle Res Cell Motil* 12: 171-182.

Duan, C., M.D. Delp, D.A. Hayes, P.D. Delp and R.B. Armstrong. 1990. Rat skeletal muscle mitochondrial [Ca²⁺] and injury from downhill walking. *J Appl Physiol* 68: 1241-1251.

Duncan, C.J. and M.J. Jackson. 1987. Different mechanisms mediate structural changes and intracellular enzyme efflux following damage to skeletal muscle. *J Cell Sci* 87 (Pt 1): 183-188.

Faulkner, J.A., D.A. Jones and J.M. Round. 1989. Injury to skeletal muscles of mice by forced lengthening during contractions. *Q J Exp Physiol* 74: 661-670.

Feasson, L., D. Stockholm, D. Freyssenet, I. Richard, S. Duguez, J.S. Beckmann and C. Denis. 2002. Molecular adaptations of neuromuscular disease-associated proteins in response to eccentric exercise in human skeletal muscle. *J Physiol* 543: 297-306.

Friden, J., M. Sjostrom and B. Ekblom. 1983. Myofibrillar damage following intense eccentric exercise in man. *Int J Sports Med* 4: 170-176.

Fritz, V.K. and W.T. Stauber. 1988. Characterization of muscles injured by forced lengthening. II. Proteoglycans. *Med Sci Sports Exerc* 20: 354-361.

Hamer, P.W., J.M. McGeachie, M.J. Davies and M.D. Grounds. 2002. Evans Blue Dye as an in vivo marker of myofibre damage: Optimising parameters for detecting initial myofibre membrane permeability. *J Anat* 200: 69-79.

Ingalls, C.P., G.L. Warren, J.H. Williams, C.W. Ward and R.B. Armstrong. 1998. E-C coupling failure in mouse EDL muscle after in vivo eccentric contractions. *J Appl Physiol* 85: 58-67.

Jones, D.A., D.J. Newham, J.M. Round and S.E. Tolfree. 1986. Experimental human muscle damage: Morphological changes in relation to other indices of damage. *J Physiol* 375: 435-448.

Koh, T.J. 2002. Do small heat shock proteins protect skeletal muscle from injury? *Exerc Sport Sci Rev* 30: 117-121.

Koh, T.J. and J. Escobedo. 2004. Cytoskeletal disruption and small heat shock protein translocation immediately after lengthening contractions. *Am J Physiol Cell Physiol* 286: C713-722.

Komulainen, J., T.E. Takala, H. Kuipers and M.K. Hesselink. 1998. The disruption of myofibre structures in rat skeletal muscle after forced lengthening contractions. *Pflugers Arch* 436: 735-741.

Lieber, R.L., L.E. Thornell and J. Friden. 1996. Muscle cytoskeletal disruption occurs within the first 15 min of cyclic eccentric contraction. *J Appl Physiol* 80: 278-284.

Lieber, R.L., T.M. Woodburn and J. Friden. 1991. Muscle damage induced by eccentric contractions of 25% strain. *J Appl Physiol* 70: 2498-2507.

Lovering, R.M. and P.G. De Deyne. 2004. Contractile function, sarcolemma integrity, and the loss of dystrophin after skeletal muscle eccentric contraction-induced injury. *Am J Physiol Cell Physiol* 286: C230-238.

Macpherson, P.C., R.G. Dennis and J.A. Faulkner. 1997. Sarcomere dynamics and contraction-induced injury to maximally activated single muscle fibres from soleus muscles of rats. *J Physiol* 500 (Pt 2): 523-533.

McArdle, A., D. Pattwell, A. Vasilaki, R.D. Griffiths and M.J. Jackson. 2001. Contractile activity-induced oxidative stress: Cellular origin and adaptive responses. *Am J Physiol Cell Physiol* 280: C621-C627.

McCully, K.K. and J.A. Faulkner. 1985. Injury to skeletal muscle fibers of mice following lengthening contractions. *J Appl Physiol* 59: 119-126.

McNeil, P.L. and R. Khakee. 1992. Disruptions of muscle fiber plasma membranes. Role in exercise-induced damage. *Am J Pathol* 140: 1097-1109.

Morgan, D.L. 1990. New insights into the behavior of muscle during active lengthening. *Biophys J* 57: 209-221.

Morgan, D.L. and D.G. Allen. 1999. Early events in stretch-induced muscle damage. *J Appl Physiol* 87: 2007-2015.

Newham, D.J., G. McPhail, K.R. Mills and R.H. Edwards. 1983. Ultrastructural changes after concentric and eccentric contractions of human muscle. *J Neurol Sci* 61: 109-122.

Patel, T.J., D. Cuizon, O. Mathieu-Costello, J. Friden and R.L. Lieber. 1998. Increased oxidative capacity does not protect skeletal muscle fibers from eccentric contraction-induced injury. *Am J Physiol* 274: R1300-1308.

Pizza, F.X., T.J. Koh, S.J. McGregor and S.V. Brooks. 2002. Muscle inflammatory cells after passive stretches, isometric contractions, and lengthening contractions. *J Appl Physiol* 92: 1873-1878.

Takekura, H., N. Fujinami, T. Nishizawa, H. Ogasawara and N. Kasuga. 2001. Eccentric exercise-induced morphological changes in the membrane systems involved in excitation-contraction coupling in rat skeletal muscle. *J Physiol* 533: 571-583.

Talbot, J.A. and D.L. Morgan. 1998. The effects of stretch parameters on eccentric exercise-induced damage to toad skeletal muscle. *J Muscle Res Cell Motil* 19: 237-245.

Warren, G.L., D.A. Hayes, D.A. Lowe and R.B. Armstrong. 1993. Mechanical factors in the initiation of eccentric contraction-induced injury in rat soleus muscle. *J Physiol* 464: 457-475.

Warren, G.L., D.A. Lowe and R.B. Armstrong. 1999. Measurement tools used in the study of eccentric contraction-induced injury. *Sports Med* 27: 43-59.

Warren, G.L., D.A. Lowe, D.A. Hayes, C.J. Karwoski, B.M. Prior and R.B. Armstrong. 1993. Excitation failure in eccentric contraction-induced injury of mouse soleus muscle. *J Physiol* 468: 487-499.

Yeung, E.W., C.D. Balnave, H.J. Ballard, J.P. Bourreau and D.G. Allen. 2002. Development of T-tubular vacuoles in eccentrically damaged mouse muscle fibres. *J Physiol* 540: 581-592.

Zerba, E., T.E. Komorowski and J.A. Faulkner. 1990. Free radical injury to skeletal muscles of young, adult, and old mice. *Am J Physiol* 258: C429-C435.

Chapter 2

Akimoto, T., M. Furudate, M. Saitoh, K. Sugiura, T. Waku, T. Akama and I. Kono. 2002. Increased plasma concentrations of intercellular adhesion molecule-1 after strenuous exercise associated with muscle damage. *European Journal of Applied Physiology* 86: 185-190.

Apple, F.S. and M. Rhodes. 1988. Enzymatic estimation of skeletal muscle damage by analysis of changes in serum creatine kinase. *Journal of Applied Physiology* 65: 2598-2600.

Armand, A.S., T. Launay, B.D. Gaspera, F. Charbonnier, C.L. Gallien and C. Chanoine. 2003. Effects of eccentric treadmill running on mouse soleus: Degeneration/regeneration studied with Myf-5 and MyoD probes. *Acta Physiologica Scandinavica* 179: 75-84.

Armstrong, R.B., M.H. Laughlin, L. Rome and C.R. Taylor. 1983. Metabolism of rats running up and down an incline. *Journal of Applied Physiology* 55: 518-521.

Armstrong, R.B., R.W. Ogilvie and J.A. Schwane. 1983. Eccentric exercise-induced injury to rat skeletal muscle. *Journal of Applied Physiology* 54: 80-93.

Babul, S., E.C. Rhodes, J.E. Taunton and M. Lepawsky. 2003. Effects of intermittent exposure to hyperbaric oxygen for the treatment of an acute soft tissue injury. *Clinical Journal of Sport Medicine* 13: 138-147.

Balnave, C.D. and D.G. Allen. 1995. Intracellular calcium and force in single mouse muscle fibres following repeated contractions with stretch. *Journal of Physiology* 488: 25-36.

Balnave, C.D., D.F. Davey and D.G. Allen. 1997. Distribution of sarcomere length and intracellular calcium in mouse skeletal muscle following stretch-induced injury. *Journal of Physiology* 502: 649-659.

Balnave, C.D. and M.W. Thompson. 1993. Effect of training on eccentric exercise-induced muscle damage. *Journal of Applied Physiology* 75: 1545-1551.

Barash, I.A., L. Mathew, A.F. Ryan, J. Chen and R.L. Lieber. 2004. Rapid muscle-specific gene expression changes after a single bout of eccentric contractions in the mouse. *American Journal of Physiology: Cell Physiology* 286: C355-364.

Beiras-Fernandez, A., E. Thein, D. Chappell and C. Hammer. 2003. Morphological assessment of reperfusion injury in a concordant xenogeneic primate model. *Annals of Transplantation* 8: 50-52.

Benz, R.J., J. Friden and R.L. Lieber. 1998. Simultaneous stiffness and force measurements reveal subtle injury to rabbit soleus muscles. *Molecular and Cellular Biochemistry* 179: 147-158.

Best, T.M., R.P. McCabe, D. Corr and R. Vanderby, Jr. 1998. Evaluation of a new method to create a standardized muscle stretch injury. *Medicine and Science in Sports and Exercise* 30: 200-205.

Black, J.D. and E.D. Stevens. 2001. Passive stretching does not protect against acute contraction-induced injury in mouse EDL muscle. *Journal of Muscle Research and Cell Motility* 22: 301-310.

Blais, C., Jr., A. Adam, D. Massicotte and F. Peronnet. 1999. Increase in blood bradykinin concentration after eccentric weight-training exercise in men. *Journal of Applied Physiology* 87: 1197-1201.

Bloomer, R.J., A.H. Goldfarb, M.J. McKenzie, T. You and L. Nguyen. 2004. Effects of antioxidant therapy in women exposed to eccentric exercise. *International Journal of Sport Nutrition and Exercise Metabolism* 14: 377-388.

Brockett, C.L., D.L. Morgan, J.E. Gregory and U. Proske. 2002. Damage to different motor units from active lengthening of the medial gastrocnemius muscle of the cat. *Journal of Applied Physiology* 92: 1104-1110.

Brooks, S.V. and J.A. Faulkner. 1990. Contraction-induced injury: Recovery of skeletal muscles in young and old mice. *American Journal of Physiology* 258: C436-442.

Brooks, S.V. and J.A. Faulkner. 1996. The magnitude of the initial injury induced by stretches of maximally activated muscle fibres of mice and rats increases in old age. *Journal of Physiology* 497: 573-580.

Brooks, S.V. and J.A. Faulkner. 2001. Severity of contraction-induced injury is affected by velocity only during stretches of large strain. *Journal of Applied Physiology* 91: 661-666.

Brooks, S.V., E. Zerba and J.A. Faulkner. 1995. Injury to muscle fibres after single stretches of passive and maximally stimulated muscles in mice. *Journal of Physiology* 488: 459-469.

Brown, S., S. Day and A. Donnelly. 1999. Indirect evidence of human skeletal muscle damage and collagen breakdown after eccentric muscle actions. *Journal of Sports Sciences* 17: 397-402.

Burkholder, T.J. and R.L. Lieber. 2001. Sarcomere length operating range of vertebrate muscles during movement. *Journal of Experimental Biology* 204: 1529-1536.

Butterfield, T.A. and W. Herzog. 2005. Quantification of muscle fiber strain during in-vivo repetitive stretch-shortening cycles. *Journal of Applied Physiology* 99: 593-602.

Byrne, C. and R. Eston. 2002a. The effect of exercise-induced muscle damage on isometric and dynamic knee extensor strength and vertical jump performance. *Journal of Sports Sciences* 20: 417-425.

Byrne, C. and R. Eston. 2002b. Maximal-intensity isometric and dynamic exercise performance after eccentric muscle actions. *Journal of Sports Sciences* 20: 951-959.

Byrne, C., R.G. Eston and R.H. Edwards. 2001. Characteristics of isometric and dynamic strength loss following eccentric exercise-induced muscle damage. *Scandinavian Journal of Medicine and Science in Sports* 11: 134-140.

Byrnes, W.C., P.M. Clarkson, J.S. White, S.S. Hsieh, P.N. Frykman and R.J. Maughan. 1985. Delayed onset muscle soreness following repeated bouts of downhill running. *Journal of Applied Physiology* 59: 710-715.

Cannon, J.G., M.A. Fiatarone, R.A. Fielding and W.J. Evans. 1994. Aging and stress-induced changes in complement activation and neutrophil mobilization. *Journal of Applied Physiology* 76: 2616-2620.

Carter, G.T., N. Kikuchi, R.T. Abresch, S.A. Walsh, S.J. Horasek and W.M. Fowler, Jr. 1994. Effects of exhaustive concentric and eccentric exercise on murine skeletal muscle. *Archives of Physical Medicine and Rehabilitation* 75: 555-559.

Carter, G.T., N. Kikuchi, S.J. Horasek and S.A. Walsh. 1994. The use of fluorescent dextrans as a marker of sarcolemmal injury. *Histology and Histopathology* 9: 443-447.

Chan, Y.S., Y. Li, W. Foster, T. Horaguchi, G. Somogyi, F.H. Fu and J. Huard. 2003. Antifibrotic effects of suramin in injured skeletal muscle after laceration. *Journal of Applied Physiology* 95: 771-780.

Chen, T.C. and S.S. Hsieh. 2001. Effects of a 7-day eccentric training period on muscle damage and inflammation. *Medicine and Science in Sports and Exercise* 33: 1732-1738.

Chen, Y.W., M.J. Hubal, E.P. Hoffman, P.D. Thompson and P.M. Clarkson. 2003. Molecular responses of human muscle to eccentric exercise. *Journal of Applied Physiology* 95: 2485-2494.

Child, R.B., J.M. Saxton and A.E. Donnelly. 1998. Comparison of eccentric knee extensor muscle actions at two muscle lengths on indices of damage and angle-specific force production in humans. *Journal of Sports Sciences* 16: 301-308.

Childers, M.K. and K.S. McDonald. 2004. Regulatory light chain phosphorylation increases eccentric contraction-induced injury in skinned fast-twitch fibers. *Muscle and Nerve* 29: 313-317.

Childs, A., C. Jacobs, T. Kaminski, B. Halliwell and C. Leeuwenburgh. 2001. Supplementation with vitamin C and N-acetyl-cysteine increases oxidative stress in humans after an acute muscle injury induced by eccentric exercise. *Free Radical Biology and Medicine* 31: 745-753.

Chleboun, G.S., J.N. Howell, R.R. Conatser and J.J. Giesey. 1998. Relationship between muscle swelling and stiffness after eccentric exercise. *Medicine and Science in Sports and Exercise* 30: 529-535.

Clarkson, P.M., W.C. Byrnes, K.M. McCormick, L.P. Turcotte and J.S. White. 1986. Muscle soreness and serum creatine kinase activity following isometric, eccentric, and concentric exercise. *International Journal of Sports Medicine* 7: 152-155.

Clarkson, P.M. and C. Ebbeling. 1988. Investigation of serum creatine kinase variability after muscle-damaging exercise. *Clinical Science* 75: 257-261.

Clarkson, P.M., W. Kroll and A.M. Melchionda. 1982. Isokinetic strength, endurance, and fiber type composition in elite American paddlers. *European Journal of Applied Physiology* 48: 67-76.

Clarkson, P.M., K. Nosaka and B. Braun. 1992. Muscle function after exercise-induced muscle damage and rapid adaptation. *Medicine and Science in Sports and Exercise* 24: 512-520.

Corr, D.T., G.E. Leverson, R. Vanderby, Jr. and T.M. Best. 2003. A nonlinear rheological assessment of muscle recovery from eccentric stretch injury. *Medicine and Science in Sports and Exercise* 35: 1581-1588.

Crisco, J.J., K.D. Hentel, W.O. Jackson, K. Goehner and P. Jokl. 1996. Maximal contraction lessens impact response in a muscle contusion model. *Journal of Biomechanics* 29: 1291-1296.

Crisco, J.J., P. Jokl, G.T. Heinen, M.D. Connell and M.M. Panjabi. 1994. A muscle contusion injury model. Biomechanics, physiology, and histology. *American Journal of Sports Medicine* 22: 702-710.

Cummins, P., A. Young, M.L. Auckland, C.A. Michie, P.C. Stone and B.J. Shepstone. 1987. Comparison of serum cardiac specific troponin-I with creatine kinase, creatine kinase-MB isoenzyme, tropomyosin, myoglobin and C-reactive protein release in marathon runners: Cardiac or skeletal muscle trauma? *European Journal of Clinical Investigation* 17: 317-324.

Cutlip, R.G., K.B. Geronilla, B.A. Baker, M.L. Kashon, G.R. Miller and A.W. Schopper. 2004. Impact of muscle length during stretch-shortening contractions on real-time and temporal muscle performance measures in rats in vivo. *Journal of Applied Physiology* 96: 507-516.

Day, S.H., A.E. Donnelly, S.J. Brown and R.B. Child. 1998. Electromyogram activity and mean power frequency in exercise-damaged human muscle. *Muscle and Nerve* 21: 961-963.

Deschenes, M.R., R.E. Brewer, J.A. Bush, R.W. McCoy, J.S. Volek and W.J. Kraemer. 2000. Neuromuscular disturbance outlasts other symptoms of exercise-induced muscle damage. *Journal of the Neurological Sciences* 174: 92-99.

Donnelly, A.E., P.M. Clarkson and R.J. Maughan. 1992. Exercise-induced muscle damage: Effects of light exercise on damaged muscle. *European Journal of Applied Physiology* 64: 350-353.

Donnelly, A.E., R.J. Maughan and P.H. Whiting. 1990. Effects of ibuprofen on exercise-induced muscle soreness and indices of muscle damage. *British Journal of Sports Medicine* 24: 191-195.

Duan, C., M.D. Delp, D.A. Hayes, P.D. Delp and R.B. Armstrong. 1990. Rat skeletal muscle mitochondrial [Ca^{2+}] and injury from downhill walking. *Journal of Applied Physiology* 68: 1241-1251.

Duarte, J.A., J.F. Magalhaes, L. Monteiro, A. Almeida-Dias, J.M. Soares and H.J. Appell. 1999. Exercise-induced signs of muscle overuse in children. *International Journal of Sports Medicine* 20: 103-108.

Duncan, C.J. 1988. The role of phospholipase A2 in calcium-induced damage in cardiac and skeletal muscle. *Cell and Tissue Research* 253: 457-462.

Edgerton, V.R., J.L. Smith and D.R. Simpson. 1975. Muscle fibre type populations of human leg muscles. *Histochemical Journal* 7: 259-266.

Edstrom, L. and B. Nystrom. 1969. Histochemical types and sizes of fibres in normal human muscles. A biopsy study. *Acta Neurologica Scandinavica* 45: 257-269.

Elder, G.C., K. Bradbury and R. Roberts. 1982. Variability of fiber type distributions within human muscles. *Journal of Applied Physiology* 53: 1473-1480.

Eston, R.G., S. Finney, S. Baker and V. Baltzopoulos. 1996. Muscle tenderness and peak torque changes after downhill running following a prior bout of isokinetic eccentric exercise. *Journal of Sports Sciences* 14: 291-299.

Eston, R. and D. Peters. 1999. Effects of cold water immersion on the symptoms of exercise-induced muscle damage. *Journal of Sports Sciences* 17: 231-238.

Evans, R.K., K.L. Knight, D.O. Draper and A.C. Parcell. 2002. Effects of warm-up before eccentric exercise on indirect markers of muscle damage. *Medicine and Science in Sports and Exercise* 34: 1892-1899.

Faulkner, J.A., D.A. Jones and J.M. Round. 1989. Injury to skeletal muscles of mice by forced lengthening during contractions. *Quarterly Journal of Experimental Physiology* 74: 661-670.

Feasson, L., D. Stockholm, D. Freyssenet, I. Richard, S. Duguez, J.S. Beckmann and C. Denis. 2002. Molecular adaptations of neuromuscular disease-associated proteins in response to eccentric exercise in human skeletal muscle. *Journal of Physiology* 543: 297-306.

Fink, E., D. Fortin, B. Serrurier, R. Ventura-Clapier and A.X. Bigard. 2003. Recovery of contractile and metabolic phenotypes in regenerating slow muscle after notexin-induced or crush injury. *Journal of Muscle Research and Cell Motility* 24: 421-429.

Fish, J.S., N.H. McKee, W.M. Kuzon, Jr. and M.J. Plyley. 1993. The effect of hypothermia on changes in isometric contractile function in skeletal muscle after tourniquet ischemia. *Journal of Hand Surgery* 18: 210-217.

Foley, J.M., R.C. Jayaraman, B.M. Prior, J.M. Pivarnik and R.A. Meyer. 1999. MR measurements of muscle damage and adaptation after eccentric exercise. *Journal of Applied Physiology* 87: 2311-2318.

Folland, J.P., J. Chong, E.M. Copeman and D.A. Jones. 2001. Acute muscle damage as a stimulus for training-induced gains in strength. *Medicine and Science in Sports and Exercise* 33: 1200-1205.

Foster, W., Y. Li, A. Usas, G. Somogyi and J. Huard. 2003. Gamma interferon as an antifibrosis agent in skeletal muscle. *Journal of Orthopaedic Research* 21: 798-804.

Friden, J. and R.L. Lieber. 1998. Segmental muscle fiber lesions after repetitive eccentric contractions. *Cell and Tissue Research* 293: 165-171.

Friden, J., R.L. Lieber and L.E. Thornell. 1991. Subtle indications of muscle damage following eccentric contractions. *Acta Physiologica Scandinavica* 142: 523-524.

Friden, J., P.N. Sfakianos, A.R. Hargens and W.H. Akeson. 1988. Residual muscular swelling after repetitive eccentric contractions. *Journal of Orthopaedic Research* 6: 493-498.

Friden, J., M. Sjostrom and B. Ekblom. 1983. Myofibrillar damage following intense eccentric exercise in man. *International Journal of Sports Medicine* 4: 170-176.

Fritz, V.K. and W.T. Stauber. 1988. Characterization of muscles injured by forced lengthening. II. Proteoglycans. *Medicine and Science in Sports and Exercise* 20: 354-361.

Fukushima, K., N. Badlani, A. Usas, F. Riano, F. Fu and J. Huard. 2001. The use of an antifibrosis agent to improve muscle recovery after laceration. *American Journal of Sports Medicine* 29: 394-402.

Gambaryan, P.P. 1974. *How mammals run: Anatomical adaptations.* New York: Wiley.

Garrett, W.E., Jr., M.R. Safran, A.V. Seaber, R.R. Glisson and B.M. Ribbeck. 1987. Biomechanical comparison of stimulated and nonstimulated skeletal muscle pulled to failure. *American Journal of Sports Medicine* 15: 448-454.

Geronilla, K.B., G.R. Miller, K.F. Mowrey, J.Z. Wu, M.L. Kashon, K. Brumbaugh, J. Reynolds, A. Hubbs and R.G. Cutlip. 2003. Dynamic force responses of skeletal muscle during stretch-shortening cycles. *European Journal of Applied Physiology* 90: 144-153.

Gibala, M.J., J.D. MacDougall, M.A. Tarnopolsky, W.T. Stauber and A. Elorriaga. 1995. Changes in human skeletal muscle ultrastructure and force production after acute resistance exercise. *Journal of Applied Physiology* 78: 702-708.

Gillette, J.H. and J.L. Mitchell. 1991. Ornithine decarboxylase: a biochemical marker of repair in damaged tissue. *Life Sciences* 48: 1501-1510.

Gleeson, N., R. Eston, V. Marginson and M. McHugh. 2003. Effects of prior concentric training on eccentric exercise induced muscle damage. *British Journal of Sports Medicine* 37: 119-125.

Goetsch, S.C., T.J. Hawke, T.D. Gallardo, J.A. Richardson and D.J. Garry. 2003. Transcriptional profiling and regulation of the extracellular matrix during muscle regeneration. *Physiological Genomics* 14: 261-271.

Gosselin, L.E. 2000. Attenuation of force deficit after lengthening contractions in soleus muscle from trained rats. *Journal of Applied Physiology* 88: 1254-1258.

Gosselin, L.E. and H. Burton. 2002. Impact of initial muscle length on force deficit following lengthening contractions in mammalian skeletal muscle. *Muscle and Nerve* 25: 822-827.

Gregorevic, P., G.S. Lynch and D.A. Williams. 2000. Hyperbaric oxygen improves contractile function of regenerating rat skeletal muscle after myotoxic injury. *Journal of Applied Physiology* 89: 1477-1482.

Gulbin, J.P. and P.T. Gaffney. 2002. Identical twins are discordant for markers of eccentric exercise-induced muscle damage. *International Journal of Sports Medicine* 23: 471-476.

Hasselman, C.T., T.M. Best, A.V. Seaber and W.E. Garrett, Jr. 1995. A threshold and continuum of injury during active stretch of rabbit skeletal muscle. *American Journal of Sports Medicine* 23: 65-73.

Hayashi, M., H. Hirose, E. Sasaki, S. Senga, S. Murakawa, Y. Mori, Y. Furusawa and M. Banodo. 1998. Evaluation of ischemic damage in the skeletal muscle with the use of electrical properties. *Journal of Surgical Research* 80: 266-271.

Hortobagyi, T., J. Houmard, D. Fraser, R. Dudek, J. Lambert and J. Tracy. 1998. Normal forces and myofibrillar disruption after repeated eccentric exercise. *Journal of Applied Physiology* 84: 492-498.

Howatson, G. and K.A. Van Someren. 2003. Ice massage. Effects on exercise-induced muscle damage. *Journal of Sports Medicine and Physical Fitness* 43: 500-505.

Howell, J.N., G. Chleboun and R. Conatser. 1993. Muscle stiffness, strength loss, swelling and soreness following exercise-induced injury in humans. *Journal of Physiology* 464: 183-196.

Huda, R., D.R. Solanki and M. Mathru. 2004. Inflammatory and redox responses to ischaemia/reperfusion in human skeletal muscle. *Clinical Science* 107: 497-503.

Huda, R., L.A. Vergara, D.R. Solanki, E.R. Sherwood and M. Mathru. 2004. Selective activation of protein kinase C delta in human neutrophils following ischemia reperfusion of skeletal muscle. *Shock* 21: 500-504.

Hughes, W. and L.E. Gosselin. 2002. Impact of endurance concentric contraction training on acute force deficit following in vitro lengthening contractions. *European Journal of Applied Physiology* 87: 283-289.

Huijing, P.A. 1999. Muscle as a collagen fiber reinforced composite: A review of force transmission in muscle and whole limb. *Journal of Biomechanics* 32: 329-345.

Huijing, P.A. and G.C. Baan. 2001. Extramuscular myofascial force transmission within the rat anterior tibial compartment: Proximo-distal differences in muscle force. *Acta Physiologica Scandinavica* 173: 297-311.

Huijing, P.A., H. Maas and G.C. Baan. 2003. Compartmental fasciotomy and isolating a muscle from neighboring muscles interfere with myofascial force transmission within the rat anterior crural compartment. *Journal of Morphology* 256: 306-321.

Hunter, K.D. and J.A. Faulkner. 1997. Pliometric contraction-induced injury of mouse skeletal muscle: Effect of initial length. *Journal of Applied Physiology* 82: 278-283.

Ide, M., F. Tajima, K. Furusawa, T. Mizushima and H. Ogata. 1999. Wheelchair marathon racing causes striated muscle distress in individuals with spinal cord injury. *Archives of Physical Medicine and Rehabilitation* 80: 324-327.

Ingalls, C.P., G.L. Warren and R.B. Armstrong. 1998. Dissociation of force production from MHC and actin contents in muscles injured by eccentric contractions. *Journal of Muscle Research and Cell Motility* 19: 215-224.

Ingalls, C.P., G.L. Warren, D.A. Lowe, D.B. Boorstein and R.B. Armstrong. 1996. Differential effects of anesthetics on in vivo skeletal muscle contractile function in the mouse. *Journal of Applied Physiology* 80: 332-340.

Ingalls, C.P., G.L. Warren, J.H. Williams, C.W. Ward and R.B. Armstrong. 1998. E-C coupling failure in mouse EDL muscle after in vivo eccentric contractions. *Journal of Applied Physiology* 85: 58-67.

Ingalls, C.P., G.L. Warren, J.Z. Zhang, S.L. Hamilton and R.B. Armstrong. 2004. Dihydropyridine and ryanodine receptor binding after eccentric contractions in mouse skeletal muscle. *Journal of Applied Physiology* 96: 1619-1625.

Jackson, M.J., D.A. Jones and R.H. Edwards. 1984. Experimental skeletal muscle damage: The nature of the calcium-activated degenerative processes. *European Journal of Clinical Investigation* 14: 369-374.

Jamurtas, A.Z., V. Theocharis, T. Tofas, A. Tsiokanos, C. Yfanti, V. Paschalis, Y. Koutedakis and K. Nosaka. 2005. Comparison between leg and arm eccentric exercises of the same relative intensity on indices of muscle damage. *European Journal of Applied Physiology* 95: 179-185.

Johnson, M.A., J. Polgar, D. Weightman and D. Appleton. 1973. Data on the distribution of fibre types in thirty-six human muscles. An autopsy study. *Journal of the Neurological Sciences* 18: 111-129.

Jones, D.A., M.J. Jackson, G. McPhail and R.H. Edwards. 1984. Experimental mouse muscle damage: The importance of external calcium. *Clinical Science* 66: 317-322.

Jones, D.A., D.J. Newham, J.M. Round and S.E. Tolfree. 1986. Experimental human muscle damage: Morphological changes in relation to other indices of damage. *Journal of Physiology* 375: 435-448.

Kasperek, G.J. and R.D. Snider. 1985a. Increased protein degradation after eccentric exercise. *European Journal of Applied Physiology* 54: 30-34.

Kasperek, G.J. and R.D. Snider. 1985b. The susceptibility to exercise-induced muscle damage increases as rats grow larger. *Experientia* 41: 616-617.

Kaufman, T., J.R. Burke, J.M. Davis and J.L. Durstine. 2001. Exercise-induced neuromuscular dysfunction under reflex conditions. *European Journal of Applied Physiology* 84: 510-520.

Kern, D.S., J.G. Semmler and R.M. Enoka. 2001. Long-term activity in upper- and lower-limb muscles of humans. *Journal of Applied Physiology* 91: 2224-2232.

Kilmer, D.D., S.G. Aitkens, N.C. Wright and M.A. McCrory. 2001. Response to high-intensity eccentric muscle contractions in persons with myopathic disease. *Muscle and Nerve* 24: 1181-1187.

Kirk, S., J. Oldham, R. Kambadur, M. Sharma, P. Dobbie and J. Bass. 2000. Myostatin regulation during skeletal muscle regeneration. *Journal of Cellular Physiology* 184: 356-363.

Koh, T.J. and J. Escobedo. 2004. Cytoskeletal disruption and small heat shock protein translocation immediately after lengthening contractions. *American Journal of Physiology: Cell Physiology* 286: C713-722.

Koller, A., J. Mair, W. Schobersberger, T. Wohlfarter, C. Haid, M. Mayr, B. Villiger, W. Frey and B. Puschendorf. 1998. Effects of prolonged strenuous endurance exercise on plasma myosin heavy chain fragments and other muscular proteins. Cycling vs running. *Journal of Sports Medicine and Physical Fitness* 38: 10-17.

Komi, P.V. and J.T. Viitasalo. 1977. Changes in motor unit activity and metabolism in human skeletal muscle during and after repeated eccentric and concentric contractions. *Acta Physiologica Scandinavica* 100: 246-254.

Komulainen, J., S.O. Koskinen, R. Kalliokoski, T.E. Takala and V. Vihko. 1999. Gender differences in skeletal muscle fibre damage after eccentrically biased downhill running in rats. *Acta Physiologica Scandinavica* 165: 57-63.

Komulainen, J., J. Kytola and V. Vihko. 1994. Running-induced muscle injury and myocellular enzyme release in rats. *Journal of Applied Physiology* 77: 2299-2304.

Komulainen, J., T.E. Takala, H. Kuipers and M.K. Hesselink. 1998. The disruption of myofibre structures in rat skeletal muscle after forced lengthening contractions. *Pflügers Archiv: European Journal of Physiology* 436: 735-741.

Komulainen, J. and V. Vihko. 1994. Exercise-induced necrotic muscle damage and enzyme release in the four days following prolonged submaximal running in rats. *Pflugers Archiv: European Journal of Physiology* 428: 346-351.

Komulainen, J. and V. Vihko. 1995. Training-induced protection and effect of terminated training on exercise-induced damage and water content in mouse skeletal muscles. *International Journal of Sports Medicine* 16: 293-297.

Kuang, W., H. Xu, J.T. Vilquin and E. Engvall. 1999. Activation of the lama2 gene in muscle regeneration: Abortive regeneration in laminin alpha2-deficiency. *Laboratory Investigation* 79: 1601-1613.

Kuipers, H., J. Drukker, P.M. Frederik, P. Geurten and G. van Kranenburg. 1983. Muscle degeneration after exercise in rats. *International Journal of Sports Medicine* 4: 45-51.

Kyparos, A., C. Matziari, M. Albani, G. Arsos and A. Deligiannis. 2001. A decrease in soleus muscle force generation in rats after downhill running. *Canadian Journal of Applied Physiology* 26: 323-335.

Lambert, M.I., P. Marcus, T. Burgess and T.D. Noakes. 2002. Electro-membrane microcurrent therapy reduces signs and symptoms of muscle damage. *Medicine and Science in Sports and Exercise* 34: 602-607.

Lannergren, J., H. Westerblad and J.D. Bruton. 2002. Dynamic vacuolation in skeletal muscle fibres after fatigue. *Cell Biology International* 26: 911-920.

Li, M., K. Chan, D. Cai, P. Leung, C. Cheng, K. Lee and K.K. Lee. 2000. Identification and purification of an intrinsic human muscle myogenic factor that enhances muscle repair and regeneration. *Archives of Biochemistry and Biophysics* 384: 263-268.

Lieber, R.L. and J. Friden. 1988. Selective damage of fast glycolytic muscle fibres with eccentric contraction of the rabbit tibialis anterior. *Acta Physiologica Scandinavica* 133: 587-588.

Lieber, R.L., M.C. Schmitz, D.K. Mishra and J. Friden. 1994. Contractile and cellular remodeling in rabbit skeletal muscle after cyclic eccentric contractions. *Journal of Applied Physiology* 77: 1926-1934.

Lieber, R.L., L.E. Thornell and J. Friden. 1996. Muscle cytoskeletal disruption occurs within the first 15 min of cyclic eccentric contraction. *Journal of Applied Physiology* 80: 278-284.

Lieber, R.L., T.M. Woodburn and J. Friden. 1991. Muscle damage induced by eccentric contractions of 25% strain. *Journal of Applied Physiology* 70: 2498-2507.

Lovering, R.M. and P.G. De Deyne. 2004. Contractile function, sarcolemma integrity, and the loss of dystrophin after skeletal muscle eccentric contraction-induced injury. *American Journal of Physiology: Cell Physiology* 286: C230-238.

Lowe, D.A., G.L. Warren, D.A. Hayes, M.A. Farmer and R.B. Armstrong. 1994. Eccentric contraction-induced injury of mouse soleus muscle: Effect of varying $[Ca^{2+}]_o$. *Journal of Applied Physiology* 76: 1445-1453.

Lowe, D.A., G.L. Warren, D.A. Hayes, B. Prior and R.B. Armstrong. 1993. Running downhill does not injure mouse soleus muscles. *International Journal of Sports Medicine* 14: 104-105.

Lowe, D.A., G.L. Warren, C.P. Ingalls, D.B. Boorstein and R.B. Armstrong. 1995. Muscle function and protein metabolism after initiation of eccentric contraction-induced injury. *Journal of Applied Physiology* 79: 1260-1270.

Lynch, G.S., C.J. Fary and D.A. Williams. 1997. Quantitative measurement of resting skeletal muscle $[Ca^{2+}]_i$ following acute and long-term downhill running exercise in mice. *Cell Calcium* 22: 373-383.

Lynch, G.S. and J.A. Faulkner. 1998. Contraction-induced injury to single muscle fibers: Velocity of stretch does not influence the force deficit. *American Journal of Physiology* 275: C1548-1554.

Lynch, G.S., J.A. Rafael, J.S. Chamberlain and J.A. Faulkner. 2000. Contraction-induced injury to single permeabilized muscle fibers from mdx, transgenic mdx, and control mice. *American Journal of Physiology: Cell Physiology* 279: C1290-1294.

Maas, H., G.C. Baan and P.A. Huijing. 2001. Intermuscular interaction via myofascial force transmission: Effects of tibialis anterior and extensor hallucis longus length on force transmission from rat extensor digitorum longus muscle. *Journal of Biomechanics* 34: 927-940.

Macpherson, P.C., R.G. Dennis and J.A. Faulkner. 1997. Sarcomere dynamics and contraction-induced injury to maximally activated single muscle fibres from soleus muscles of rats. *Journal of Physiology* 500: 523-533.

Macpherson, P.C., M.A. Schork and J.A. Faulkner. 1996. Contraction-induced injury to single fiber segments from fast and slow muscles of rats by single stretches. *American Journal of Physiology* 271: C1438-1446.

Mair, J., A. Koller, E. Artner-Dworzak, C. Haid, K. Wicke, W. Judmaier and B. Puschendorf. 1992. Effects of exercise on plasma myosin heavy chain fragments and MRI of skeletal muscle. *Journal of Applied Physiology* 72: 656-663.

Mair, J., M. Mayr, E. Muller, A. Koller, C. Haid, E. Artner-Dworzak, C. Calzolari, C. Larue and B. Puschendorf. 1995. Rapid adaptation to eccentric exercise-induced muscle damage. *International Journal of Sports Medicine* 16: 352-356.

Mair, S.D., A.V. Seaber, R.R. Glisson and W.E. Garrett Jr. 1996. The role of fatigue in susceptibility to acute muscle strain injury. *American Journal of Sports Medicine* 24: 137-143.

Matin, P., G. Lang, R. Carretta and G. Simon. 1983. Scintigraphic evaluation of muscle damage following extreme exercise: Concise communication. *Journal of Nuclear Medicine* 24: 308-311.

McArdle, A., W.H. Dillmann, R. Mestril, J.A. Faulkner and M.J. Jackson. 2004. Overexpression of HSP70 in mouse skeletal muscle protects against muscle damage and age-related muscle dysfunction. *FASEB Journal* 18: 355-357.

McBride, T. 2000. Increased depolarization, prolonged recovery and reduced adaptation of the resting membrane potential in aged rat skeletal muscles following eccentric contractions. *Mechanisms of Ageing and Development* 115: 127-138.

McBride, T.A., B.W. Stockert, F.A. Gorin and R.C. Carlsen. 2000. Stretch-activated ion channels contribute to membrane depolarization after eccentric contractions. *Journal of Applied Physiology* 88: 91-101.

McCall, K.E. and C.J. Duncan. 1989. Independent pathways causing cellular damage in mouse soleus muscle under hypoxia. *Comparative Biochemistry and Physiology. A, Comparative Physiology* 94: 799-804.

McCully, K.K. and J.A. Faulkner. 1985. Injury to skeletal muscle fibers of mice following lengthening contractions. *Journal of Applied Physiology* 59: 119-126.

McCully, K.K. and J.A. Faulkner. 1986. Characteristics of lengthening contractions associated with injury to skeletal muscle fibers. *Journal of Applied Physiology* 61: 293-299.

McHugh, M.P. 2003. Recent advances in the understanding of the repeated bout effect: The protective effect against muscle damage from a single bout of eccentric exercise. *Scandinavian Journal of Medicine and Science in Sports* 13: 88-97.

McNeil, P.L. and R. Khakee. 1992. Disruptions of muscle fiber plasma membranes. Role in exercise-induced damage. *American Journal of Pathology* 140: 1097-1109.

Menetrey, J., C. Kasemkijwattana, F.H. Fu, M.S. Moreland and J. Huard. 1999. Suturing versus immobilization of a muscle laceration. A morphological and functional study in a mouse model. *American Journal of Sports Medicine* 27: 222-229.

Michaut, A., M. Pousson, N. Babault and J. Van Hoecke. 2002. Is eccentric exercise-induced torque decrease contraction type dependent? *Medicine and Science in Sports and Exercise* 34: 1003-1008.

Michaut, A., M. Pousson, Y. Ballay and J. Van Hoecke. 2001. Short-term changes in the series elastic component after an acute eccentric exercise of the elbow flexors. *European Journal of Applied Physiology* 84: 569-574.

Mishra, D.K., J. Friden, M.C. Schmitz and R.L. Lieber. 1995. Anti-inflammatory medication after muscle injury. A treatment resulting in short-term improvement but subsequent loss of muscle function. *Journal of Bone and Joint Surgery* 77: 1510-1519.

Morgan, D.L., D.R. Claflin and F.J. Julian. 1996. The effects of repeated active stretches on tension generation and myoplasmic calcium in frog single muscle fibres. *Journal of Physiology* 497: 665-674.

Morgan, D.L., J.E. Gregory and U. Proske. 2004. The influence of fatigue on damage from eccentric contractions in the gastrocnemius muscle of the cat. *Journal of Physiology* 561: 841-850.

Newham, D.J., D.A. Jones and P.M. Clarkson. 1987. Repeated high-force eccentric exercise: Effects on muscle pain and damage. *Journal of Applied Physiology* 63: 1381-1386.

Newham, D.J., D.A. Jones, G. Ghosh and P. Aurora. 1988. Muscle fatigue and pain after eccentric contractions at long and short length. *Clinical Science* 74: 553-557.

Newham, D.J., D.A. Jones, S.E. Tolfree and R.H. Edwards. 1986. Skeletal muscle damage: a study of isotope uptake, enzyme efflux and pain after stepping. *European Journal of Applied Physiology* 55: 106-112.

Newham, D.J., G. McPhail, K.R. Mills and R.H. Edwards. 1983. Ultrastructural changes after concentric and eccentric contractions of human muscle. *Journal of the Neurological Sciences* 61: 109-122.

Newton, M., K. Nosaka and P. Sacco. 2000. Relationship between force and histological changes in mouse anterior crural muscles following eccentric exercise. *Basic and Applied Myology* 10: 225-229.

Noakes, T.D. and J.W. Carter. 1982. The responses of plasma biochemical parameters to a 56-km race in novice and experienced ultra-marathon runners. *European Journal of Applied Physiology* 49: 179-186.

Noakes, T.D., G. Kotzenberg, P.S. McArthur and J. Dykman. 1983. Elevated serum creatine kinase MB and creatine kinase BB-isoenzyme fractions after ultra-marathon running. *European Journal of Applied Physiology* 52: 75-79.

Nosaka, K. and P.M. Clarkson. 1992. Changes in plasma zinc following high force eccentric exercise. *International Journal of Sport Nutrition* 2: 175-184.

Nosaka, K. and P.M. Clarkson. 1996a. Changes in indicators of inflammation after eccentric exercise of the elbow flexors. *Medicine and Science in Sports and Exercise* 28: 953-961.

Nosaka, K. and P.M. Clarkson. 1996b. Variability in serum creatine kinase response after eccentric exercise of the elbow flexors. *International Journal of Sports Medicine* 17: 120-127.

Nosaka, K. and P.M. Clarkson. 1997. Influence of previous concentric exercise on eccentric exercise-induced muscle damage. *Journal of Sports Sciences* 15: 477-483.

Nosaka, K. and M. Newton. 2002a. Concentric or eccentric training effect on eccentric exercise-induced muscle damage. *Medicine and Science in Sports and Exercise* 34: 63-69.

Nosaka, K. and M. Newton. 2002b. Difference in the magnitude of muscle damage between maximal and submaximal eccentric loading. *Journal of Strength and Conditioning Research* 16: 202-208.

Nosaka, K., M. Newton and P. Sacco. 2002. Delayed-onset muscle soreness does not reflect the magnitude of eccentric exercise-induced muscle damage. *Scandinavian Journal of Medicine and Science in Sports* 12: 337-346.

Nosaka, K. and K. Sakamoto. 2001. Effect of elbow joint angle on the magnitude of muscle damage to the elbow flexors. *Medicine and Science in Sports and Exercise* 33: 22-29.

Nosaka, K., K. Sakamoto, M. Newton and P. Sacco. 2001. The repeated bout effect of reduced-load eccentric exercise on elbow flexor muscle damage. *European Journal of Applied Physiology* 85: 34-40.

Ogilvie, R.W., R.B. Armstrong, K.E. Baird and C.L. Bottoms. 1988. Lesions in the rat soleus muscle following eccentrically biased exercise. *American Journal of Anatomy* 182: 335-346.

Pachori, A.S., L.G. Melo, M.L. Hart, N. Noiseux, L. Zhang, F. Morello, S.D. Solomon, G.L. Stahl, R.E. Pratt and V.J. Dzau. 2004. Hypoxia-regulated therapeutic gene as a preemptive treatment strategy against ischemia/reperfusion tissue injury. *Proceedings of the National Academy of Sciences of the United States of America* 101: 12282-12287.

Paddon-Jones, D. and P.J. Abernethy. 2001. Acute adaptation to low volume eccentric exercise. *Medicine and Science in Sports and Exercise* 33: 1213-1219.

Paddon-Jones, D., A. Keech and D. Jenkins. 2001. Short-term beta-hydroxy-beta-methylbutyrate supplementation does not reduce symptoms of eccentric muscle damage. *International Journal of Sport Nutrition and Exercise Metabolism* 11: 442-450.

Paddon-Jones, D., M. Muthalib and D. Jenkins. 2000. The effects of a repeated bout of eccentric exercise on indices of muscle damage and delayed onset muscle soreness. *Journal of Science and Medicine in Sport* 3: 35-43.

Paddon-Jones, D.J. and B.M. Quigley. 1997. Effect of cryotherapy on muscle soreness and strength following eccentric exercise. *International Journal of Sports Medicine* 18: 588-593.

Parikh, S., D.L. Morgan, J.E. Gregory and U. Proske. 2004. Low-frequency depression of tension in the cat gastrocnemius muscle after eccentric exercise. *Journal of Applied Physiology* 97: 1195-1202.

Patel, T.J., D. Cuizon, O. Mathieu-Costello, J. Friden and R.L. Lieber. 1998. Increased oxidative capacity does not protect skeletal muscle fibers from eccentric contraction-induced injury. *American Journal of Physiology* 274: R1300-1308.

Patel, T.J., R. Das, J. Friden, G.J. Lutz and R.L. Lieber. 2004. Sarcomere strain and heterogeneity correlate with injury to frog skeletal muscle fiber bundles. *Journal of Applied Physiology* 97: 1803-1813.

Pavlath, G.K., D. Thaloor, T.A. Rando, M. Cheong, A.W. English and B. Zheng. 1998. Heterogeneity among muscle precursor cells in adult skeletal muscles with differing regenerative capacities. *Developmental Dynamics* 212: 495-508.

Peters, D., I.A. Barash, M. Burdi, P.S. Yuan, L. Mathew, J. Friden and R.L. Lieber. 2003. Asynchronous functional, cellular and transcriptional changes after a bout of eccentric exercise in the rat. *Journal of Physiology* 553: 947-957.

Phillips, T., A.C. Childs, D.M. Dreon, S. Phinney and C. Leeuwenburgh. 2003. A dietary supplement attenuates IL-6 and CRP after eccentric exercise in untrained males. *Medicine and Science in Sports and Exercise* 35: 2032-2037.

Pizza, F.X., J.M. Peterson, J.H. Baas and T.J. Koh. 2005. Neutrophils contribute to muscle injury and impair its resolution after lengthening contractions in mice. *Journal of Physiology* 562: 899-913.

Ploutz-Snyder, L.L., E.L. Giamis, M. Formikell and A.E. Rosenbaum. 2001. Resistance training reduces susceptibility to eccentric exercise-induced muscle dysfunction in older women. *Journals of Gerontology Series A: Biological Sciences and Medical Sciences* 56: B384-390.

Publicover, S.J., C.J. Duncan and J.L. Smith. 1978. The use of A23187 to demonstrate the role of intracellular calcium in causing ultrastructural damage in mammalian muscle. *Journal of Neuropathology and Experimental Neurology* 37: 554-557.

Qiu, F.H., K. Wada, G.L. Stahl and C.N. Serhan. 2000. IMP and AMP deaminase in reperfusion injury down-regulates neutrophil recruitment. *Proceedings of the National Academy of Sciences of the United States of America* 97: 4267-4272.

Rathbone, C.R., J.C. Wenke, G.L. Warren and R.B. Armstrong. 2003. Importance of satellite cells in the strength recovery after eccentric contraction-induced muscle injury. *American Journal of Physiology: Regulatory, Integrative, and Comparative Physiology* 285: R1490-1495.

Rawson, E.S., B. Gunn and P.M. Clarkson. 2001. The effects of creatine supplementation on exercise-induced muscle damage. *Journal of Strength and Conditioning Research* 15: 178-184.

Rinard, J., P.M. Clarkson, L.L. Smith and M. Grossman. 2000. Response of males and females to high-force eccentric exercise. *Journal of Sports Sciences* 18: 229-236.

Rowlands, A.V., R.G. Eston and C. Tilzey. 2001. Effect of stride length manipulation on symptoms of exercise-induced muscle damage and the repeated bout effect. *Journal of Sports Sciences* 19: 333-340.

Sacco, P., D.A. Jones, J.R. Dick and G. Vrbova. 1992. Contractile properties and susceptibility to exercise-induced damage of normal and mdx mouse tibialis anterior muscle. *Clinical Science* 82: 227-236.

Sacheck, J.M., P.E. Milbury, J.G. Cannon, R. Roubenoff and J.B. Blumberg. 2003. Effect of vitamin E and eccentric exercise on selected biomarkers of oxidative stress in young and elderly men. *Free Radical Biology and Medicine* 34: 1575-1588.

Safran, M.R., W.E. Garrett Jr., A.V. Seaber, R.R. Glisson and B.M. Ribbeck. 1988. The role of warmup in muscular injury prevention. *American Journal of Sports Medicine* 16: 123-129.

Sakamoto, K., K. Nosaka, S. Shimegi, H. Ohmori and S. Katsuta. 1996. Creatine kinase release from regenerated muscles after eccentric contractions in rats. *European Journal of Applied Physiology* 73: 516-520.

Salminen, A. and M. Kihlstrom. 1987. Protective effect of indomethacin against exercise-induced injuries in mouse skeletal muscle fibers. *International Journal of Sports Medicine* 8: 46-49.

Sam, M., S. Shah, J. Friden, D.J. Milner, Y. Capetanaki and R.L. Lieber. 2000. Desmin knockout muscles generate lower stress and are less vulnerable to injury compared with wild-type muscles. *American Journal of Physiology: Cell Physiology* 279: C1116-1122.

Savage, K.J. and P.M. Clarkson. 2002. Oral contraceptive use and exercise-induced muscle damage and recovery. *Contraception* 66: 67-71.

Sayers, S.P. and P.M. Clarkson. 2003. Short-term immobilization after eccentric exercise. Part II: Creatine kinase and myoglobin. *Medicine and Science in Sports and Exercise* 35: 762-768.

Sayers, S.P., C.A. Knight and P.M. Clarkson. 2003. Neuromuscular variables affecting the magnitude of force loss after eccentric exercise. *Journal of Sports Sciences* 21: 403-410.

Sayers, S.P., B.T. Peters, C.A. Knight, M.L. Urso, J. Parkington and P.M. Clarkson. 2003. Short-term immobilization after eccentric exercise. Part I: Contractile properties. *Medicine and Science in Sports and Exercise* 35: 753-761.

Sbriccoli, P., F. Felici, A. Rosponi, A. Aliotta, V. Castellano, C. Mazza, M. Bernardi and M. Marchetti. 2001. Exercise induced muscle damage and recovery assessed by means of linear and non-linear sEMG analysis and ultrasonography. *Journal of Electromyography and Kinesiology* 11: 73-83.

Schneider, B.S., H. Sannes, J. Fine and T. Best. 2002. Desmin characteristics of CD11b-positive fibers after eccentric contractions. *Medicine and Science in Sports and Exercise* 34: 274-281.

Schwane, J.A. and R.B. Armstrong. 1983. Effect of training on skeletal muscle injury from downhill running in rats. *Journal of Applied Physiology* 55: 969-975.

Segal, S.S. and J.A. Faulkner. 1985. Temperature-dependent physiological stability of rat skeletal muscle in vitro. *American Journal of Physiology* 248: C265-270.

Serrao, F.V., B. Foerster, S. Spada, M.M. Morales, V. Monteiro-Pedro, A. Tannus and T.F. Salvini. 2003. Functional changes of human quadriceps muscle injured by eccentric exercise. *Brazilian Journal of Medical and Biological Research* 36: 781-786.

Shamsadeen, N.M. and C.J. Duncan. 1989. Effects of anoxia on cellular damage in the incubated mouse diaphragm. *Tissue and Cell* 21: 857-861.

Shave, R., E. Dawson, G. Whyte, K. George, D. Ball, P. Collinson and D. Gaze. 2002. The cardiospecificity of the third-generation cTnT assay after exercise-induced muscle damage. *Medicine and Science in Sports and Exercise* 34: 651-654.

Shimomura, Y., M. Suzuki, S. Sugiyama, Y. Hanaki and T. Ozawa. 1991. Protective effect of coenzyme Q10 on exercise-induced muscular injury. *Biochemical and Biophysical Research Communications* 176: 349-355.

Shiotani, A., M. Fukumura, M. Maeda, X. Hou, M. Inoue, T. Kanamori, S. Komaba, K. Washizawa, S. Fujikawa, T. Yamamoto, C. Kadono, K. Watabe, H. Fukuda, K. Saito, Y. Sakai, Y. Nagai, J. Kanzaki and M. Hasegawa. 2001. Skeletal muscle regeneration after insulin-like growth factor I gene transfer by recombinant Sendai virus vector. *Gene Therapy* 8: 1043-1050.

Siekmann, L., R. Bonora, C.A. Burtis, F. Ceriotti, P. Clerc-Renaud, G. Ferard et al. 2002. IFCC primary reference procedures for the measurement of catalytic activity concentrations of enzymes at 37 degrees C. Part 1. The concept of reference procedures for the measurement of catalytic activity concentrations of enzymes. *Clinical Chemistry and Laboratory Medicine* 40: 631-634.

Smith, H.K., M.J. Plyley, C.D. Rodgers and N.H. McKee. 1997. Skeletal muscle damage in the rat hindlimb following single or repeated daily bouts of downhill exercise. *International Journal of Sports Medicine* 18: 94-100.

Snyder, A.C., D.R. Lamb, C.P. Salm, M.D. Judge, E.D. Aberle and E.W. Mills. 1984. Myofibrillar protein degradation after eccentric exercise. *Experientia* 40: 69-70.

Soeta, C., M. Suzuki, S. Suzuki, K. Naito, C. Tachi and H. Tojo. 2001. Possible role for the c-ski gene in the proliferation of myogenic cells in regenerating skeletal muscles of rats. *Development, Growth and Differentiation* 43: 155-164.

Sorichter, S., A. Koller, C. Haid, K. Wicke, W. Judmaier, P. Werner and E. Raas. 1995. Light concentric exercise and heavy eccentric muscle loading: Effects on CK, MRI and markers of inflammation. *International Journal of Sports Medicine* 16: 288-292.

Sorichter, S., J. Mair, A. Koller, W. Gebert, D. Rama, C. Calzolari, E. Artner-Dworzak and B. Puschendorf. 1997. Skeletal troponin I as a marker of exercise-induced muscle damage. *Journal of Applied Physiology* 83: 1076-1082.

Sorichter, S., J. Mair, A. Koller, E. Muller, C. Kremser, W. Judmaier, C. Haid, C. Calzolari and B. Puschendorf. 2001. Creatine kinase, myosin heavy chains and magnetic resonance imaging after eccentric exercise. *Journal of Sports Sciences* 19: 687-691.

Sorichter, S., J. Mair, A. Koller, E. Muller, C. Kremser, W. Judmaier, C. Haid, D. Rama, C. Calzolari and B. Puschendorf. 1997. Skeletal muscle troponin I release and magnetic resonance imaging signal intensity changes after eccentric exercise-induced skeletal muscle injury. *Clinica Chimica Acta* 262: 139-146.

Sotiriadou, S., A. Kyparos, V. Mougios, C. Trontzos, G. Sidiras and C. Matziari. 2003. Estrogen effect on some enzymes in female rats after downhill running. *Physiological Research* 52: 743-748.

Staron, R.S., F.C. Hagerman, R.S. Hikida, T.F. Murray, D.P. Hostler, M.T. Crill, K.E. Ragg and K. Toma. 2000. Fiber type composition of the vastus lateralis muscle of young men and women. *Journal of Histochemistry and Cytochemistry* 48: 623-629.

Stauber, W.T., P.M. Clarkson, V.K. Fritz and W.J. Evans. 1990. Extracellular matrix disruption and pain after eccentric muscle action. *Journal of Applied Physiology* 69: 868-874.

Stauber, W.T., K.K. Knack, G.R. Miller and J.G. Grimmett. 1996. Fibrosis and intercellular collagen connections from four weeks of muscle strains. *Muscle and Nerve* 19: 423-430.

Stauber, W.T. and M.E. Willems. 2002. Prevention of histopathologic changes from 30 repeated stretches of active rat skeletal muscles by long inter-stretch rest times. *European Journal of Applied Physiology* 88: 94-99.

St Pierre Schneider, B., L.A. Correia and J.G. Cannon. 1999. Sex differences in leukocyte invasion in injured murine skeletal muscle. *Research in Nursing and Health* 22: 243-250.

Strickler, T., T. Malone and W.E. Garrett. 1990. The effects of passive warming on muscle injury. *American Journal of Sports Medicine* 18: 141-145.

Stupka, N., M.A. Tarnopolsky, N.J. Yardley and S.M. Phillips. 2001. Cellular adaptation to repeated eccentric exercise-induced muscle damage. *Journal of Applied Physiology* 91: 1669-1678.

Summan, M., M. McKinstry, G.L. Warren, T. Hulderman, D. Mishra, K. Brumbaugh, M.I. Luster and P.P. Simeonova. 2003. Inflammatory mediators and skeletal muscle injury: A DNA microarray analysis. *Journal of Interferon and Cytokine Research* 23: 237-245.

Takekura, H., N. Fujinami, T. Nishizawa, H. Ogasawara and N. Kasuga. 2001. Eccentric exercise-induced morphological changes in the membrane systems involved in excitation-contraction coupling in rat skeletal muscle. *Journal of Physiology* 533: 571-583.

Talbot, J.A. and D.L. Morgan. 1998. The effects of stretch parameters on eccentric exercise-induced damage to toad skeletal muscle. *Journal of Muscle Research and Cell Motility* 19: 237-245.

Taylor, C.R., S.L. Caldwell and V.J. Rowntree. 1972. Running up and down hills: Some consequences of size. *Science* 178: 1096-1097.

Teague, B.N. and J.A. Schwane. 1995. Effect of intermittent eccentric contractions on symptoms of muscle microinjury. *Medicine and Science in Sports and Exercise* 27: 1378-1384.

Thompson, H.S., P.M. Clarkson and S.P. Scordilis. 2002. The repeated bout effect and heat shock proteins: Intramuscular HSP27 and HSP70 expression following two bouts of eccentric exercise in humans. *Acta Physiologica Scandinavica* 174: 47-56.

Thompson, H.S., E.B. Maynard, E.R. Morales and S.P. Scordilis. 2003. Exercise-induced HSP27, HSP70 and MAPK responses in human skeletal muscle. *Acta Physiologica Scandinavica* 178: 61-72.

Tiidus, P.M., D. Holden, E. Bombardier, S. Zajchowski, D. Enns and A. Belcastro. 2001. Estrogen effect on post-exercise skeletal muscle neutrophil infiltration and calpain activity. *Canadian Journal of Physiology and Pharmacology* 79: 400-406.

Tomiya, A., T. Aizawa, R. Nagatomi, H. Sensui and S. Kokubun. 2004. Myofibers express IL-6 after eccentric exercise. *American Journal of Sports Medicine* 32: 503-508.

Tsivitse, S.K., T.J. McLoughlin, J.M. Peterson, E. Mylona, S.J. McGregor and F.X. Pizza. 2003. Downhill running in rats: Influence on neutrophils, macrophages, and MyoD+ cells in skeletal muscle. *European Journal of Applied Physiology* 90: 633-638.

Uchiyama, Y., T. Tamaki and H. Fukuda. 2001. Relationship between functional deficit and severity of experimental fast-strain injury of rat skeletal muscle. *European Journal of Applied Physiology* 85: 1-9.

Ullman, M. and A. Oldfors. 1991. Skeletal muscle regeneration in young rats is dependent on growth hormone. *Journal of the Neurological Sciences* 106: 67-74.

Varejao, A.S., A.M. Cabrita, M.F. Meek, J. Bulas-Cruz, R.C. Gabriel, V.M. Filipe, P. Melo-Pinto and D.A. Winter. 2002. Motion of the foot and ankle during the stance phase in rats. *Muscle and Nerve* 26: 630-635.

Vignaud, A., P. Noirez, S. Besse, M. Rieu, D. Barritault and A. Ferry. 2003. Recovery of slow skeletal muscle after injury in the senescent rat. *Experimental Gerontology* 38: 529-537.

Vihko, V., J. Rantamaki and A. Salminen. 1978. Exhaustive physical exercise and acid hydrolase activity in mouse skeletal muscle. A histochemical study. *Histochemistry* 57: 237-249.

Warren, G.L., J.M. Fennessy and M.L. Millard-Stafford. 2000. Strength loss after eccentric contractions is unaffected by creatine supplementation. *Journal of Applied Physiology* 89: 557-562.

Warren, G.L., D.A. Hayes, D.A. Lowe and R.B. Armstrong. 1993. Mechanical factors in the initiation of eccentric contraction-induced injury in rat soleus muscle. *Journal of Physiology* 464: 457-475.

Warren, G.L., D.A. Hayes, D.A. Lowe, B.M. Prior and R.B. Armstrong. 1993. Materials fatigue initiates eccentric contraction-induced injury in rat soleus muscle. *Journal of Physiology* 464: 477-489.

Warren, G.L., D.A. Hayes, D.A. Lowe, J.H. Williams and R.B. Armstrong. 1994. Eccentric contraction-induced injury in normal and hindlimb-suspended mouse soleus and EDL muscles. *Journal of Applied Physiology* 77: 1421-1430.

Warren, G.L., K.M. Hermann, C.P. Ingalls, M.R. Masselli and R.B. Armstrong. 2000. Decreased EMG median frequency during a second bout of eccentric contractions. *Medicine and Science in Sports and Exercise* 32: 820-829.

Warren, G.L., T. Hulderman, N. Jensen, M. McKinstry, M. Mishra, M.I. Luster and P.P. Simeonova. 2002. Physiological role of tumor necrosis factor alpha in traumatic muscle injury. *FASEB Journal* 16: 1630-1632.

Warren, G.L., T. Hulderman, D. Mishra, X. Gao, L. Millecchia, L. O'Farrell, W.A. Kuziel and P.P. Simeonova. 2005. Chemokine receptor CCR2 involvement in skeletal muscle regeneration. *FASEB Journal* 19: 413-415.

Warren, G.L., C.P. Ingalls and R.B. Armstrong. 1998. A stimulating nerve cuff for chronic in vivo measurements of torque produced about the ankle in the mouse. *Journal of Applied Physiology* 84: 2171-2176.

Warren, G.L., C.P. Ingalls and R.B. Armstrong. 2002. Temperature dependency of force loss and Ca²⁺ homeostasis in mouse EDL muscle after eccentric contractions. *American Journal of Physiology: Regulatory, Integrative, and Comparative Physiology* 282: R1122-1132.

Warren, G.L., C.P. Ingalls, D.A. Lowe and R.B. Armstrong. 2001. Excitation-contraction uncoupling: Major role in contraction-induced muscle injury. *Exercise and Sport Sciences Reviews* 29: 82-87.

Warren, G.L., C.P. Ingalls, S.J. Shah and R.B. Armstrong. 1999. Uncoupling of in vivo torque production from EMG in mouse muscles injured by eccentric contractions. *Journal of Physiology* 515: 609-619.

Warren, G.L., D.A. Lowe and R.B. Armstrong. 1999. Measurement tools used in the study of eccentric contraction-induced injury. *Sports Medicine* 27: 43-59.

Warren, G.L., D.A. Lowe, D.A. Hayes, M.A. Farmer and R.B. Armstrong. 1995. Redistribution of cell membrane probes following contraction-induced injury of mouse soleus muscle. *Cell and Tissue Research* 282: 311-320.

Warren, G.L., D.A. Lowe, D.A. Hayes, C.J. Karwoski, B.M. Prior and R.B. Armstrong. 1993. Excitation failure in eccentric contraction-induced injury of mouse soleus muscle. *Journal of Physiology* 468: 487-499.

Warren, G.L., D.A. Lowe, C.L. Inman, O.M. Orr, H.A. Hogan, S.A. Bloomfield and R.B. Armstrong. 1996. Estradiol effect on anterior crural muscles-tibial bone relationship and susceptibility to injury. *Journal of Applied Physiology* 80: 1660-1665.

Warren, G.L., L. O'Farrell, M. Summan, T. Hulderman, D. Mishra, M.I. Luster, W.A. Kuziel and P.P. Simeonova. 2004. Role of CC chemokines in skeletal muscle functional restoration after injury. *American Journal of Physiology: Cell Physiology* 286: C1031-1036.

Warren, G.L., 3rd, J.H. Williams, C.W. Ward, H. Matoba, C.P. Ingalls, K.M. Hermann and R.B. Armstrong. 1996. Decreased contraction economy in mouse EDL muscle injured by eccentric contractions. *Journal of Applied Physiology* 81: 2555-2564.

Warren, J.A., R.R. Jenkins, L. Packer, E.H. Witt and R.B. Armstrong. 1992. Elevated muscle vitamin E does not attenuate eccentric exercise-induced muscle injury. *Journal of Applied Physiology* 72: 2168-2175.

Whitehead, N.P., D.L. Morgan, J.E. Gregory and U. Proske. 2003. Rises in whole muscle passive tension of mammalian muscle after eccentric contractions at different lengths. *Journal of Applied Physiology* 95: 1224-1234.

Whitehead, N.P., N.S. Weerakkody, J.E. Gregory, D.L. Morgan and U. Proske. 2001. Changes in passive tension of muscle in humans and animals after eccentric exercise. *Journal of Physiology* 533: 593-604.

Willems, M.E., G.R. Miller and W.T. Stauber. 2001. Force deficits after stretches of activated rat muscle-tendon complex with reduced collagen cross-linking. *European Journal of Applied Physiology* 85: 405-411.

Willems, M.E. and W.T. Stauber. 2000. Force output during and following active stretches of rat plantar flexor muscles: Effect of velocity of ankle rotation. *Journal of Biomechanics* 33: 1035-1038.

Willems, M.E. and W.T. Stauber. 2002a. Effect of contraction history on torque deficits by stretches of active rat skeletal muscles. *Canadian Journal of Applied Physiology* 27: 323-335.

Willems, M.E. and W.T. Stauber. 2002b. Force deficits by stretches of activated muscles with constant or increasing velocity. *Medicine and Science in Sports and Exercise* 34: 667-672.

Wojcik, J.R., J. Walber-Rankin, L.L. Smith and F.C. Gwazdauskas. 2001. Comparison of carbohydrate and milk-based beverages on muscle damage and glycogen following exercise. *International Journal of Sport Nutrition and Exercise Metabolism* 11: 406-419.

Wood, S.A., D.L. Morgan and U. Proske. 1993. Effects of repeated eccentric contractions on structure and mechanical properties of toad sartorius muscle. *American Journal of Physiology* 265: C792-800.

Yasuda, T., K. Sakamoto, K. Nosaka, M. Wada and S. Katsuta. 1997. Loss of sarcoplasmic reticulum membrane integrity after eccentric contractions. *Acta Physiologica Scandinavica* 161: 581-582.

Yeung, E.W., H.J. Ballard, J.P. Bourreau and D.G. Allen. 2003. Intracellular sodium in mammalian muscle fibers after eccentric contractions. *Journal of Applied Physiology* 94: 2475-2482.

Yeung, E.W., C.D. Balnave, H.J. Ballard, J.P. Bourreau and D.G. Allen. 2002. Development of T-tubular vacuoles in eccentrically damaged mouse muscle fibres. *Journal of Physiology* 540: 581-592.

Yeung, E.W., J.P. Bourreau, D.G. Allen and H.J. Ballard. 2002. Effect of eccentric contraction-induced injury on force and intracellular pH in rat skeletal muscles. *Journal of Applied Physiology* 92: 93-99.

Yu, J.G., C. Malm and L.E. Thornell. 2002. Eccentric contractions leading to DOMS do not cause loss of desmin nor fibre necrosis in human muscle. *Histochemistry and Cell Biology* 118: 29-34.

Zhang, J. and G.K. Dhoot. 1998. Localized and limited changes in the expression of myosin heavy chains in injured skeletal muscle fibers being repaired. *Muscle and Nerve* 21: 469-481.

Zhao, P., S. Iezzi, E. Carver, D. Dressman, T. Gridley, V. Sartorelli and E.P. Hoffman. 2002. Slug is a novel downstream target of MyoD. Temporal profiling in muscle regeneration. *Journal of Biological Chemistry* 277: 30091-30101.

Chapter 3

Abrams, G.D. 1997. Response of the body to injury: Inflammation and repair. In: S.A. Price and L.M. Wilson (Eds.), *Pathophysiology: Clinical concepts of disease processes* (5th ed.), pp. 38-58. St. Louis: Mosby.

Armstrong, R.B. 1984. Mechanisms of exercise-induced delayed onset muscular soreness: A brief review. *Medicine and Science in Sports and Exercise* 16:529-538.

Asp, S., J.R. Daugaard, S. Kristiansen, B. Kiends, and E.A. Richter. 1996. Eccentric exercise decreases maximal insulin action in humans: Muscle and systemic effects. *Journal of Physiology* 493:891-898.

Asp, S., J.R. Daugaard, and E.A. Richter. 1995. Eccentric exercise decreases glucose transporter GLUT4 protein in human skeletal muscle. *Journal of Physiology* 482 (Pt 3):705-712.

Avela, J., H. Kyrolainen, P.V. Komi, and D. Rama. 1999. Reduced reflex sensitivity persists several days after long-lasting stretch-shortening cycle exercise. *Journal of Applied Physiology* 86:1292-1300.

Baldwin, A.C., S.W. Stevenson, and G.A. Dudley. 2001. Nonsteroidal anti-inflammatory therapy after eccentric exercise in healthy older individuals. *Journals of Gerontology* 56:M510-513.

Balnave, C.D., and M.W. Thompson. 1993. Effect of training on eccentric exercise-induced muscle damage. *Journal of Applied Physiology* 75:1545-1551.

Behm, D.G., K.M. Baker, R. Kelland, and J. Lomond. 2001. The effect of muscle damage on strength and fatigue deficits. *Journal of Strength and Conditioning Research* 15:255-263.

Benz, R.J., J. Friden, and R.L. Lieber. 1998. Simultaneous stiffness and force measurements reveal subtle injury to rabbit soleus muscles. *Molecular and Cellular Biochemistry* 179:147-158.

Bigland-Ritchie, B., F. Furbush, and J.J. Woods. 1986. Fatigue of intermittent submaximal voluntary contractions: Central and peripheral factors. *Journal of Applied Physiology* 61:421-429.

Bobbert, M.F., A.P. Hollander, and P.A. Huijing. 1986. Factors in delayed onset muscular soreness of man. *Medicine and Science in Sports and Exercise* 18:75-81.

Bulbulian, R., and D.K. Bowles. 1992. Effect of downhill running on motoneuron pool excitability. *Journal of Applied Physiology* 73:968-973.

Busch, W.A., M.H. Stromer, D.E. Goll, and A. Suzuki. 1972. Ca^{2+}-specific removal of Z-lines from rabbit skeletal muscle. *Journal of Cell Biology* 52:367-381.

Butterfield, T.A., and W.A. Herzog. 2006. Is the force-length relationship a useful indicator of contractile element damage following eccentric exercise? *Journal of Biomechanics* 38:1932-1937.

Byrnes, W.C., P.M. Clarkson, J.S. White, S.S. Hsieh, P.N. Frykman, and R.J. Maughan. 1985. Delayed onset muscle soreness following repeated bouts of downhill running. *Journal of Applied Physiology* 59:710-715.

Cannon, J.G., M.A. Fiatarone, R.A. Fielding, and W.J. Evans. 1994. Aging and stress-induced changes in complement activation and neutrophil mobilization. *Journal of Applied Physiology* 76:2616-2620.

Cannon, J.G., R.A. Fielding, M.A. Fiatarone, S.F. Orencole, C.A. Dinarello, and W.J. Evans. 1989. Increased interleukin 1 in human skeletal muscle after exercise. *American Journal of Physiology* 257: R451-R455.

Cannon, J.G., S.N. Meydani, M.A. Fiatarone, R.A. Fielding, M. Meydani, S. Farhangmehr, S.F. Orencole, J.B. Blumberg, and W.J. Evans. 1991. Acute phase response in exercise II. Associations between vitamin E, cytokines, and muscle proteolysis. *American Journal of Physiology* 260:1235-1240.

Child, R.B., J.M. Saxton, and A.E. Donnelly. 1998. Comparison of eccentric knee extensor muscle actions at two muscle lengths on indices of damage and angle-specific force production in humans. *Journal of Sports Sciences* 16:301-308.

Chin, E.R., C.D. Balnave, and D.G. Allen. 1997. Role of intracellular calcium and metabolites in low-frequency fatigue of mouse skeletal muscle. *American Journal of Physiology* 272:C550-559.

Chleboun, G.S., J.N. Howell, H.L. Baker, T.N. Ballard, J.L. Graham, H.L. Hallman, L.E. Perkins, J.H. Schauss, and R.R. Conatser. 1995. Intermittent pneumatic compression effect on eccentric exercise-induced swelling, stiffness, and strength loss. *Archives of Physical Medicine and Rehabilitation* 76:744-749.

Chleboun, G.S., J.N. Howell, R.R. Conatser, and J.J. Giesey. 1998. Relationship between muscle swelling and stiffness after eccentric exercise. *Medicine and Science in Sports and Exercise* 30:529-535.

Clarkson, P.M., and M.J. Hubal. 2001. Exercise-induced muscle damage in humans. *American Journal of Physical Medicine and Rehabilitation* 81:(11 Suppl.)S52-S69.

Clarkson, P.M., K. Nosaka, and B. Braun. 1992. Muscle function after exercise-induced muscle damage and rapid adaptation. *Medicine and Science in Sports and Exercise* 24:512-520.

Clarkson, P.M., and I. Tremblay. 1988. Exercise-induced muscle damage, repair, and adaptation in humans. *Journal of Applied Physiology* 65:1-6.

Del Aguila, L.F., R.K. Krishnan, J.S. Ulbrecht, P.A. Farrell, P.H. Correll, C.H. Lang, J.R. Zierath, and J.P. Kirwan. 2000. Muscle damage impairs insulin stimulation of IRS-1, PI 3-kinase, and Akt-kinase in human skeletal muscle. *American Journal of Physiology: Endocrinology and Metabolism* 279:E206-E212.

Deschenes, M.R., R.E. Brewer, J.A. Bush, R.W. McCoy, J.S. Volek, and W.J. Kraemer. 2000. Neuromuscular disturbance outlasts other symptoms of exercise-induced muscle damage. *Journal of Neurological Sciences* 174:92-99.

Duan, C., M.D. Delp, D.A. Hayes, P.D. Delp, and R.B. Armstrong. 1990. Rat skeletal muscle mitochondria (Ca^{2+}) and injury from downhill walking. *Journal of Applied Physiology* 68:1241-1251.

Duncan, C.J. 1978. Role of intracellular calcium in promoting muscle damage: A strategy for controlling the dystrophic condition. *Experientia* 34:1531-1535.

Ebbeling, C.B., and P.M. Clarkson. 1990. Muscle adaptation prior to recovery following eccentric exercise. *European Journal of Applied Physiology* 60:26-31.

Edwards, R.H., D.K. Hill, D.A. Jones, and P.A. Merton. 1977. Fatigue of long duration in human skeletal muscle after exercise. *Journal of Physiology* 272:769-778.

Edwards, R.H.T., K.R. Mills, and D.J. Newham. 1981. Measurement of severity and distribution of experimental muscle tenderness. *Journal of Physiology* 317:1P-2P.

Evans, G.F., R.G. Haller, P.S. Wyrick, R.W. Parkey, and J.L. Fleckenstein. 1998. Submaximal delayed-onset muscle soreness: Correlations between MR imaging findings and clinical measures. *Radiology* 208:815-820.

Evans, W.J., and J.G. Cannon. 1991. The metabolic effects of exercise-induced muscle damage. *Exercise and Sport Sciences Reviews* 19:99-125.

Evans, W.J., C.A. Meredith, J.G. Cannon, C.A. Dinarello, W.R. Frontera, V.A. Hughes, B.H. Jones, and H.G. Knuttgen. 1986. Metabolic changes following eccentric exercise in trained and untrained men. *Journal of Applied Physiology* 61:1864-1868.

Fielding, R.A., T.J. Manfredi, W. Ding, M.A. Fiatarone, W.J. Evans, and J.G. Cannon. 1993. Acute phase response in exercise III. Neutrophil and IL-β accumulation in skeletal muscle. *American Journal of Physiology* 265:R166-R172.

Foley, J.M., R.C. Jayaraman, B.M. Prior, J.M. Pivarnik, and R.A. Meyer. 1999. MR measurements of muscle damage and adaptation after eccentric exercise. *Journal of Applied Physiology* 87:2311-2318.

Friden, J., J. Seger, and B. Ekblom. 1988. Sublethal muscle fiber injuries after high-tension anaerobic exercise. *European Journal of Applied Physiology* 57:360-368.

Friden J., P.N. Sfakianos, and A.R. Hargens. 1986. Muscle soreness and intramuscular fluid pressure: Comparison between eccentric and concentric load. *Journal of Applied Physiology* 61:2175-2179.

Friden, J., J. Sjostrom, and B. Ekblom. 1981. A morphological study of delayed muscle soreness. *Experientia* 37:506-507.

Friden, J., J. Sjostrom, and B. Ekblom. 1983. Myofibrillar damage following intense eccentric exercise in man. *International Journal of Sports Medicine* 4:170-176.

Friden, J., and R.L. Lieber. 1992. Structural and mechanical basis of exercise-induced muscle injury. *Medicine and Science in Sports and Exercise* 24:521-530.

Friden, J., and R.L. Lieber. 1996. Ultrastructural evidence for loss of calcium homeostasis in exercised skeletal muscle. *Acta Physiologica Scandinavica* 158:381-382.

Friden, J., and R.L. Lieber. 1998. Segmental muscle fiber lesions after repetitive eccentric contractions. *Cell and Tissue Research* 293:165-171.

Gibala, M.J., J.D. MacDougall, M.A. Tarnopolsky, W.T. Stauber, and A. Elorriaga. 1995. Changes in human skeletal muscle ultrastructure and force production after acute resistance exercise. *Journal of Applied Physiology* 78:702-708.

Harrison, B.C., D. Robinson, B.J. Davison, B. Foley, E. Seda, and W.C. Byrnes. 2001. Treatment of exercise induced muscle injury via hyperbaric oxygen therapy. *Medicine and Science in Sports and Exercise* 33:36-42.

Hill, C.A., M.W. Thompson, P.A. Ruell, J.M. Thom, and M.J. White. 2001. Sarcoplasmic reticulum function and muscle contractile character following fatiguing exercise in humans. *Journal of Physiology* 531:871-878.

Hirose, L., K. Nosaka, M. Newton, A. Lavender, M. Kano, J. Peake, and K. Suzuki. 2004. Changes in inflammatory mediators following eccentric exercise of the elbow flexors. *Exercise Immunology Review* 10:75-90.

Hough, T. 1902. Ergographic studies in muscular soreness. *American Journal of Physiology* 7:76-92.

Howell, J.N., G. Chleboun, and R. Conatser. 1993. Muscle stiffness, strength loss, swelling and soreness following exercise-induced injury in humans. *Journal of Physiology* 464:183-196.

Ingalls, C.P., G.L. Warren, and R.B. Armstrong. 1998. Dissociation of force production from MHC and actin contents in muscles injured by eccentric contractions. *Journal of Muscle Research and Cell Motility* 19:215-224.

Jones, D.A. 1996. High-and low-frequency fatigue revisited. *Acta Physiologica Scandinavica* 156:265-270.

Jones, D.A., D.J. Newham, and P.M. Clarkson. 1987. Skeletal muscle stiffness and pain following eccentric exercise of the elbow flexors. *Pain* 30:233-242.

Jones, D.A., D.J. Newham, J.M. Round, and S.E.J. Tollfree. 1986. Experimental human muscle damage: Morphological changes in relation to other indices of damage. *Journal of Physiology* 375:435-448.

Jones, D.A., D.J. Newham, and C. Torgan. 1989. Mechanical influences on long-lasting human muscle fatigue and delayed-onset pain. *Journal of Physiology* 412:415-427.

Kent-Braun, J.A. 1999. Central and peripheral contributions to muscle fatigue in humans during sustained maximal effort. *European Journal of Applied Physiology* 80:57-63.

Kent-Braun, J.A., and R. Le Blanc. 1996. Quantitation of central activation failure during maximal voluntary contractions in humans. *Muscle and Nerve* 19:861-869.

King, D.S., T.L. Feltmeyer, P.J. Baldus, R.L. Sharp, and J. Nespor. 1993. Effects of eccentric exercise on insulin secretion and action in humans. *Journal of Applied Physiology* 75:2151-2156.

Kirwan, J.P., R.C. Hickner, K.E. Yarasheski, W.M. Kohrt, B.V. Wiethop, and J.O. Holloszy. 1992. Eccentric exercise induces transient insulin resistance in healthy individuals. *Journal of Applied Physiology* 72:2197-2202.

Kraemer, W.J., J.A. Bush, R.B. Wickham, C.R. Denegar, A.L. Gomez, L.A. Gotshalk, N.D. Duncan, J.S. Volek, M. Putukian, and W.J. Sebastianelli. 2001. Influence of compression therapy on symptoms following soft tissue injury from maximal eccentric exercise. *Journal of Orthopedic and Sports Physical Therapy* 31:282-290.

Kristiansen, S., J. Jones, A. Handberg, G.L. Dohm, and E.A. Richter. 1997. Eccentric contractions decrease glucose transporter transcription rate, mRNA, and protein in skeletal muscle. *American Journal of Physiology* 272:C1734-C1738.

Kuipers, H., E. Janssen, H. Keizer, and S. Verstappen. 1985. Serum CPK and amount of muscle damage in rats. *Medicine and Science in Sports and Exercise* 17:195.

Leger, A.B., and T.E. Milner. 2000a. The effect of eccentric exercise on intrinsic and reflex stiffness in the human hand. *Clinical Biomechanics* (Bristol, Avon) 15:574-582.

Leger, A.B., and T.E. Milner. 2000b. Passive and active wrist joint stiffness following eccentric exercise. *European Journal of Applied Physiology* 82:472-479.

Lieber, R.L., L.-E. Thornell, and J. Friden. 1996. Muscle cytoskeletal disruption occurs within the first 15 min of cyclic eccentric contractions. *Journal of Applied Physiology* 80:278-284.

Loscher, W.N., and M.M. Nordlund. 2002. Central fatigue and motor cortical excitability during repeated shortening and lengthening actions. *Muscle and Nerve* 25:864-872.

Lynch, G.S., C.J. Fary, and D.A. Williams. 1997. Quantitative measurement of resting skeletal muscle [Ca^{2+}]i following acute and long-term downhill running exercise in mice. *Cell Calcium* 22:373-383.

MacIntyre, D.L., W.D. Reid, D.M. Lyster, and D.C. McKenzie. 2000. Different effects of strenuous eccentric exercise on the accumulation of neutrophils in muscle in women and men. *European Journal of Applied Physiology* 81:47-53.

MacIntyre, D.L., W.D. Reid, D.M. Lyster, I.J. Szasz, and D.C. McKenzie. 1996. Presence of WBC, decreased strength, and delayed soreness in muscles after eccentric exercise. *Journal of Applied Physiology* 80:1006-1013.

Mair, J., A. Koller, E. Artner-Dworzak, C. Haid, K. Wicke, W. Judmaier, and B. Puschendorf. 1992. Effects of exercise on plasma myosin heavy chain fragments and MRI of skeletal muscle. *Journal of Applied Physiology* 72:656-663.

Malm, C., R. Lenkai, and B. Sjodin. 1999. Effect of eccentric exercise on the immune system in men. *Journal of Applied Physiology* 86:461-468.

Malm, C., P. Nyberg, M. Engstrom, B. Sjodin, R. Lenkei, B. Ekblom, and I. Lundberg. 2000. Immunological changes in human skeletal muscle and blood after eccentric exercise and multiple biopsies. *Journal of Physiology* 15:243-262.

Manfredi, T.G., R.A. Fielding, K.P. O'Reilly, C.N. Meredith, H.-Y. Lee, and W.J. Evans. 1991. Plasma creatine kinase activity and exercise-induced muscle damage in older men. *Medicine and Science in Sports and Exercise* 23:1028-1034.

Martin, V., G.Y. Millet, G. Lattier, and L. Perrod. 2004. Effects of recovery modes after knee extensor muscles eccentric contractions. *Medicine and Science in Sports and Exercise* 36:1907-1915.

Mishra, D.K., J. Friden, M.C. Schmitz, and R.L. Lieber. 1995. Anti-inflammatory medication after muscle injury. A treatment resulting in short-term improvement but subsequent loss of muscle function. *Journal of Bone Joint Surgery* 77:1510-1519.

Newham, D.J., D.A. Jones, and R.H. Edwards. 1986. Plasma creatine kinase changes after eccentric and concentric contractions. *Muscle and Nerve* 9:59-63.

Newham, D.J., G. McPhail, K.R. Mills, and R.H.T. Edwards. 1983. Ultrastructural changes after concentric and eccentric contractions of human muscle. *Journal of Neurological Sciences* 61:109-122.

Newham, D.J., K.R. Mills, B.M. Quigley, and R.H.T. Edwards. 1982. Muscle pain and tenderness after exercise. *Australian Journal of Sports Medicine and Exercise Science* 14:910-913.

Newham, D.J., K.R. Mills, B.M. Quigley, and R.H. Edwards. 1983. Pain and fatigue after concentric and eccentric muscle contractions. *Clinical Sciences* (London) 64:55-62.

Nosaka, K., and P.M. Clarkson. 1994. Effect of eccentric exercise on plasma enzyme activities previously elevated by eccentric exercise. *European Journal of Applied Physiology* 69:492-497.

Nosaka, K., and P.M. Clarkson. 1996a. Changes in indicators of inflammation after eccentric exercise of the elbow flexors. *Medicine and Science in Sports and Exercise* 28:953-961.

Nosaka, K., and P.M. Clarkson. 1996b. Variability in serum creatine kinase response after eccentric exercise of the elbow flexors. *International Journal of Sports Medicine* 17:120-127.

Nosaka, K., and K. Sakamoto. 2001. Effect of elbow joint angle on the magnitude of muscle damage to the elbow flexors. *Medicine and Science in Sports and Exercise* 33:22-29.

Nurenberg, P., C.J. Giddings, J. Stray-Gundersen, J.L. Fleckenstein, W.J. Gonyea, and R.M. Peshock. 1992. MR imaging-guided muscle biopsy for correlation of increased signal intensity with ultrastructural change and delayed-onset muscle soreness after exercise. *Radiology* 184:865-869.

O'Reilly, K.P., M.J. Warhol, R.A. Fielding, W.R. Frontera, C.N. Meredith, and W.J. Evans. 1987. Eccentric exercise-induced muscle damage impairs muscle glycogen repletion. *Journal of Applied Physiology* 63:252-256.

Pearce, A.J., P. Sacco, M.L. Byrnes, G.W. Thickbroom, and F.L. Mastaglia. 1998. The effects of eccentric exercise on neuromuscular function of the biceps brachii. *Journal of Science and Medicine in Sport* 1:236-244.

Pizza, F.X., D. Cavender, A. Stockard, H. Baylies, and A. Beighle. 1999. Anti-inflammatory doses of ibuprofen: Effect on neutrophils and exercise-induced muscle injury. *International Journal of Sports Medicine* 20:98-102.

Pizza, F.X., J.B. Mitchell, B.H. Davis, R.D. Starling, R.W. Holtz, and N. Bigelow. 1995. Exercise-induced muscle damage: Effect on circulating leukocyte and lymphocyte subsets. *Medicine and Science in Sports and Exercise* 27:363-370.

Ploutz-Snyder, L.L., S. Nyren, T.G. Cooper, E.J. Potchen, and R.A. Meyer. 1997. Different effects of exercise and edema on T2 relaxation in skeletal muscle. *Magnetic Resonance in Medicine* 37:676-682.

Proske, U., J.E. Gregory, D.L. Morgan, P. Percival, N.S. Weerakkody, and B.J. Canny. 2004. Force matching errors following eccentric exercise. *Human Movement Science* 23:365-378.

Proske, U., N.S. Weerakkody, P. Percival, D.L. Morgan, J.E. Gregory, and B.J. Canny. 2003. Force-matching errors after eccentric exercise attributed to muscle soreness. *Clinical and Experimental Pharmacology and Physiology* 30:576-579.

Rodenburg, J.B., D. Steenbeek, P. Schiereck, and P.R. Bar. 1994. Warm-up, stretching and massage diminish harmful effects of eccentric exercise. *International Journal of Sports Medicine* 15:414-419.

Roth, S.M., G.F. Martel, and M.A. Rogers. 2000. Muscle biopsy and muscle fiber hypercontraction: A brief review. *European Journal of Applied Physiology* 83:239-245.

Round, J.M., D.A. Jones, and G. Cambridge. 1987. Cellular infiltrates in human skeletal muscle: Exercise induced damage as a model for inflammatory muscle disease? *Journal of Neurological Science* 82:1-11.

Sargeant, A.J., and P. Dolan. 1987. Human muscle function following prolonged eccentric exercise. *European Journal of Applied Physiology* 56:704-711.

Saxton, J.M., P.M. Clarkson, R. James, M. Miles, M. Westerfer, S. Clark, and A.E. Donnelly. 1995. Neuromuscular dysfunction following eccentric exercise. *Medicine and Science in Sports and Exercise* 27:1185-1193.

Sayers, S.P., and P.M. Clarkson. 2001. Force recovery after eccentric exercise in males and females. *European Journal of Applied Physiology* 84:122-126.

Sayers, S.P., and P.M. Clarkson. 2003. Short-term immobilization after eccentric exercise. Part II. CK and myoglobin. *Medicine and Science in Sports and Exercise* 35:762-768.

Sayers, S.P., P.M. Clarkson, P.A. Rouzier, and G. Kamen. 1999. Adverse events associated with eccentric exercise protocols: Six case studies. *Medicine and Science in Sports and Exercise* 31:1697-1702.

Sayers, S.P., C.A. Knight, P.M. Clarkson, E.H. van Wegen, and G. Kamen. 2001. Effect of ketoprofen on muscle function and sEMG after eccentric exercise. *Medicine and Science in Sports and Exercise* 33:702-710.

Sayers, S.P., B.T. Peters, C.A. Knight, M.L. Urso, J. Parkington, and P.M. Clarkson. 2003. Short-term immobilization after eccentric exercise. Part I: Contractile properties. *Medicine and Science in Sports and Exercise* 35:753-761.

Shellock, F.G., T. Fukunaga, J.H. Mink, and V.R. Edgerton. 1991. Exertional muscle injury: Evaluation of concentric versus eccentric actions with serial MR imaging. *Radiology* 179:659-664.

Smith, L.L., A. Anwar, M. Fragen, C. Rananto, R. Johnson, and D. Holbert. 2000. Cytokines and cell adhesion molecules associated with high-intensity eccentric exercise. *European Journal of Applied Physiology* 82:61-67.

Smith, L.L., M. McCammon, S. Smith, M. Chamness, R.G. Israel, and K.F. O'Brien. 1989. White blood cell response to uphill walking and downhill jogging at similar metabolic loads. *Journal of Applied Physiology* 58:833-837.

Stauber, W.T., P.M. Clarkson, V.K. Fritz, and W.J. Evans. 1990. Extracellular matrix disruption and pain after eccentric muscle action. *Journal of Applied Physiology* 69:868-874.

Stupka, N., S. Lowther, K. Chorneydo, J.M. Bourgeoise, C. Hogben, and M. Tarnopolsky. 2000. Gender differences in muscle inflammation after eccentric exercise. *Journal of Applied Physiology* 89:2325-2332.

Takahashi, H., S. Kuno, T. Miyamoto, H. Yoshioko, M. Inaki, H. Akima, S. Katsuda, I. Anno, and Y. Itai. 1994. Changes in magnetic resonance images in human skeletal muscle after eccentric exercise. *European Journal of Applied Physiology* 69:408-413.

Takekura, H., N. Fujinami, T. Nishizawa, H. Ogasawara, and N. Kasuga. 2001. Eccentric exercise-induced morphological changes in the membrane systems involved in excitation-contraction coupling in rat skeletal muscle. *Journal of Physiology* 533:571-583.

Tidball, J.G. 1995. Inflammatory cell response to acute muscle injury. *Medicine and Science in Sports and Exercise* 27:1022-1032.

Van der Meulen, J.H., H. Kuipers, and J. Drukker. 1991. Relationship between exercise-induced muscle damage and enzyme release in rats. *Journal of Applied Physiology* 71:999-1004.

Walsh, L.D., C.W. Hesse, D.L. Morgan, and U. Proske. 2004. Human forearm position sense after fatigue of elbow flexor muscles. *Journal of Physiology* 558:705-715.

Warren, G.L., D.A. Lowe, and R.B. Armstrong. 1999. Measurement tools used in the study of eccentric contraction-induced injury. *Sports Medicine* 27:43-59.

Warren, G.L., D.A. Lowe, D.A. Hayes, M.A. Farmer, and R.B. Armstrong. 1995. Redistribution of cell membrane probes following contraction-induced injury of mouse soleus muscle. *Cell Tissue Research* 282:311-320.

Warren, G.L., D.A. Lowe, D.A. Hayes, C.J. Karwoski, B.M. Prior, and R.B. Armstrong. 1993. Excitation failure in eccentric contraction-induced injury of mouse soleus muscle. *Journal of Physiology* 468:487-499.

Weerakkody, N.S., P. Percival, M.W. Hickey, D.L. Morgan, J.E. Gregory, B.J. Canny, and U. Proske. 2003. Effects of local pressure and vibration on muscle pain from eccentric exercise and hypertonic saline. *Pain* 105:425-435.

Weerakkody, N.S., N.P. Whitehead, B.J. Canny, J.E. Gregory, and U. Proske. 2001. Large-fiber mechanoreceptors contribute to muscle soreness after eccentric exercise. *Journal of Pain* 2:209-219.

Westerblad, H., S. Duty, and D.G. Allen. 1993. Intracellular calcium concentration during low-frequency fatigue in isolated single fibers of mouse skeletal muscle. *Journal of Applied Physiology* 75:382-388.

Yasuda, T., K. Sakamoto, K. Nosaka, M. Wada, and S. Katsuta. 1997. Loss of sarcoplasmic reticulum membrane integrity after eccentric exercise. *Acta Physiologica Scandinavica* 161:581-582.

Yu, J.-G., L. Carlsson, and L.E. Thornell. 2004. Evidence for myofibril remodeling as opposed to myofibril damage in human muscles with DOMS: An ultrastructural and immuno-electron microscopic study. *Histochemistry and Cell Biology* 121:219-227.

Yu, J.-G., D.O. Furst, and L.E. Thornell. 2003. The mode of myofibril remodeling in human skeletal muscle affected by DOMS induced by eccentric contractions. *Histochemistry and Cell Biology* 119: 383-393.

Yu, J.-G., C. Malm, and L.E. Thornell. 2002. Eccentric contractions leading to DOMS do not cause loss of desmin nor fibre necrosis in human muscle. *Histochemistry and Cell Biology* 118:29-34.

Yu, J.-G., and L.E. Thornell. 2002. Desmin and actin alterations in human muscles affected by delayed onset muscle soreness: A high resolution immunocytochemical study. *Histochemistry and Cell Biology* 118:171-179.

Chapter 4

Baker, W., B.A. Schneider, A. Kulkarni, G. Sloan, R. Schaub, J. Sypek, and J.G. Cannon. 2004. P-selectin inhibition suppresses muscle regeneration following injury. *J Leukoc Biol* 76 (2):352-358.

Beaton, L.J., M.A. Tarnopolsky, and S.M. Phillips. 2002. Contraction-induced muscle damage in humans following calcium channel blocker administration. *J Physiol* 544 (Pt 3):849-859.

Belcastro, A.N., L.D. Shewchuk, and D.A. Raj. 1998. Exercise-induced muscle injury: A calpain hypothesis. *Mol Cell Biochem* 179 (1-2):135-145.

Bondesen, B.A., S.T. Mills, K.M. Kegley, and G.K. Pavlath. 2004. The COX-2 pathway is essential during early stages of skeletal muscle regeneration. *Am J Physiol Cell Physiol* 287 (2):C475-483.

Brickson, S., L.L. Ji, K. Schell, R. Olabisi, B. St Pierre Schneider, and T.M. Best. 2003. M1/70 attenuates blood-borne neutrophil oxidants, activation, and myofiber damage following stretch injury. *J Appl Physiol* 95 (3):969-976.

Cantini, M., and U. Carraro. 1995. Macrophage-released factor stimulates selectively myogenic cells in primary muscle culture. *J Neuropathol Exp Neurol* 54 (1):121-128.

Cantini, M., E. Giurisato, C. Radu, S. Tiozzo, F. Pampinella, D. Senigaglia, G. Zaniolo, F. Mazzoleni, and L. Vitiello. 2002. Macrophage-secreted myogenic factors: A promising tool for greatly enhancing the proliferative capacity of myoblasts in vitro and in vivo. *Neurol Sci* 23 (4):189-194.

Cantini, M., M.L. Massimino, E. Rapizzi, K. Rossini, C. Catani, L. Dalla Libera, and U. Carraro. 1995. Human satellite cell proliferation in vitro is regulated by autocrine secretion of IL-6 stimulated by a soluble factor(s) released by activated monocytes. *Biochem Biophys Res Commun* 216 (1):49-53.

Cassatella, M.A. 1999. Neutrophil-derived proteins: Selling cytokines by the pound. *Advances in Immunology* 73: 369-509.

Chazaud, B., C. Sonnet, P. Lafuste, G. Bassez, A.C. Rimaniol, F. Poron, F.J. Authier, P.A. Dreyfus, and R.K. Gherardi. 2003. Satellite cells attract monocytes and use macrophages as a support to escape apoptosis and enhance muscle growth. *J Cell Biol* 163 (5):1133-1143.

Doyonnas, R., M.A. LaBarge, A. Sacco, C. Charlton, and H.M. Blau. 2004. Hematopoietic contribution to skeletal muscle regeneration by myelomonocytic precursors. *Proc Natl Acad Sci USA* 101 (37):13507-13512.

Echeverri, K., and E.M. Tanaka. 2002. Mechanisms of muscle dedifferentiation during regeneration. *Semin Cell Dev Biol* 13 (5):353-360.

Farges, M.C., D. Balcerzak, B.D. Fisher, D. Attaix, D. Bechet, M. Ferrara, and V.E. Baracos. 2002. Increased muscle proteolysis after local trauma mainly reflects macrophage-associated lysosomal proteolysis. *Am J Physiol Endocrinol Metab* 282 (2):E326-335.

Faulkner, J.A., D.A. Jones, and J.M. Round. 1989. Injury to skeletal muscles of mice by forced lengthening during contractions. *Q J Exp Physiol* 74 (5):661-670.

Frenette, J., B. Cai, and J.G. Tidball. 2000. Complement activation promotes muscle inflammation during modified muscle use. *Am J Pathol* 156 (6):2103-2110.

Frenette, J., N. Chbinou, C. Godbout, D. Marsolais, and P.S. Frenette. 2003. Macrophages, not neutrophils, infiltrate skeletal muscle in mice deficient in P/E selectins after mechanical reloading. *Am J Physiol Regul Integr Comp Physiol* 285 (4): R727-732.

Frenette, J., M. St-Pierre, C.H. Cote, E. Mylona, and F.X. Pizza. 2002. Muscle impairment occurs rapidly and precedes inflammatory cell accumulation after mechanical loading. *Am J Physiol Regul Integr Comp Physiol* 282 (2):R351-357.

Grounds, M.D., and M.J. Davies. 1996. Chemotaxis in myogenesis. *Basic Appl Myol* 6:469-483.

Grounds, M.D., and J.K. McGeachie. 1989. A comparison of muscle precursor replication in crush-injured skeletal muscle of Swiss and BALBc mice. *Cell Tissue Res* 255 (2):385-391.

Hawke, T.J., and D.J. Garry. 2001. Myogenic satellite cells: Physiology to molecular biology. *J Appl Physiol* 91 (2):534-551.

Hellsten, Y., U. Frandsen, N. Orthenblad, B. Sjodin, and E.A. Richter. 1997. Xanthine oxidase in human skeletal muscle following eccentric exercise: A role in inflammation. *J Physiol* 498 (Pt 1):239-248.

Koh, T.J., J.M. Peterson, F.X. Pizza, and S.V. Brooks. 2003. Passive stretches protect skeletal muscle of adult and old mice from lengthening contraction-induced injury. *J Gerontol A Biol Sci Med Sci* 58 (7):592-597.

Kuschel, R., M.H. Deininger, R. Meyermann, A. Bornemann, Z. Yablonka-Reuveni, and H.J. Schluesener. 2000. Allograft inflammatory factor-1 is expressed by macrophages in injured skeletal muscle and abrogates proliferation and differentiation of satellite cells. *J Neuropathol Exp Neurol* 59 (4):323-332.

Kvietys, P.R., and M. Sandig. 2001. Neutrophil diapedesis: Paracellular or transcellular? *News Physiol Sci* 16:15-19.

Lapointe, B.M., J. Frenette, and C.H. Cote. 2002. Lengthening contraction-induced inflammation is linked to secondary damage but devoid of neutrophil invasion. *J Appl Physiol* 92 (5):1995-2004.

Lowe, D.A., G.L. Warren, C.P. Ingalls, D.B. Boorstein, and R.B. Armstrong. 1995. Muscle function and protein metabolism after initiation of eccentric contraction-induced injury. *J Appl Physiol* 79 (4):1260-1270.

MacIntyre, D.L., S. Sorichter, J. Mair, A. Berg, and D.C. McKenzie. 2001. Markers of inflammation and myofibrillar proteins following eccentric exercise in humans. *Eur J Appl Physiol* 84 (3):180-186.

Massimino, M.L., E. Rapizzi, M. Cantini, L.D. Libera, F. Mazzoleni, P. Arslan, and U. Carraro. 1997. ED2+ macrophages increase selectively myoblast proliferation in muscle cultures. *Biochem Biophys Res Commun* 235 (3):754-759.

Matsukawa, A., C.M. Hogaboam, N.W. Lukacs, and S.L. Kunkel. 2000. Chemokines and innate immunity. *Rev Immunogenet* 2 (3):339-258.

McLoughlin, T.J., E. Mylona, T.A. Hornberger, K.A. Esser, and F.X. Pizza. 2003. Inflammatory cells in rat skeletal muscle are elevated after electrically stimulated contractions. *J Appl Physiol* 94 (3):876-882.

McLoughlin, T.J., S.K. Tsivitse, J.A. Edwards, B.A. Aiken, and F.X. Pizza. 2003. Deferoxamine reduces and nitric oxide synthase inhibition increases neutrophil-mediated myotube injury. *Cell Tissue Res* 313 (3):313-319.

Meerschaert, J., and M.B. Furie. 1995. The adhesion molecules used by monocytes for migration across endothelium include CD11a/CD18, CD11b/CD18, and VLA-4 on monocytes and ICAM-1, VCAM-1, and other ligands on endothelium. *J Immunol* 154 (8):4099-4112.

Merly, F., L. Lescaudron, T. Rouaud, F. Crossin, and M.F. Gardahaut. 1999. Macrophages enhance muscle satellite cell proliferation and delay their differentiation. *Muscle Nerve* 22 (6):724-732.

Mitchell, C.A., J.K. McGeachie, and M.D. Grounds. 1992. Cellular differences in the regeneration of murine skeletal muscle: A quantitative histological study in SJL/J and BALB/c mice. *Cell Tissue Res* 269 (1):159-166.

Morozov, V.I., T.N. Usenko, and V.A. Rogozkin. 2001. Neutrophil antiserum response to decrease in proteolytic activity in loaded rat muscle. *Eur J Appl Physiol* 84 (3):195-200.

Nagaraju, K. 2001. Immunological capabilities of skeletal muscle cells. *Acta Physiol Scand* 171 (3):215-223.

Nguyen, H.X., and J.G. Tidball. 2003. Interactions between neutrophils and macrophages promote macrophage killing of rat muscle cells in vitro. *J Physiol* 547 (Pt 1):125-132.

Padgett, E.L., and S.B. Pruett. 1992. Evaluation of nitrite production by human monocyte-derived macrophages. *Biochem Biophys Res Commun* 186 (2):775-781.

Papadimitriou, J.M., T.A. Robertson, C.A. Mitchell, and M.D. Grounds. 1990. The process of new plasmalemma formation in focally injured skeletal muscle fibers. *J Struct Biol* 103 (2):124-134.

Patel, K.D., S.L. Cuvelier, and S. Wiehler. 2002. Selectins: Critical mediators of leukocyte recruitment. *Semin Immunol* 14 (2):73-81.

Peterson, J.M., T.A. Trappe, E. Mylona, F. White, C.P. Lambert, W.J. Evans, and F.X. Pizza. 2003. Ibuprofen and acetaminophen: Effect on muscle inflammation after eccentric exercise. *Med Sci Sports Exerc* 35 (6):892-896.

Pizza, F.X., T.J. Koh, S.J. McGregor, and S.V. Brooks. 2002. Muscle inflammatory cells after passive stretches, isometric contractions, and lengthening contractions. *J Appl Physiol* 92 (5):1873-1878.

Pizza, F.X., T.J. McLoughlin, S.J. McGregor, E.P. Calomeni, and W.T. Gunning. 2001. Neutrophils injure cultured skeletal myotubes. *Am J Physiol Cell Physiol* 281 (1):C335-341.

Pizza, F.X., J.M. Peterson, J.H. Baas, and T.J. Koh. 2005. Neutrophils contribute to muscle injury and impair its resolution after lengthening contractions in mice. *J Physiol* 562 (Pt 3):899-913.

Raj, D.A., T.S. Booker, and A.N. Belcastro. 1998. Striated muscle calcium-stimulated cysteine protease (calpain-like) activity promotes myeloperoxidase activity with exercise. *Pflugers Arch* 435 (6):804-809.

Roberts, P., J.K. McGeachie, and M.D. Grounds. 1997. The host environment determines strain-specific differences in the timing of skeletal muscle regeneration: Cross-transplantation studies between SJL/J and BALB/c mice. *J Anat* 191 (Pt 4):585-594.

Robertson, T.A., M.D. Grounds, and J.M. Papadimitriou. 1992. Elucidation of aspects of murine skeletal muscle regeneration using local and whole body irradiation. *J Anat* 181 (Pt 2):265-276.

Robertson, T.A., M.A. Maley, M.D. Grounds, and J.M. Papadimitriou. 1993. The role of macrophages in skeletal muscle regeneration with particular reference to chemotaxis. *Exp Cell Res* 207 (2):321-331.

Sandri, M., C. Sandri, B. Brun, E. Giurisato, M. Cantini, K. Rossini, C. Destro, P. Arslan, and U. Carraro. 2001. Inhibition of fasL sustains phagocytic cells and delays myogenesis in regenerating muscle fibers. *J Leukoc Biol* 69 (3):482-489.

Seale, P., and M.A. Rudnicki. 2000. A new look at the origin, function, and "stem-cell" status of muscle satellite cells. *Dev Biol* 218 (2):115-124.

Simon, H.U. 2003. Neutrophil apoptosis pathways and their modifications in inflammation. *Immunol Rev* 193:101-110.

Stupka, N., M.A. Tarnopolsky, N.J. Yardley, and S.M. Phillips. 2001. Cellular adaptation to repeated eccentric exercise-induced muscle damage. *J Appl Physiol* 91 (4):1669-1678.

Summan, M., M. McKinstry, G.L. Warren, T. Hulderman, D. Mishra, K. Brumbaugh, M.I. Luster, and P.P. Simeonova. 2003. Inflammatory mediators and skeletal muscle injury: A DNA microarray analysis. *J Interferon Cytokine Res* 23 (5):237-245.

Teixeira, C.F., S.R. Zamuner, J.P. Zuliani, C.M. Fernandes, M.A. Cruz-Hofling, I. Fernandes, F. Chaves, and J.M. Gutierrez. 2003. Neutrophils do not contribute to local tissue damage, but play a key role in skeletal muscle regeneration, in mice injected with Bothrops asper snake venom. *Muscle Nerve* 28 (4):449-459.

Tidball, J.G., E. Berchenko, and J. Frenette. 1999. Macrophage invasion does not contribute to muscle membrane injury during inflammation. *J Leukoc Biol* 65 (4):492-498.

Tsivitse, S.K., T.J. McLoughlin, J.M. Peterson, E. Mylona, S.J. McGregor, and F.X. Pizza. 2003. Downhill running in rats: Influence on neutrophils, macrophages, and MyoD+ cells in skeletal muscle. *Eur J Appl Physiol* 90 (5-6):633-638.

Tsivitse, S.K., E. Mylona, J.M. Peterson, W.T. Gunning, and F.X. Pizza. 2005. Mechanical loading and injury induce human myotubes to release neutrophil chemoattractants. *Am J Physiol Cell Physiol* 288 (3):C721-729.

Walzog, B., and P. Gaehtgens. 2000. Adhesion molecules: The path to a new understanding of acute inflammation. *News Physiol Sci* 15:107-113.

Warren, G.L., T. Hulderman, N. Jensen, M. McKinstry, M. Mishra, M.I. Luster, and P.P. Simeonova. 2002. Physiological role of tumor necrosis factor alpha in traumatic muscle injury. *FASEB J* 16 (12):1630-1632.

Warren, G.L., T. Hulderman, D. Mishra, X. Gao, L. Millecchia, L. O'Farrell, W.A. Kuziel, and P.P. Simeonova. 2005. Chemokine receptor CCR2 involvement in skeletal muscle regeneration. *FASEB J* 19 (3):413-415.

Warren, G.L., L. O'Farrell, M. Summan, T. Hulderman, D. Mishra, M.I. Luster, W.A. Kuziel, and P.P. Simeonova. 2004. Role of CC chemokines in skeletal muscle functional restoration after injury. *Am J Physiol Cell Physiol* 286 (5): C1031-1036.

Chapter 5

Abraham, W.M. 1977. Factors in delayed muscle soreness. *Medicine and Science in Sports* 9: 11-20.

Armstrong, R.B. 1984. Mechanisms of exercise-induced delayed onset muscular soreness: A brief review. *Medicine and Science in Sports and Exercise* 16: 529-538.

Armstrong, R.B., G.L. Warren, and J.A. Warren. 1991. Mechanisms of exercise induced muscle fibre injury. *Sports Medicine* 12: 184-207.

Bär, P.R.D., J.P. Reijneveld, J.H.J. Wokke, S.C.J.M. Jacobs, and A.L. Bootsma. 1997. Muscle damage induced by exercise: Nature, prevention and repair. In *Muscle damage,* ed. S. Salmons, 1-27. Oxford, UK: Oxford University Press.

Brown, S., R.B. Child, S. Day, and A. Donnelly. 1997. Exercise-induced skeletal muscle damage and adaptation following repeated bouts of eccentric muscle contraction. *Journal of Sports Science* 15: 215-222.

Brown, S., S. Day, and A. Donnelly. 1999. Indirect evidence of human skeletal muscle damage and collagen breakdown after eccentric muscle actions. *Journal of Sports Science* 17: 397-402.

Calzá, L. 2001. Neurochemistry of pain circuits: Physiological versus pathological pain. In *Neuroscience: Focus on acute and chronic pain,* ed. M.A. Tiendo. New York, NY: Springer-Verlag .

Chen, T.C. 2003. Effects of a second bout of maximal eccentric exercise on muscle damage and electromyographic activity. *European Journal of Applied Physiology* 89: 115-121.

Chen, T.C., K. Nosaka, and P. Sacco. 2007. Intensity of eccentric exercise, shift of optimum angle, and the magnitude of repeated-bout effect. *Journal of Applied Physiology* 102: 992-999.

Cheung, K., P.A. Hume, and L. Maxwell. 2003. Delayed onset muscle soreness. Treatment strategies and performance factors. *Sports Medicine* 33: 145-164.

Clarkson, P.M., K. Nosaka, and B. Braun. 1992. Muscle function after exercise-induced muscle damage and rapid adaptation. *Medicine and Science in Sports and Exercise* 24: 512-520.

Clarkson, P.M., and I. Tremblay. 1988. Exercise-induced muscle damage, repair, and adaptation in humans. *Journal of Applied Physiology* 65: 1-6.

Cleak, M.J., and R.G. Eston. 1992. Delayed onset muscle soreness: Mechanisms and management. *Journal of Sports Science* 10: 325-341.

Ebbeling, C.B., and P.M. Clarkson. 1989. Exercise-induced muscle damage and adaptation. *Sports Medicine* 7: 207-234.

Eston, R., S. Finney, S. Baker, and V. Baltzpoulos. 1996. Muscle tenderness and peak torque changes after downhill running following a prior bout of isokinetic eccentric exercise. *Journal of Sports Science* 14: 291-299.

Faulkner, J.A., A.V. Brooks, and J.A. Opiteck. 1993. Injury to skeletal muscle fibers during contractions: Conditions of occurrence and prevention. *Physical Therapy* 73: 911-921.

Foley, J.M., R.C. Jayaraman, B.M. Prior, J.M. Pivarnik, and R.A. Meyer. 1999. MR measurements of muscle damage and adaptation after eccentric exercise. *Journal of Applied Physiology* 87: 2311-2318.

Friden, J., M. Sjöström, and B. Ekblom. 1981. A morphological study of delayed muscle soreness. *Experientia* 37: 506-507.

Friden, J., M. Sjöström, and B. Ekblom. 1984. Myofibrillar damage following intense eccentric exercise in men. *International Journal of Sports Medicine* 4: 170-176.

Friden, J., and R.L. Lieber. 2001. Eccentric exercise-induced injuries to contractile and cytoskeletal muscle fibre components. *Acta Physiologica Scandinavica* 171: 321-326.

Gibala, M.J., J.D. MacDougall, W.T. Stauber, and A. Elorriaga. 1995. Changes in human skeletal muscle ultrastructure and force production after acute resistance exercise. *Journal of Applied Physiology* 78: 702-708.

Guyton, A.C., and J.E. Hall. *Textbook of medical physiology.* 9th ed. Philadelphia, PA: Saunders.

Hargreaves, K.M., J.A. Buckwalter, and S.L. Gordon 1989. Mechanisms of pain sensation resulting from inflammation. In *Sports-induced inflammation: clinical and basic science concepts,* ed. W.B. Leadbetter et al., 383-392. Park Ridge, IL: American Academy of Orthopaedic Surgeons.

Hirose, L., K. Nosaka, M. Newton, A. Lavender, M. Kano, J. Peake, and K. Suzuki. 2004. Changes in inflammatory mediators following eccentric exercise of the elbow flexors. *Exercise Immunology Review* 10: 75-90.

Hough, T. 1902. Ergographic studies in muscular soreness. *American Journal of Physiology* 7: 76-92.

Howatson, G., and K.A. van Someren. 2007. Evidence of a contralateral repeated bout effect after maximal eccentric contractions. *European Journal of Applied Physiology* 2007 May 30; [Epub ahead of print]

Howell, J.N., G. Chleboun, and R. Conatser. 1993. Muscle stiffness, strength loss, swelling and soreness following exercise-induced injury in humans. *Journal of Physiology* 464: 183-196.

Ingalls, C.P., J.C. Wenke, T. Nofal, and R.B. Armstrong. 2004. Adaptation to lengthening contraction-induced injury in mouse muscle. *Journal of Applied Physiology* 97: 1067-1076.

Jamurtas, A.Z., V. Theocharis, T. Tofas, A. Tsiokanos, C. Yfanti, V. Paschalis, Y. Koutedakis, and K. Nosaka. 2005. Comparison between leg and arm eccentric exercise of the same relative intensity on indices of muscle damage. *European Journal of Applied Physiology,* 95: 179-185

Jones, D.A., D.J. Newham, and P.M. Clarkson. 1987. Skeletal muscle stiffness and pain following eccentric exercise of the elbow flexors. *Pain* 30: 233-242.

Jones, D.A., D.J. Newham, J.M. Round, and S.E.J. Tolfree. 1986. Experimental human muscle damage: Morphological changes in relation to other indices of damage. *Journal of Physiology* 375: 435-448.

Jones, D.A., D.J. Newham, and C. Torgan. 1989. Mechanical influences on long-lasting human muscle fatigue and delayed-onset pain. *Journal of Physiology* 412: 415-427.

Jones, D.A., and J.M. Round. 1990. *Skeletal muscle in health and disease.* Manchester, UK: Manchester University Press.

Kingsley, R.E. 2002. *Concise text of neuroscience.* 2nd ed. Baltimore: Lippincott, Williams & Wilkins.

Koh, T.J. 2002. Do small heat shock proteins protect skeletal muscle from injury? *Exercise and Sport Sciences Reviews* 30: 117-121.

Koh, T.J., and S.V. Brooks. 2001. Lengthening contractions are not required to induce protection from contraction-induced muscle injury. *American Journal of Physiology: Regulatory, Integrative, and Comparative Physiology* 281: R155-R161.

Koltyn, K.F. 2000. Analgesia following exercise. *Sports Medicine* 29: 85-98.

Kuipers, H. 1994. Exercise-induced muscle damage. *International Journal of Sports Medicine* 15: 132-135.

Lavender, A.P., and K. Nosaka. 2006. Changes in steadiness of isometric force following eccentric and concentric exercise. *European Journal of Applied Physiology* 96: 235-240.

Lavender, A.P., and K. Nosaka. 2007. A light load eccentric exercise confers protection against a subsequent bout of more demanding eccentric exercise. *Journal of Science and Medicine in Sport,* in press

Lieber, R.L., and J. Friden. 2002. Morphologic and mechanical basis of delayed-onset muscle soreness. *Journal of American Academy of Orthopedic Surgery* 10: 67-73.

MacIntyre, D.L., W.D. Reid, and D.C. McKenzie. 1995. Delayed muscle soreness. The inflammatory response to muscle injury and its clinical implications. *Sports Medicine* 20: 24-40.

Marchettini, P. 1993. Muscle pain: Animal and human experimental and clinical studies. *Muscle and Nerve* 16: 1033-1039.

McArdle, F., S. Spiers, H. Aldemir, A. Vasilaki, A. Beaver, L. Iwanejko, A. McArdle, and M.J. Jackson. 2004. Preconditioning of skeletal muscle against contraction-induced damage: The role of adaptations to oxidants in mice. *Journal of Physiology* 561: 233-244.

McHugh, M.P. 2003. Recent advances in the understanding of the repeated bout effect: The protective effect against muscle damage from a single bout of eccentric exercise. *Scandinavian Journal of Medicine and Science in Sports* 13: 88-97.

McHugh, M.P., and S. Pasiakos. 2004. The role of exercising muscle length in the protective adaptation to a single bout of eccentric exercise. *European Journal of Physiology,* 93: 286-293

Melzack, R. 1982. Recent concept of pain. *Journal of Medicine* 13: 147-160.

Mense, S. 1993. Nociception from skeletal muscle in relation to clinical muscle pain. *Pain* 54: 241-289.

Merskey, H., and N. Bogduk. 1994. Classification of chronic pain. In *Definition of chronic pain syndromes and definition of pain terms.* 2nd ed. Seattle: International Association for the Study of Pain.

Mikkelsen, U.R., H. Gissel, A. Fredsted, H. Gissel, and T. Clausen. 2006. Excitation-induced cell damage and β_2-adrenoceptor agonist stimulated force recovery in rat skeletal muscle. *American Journal of Physiology: Regulative, and Comparative Physiology* 290: R265-R272.

Miles, M.P., and P.M. Clarkson. 1994. Exercise-induced muscle pain, soreness, and cramps. *Journal of Sports Medicine and Physical Fitness* 34: 203-216.

Millan, M.J. 1999. The induction of pain: An integrative review. *Progress in Neurobiology* 57: 1-164.

Newham, D.J. 1988. The consequences of eccentric contractions and their relationship to delayed onset muscle pain. *European Journal of Applied Physiology* 57: 353-359.

Newham, D.J., D.A. Jones, and P.M. Clarkson. 1987. Repeated high-force eccentric exercise: Effects on muscle pain and damage. *Journal of Applied Physiology* 63: 1381-1386.

Newham, D.J., K.R. Mills, B.M. Quigley, and R.H.T. Edwards. 1983. Pain and fatigue after concentric and eccentric muscle contractions. *Clinical Science* 64: 55-62.

Newton, M., G.T. Morgan, P. Sacco, D. Chapman, and K. Nosaka. 20% Comparison between resistance-trained and untrained men for responses to a bout of strenuous eccentric exercise of the elbow flexors. *Journal of Strength and Conditioning Research,* in press

Newton, M., P. Sacco, D. Chapman, and K. Nosaka. 20% Comparison between arms for responses to a bout of maximal eccentric exercise of the elbow flexors. *European Journal of Applied Physiology,* in press.

Nosaka, K., and P.M. Clarkson. 1996. Changes in indicators of inflammation after eccentric exercise of the elbow flexors. *Medicine and Science in Sports and Exercise* 28: 953-961.

Nosaka, K., P.M. Clarkson, M.E. McGuiggin, and J.M. Byrne. 1991. Time course of muscle adaptation after high force eccentric exercise. *European Journal of Applied Physiology* 63: 70-76.

Nosaka, K., M. Muthalib, A. Lavender, and P.B. Laursen. 2007. Attenuation of muscle damage by preconditioning with muscle hyperthermia 1 day prior to eccentric exercise. *European Journal of Applied Physiology* 99: 183-192.

Nosaka, K., and M. Newton. 2002a. Concentric or eccentric training on eccentric exercise-induced muscle damage. *Medicine and Science in Sports and Exercise* 34: 63-69.

Nosaka, K., and M. Newton. 2002b. Differences in the magnitude of muscle damage between maximal and submaximal eccentric loading. *Journal of Strength and Conditioning Research* 16: 202-208.

Nosaka, K., and M. Newton. 2002c. Is recovery from muscle damage retarded by a subsequent bout of eccentric exercise inducing larger decreases in force? *Journal of Science and Medicine in Sport* 5: 204-208.

Nosaka, K., and M. Newton. 2002d. Repeated eccentric exercise bouts do not exacerbate muscle damage and repair. *Journal of Strength and Conditioning Research* 16: 117-122.

Nosaka, K., M. Newton, and P. Sacco. 2002a. DOMS does not reflect the magnitude of eccentric exercise-induced muscle damage. *Scandinavian Journal of Medicine and Science in Sports* 12: 337-346.

Nosaka, K., M. Newton, and P. Sacco. 2002b. Responses of human elbow flexor muscles to electrically stimulated forced lengthening exercise. *Acta Physiologica Scandinavica* 174: 137-145.

Nosaka, K., M.J. Newton, and P. Sacco. 2005. Attenuation of protective effect against eccentric exercise-induced muscle damage. *Canadian Journal of Applied Physiology* 30: 529-542.

Nosaka K., M. Newton, P. Sacco, D. Chapman, and A. Lavender. 2005b. Partial protection against muscle damage by eccentric actions at short muscle lengths. *Medicine and Science in Sports and Exercise* 37: 746-753.

Nosaka, K., and K. Sakamoto. 2001. Effect of elbow joint angle on the magnitude of muscle damage to the elbow flexors. *Medicine and Science in Sports and Exercise* 33: 22-29.

Nosaka, K., K. Sakamoto, M. Newton, and P. Sacco. 2001a. How long does the protective effect of eccentric-induced muscle damage last? *Medicine and Science in Sports and Exercise* 33: 1490-1495.

Nosaka, K., K. Sakamoto, M. Newton, and P. Sacco. 2001b. The repeated bout effect of reduced-load eccentric exercise on elbow flexor muscle damage. *European Journal of Applied Physiology* 85: 34-40.

Nurenberg, P., C.J. Giddings, J. Stray-Gundersen, J. Fleckenstein, W. Gonyea, and R.M. Peshock. 1992. MR imaging-guided muscle biopsy for correlation of increased signal intensity with ultrastructural change and delayed-onset muscle soreness after exercise. *Radiology* 184: 865-869.

O'Connor, P.J., and D.B. Cook. 1999. Exercise and pain: The neurobiology, measurement, and laboratory study of pain in relation to exercise in humans. *Exercise and Sport Sciences Reviews* 27: 119-166.

Ohbach, R., and E.N. Gale. 1989. Pressure pain threshold, clinical assessment, and differential diagnosis: Reliability and validity in patient with myogenic pain. *Pain* 39: 157-169.

Ohnhaus, E.E., and R. Adler. 1975. Methodological problems in the measurement of pain: A comparison between the verbal rating scale and the visual analogue scale. *Pain* 1: 379-384.

Philippou, A., G.C. Bogdanis, A.M. Nevill, and M. Maridaki. 2004. Changes in the angle-force curve of human elbow flexors following eccentric and isometric exercise. *European Journal of Applied Physiology* 93: 237-244.

Pizza, F.X., D. Cavender, A. Stockard, H. Baylies, and A. Beighle. 1999. Anti-inflammatory doses of ibuprofen: Effect on neutrophils and exercise-induced muscle injury. *International Journal of Sports Medicine* 20: 98-102.

Proske, U., and D.L. Morgan. 2001. Muscle damage from eccentric exercise: mechanism, mechanical signs, and adaptation and clinical applications. *Journal of Physiology* 537: 333-345.

Rodenburg, J.B., P.R. Bär, and R.W. De Boer. 1993. Relationship between muscle soreness and biochemical and functional outcomes of eccentric exercise. *Journal of Applied Physiology* 74: 2976-2983.

Safran, M.R., A.V. Seaber, and W.E. Garrett Jr. 1989. Warm-up and muscular injury prevention. *Sports Medicine* 8: 239-249.

Saxon, J.M., and A.E. Donnelly. 1995. Light concentric exercise during recovery from exercise-induced muscle damage. *International Journal of Sports Medicine* 16: 347-351.

Smith, L.L. 1991. Acute inflammation: The underlying mechanism in delayed onset muscle soreness? *Medicine and Science in Sports and Exercise* 23: 542-551.

Stauber, W.T., and C.A. Smith. 1998. Cellular responses in exertion-induced skeletal muscle injury. *Molecular and Cellular Biochemistry* 179: 189-196.

Talag, T.S. 1973. Residual muscular soreness as influenced by concentric, eccentric, and static contractions. *Research Quarterly* 44: 458-469.

Unruh, A.M., J. Strong, and A. Wright. 2002. Introduction to pain. In *Pain: A textbook for therapists,* ed. J. Strong, A. M. Unruh, A. Wright, and G.D. Baxter, 3-11. London, UK: Churchill Livingstone.

Warren, G.L., D.A. Lowe, and R.B. Armstrong. 1999. Measurement tools used in the study of eccentric contraction-induced injury. *Sports Medicine* 27: 43-59.

Weerakkody, N.S., N.P. Whitehead, B.J. Canny, J.E. Gregory, and U. Proske. 2001. Large-fiber mechanoreceptors contribute to muscle soreness after eccentric exercise. *Journal of Pain* 2: 209-219.

Whitehead, N.P., T.J. Allen, D.L. Morgan, and U. Proske. 1998. Damage to human muscle from eccentric exercise after training with concentric exercise. *Journal of Physiology* 512: 615-620.

Yu, J-G., C. Malm, and L-E. Thornell. 2002. Eccentric contractions leading to DOMS do not cause loss of desmin nor fibre necrosis in human muscle. *Histochemistry and Cell Biology* 118: 29-34.

Yu, J-G., and L-E. Thornell. 2002. Desmin and actin alterations in human muscles affected by delayed onset muscle soreness: A high resolution immunocytochemical study. *Histochemistry and Cell Biology* 118: 171-179.

Zainuddin, Z., P. Sacco, M. Newton, and K. Nosaka. 2006.

Light concentric exercise has a temporarily analgesic effect on DOMS but no effect on recovery from eccentric exercise. *Canadian Journal of Applied Physiology* 31: 126-134.

Chapter 6

Allen, R.E., and L.K. Boxhorn. 1989. Regulation of skeletal muscle myogenic satellite cell proliferation and differentiation by transforming growth factor-beta, insulin-like growth factor I, and fibroblast growth factor. *Journal of Cellular Physiology* 138: 311-315.

Allen, R.E., S.M. Sheehan, R.G. Taylor, T.L. Kendall, and G.M. Rice. 1995. Hepatocyte growth factor activates quiescent skeletal muscle satellite cells in vitro. *Journal of Cellular Physiology* 165: 307-312.

Appelbaum, F.R. 1996. The use of bone marrow and peripheral blood stem cell transplantation in the treatment of cancer. *Cancer Journal for Clinicians* 46: 142-164.

Armand, A.S., P. Pariset, I. Laziz, T. Launay, F. Fiore, B.D. Gaspera, D. Birnbaum, F. Charbonnier, and C. Chanoine. 2005. FGF6 regulates muscle differentiation through a calcineurin-dependent pathway in regenerating soleus of adult mice. *Journal of Cellular Physiology* 204: 297-308.

Austin, L., and A.W. Burgess. 1991. Stimulation of myoblast proliferation in culture by leukaemia inhibitory factor and other cytokines. *Journal of the Neurological Sciences* 101: 193-197.

Barton-Davis, E.R., D.I. Shoturma, A. Musaro, N. Rosenthal, and H.L. Sweeney. 1998. Viral mediated expression of insulin-like growth factor I blocks the aging-related loss of skeletal muscle function. *Proceedings of the National Academy of Sciences of the United States of America* 95: 15603-15607.

Bell, C.D., and P.E. Conen. 1968. Histopathological changes in Duchenne muscular dystrophy. *Journal of the Neurological Sciences* 7: 529-544.

Bernasconi, P., E. Torchiana, P. Confalonieri, R. Brugnoni, R. Barresi, M. Mora, F. Cornelio, L. Morandi, and R. Mantegazza. 1999. Expression of transforming growth factor-beta 1 in dystrophic patient muscles correlates with fibrosis. Pathogenetic role of a fibrogenic cytokine. *Journal of Clinical Investigation* 96: 1137-1144.

Bischoff, R. 1986. A satellite cell mitogen from crushed adult muscle. *Developmental Biology* 115: 140-147.

Bischoff, R. 1989. Analysis of muscle regeneration using single myofibres in culture. *Medicine and Science in Sports and Exercise* (Suppl. 5): 164-172.

Bischoff, R. 1997. Chemotaxis of skeletal muscle myogenic satellite cells. *Developmental Dynamics* 208: 505-515.

Bondesen, B.A., S.T. Mills, K.M. Kegley, and G.K. Pavlath. 2004. The COX-2 pathway is essential during early stages of skeletal muscle regeneration. *American Journal of Physiology* 287: C475-C483.

Burghes, A.H., C. Logan, X. Hu, B. Belfall, R.G. Worton, and P.N. Ray. 1987. A cDNA clone from the Duchenne/Becker muscular dystrophy gene. *Nature* 328: 434-437.

Cantini, M., and F. Carraro. 1996. Control of cell proliferation by macrophage-myoblast interactions. *Basic and Applied Myology* 6: 485-489.

Carlson, B.M., and J.A. Faulkner. 1989. Muscle transplantation between young and old rats: Age of host determines recovery. *American Journal of Physiology* 256: C1262-C1266.

Chakravarthy, M.V., T.W. Abraha, R.J. Schwartz, M.L. Fiorotto, and F.W. Booth. 2000. Insulin-like growth factor-I extends in vitro replicative life span of skeletal muscle myogenic satellite cells by enhancing G1/S cell cycle progression via the activation of phosphatidylinositol 3'-kinase/Akt signaling pathway. *Journal of Biological Chemistry* 275: 35942-35952.

Chakravarthy, M.V., B.S. Davis, and F.W. Booth. 2000. IGF-I restores satellite cell proliferative potential in immobilized old skeletal muscle. *Journal of Applied Physiology* 89: 1365-1379.

Charge, S.B., and M.A. Rudnicki. 2004. Cellular and molecular regulation of muscle regeneration. *Physiological Review* 84(1) 209-238.

Clegg, C.H., T.A. Linkhart, B.B. Olwin, and S.D. Hauschka. 1987. Growth factor control of skeletal muscle differentiation: commitment to terminal differentiation occurs in G1 phase and is repressed by fibroblast growth factor. *Journal of Cell Biology* 105(2): 949-956.

Coggan, A.R., R.J. Spina, D.S. King, M.A. Rogers, M. Brown, P.M. Nemeth, and J.O. Holloszy. 1992. Histochemical and enzymatic comparison of the gastrocnemius muscle of young and elderly men and women. *Journals of Gerontology Series B: Psychological Sciences and Social Sciences* 47: B71-B76.

Coleman, M.E., F. Demayo, K.C. Yin, H.M. Lee, R. Geske, C. Montgomery, and R.J. Schwartz. 1995. Myogenic vector expression of insulin-like growth factor I stimulates muscle cell differentiation and muscle fibre hypertrophy in transgenic mice. *Journal of Biological Chemistry* 270: 12109-12116.

Conboy, I.M., M.J. Conboy, A.J. Wagers, E.R. Girma, I.L. Weissman, and T.A. Rando. 2005. Rejuvenation of aged progenitor cells by exposure to a young systemic environment. *Nature* 433: 760-764.

Cornelison, D.D., B.B. Olwin, M.A. Rudnicki, and B.J. Wold. 2000. MyoD(-/-) satellite cells in single-fibre culture are differentiation defective and MRF4 deficient. *Developmental Biology* 224: 122-137.

Cossu, G., and F. Mavilio. 2000. Myogenic stem cells for the therapy of primary myopathies: Wishful thinking or therapeutic perspective? *Journal of Clinical Investigation* 105: 1669-1674.

Danon, D., M.A. Kowatch, and G.S. Roth. 1989. Promotion of wound repair in old mice by local injection of macrophages. *Proceedings of the National Academy of Sciences of the United States of America* 86: 2018-2020.

Darr, K.C., and E. Schultz. 1989. Hindlimb suspension suppresses muscle growth and satellite cell proliferation. *Journal of Applied Physiology* 67: 1827-1834.

Dasarathy, S., M. Dodig, S.M. Muc, S.C. Kalhan, and A.J. McCullough. 2004. Skeletal muscle atrophy is associated with an increased expression of myostatin and impaired satellite cell function in the portacaval anastamosis rat. *American Journal of Physiology* 287: G1124-1130.

Decary, S., C.B. Hamida, V. Mouly, J.P. Barbet, F. Hentati, and G.S. Butler-Browne. 2000. Shorter telomeres in dystrophic muscle consistent with extensive regeneration in young children. *Neuromuscular Disorders* 10: 113-120.

deLapeyriere, O., V. Ollendorff, J. Planche, M.O. Ott, S. Pizette, F. Coulier, and D. Birnbaum. 1993. Expression of the FGF6 gene is restricted to developing skeletal muscle in the mouse embryo. *Development* 118: 601-611.

Dodson, M.V., and R.E. Allen. 1987. Interaction of multiplication stimulating activity/rat insulin-like growth factor II with skeletal muscle satellite cells during aging. *Mechanisms of Ageing and Development* 39: 121-128.

Doumit, M.E., D.R. Cook, and R.A. Merkel. 1993. Fibroblast growth factor, epidermal growth factor, insulin-like growth factors, and platelet-derived growth factor-BB stimulate proliferation of clonally derived porcine myogenic satellite cells. *Journal of Cellular Physiology* 157: 326-332.

Faulkner, J.A., S.V. Brooks, and E. Zerba. 1995. Muscle atrophy and weakness with aging: Contraction-induced injury as an underlying mechanism. *Journals of Gerontology Series A: Biological Sciences and Medical Sciences* 50: 124-129.

Fiore, F., A. Sebille, and D. Birnbaum. 2000. Skeletal muscle regeneration is not impaired in Fgf6 -/- mutant mice. *Biochemical and Biophysical Research Communications* 272: 138-143.

Florini, J.R., K.A. Magri, D.Z. Ewton, P.L. James, K. Grindstaff, and P.S. Rotwein. 1991. "Spontaneous" differentiation of skeletal myoblasts is dependent upon autocrine secretion of insulin-like growth factor-II. *Journal of Biological Chemistry* 266: 15917-15923.

Floss, T., H.H. Arnold, and T. Braun. 1997. A role for FGF-6 in skeletal muscle regeneration. *Genes and Development* 11: 2040-2051.

Fulle, S., S. DiDonna, C. Puglielli, T. Pietrangelo, S. Beccafico, R. Bellomo, F. Protasi, and G. Fano. 2005. Age-dependent imbalance of the antioxidative system in human satellite cells. *Experimental Gerontology* 40: 189-197.

Gal-Levi, R., Y. Leshem, S. Aoki, T. Nakamura, and O. Halevy. 1998. Hepatocyte growth factor plays a dual role in regulating skeletal muscle myogenic satellite cell proliferation and differentiation. *Biochimica et Biophysica Acta* 1402: 39-51.

Gayan-Ramirez, G., F. Vanderhoydonc, G. Verhoeven, and M. Decramer. 1999. Acute treatment with corticosteroids decreases IGF-1 and IGF-2 expression in the rat diaphragm and gastrocnemius. *American Journal of Respiratory and Critical Care Medicine* 159: 283-289.

Gibson, M.C., and E. Schultz. 1983. Age-related differences in absolute numbers of skeletal muscle satellite cells. *Muscle and Nerve* 6: 574-580.

Goldspink, G., and S.D. Harridge. 2004. Growth factors and muscle ageing. *Experimental Gerontology* 39: 1433-1438.

Grande, J.P., D.C. Melder, and A.R. Zinsmeister. 1997. Modulation of collagen gene expression by cytokines: Stimulatory effect of transforming growth factor-beta1, with divergent effects of epidermal growth factor and tumor necrosis factor-alpha on collagen type I and collagen type IV. *Journal of Laboratory and Clinical Medicine* 130: 476-486.

Gregorevic, P., D.R. Plant, and G.S. Lynch. 2004. Administration of insulin-like growth factor-I improves fatigue resistance of skeletal muscles from dystrophic mdx mice. *Muscle and Nerve* 30: 295-304.

Grounds, M.D. 1998. Age-associated changes in the response of skeletal muscle cells to exercise and regeneration. *Annals of the New York Academy of Sciences* 854: 78-91.

Grounds, M.D. 1999. Muscle regeneration: Molecular aspects and therapeutic implications. *Current Opinion in Neurology* 12: 535-543.

Grounds, M.D., J.D. White, N. Rosenthal, and M.A. Bogoyevitch. 2002. The role of stem cells in skeletal and cardiac muscle repair. *Journal of Histochemistry and Cytochemistry* 50: 589-610.

Gussoni, E., G.K. Pavlath, A.M. Lanctot, K.R. Sharma, R.G. Miller, L. Steinman, and H.M. Blau. 1992. Normal dystrophin transcripts detected in Duchenne muscular dystrophy patients after myoblast transplantation. *Nature* 356: 435-438.

Hartel, J.V., J.A. Granchelli, M.S. Hudecki, C.M. Pollina, and L.E. Gosselin. 2001. Impact of prednisone on TGF-beta1 and collagen in diaphragm muscle from mdx mice. *Muscle and Nerve* 24: 428-432.

Hasty, P. 2001. The impact energy metabolism and genome maintenance have on longevity and senescence: Lessons from yeast to mammals. *Mechanisms of Ageing and Development* 122: 1651-1662.

Hasty, P., A. Bradley, J.H. Morris, D.G. Edmondson, J.M. Venuti, E.N. Olson, and W.H. Klein. 1993. Muscle deficiency and neonatal death in mice with a targeted mutation in the myogenin gene. *Nature* 364: 501-506.

Haugk, K.L., R.A. Roeder, M.J. Garber, and G.T. Schelling. 1995. Regulation of muscle cell proliferation by extracts from crushed muscle. *Journal of Animal Science* 73: 1972-1981.

Hawke, T.J. 2005. Muscle stem cells and exercise training. *Exercise and Sport Sciences Reviews* 33: 63-68.

Hawke, T.J., and D.J. Garry. 2001. Myogenic satellite cells: Physiology to molecular biology. *Journal of Applied Physiology* 9: 534-551.

Hawke, T.J., A.P. Meeson, N. Jiang, S. Graham, K. Hutcheson, J.M. DiMaio, and D.J. Garry. 2003. p21 is essential for normal myogenic progenitor cell function in regenerating skeletal muscle. *American Journal of Physiology* 285: C1019-1027.

Heslop, L., J.E. Morgan, and T.A. Partridge. 2000. Evidence for a myogenic stem cell that is exhausted in dystrophic muscle. *Journal of Cell Science* 113: 2299-2308.

Hibi, M., K. Nakajima, and T. Hirano. 1996. IL-6 cytokine family and signal transduction: A model of the cytokine system. *Journal of Molecular Medicine* 74: 1-12.

Hill, M., and G. Goldspink. 2003. Expression and splicing of the insulin-like growth factor gene in rodent muscle is associated with muscle satellite (stem) cell activation following local tissue damage. *Journal of Physiology* 549: 409-418.

Hodgetts, S.I., and M.D. Grounds. 2003. Irradiation of dystrophic host tissue prior to myoblast transfer therapy enhances initial (but not long-term) survival of donor myoblasts. *Journal of Cell Science* 116: 4131-4146.

Huard, J., J.P. Bouchard, R. Roy, F. Malouin, G. Dansereau, C. Labrecque, N. Albert, C.L. Richards, B. Lemieux, and J.P. Tremblay. 1992. Human myoblast transplantation: Preliminary results of 4 cases. *Muscle and Nerve* 15: 550-560.

Ignotz, R.A., and J. Massague. 1986. Transforming growth factor-beta stimulates the expression of fibronectin and collagen and their incorporation into the extracellular matrix. *Journal of Biological Chemistry* 261: 4337-4345.

Kadi, F., N. Charifi, C. Denis, and J. Lexell. 2004. Satellite cells and myonuclei in young and elderly women and men. *Muscle and Nerve* 29: 120-127.

Karpati, G., D. Ajdukovic, D. Arnold, R.B. Gledhill, R. Guttmann, P. Holland, P.A. Koch, E. Shoubridge, D. Spence, and M. Vanasse. 1993. Myoblast transfer in Duchenne muscular dystrophy. *Annals of Neurology* 34: 8-17.

Kirk, S.P., J.M. Oldham, F. Jeanplong, and J.J. Bass. 2003. Insulin-like growth factor-II delays early but enhances late regeneration of skeletal muscle. *Journal of Histochemistry and Cytochemistry* 51: 1611-1620.

Kirk, S., J. Oldham, R. Kambadur, M. Sharma, P. Dobbie, and J. Bass. 2000. Myostatin regulation during skeletal muscle regeneration. *Journal of Cellular Physiology* 184: 356-363.

Kraemer, W.J., K. Hakkinen, R.U. Newton, B.C. Nindl, J.S. Volek, M. McCormick, L.A. Gotshalk, S.E. Gordon, S.J. Fleck, W.W. Campbell, M. Putukian, and W.J. Evans. 1999. Effects of heavy-resistance training on hormonal response patterns in younger vs. older men. *Journal of Applied Physiology* 87: 982-992.

Kurek, J.B., J.J. Bower, M. Romanella, F. Koentgen, M. Murphy, and L. Austin. 1997. The role of leukemia inhibitory factor in skeletal muscle regeneration. *Muscle and Nerve* 20: 815-822.

Kuschel, R., Z. Yablonka-Reuveni, and A. Bornemann. 1999. Satellite cells on isolated myofibres from normal and denervated adult rat muscle. *Journal of Histochemistry and Cytochemistry* 47: 1375-1384.

Lange, K.H., J.L. Andersen, N. Beyer, F. Isaksson, B. Larsson, M.H. Rasmussen, A. Juul, J. Bulow, and M. Kjaer. 2002. GH administration changes myosin heavy chain isoforms in skeletal muscle but does not augment muscle strength or hypertrophy, either alone or combined with resistance exercise training in healthy elderly men. *Journal of Clinical Endocrinology and Metabolism* 87: 513-523.

LeRoith, D., M. McGuinness, J. Shemer, B. Stannard, F. Lanau, T.N. Faria, H. Kato, H. Werner, M. Adamo, and C.T. Roberts. 1992. Insulin-like growth factors. *Biological Signals* 1: 173-181.

Lescaudron, L., E. Peltekian, J. Fontaine-Perus, D. Paulin, M. Zampieri, L. Garcia, and E. Parrish. 1999. Blood borne macrophages are essential for the triggering of muscle regeneration following muscle transplant. *Neuromuscular Disorders* 9: 72-80.

Lu, D.X., S.K. Huang, and B.M. Carlson. 1997. Electron microscopic study of long-term denervated rat skeletal muscle. *Anatomical Record* 248: 355-365.

Machida, S., and F.W. Booth. 2004. Insulin-like growth factor 1 and muscle growth: Implication for satellite cell proliferation. *Proceedings of the Nutrition Society* 63: 337-340.

Martin, J.F., L. Li, and E.N. Olson. 1992. Repression of myogenin function by TGF-β1 is targeted at the basic helix-loop-helix motif and is independent of E2A products. *Journal of Biological Chemistry* 267: 10956-10960.

Marshall, P.A., P.E. Williams, and G. Goldspink. 1989. Accumulation of collagen and altered fiber-type ratios as indicators of abnormal muscle gene expression in the mdx dystrophic mouse. *Muscle and Nerve* 12: 528-537.

Mauro, A. 1961. Satellite cell of skeletal muscle fibres. *Journal of Biophysical and Biochemical Cytology* 9: 493-498.

McCroskery, S., M. Thomas, L. Maxwell, M. Sharma, and R. Kambadur. 2003. Myostatin negatively regulates satellite cell activation and self-renewal. *Journal of Cell Biology* 162: 1135-1147.

McGeachie, J.K. 1989. Sustained cell proliferation in denervated skeletal muscle of mice. *Cell and Tissue Research* 257: 455-457.

McKoy, G., W. Ashley, J. Mander, S.Y. Yang, N. Williams, B. Russell, and G. Goldspink. 1999. Expression of insulin growth factor-1 splice variants and structural genes in rabbit skeletal muscle induced by stretch and stimulation. *Journal of Physiology* 516: 583-592.

McPherron, A.C., and S. Lee. 1997. Double muscling in cattle due to mutations in the myostatin gene. *Proceedings of the National Academy of Sciences of the United States of America* 94: 12457-12461.

Meeson, A.P., T.J. Hawke, S. Graham, N. Jiang, J. Eltermann, K. Hutcheson, J.M. DiMaio, T. Gallardo, and D.J. Garry. 2004. Cellular and molecular regulation of skeletal muscle SP cells. *Stem Cells* 22: 1305-1320.

Melone, M.A., G. Peluso, U. Galderisi, O. Petillo, and R. Cotrufo. 2000. Increased expression of IGF-binding protein-5 in Duchenne muscular dystrophy (DMD) fibroblasts correlates with the fibroblast-induced downregulation of DMD myoblast growth: An in vitro analysis. *Journal of Cellular Physiology* 185: 143-153.

Menasche, P. 2003. Skeletal muscle satellite cell transplantation. *Cardiovascular Research* 58: 351-357.

Monaco, A.P., R.L. Neve, C. Colletti-Feener, C.J. Bertelson, D.M. Kurnit, and L.M. Kunkel. 1986. Isolation of candidate cDNAs for portions of the Duchenne muscular dystrophy gene. *Nature* 323: 646-650.

Mozdziak, P.E., P.M. Pulvermacher, and E. Schultz. 2000. Unloading of juvenile muscle results in a reduced muscle size 9 wk after reloading. *Journal of Applied Physiology* 88: 158-164.

Nathan, C.F. 1987. Secretory products of macrophages. *Journal of Clinical Investigation* 79: 319-326.

Nnodim, J.O. 2000. Satellite cell numbers in senile rat levator ani muscle. *Mechanisms of Ageing and Development* 112: 99-111.

Olguin, H.C., and B.B. Olwin. 2004. Pax-7 up-regulation inhibits myogenesis and cell cycle progression in satellite cells: A potential mechanism for self-renewal. *Developmental Biology* 275: 375-388.

Owino, V., S.Y. Yang, and G. Goldspink. 2001. Age-related loss of skeletal muscle function and the inability to express the autocrine form of insulin-like growth factor-I (MGF) in response to mechanical overload. *FEBS Letters* 505: 259-263.

Parkhouse, W.S., D.C. Coupland, C. Li, and K.J. Vanderhoek. 2000. IGF-1 bioavailability is increased by resistance training in older women with low bone mineral density. *Mechanisms of Ageing and Development* 113: 75-83.

Patapoutian, A., J.K. Yoon, J.H. Miner, S. Wang, K. Stark, and B. Wold. 1995. Disruption of the mouse MRF4 gene identifies multiple waves of myogenesis in the myotome. *Development* 121: 3347-3358.

Pennica, D., K.J. Shaw, T.A. Swanson, M.W. Moore, D.L. Shelton, K.A. Zioncheck, A. Rosenthal, T. Taga, N.F. Paoni, and W.I. Wood. 1995. Cardiotrophin-1. Biological activities and binding to the leukemia inhibitory factor receptor/gp130 signaling complex. *Journal of Biological Chemistry* 270: 10915-10922.

Proctor, D.N., P. Balagopal, and K.S. Nair. 1998. Age-related sarcopenia in humans is associated with reduced synthetic rates of specific muscle proteins. *Journal of Nutrition* 128 (Suppl. 2): 351S-355.

Rawls, A., J.H. Morris, M. Rudnicki, T. Braun, H.H. Arnold, W.H. Klein, and E.N. Olson. 1995. Myogenin's functions do not overlap with those of MyoD or Myf-5 during mouse embryogenesis. *Developmental Biology* 172: 37-50.

Rawls, A., M.R. Valdez, W. Zhang, J. Richardson, W.H. Klein, and E.N. Olson. 1998. Overlapping functions of the myogenic bHLH genes MRF4 and MyoD revealed in double mutant mice. *Development* 125: 2349-2358.

Reardon, K.A., J. Davis, R.M. Kapsa, P. Choong, and E. Byrne. 2001. Myostatin, insulin-like growth factor-1, and leukemia inhibitory factor mRNAs are upregulated in chronic human disuse muscle atrophy. *Muscle and Nerve* 24: 893-899.

Renault, V., G. Piron-Hamelin, C. Forestier, S. DiDonna, S. Decary, F. Hentati, G. Saillant, G.S. Butler-Browne, and V. Mouly. 2000. Skeletal muscle regeneration and the mitotic clock. *Experimental Gerontology* 35: 711-719.

Renault, V., L.E. Thornell, P.O. Eriksson, G. Butler-Browne, and V. Mouly. 2002. Regenerative potential of human skeletal muscle during aging. *Aging Cell* 1: 132-139.

Rodrigues, A., and H. Schmalbruch. 1995. Satellite cells and myonuclei in long-term denervated rat muscles. *Anatomical Record* 243: 430-437.

Roth, S.M., G.F. Martel, R.E. Ferrell, E.J. Metter, B.F. Hurley, and M.A. Rogers. 2003. Myostatin gene expression is reduced in humans with heavy-resistance strength training: A brief communication. *Experimental Biology and Medicine* 228: 706-709.

Roth, S.M., G.F. Martel, F.M. Ivey, J.T. Lemmer, E.J. Metter, B.F. Hurley, and M.A. Rogers. 2000. Skeletal muscle satellite cell populations in healthy young and older men and women. *Anatomical Record* 260: 351-358.

Roth, S.M., G.F. Martel, F.M. Ivey, J.T. Lemmer, B.L. Tracy, E.J. Metter, B.F. Hurley, and M.A. Rogers. 2001. Skeletal muscle satellite cell characteristics in young and older men and women after heavy resistance strength training. *Journals of Gerontology Series A: Biological Sciences and Medical Sciences* 56: B240-B247.

Rudnicki, M.A., P.N. Schnegelsberg, R.H. Stead, T. Braun, H.H. Arnold, and R. Jaenisch. 1993. MyoD or Myf-5 is required for the formation of skeletal muscle. *Cell* 75: 1351-1359.

Sajko, S., L. Kubinova, E. Cvetko, M. Kreft, A. Wernig, and I. Erzen. 2004. Frequency of M-cadherin-stained satellite cells declines in human muscles during aging. *Journal of Histochemistry and Cytochemistry* 52: 179-185.

Sakuma, K., K. Watanabe, M. Sano, S. Kitajima, K. Sakamoto, I. Uramoto, and T. Totsuka. 2000. The adaptive response of transforming growth factor-β 2 and -β RII in the overloaded, regenerating and denervated muscles of rats. *Acta Neuropathologica* 99: 177-185.

Schuelke M., K.R. Wagner, L.E. Stolz, C. Hübner, T. Riebel, W. Kömen, T. Braun, J.F. Tobin, and S. Lee. 2004. Myostatin mutation associated with gross muscle hypertrophy in a child. *New England Journal of Medicine* 350: 2682-2688.

Schultz, E., and B.H. Lipton. 1982. Skeletal muscle satellite cells: Changes in proliferation potential as a function of age. *Mechanisms of Ageing and Development* 20: 377-383.

Sheehan, S.M., and R.E. Allen. 1999. Skeletal muscle myogenic satellite cell proliferation in response to members of the fibroblast growth factor family and hepatocyte growth factor. *Journal of Cellular Physiology* 181: 499-506.

Singh, M.A., W. Ding, Y.J. Manfredi, G.S. Solares, E.F. O'Neill, K.M. Clements, N.D. Ryan, J.J. Kehayias, R.A. Fielding, and W.J. Evans. 1999. Insulin-like growth factor I in skeletal muscle after weight-lifting exercise in frail elders. *American Journal of Physiology* 277: E135-143.

Singleton, J.R., and E.L. Feldman. 2001. Insulin-like growth factor-I in muscle metabolism and myotherapies. *Neurobiology of Disease* 8: 541-554.

Skuk, D., B. Roy, M. Goulet, P. Chapdelaine, J.P. Bouchard, R. Roy, F.J. Dugre, J.G. Lachance, L. Deschenes, S. Helene, M. Sylvain, and J.P. Tremblay. 2004. Dystrophin expression in myofibers of Duchenne muscular dystrophy patients following intramuscular injections of normal myogenic cells. *Molecular Therapeutics* 9: 475-482.

Skuk, D., and J.P. Tremblay. 2003. Myoblast transplantation: The current status of a potential therapeutic tool for myopathies. *Journal of Muscle Research and Cell Motility* 24: 285-300.

Snow, M.H. 1977. The effects of aging on satellite cells in skeletal muscles of mice and rats. *Cell and Tissue Research* 185: 399-408.

Sunderland, S. 1978. Traumatized nerves, roots and ganglia: Musculoskeletal factors and neuropathological consequences. In *The neurobiologic mechanisms in manipulative therapy*, ed. I.M. Korr, 137-166. New York: Plenum Press.

Taaffe, D.R., J.A. Cauley, M. Danielson, M.C. Nevitt, T.F. Lang, D.C. Bauer, and T.B. Harris. 2001. Race and sex effects on the association between muscle strength, soft tissue, and bone mineral density in healthy elders: The Health, Aging, and Body Composition Study. *Journal of Bone Mineral Research* 16: 1343-1352.

Taaffe, D.R., I.H. Jin, T.H. Vu, A.R. Hoffman, and R. Marcus. 1996. Lack of effect of recombinant human growth hormone (GH) on muscle morphology and GH-insulin-like growth factor expression in resistance-trained elderly men. *Journal of Clinical Endocrinology and Metabolism* 81: 421-425.

Tatsumi, R., J.E. Anderson, C.J. Nevoret, O. Halevy, and R.E. Allen. 1998. HGF/SF is present in normal adult skeletal muscle and is capable of activating satellite cells. *Developmental Biology* 194: 114-128.

Tatsumi, R., A. Hattori, Y. Ikeuchi, J.E. Anderson, and R.E. Allen. 2002. Release of hepatocyte growth factor from mechanically stretched skeletal muscle myogenic satellite cells and role of ph in nitric oxide. *Molecular Biology of the Cell* 13: 2090-2098.

Thaler, F.J., and G.K. Michalopoulos. 1985. Hepatopoietin A: Partial characterization and trypsin activation of a hepatocyte growth factor. *Cancer Research* 45(6): 2545-2549.

Valdez, M.R., J.A. Richardson, W.H. Klein, and E.N. Olson. 2000. Failure of Myf5 to support myogenic differentiation without myogenin, MyoD, and MRF4. *Developmental Biology* 219: 287-298.

Vierck, J., B. O'Reilly, K. Hossner, J. Antonio, K. Byrne, L. Bucci, and M. Dodson. 2000. Satellite cell regulation following myotrauma caused by resistance exercise. *Cell Biology International* 24: 263-272.

Viguie, C.A., D.X. Lu, S.K. Huang, H. Rengen, and B.M. Carlson. 1997. Quantitative study of the effects of long-term denervation on the extensor digitorum longus muscle of the rat. *Anatomical Record* 248: 346-354.

Vincent, K.R., and R.W. Braith. 2002. Resistance exercise and bone turnover in elderly men and women. *Medicine and Science in Sports and Exercise* 34: 17-23.

Wanek, L.J., and M.H. Snow. 2000. Activity-induced fibre regeneration in rat soleus muscle. *Anatomical Record* 258: 176-185.

Whitman, M. 1998. Smads and early developmental signaling by the TGF-β superfamily. *Genes and Development* 12: 2445-2462.

Yan, Z., R.B. Biggs, and F.W. Booth. 1993. Insulin-like growth factor immunoreactivity increases in muscle after acute eccentric contractions. *Journal of Applied Physiology* 74: 410-414.

Yang, S., M. Alnaqeeb, H. Simpson, and G. Goldspink. 1996. Cloning and characterization of an IGF-I isoform expressed in skeletal muscle subjected to stretch. *Journal of Muscle Research and Cell Motility* 17: 487-495.

Yang, S.Y., and G. Goldspink. 2002. Different roles of the IGF-I Ec peptide (MGF) and mature IGF-I in myoblast proliferation and differentiation. *FEBS Letters* 522(1-3): 156-160.

Yarasheski, K.E., S. Bhasin, I. Sinha-Hikim, J. Pak-Loduca, and N.F. Gonzalez-Cadavid. 2002. Serum myostatin-immunoreactive protein is increased in 60-92 year old women and men with muscle wasting. *Journal of Nutrition, Health and Aging* 6: 343-348.

Zammit, P.S., L. Heslop, V. Hudon, J.D. Rosenblatt, S. Tajbakhsh, M.E. Buckingham, J.R. Beauchamp, and T.A. Partridge. 2002. Kinetics of myoblast proliferation show that resident satellite cells are competent to fully regenerate skeletal muscle fibres. *Experimental Cell Research* 281: 39-49.

Chapter 7

Abmayr, S., Crawford, R.W., & Chamberlain, J.S. (2004). Characterization of ARC, apoptosis repressor interacting with CARD, in normal and dystrophin-deficient skeletal muscle. *Hum Mol Genet* 13, 213-221.

Bakay, M., Zhao, P., Chen, J., & Hoffman, E.P. (2002). A web-accessible complete transcriptome of normal human and DMD muscle. *Neuromuscul Disord* 12 (Suppl 1), S125-141.

Baldi, P., & Long, A.D. (2001). A Bayesian framework for the analysis of microarray expression data: Regularized *t*-test and statistical inferences of gene changes. *Bioinformatics* 17, 509-519.

Barash, I.A., Mathew, L., Ryan, A.R., Chen, J., & Lieber, R.L. (2004). Rapid muscle-specific gene expression changes after a single bout of eccentric contractions in the mouse. *Am J Physiol Cell Physiol* 286, C355-364.

Bilello, J.A. (2005). The agony and ecstasy of "OMIC" technologies in drug development. *Curr Mol Med* 5, 39-52.

Blattner, F.R., Plunkett, G. III, Bloch, C.A., Perna, N.T., Burland, V., Riley, M., Collado-Vides, J., Glasner, J.D., Rode, C.K., Mayhew, G.F., Gregor, J., Davis, N.W., Kirkpatrick, H.A., Goeden, M.A., Rose, D.J., Mau, B., Shao, Y. (1997). The complete genome sequence of Escherichia coli K-12. *Science* 277, 1453-1474.

Boer, J.M., de Meijer, E.J., Mank, E.M., van Ommen, G.B., & den Dunnen, J.T. (2002). Expression profiling in stably regenerating skeletal muscle of dystrophin-deficient mdx mice. *Neuromuscul Disord* 12 (Suppl 1), S118-124.

Brown, M.S., & Goldstein, J.L. (1997). The SREBP pathway: Regulation of cholesterol metabolism by proteolysis of a membrane-bound transcription factor. *Cell* 89, 331-340.

Carson, J.A., Nettleton, D., & Reecy, J.M. (2002). Differential gene expression in the rat soleus muscle during early work overload-induced hypertrophy. *FASEB J* 16, 207-209.

Chen, Y.W., Hubal, M.J., Hoffman, E.P., Thompson, P.D., & Clarkson, P.M. (2003). Molecular responses of human muscle to eccentric exercise. *J Appl Physiol* 95, 2485-2494.

Chen, Y.W., Nader, G.A., Baar, K.R., Fedele, M.J., Hoffman, E.P., Esser, K.A. (2002). Response of rat muscle to acute resistance exercise defined by transcriptional and translational profiling. *J Physiol* 545, 27-41.

Chen, Y.W., Zhao, P., Borup, R., & Hoffman, E.P. (2000). Expression profiling in the muscular dystrophies: Identification of novel aspects of molecular pathophysiology. *J Cell Biol* 151, 1321-1336.

Delgado, I., Huang, X., Jones, S., Zhang, L., Hatcher, R., Gao, B., Zhang, P. (2003). Dynamic gene expression during the onset of myoblast differentiation in vitro. *Genomics* 82, 109-121.

Fehrenbach, E., Zieker, D., Niess, A.M., Moeller, E., Russwurm, S., Northoff, H. (2003). Microarray technology—the future analysis tool in exercise physiology? *Exerc Immunol Rev* 9, 58-69.

Flick, M.J., & Konieczny, S.F. (2000). The muscle regulatory and structural protein MLP is a cytoskeletal binding partner of betaI-spectrin. *J Cell Sci* 113 (Pt 9), 1553-1564.

Ge, H., Walhout, A.J., & Vidal, M. (2003). Integrating "omic" information: A bridge between genomics and systems biology. *Trends Genet* 19, 551-560.

Gibbs, R.A., Weinstock, G.M., Metzker, M.L., Muzny, D.M., Sodergren, E., et al. (2004). Genome sequence of the Brown Norway rat yields insights into mammalian evolution. *Nature* 428, 493-521.

Goffeau, A.E.A. (1997). The yeast genome directory. *Nature* 387, 5.

Goodacre, R., Vaidyanathan, S., Dunn, W.B., Harrigan, G.G., & Kell, D.B. (2004). Metabolomics by numbers: Acquiring and understanding global metabolite data. *Trends Biotechnol* 22, 245-252.

Haslett, J.N., Sanoudou, D., Kho, A.T., Bennett, R.R., Greenberg, S.A., Kohane, I.S., Beggs, A.H., Kunkel, L.M. (2002). Gene expression comparison of biopsies from Duchenne muscular dystrophy (DMD) and normal skeletal muscle. *Proc Natl Acad Sci USA* 99, 15000-15005.

Haslett, J.N., Sanoudou, D., Kho, A.T., Han, M., Bennett, R.R., Kohane, I.S., Beggs, A.H., Kunkel, L.M. (2003). Gene expression profiling of Duchenne muscular dystrophy skeletal muscle. *Neurogenetics* 4, 163-171.

Hedge, P.S., White, I.R., & Debouck, C. (2003). Interplay of transcriptomics and proteomics. *Curr Opin Biotechnol* 14, 647-651.

Heller, M.J. (2002). DNA microarray technology: Devices, systems, and applications. *Annu Rev Biomed Eng* 4, 129-153.

Howbrook, D.N., van der Valk, A.M., O'Shaughnessy, M.C., Sarker, D.K., Baker, S.C., Lloyd, A.W. (2003). Developments in microarray technologies. *Drug Discov Today* 8, 642-651.

Knöll, R., Hoshijima, M., Hoffman, H.M., Person, V., Lorenzen-Schmidt, I., Bang, M.L., Hayashi, T., Shiga, N., Yasukawa, H., Schaper, W., McKenna, W., Yokoyama, M., Schork, N.J., Omens, J.H., McCulloch, A.D., Kimura, A., Gregorio, C.C., Poller, W., Schaper, J., Schultheiss, H.P., Chien, K.R. (2002). The cardiac mechanical stretch sensor machinery involves a Z disc complex that is defective in a subset of human dilated cardiomyopathy. *Cell* 111, 943-955.

Lee, T.I., & Young, R.A. (2000). Transcription of eukaryotic protein-coding genes. *Annu Rev Genet* 34, 77-137.

Mahoney, D.J., & Tarnopolsky, M.A. (2005). Analysis of global gene expression in human skeletal muscle following eccentric exercise. Unpublished observations.

McKusick, V.A. (1997). Genomics: Structural and functional studies of genomes. *Genomics* 45, 244-249.

Mejat, A., Ravel-Chapuis, A., Vandromme, M. & Schaeffer, L. (2003). Synapse-specific gene expression at the neuromuscular junction. *Ann N Y Acad Sci* 998, 53-65.

Moran, J. L., Li, Y., Hill, A. A., Mounts, W. M. & Miller, C. P. (2002). Gene expression changes during mouse skeletal myoblast differentiation revealed by transcriptional profiling. *Physiol Genomics* 10, 103-111.

Noguchi, S., Tsukahara, T., Fujita, M., Kurokawa, R., Tachikawa, M., Toda, T., Tsujimoto, A., Arahata, K., Nishino, I. (2003). cDNA microarray analysis of individual Duchenne muscular dystrophy patients. *Hum Mol Genet* 12, 595-600.

Norrbom, J., Sundberg, C.J., Ameln, H., Kraus, W.E., Jansson, E., Gustafsson, T. (2004). PGC-1alpha mRNA expression is influenced by metabolic perturbation in exercising human skeletal muscle. *J Appl Physiol* 96, 189-194.

Porter, J.D., Khanna, S., Kaminski, H.J., Rao, J.S., Merriam, A.P., Richmonds, C.R., Leahy, P., Li, J., Guo, W., Andrade, F.H. (2002). A chronic inflammatory response dominates the skeletal muscle molecular signature in dystrophin-deficient mdx mice. *Hum Mol Genet* 11, 263-272.

Porter, J.D., Merriam, A.P., Leahy, P., Gong, B., Feuerman, J., Cheng, G., Khanna, S. (2004). Temporal gene expression profiling of dystrophin-deficient (mdx) mouse diaphragm identifies conserved and muscle group-specific mechanisms in the pathogenesis of muscular dystrophy. *Hum Mol Genet* 13, 257-269.

Porter, J.D., Merriam, A.P., Leahy, P., Gong, B., & Khanna, S. (2003). Dissection of temporal gene expression signatures of affected and spared muscle groups in dystrophin-deficient (mdx) mice. *Hum Mol Genet* 12, 1813-1821.

Preiss, T., & Hentze, M. (2003). Starting the protein synthesis machine: Eukaryotic translation initiation. *Bioessays* 25, 1201-1211.

Roth, S.M., Ferrell, R.E., Peters, D.G., Metter, E.J., Hurley, B.F., Rogers, M.A. (2002). Influence of age, sex, and strength training on human muscle gene expression determined by microarray. *Physiol Genomics* 10, 181-190.

Rouger, K., Le Cunff, M., Steenman, M., Potier, M.C., Gibelin, N., Dechesne, C.A., Leger, J.J. (2002). Global/temporal gene expression in diaphragm and hindlimb muscles of dystrophin-deficient (mdx) mice. *Am J Physiol Cell Physiol* 283, C773-784.

Shen, X., Collier, J.M., Hlaing, M., Zhang, L., Delshad, E.H., Bristow, J., Bernstein, H.S. (2003). Genome-wide examination of myoblast cell cycle withdrawal during differentiation. *Dev Dyn* 226, 128-138.

Shriver, Z., Raguram, S., & Sasisekharan, R. (2004). Glycomics: A pathway to a class of new and improved therapeutics. *Nat Rev Drug Discov* 3, 863-873.

Sterrenburg, E., Turk, R., 't Hoen, P.A., van Deutekom, J.C., Boer, J.M., van Ommen, G.J., den Dunnen, J.T. (2004). Large-scale gene expression analysis of human skeletal myoblast differentiation. *Neuromuscul Disord* 14, 507-518.

Tkatchenko, A.V., Le Cam, G., Leger, J.J., & Dechesne, C.A. (2000). Large-scale analysis of differential gene expression in the hindlimb muscles and diaphragm of mdx mouse. *Biochim Biophys Acta* 1500, 17-30.

Tkatchenko, A.V., Pietu, G., Cros, N., Gannoun-Zaki, L., Auffray, C., Léger, J.J., Dechesne, C.A. (2001). Identification of altered gene expression in skeletal muscles from Duchenne muscular dystrophy patients. *Neuromuscul Disord* 11, 269-277.

Tomczak, K.K., Marinescu, V.D., Ramoni, M.F., Sanoudou, D., Montanaro, F., Han, M., Kunkel, L.M., Kohane, I.S., Beggs, A.H. (2004). Expression profiling and identification of novel genes involved in myogenic differentiation. *FASEB J* 18, 403-405.

Tseng, B.S., Zhao, P., Pattison, J.S., Gordon, S.E., Granchelli, J.A., Madsen, R.W., Folk, L.C., Hoffman, E.P., Booth, F.W. (2002). Regenerated mdx mouse skeletal muscle shows differential mRNA expression. *J Appl Physiol* 93, 537-545.

Tusher, V.G., Tibshirani, R., & Chu, G. (2001). Significance analysis of microarrays applied to the ionizing radiation response. *Proc Natl Acad Sci USA* 98, 5116-5121.

Venter, J.C., Adams, M.D., Myers, E.W., Li, P.W., Mural, R.J., et al. (2001). The sequence of the human genome. *Science* 291, 1304-1351.

Vissing, K., Andersen, J.L., & Schjerling, P. (2005). Are exercise-induced genes induced by exercise? *FASEB J* 19, 94-96.

Waterston, R.H., Lindblad-Toh, K., Birney, E., Rogers, J., Abril, J.F., et al. (2002). Initial sequencing and comparative analysis of the mouse genome. *Nature* 420, 520-562.

Yan, Z., Choi, S., Liu, X., Zhang, M., Schageman, J.J., Lee, S.Y., Hart, R., Lin, L., Thurmond, F.A., Williams, R.S. (2003). Highly coordinated gene regulation in mouse skeletal muscle regeneration. *J Biol Chem* 278, 8826-8836.

Zambon, A.C., McDearmon, E.L., Salomonis, N., Vranizan, K.M., Johansen, K.L., Adey, D., Takahashi, J.S., Schambelan, M., Conklin, B.R. (2003). Time- and exercise-dependent gene regulation in human skeletal muscle. *Genome Biol* 4, R61.

Zhu, H., & Snyder, M. (2002). "Omic" approaches for unraveling signaling networks. *Curr Opin Cell Biol* 14, 173-179.

Chapter 8

Artavanis-Tsakonas, S., M.D. Rand, and R.J. Lake. 1999. Notch signaling: Cell fate control and signal integration in development. *Science* 284: 770-776.

Beaton, L.J., M.A. Tarnopolsky, and S.M. Phillips. 2002. Contraction-induced muscle damage in humans following calcium channel blocker administration. *Journal of Physiology* 544: 849-859.

Best, T.M., R. Fiebig, D.T. Corr, S. Brickson, and L. Ji. 1999. Free radical activity, antioxidant enzyme, and glutathione changes with muscle stretch injury in rabbits. *Journal of Applied Physiology* 87: 74-82.

Brooks, S.V., and J.A. Faulkner. 1988. Contractile properties of skeletal muscles from young, adult and aged mice. *Journal of Physiology* 404: 71-82.

Brooks, S.V., and J.A. Faulkner. 1990. Contraction-induced injury: Recovery of skeletal muscles in young and old mice. *American Journal of Physiology: Cell Physiology* 258: C436-C442.

Brooks, S.V., and J.A. Faulkner. 1991. Maximum and sustained power of extensor digitorum longus muscles from young, adult, and old mice. *Journals of Gerontology: Biological Sciences* 46: B28-B33.

Brooks, S.V., and J.A. Faulkner. 1994. Skeletal muscle weakness in old age: Underlying mechanisms. *Medicine and Science in Sports and Exercise* 26: 432-439.

Brooks, S.V., and J.A. Faulkner. 1996. The magnitude of the initial injury induced by stretches of maximally activated muscle fibres of mice and rats increases in old age. *Journal of Physiology* 497: 573-580.

Brooks, S.V., J.A. Opiteck, and J.A. Faulkner. 2001. Conditioning of skeletal muscles in adult and old mice for protection from contraction-induced injury. *Journals of Gerontology: Biological Sciences* 56: B163-B171.

Brooks, S.V., E. Zerba, and J.A. Faulkner. 1995. Injury to fibres after single stretches of passive and maximally stimulated muscles in mice. *Journal of Physiology* 488: 459-469.

Carlson, B.M., and J.A. Faulkner. 1989. Muscle transplantation between young and old rats: Age of host determines recovery. *American Journal of Physiology: Cell Physiology* 256: C1262-C1266.

Clarkson, P.M., W.C. Byrnes, K.M. McCormick, L.P. Turcotte, and J.S. White. 1986. Muscle soreness and serum creatine kinase activity following isometric, eccentric, and concentric exercise. *International Journal of Sports Medicine* 7: 152-155.

Clarkson, P.M., and M.E. Dedrick. 1988. Exercise-induced muscle damage, repair, and adaptation in old and young subjects. *Journals of Gerontology: Medical Sciences* 43: M91-M96.

Conboy, I.M., M.J. Conboy, G.M. Smythe, and T.A. Rando. 2003. Notch-mediated restoration of regenerative potential to aged muscle. *Science* 302: 1575-1577.

Conboy, I.M., M.J. Conboy, A.J. Wagers, E.R. Girma, I.L. Weissman, and T.A. Rando. 2005. Rejuvenation of aged progenitor cells by exposure to a young systemic environment. *Nature* 433: 760-764.

Dedrick, M.E., and P.M. Clarkson. 1990. The effects of eccentric exercise on motor performance in young and older women. *European Journal of Applied Physiology and Occupational Physiology* 60: 183-186.

Devor, S.T., and J.A. Faulkner. 1999. Regeneration of new fibers in muscles of old rats reduces contraction-induced injury. *Journal of Applied Physiology* 87: 750-756.

Faulkner, J.A., S.V. Brooks, and J.A. Opiteck. 1993. Injury to skeletal muscle fibers during contractions: Conditions of occurrence and prevention. *Physical Therapy* 73: 911-921.

Faulkner, J.A., S.V. Brooks, and E. Zerba. 1995. Muscle atrophy and weakness with aging: Contraction-induced injury as an underlying mechanism. *Journals of Gerontology Series A* 50A: 124-129.

Faulkner, J.A., H.J. Green, and T.P. White. 1994. Response and adaptation of skeletal muscle to changes in physical activity. In: *Physical activity, fitness, and health,* ed. C. Bouchard, R.J. Shephard, and T. Stephens. Champaign, IL: Human Kinetics, pp. 343-357.

Faulkner, J.A., D.A. Jones, and J.M. Round. 1989. Injury to skeletal muscles of mice by forced lengthening during contractions. *Quarterly Journal of Experimental Physiology* 74: 661-670.

Friden, J., M. Sjostrom, and B. Ekblom. 1983. Myofibrillar damage following intense eccentric exercise in man. *International Journal of Sports Medicine* 4: 170-176.

Grimby, G. 1995. Muscle performance and structure in the elderly as studied cross-sectionally and longitudinally. *Journals of Gerontology Series A* 50A: 17-22.

Grimby, G., and B. Saltin. 1983. The ageing muscle. *Clinical Physiology* 3: 209-218.

Hadley, E.C., M.G. Ory, R. Suzman, R. Weindruch, and L. Fried. 1993. Physical frailty: A treatable cause of dependence in old age. *Journals of Gerontology: Biological Sciences* 48: B1-B88.

Holloszy, J.O. 1995. Workshop on sarcopenia: Muscle atrophy in old age. *Journals of Gerontology Series A* 50A: 1-161.

Hough, T. 1902. Ergographic studies in muscular soreness. *American Journal of Physiology* 7: 76-92.

Jackson, M.J., and R.H.T. Edwards. 1988. Free radicals, muscle damage and muscular dystrophy. In: *Reactive oxygen species in chemistry, biology, and medicine,* ed. A. Quintanilha. New York: NY, Plenum Press, pp. 197-210.

Jackson, M.J., D.J. Jones, and R.H.T. Edwards. 1984. Experimental skeletal muscle damage: the nature of the calcium-activated degenerative processes. *European Journal of Clinical Investigation* 14: 369-374.

Julian, F.J., and D.L. Morgan. 1979a. The effect on tension of non-uniform distribution of length changes applied to frog muscle fibres. *Journal of Physiology* 293: 379-392.

Julian, F.J., and D.L. Morgan. 1979b. Intersarcomere dynamics during fixed-end tetanic contractions of frog muscle fibres. *Journal of Physiology* 293: 365-378.

Kadhiresan, V.A., C.A. Hassett, and J.A. Faulkner. 1996. Properties of single motor units in medial gastrocnemius muscles of adult and old rats. *Journal of Physiology* 493: 543-552.

Kanda, K., and K. Hashizume. 1989. Changes in properties of the medial gastrocnemius motor units in aging. *Journal of Neurophysiology* 61: 737-746.

Koh, T.J., and S.V. Brooks. 2001. Lengthening contractions are not required to induce protection from contraction-induced muscle injury. *American Journal of Physiology: Regulatory, Integrative, and Comparative Physiology* 281: R155-161.

Koh, T.J., J.M. Peterson, F.X. Pizza, and S.V. Brooks. 2003. Passive stretches protect skeletal muscle of adult and old mice from lengthening contraction-induced injury. *Journals of Gerontology: Biological Sciences* 58A: B592-B597.

Larsson, L. 1995. Motor units: Remodeling in aged animals. *Journals of Gerontology Series A* 50A: 91-95.

Larsson, L., and B. Ramamurthy. 2000. Aging-related changes in skeletal muscle: Mechanisms and interventions. *Drugs and Aging* 17: 303-316.

Lexell, J. 1995. Human aging, muscle mass, and fiber type composition. *Journals of Gerontology Series A* 50A: 11-16.

Lieber, R.L., and J. Friden. 1993. Muscle damage is not a function of muscle force by active muscle strain. *Journal of Applied Physiology* 74: 520-526.

Lieber, R.L., T.M. Woodburn, and J. Friden. 1991. Muscle damage induced by eccentric contractions of 25% strain. *Journal of Applied Physiology* 70: 2498-2507.

Macpherson, P.C.D., R.G. Dennis, and J.A Faulkner. 1997. Sarcomere dynamics and contraction-induced injury to maximally activated single muscle fibres from soleus muscles of rats. *Journal of Physiology* 500: 523-533.

Macpherson, P.C.D., A.M. Schork, and J.A. Faulkner. 1996. Contraction-induced injury to single fiber segments from fast and slow muscles of rats by single stretches. *American Journal of Physiology: Cell Physiology* 271: C1438-C1446.

Manfredi, T.G., R.A. Fielding, K.P. O'Reilly, C.N. Meredith, H.Y. Lee, and W.J. Evans. 1991. Plasma creatine kinase activity and exercise-induced muscle damage in older men. *Medicine and Science in Sports and Exercise* 23: 1028-1034.

McArdle, A., W.H. Dillmann, R. Mestril, J.A. Faulkner, and M.J. Jackson. 2004. Overexpression of HSP70 in mouse skeletal muscle protects against muscle damage and age-related muscle dysfunction. *FASEB Journal* 18: 355-357.

McArdle, A., J.H. Van der Meulen, M. Catapano, M.C. Symons, J.A. Faulkner, and M.J. Jackson. 1999. Free radical activity following contraction-induced injury to the extensor digitorum longus muscles of rats. *Free Radicals in Biology and Medicine* 26: 1085-1091.

McBride, T. 2000. Increased depolarization, prolonged recovery and reduced adaptation of the resting membrane potential in aged rat skeletal muscles following eccentric contractions. *Mechanisms of Ageing and Development* 155: 127-138.

McBride, T.A., F.A. Gorin, and R.C. Carlsen. 1995. Prolonged recovery and reduced adaptation in aged rat muscle following eccentric exercise. *Mechanisms of Ageing and Development* 83: 185-200.

McBride, T.A., B.W. Stockert, F.A. Gorin, and R.C. Carlsen. 2000. Stretch-activated ion channels contribute to membrane depolarization after eccentric contractions. *Journal of Applied Physiology* 88: 91-101.

McCord, J.M., and R.R. Roy. 1982. The pathophysiology of superoxide: Roles in inflammation and ischemia. *Canadian Journal of Physiology and Pharmacology* 60: 1346-1352.

McCully, K.K., and J.A. Faulkner. 1985. Injury to skeletal muscle fibers of mice following lengthening contractions. *Journal of Applied Physiology* 59: 119-126.

McCully, K.K., and J.A. Faulkner. 1986. Characteristics of lengthening contractions associated with injury to skeletal muscle fibers. *Journal of Applied Physiology* 61: 293-299.

Moore, D.H. 1975. A study of age group track and field records to relate age and running speed. *Nature* 253: 264-265.

Morgan, D.L. 1990. New insights into the behavior of muscle during active lengthening. *Biophysical Journal* 57: 209-221.

Newham, D.J., G. McPhail, K.R. Mills, and R.H. Edwards. 1983. Ultrastructural changes after concentric and eccentric contractions of human muscle. *Journal of the Neurological Sciences* 61: 109-122.

Official NBA Resister 1993-94 Edition. 1993. ed. A. Sachare, and M. Shimabukuro. St. Louis: MO, The Sporting News Publishing Co.

Ogilvie, R.W., R.B. Armstrong, K.E. Baird, and C.L. Bottoms. 1988. Lesions in the rat soleus muscle following eccentrically biased exercise. *American Journal of Anatomy* 182: 335-346.

Pizza, F.X., T.J. McLoughlin, S.J. McGregor, E.P. Calomeni, and W.T. Gunning. 2001. Neutrophils injure cultured skeletal myotubes. *American Journal of Physiology: Cell Physiology* 281: C335-C341.

Ploutz-Snyder, L.L., E.L. Giamis, M. Formikell, and A.E. Rosenbaum. 2001. Resistance training reduces susceptibility to eccentric exercise-induced muscle dysfunction in older women. *Journals of Gerontology: Biological Sciences* 56: B384-B390.

Rader, E.P., and J.A. Faulkner. 2006a. Effect of aging on the recovery following contraction-induced injury in muscles of female mice. *Journal of Applied Physiology* 101: 887-892.

Rader, E.P., and J.A. Faulkner. 2006b. Recovery from contraction-induced injury is impaired in weight-bearing muscles of old male mice. *Journal of Applied Physiology* 100: 656-661.

Roth, S.M., G.F. Martel, F.M. Ivey, J.T. Lemmer, E.J. Metter, B.F. Hurley, and M.A. Rogers. 2000. High volume, heavy-resistance strength training and muscle damage in young and old women. *Journal of Applied Physiology* 88: 1112-1118.

Roth, S.M., G.F. Martel, F.M. Ivey, J.T. Lemmer, B.L. Tracy, D.E. Hurlburt, E.J. Metter, B.F. Hurley, and M.A. Rogers. 1999. Ultrastructural muscle damage in young vs. older men after high volume, heavy-resistance strength training. *Journal of Applied Physiology* 86: 1833-1840.

Schultz, E., and B.H. Lipton. 1982. Skeletal muscle satellite cells: Changes in proliferation potential as a function of age. *Mechanisms of Ageing and Development* 20: 377-383.

Schulz, R., and C. Curnow. 1988. Peak performance and age among superathletes: Track and field, swimming, baseball, tennis, and golf. *Journals of Gerontology* 43: P113-P120.

Smith, L.L. 1991. Acute inflammation: The underlying mechanism in delayed onset muscle soreness? *Medicine and Science in Sports and Exercise* 23: 542-551.

Stauber, W.T., V.K. Fritz, D.W. Vogelbach, and B. Dahlmann. 1988. Characterization of muscles injured by forced lengthening. I. Cellular infiltrates. *Medicine and Science in Sports and Exercise* 20: 345-353.

Tidball, J.G. 1995. Inflammatory cell response to acute muscle injury. *Medicine and Science in Sports and Exercise* 27: 1022-1032.

Trappe, S., P. Gallagher, M. Harber, J. Carrithers, J. Fluckey, and T. Trappe. 2003. Single muscle fibre contractile properties in young and old men and women. *Journal of Physiology* 552: 47-58.

Trappe, S., D. Williamson, M. Godard, D. Porter, G. Rowden, and D. Costill. 2000. Effect of resistance training on single muscle fiber contractile function in older men. *Journal of Applied Physiology* 89: 143-152.

Van der Meulen, J.H., A. McArdle, M.J. Jackson, and J.A. Faulkner. 1997. Contraction-induced injury to the extensor digitorum longus muscles of rats: The role of vitamin E. *Journal of Applied Physiology* 83: 817-823.

Warren, G.L., D.A. Hayes, D.A. Lowe, and R.B. Armstrong. 1993. Mechanical factors in the initiation of eccentric contraction-induced injury in rat soleus muscle. *Journal of Physiology* 464: 457-475.

Zerba, E., T.E. Komorowski, and J.A. Faulkner. 1990. Free radical injury to skeletal muscles of young, adult, and old mice. *American Journal of Physiology: Cell Physiology* 258: C429-C435.

Chapter 9

Alderton, J.M., and R.A. Steinhardt. 2000a. Calcium influx through calcium leak channels is responsible for the elevated levels of calcium-dependent proteolysis in dystrophic myotubes. *Journal of Biological Chemistry* 275(13): 9452-9460.

Alderton, J.M., and R.A. Steinhardt. 2000b. How calcium influx through calcium leak channels is responsible for the elevated levels of calcium-dependent proteolysis in dystrophic myotubes. *Trends in Cardiovascular Medicine* 10: 268-272.

Allen, D.G. 2004. Skeletal muscle function: Role of ionic changes in fatigue, damage and disease. *Clinical and Experimental Pharmacology and Physiology* 31(8): 485-493.

Anzai, K., K. Ogawa, T. Ozawa, and H. Yamamoto. 2000. Oxidative modification of ion channel activity of ryanodine receptor. *Antioxidants and Redox Signaling* 2(1): 35-40.

Bakker, A.J., S.I. Head, D.A. Williams, and D.G. Stephenson. 1993. Ca^{2+} levels in myotubes grown from the skeletal muscle of dystrophic (mdx) and normal mice. *Journal of Physiology* (London) 460: 1-13.

Berchtold, M.W., H. Brinkmeier, and M. Muntener. 2000. Calcium ion in skeletal muscle: Its crucial role for muscle function, plasticity, and disease. *Physiological Review* 80(3): 1215-1265.

Bertorini, T.E., S.K. Bhattacharya, G.M. Palmieri, C.M. Chesney, D. Pifer, and B. Baker. 1982. Muscle calcium and magnesium content in Duchenne muscular dystrophy. *Neurology* 32: 1088-1092.

Bertorini, T.E., F. Cornelio, S.K. Bhattacharya, G.M. Palmieri, I. Dones, F. Dworzak, and B. Brambati. 1984. Calcium and magnesium content in fetuses at risk and prenecrotic Duchenne muscular dystrophy. *Neurology* 34: 1436-1440.

Blake, D.J., J.N. Schofield, R.A. Zuellig, D.C. Gorecki, S.R. Phelps, E.A. Barnard, Y.H. Edwards, and K.E. Davies. 1995. G-utrophin, the autosomal homologue of dystrophin Dp116, is expressed in sensory ganglia and brain. *Proceedings of the National Academy of Sciences USA* 92(9): 3697-3701.

Blake, D.J., A. Weir, E. Sarall, and K.E. Davies. 2002. Function and genetics of dystrophin and dystrophin-related proteins in muscle. *Physiological Reviews* 82: 291-232.

Bredt, D.S., and S.H. Snyder. 1994. Transient nitric oxide synthase neurons in embryonic cerebral cortical plate, sensory ganglia, and olfactory epithelium. *Neuron* 13(2): 301-313.

Burkin, D.J., and S.J. Kaufman. 1999. The $\alpha7\beta1$ integrin in muscle development and disease. *Cell Tissue Research* 296: 183-190.

Burkin, D.J., G.Q. Wallace, K.J. Nicol, D.J. Kaufman, and S.J. Kaufman. 2001. Enhanced expression of the $\alpha7\beta1$ integrin reduces muscular dystrophy and restores viability in dystrophic mice. *Journal of Cell Biology* 152(6): 1207-1218.

Byrne, E., A.J. Kornberg, and R. Kapsa. 2003. Duchenne muscular dystrophy: Hopes for the sesquicentenary. *Medical Journal of Australia* 179 (9): 463-464.

Campanelli, J.T., S.L. Roberds, K.P. Campbell, and R.H. Scheller. 1994. A role for dystrophin-associated glycoproteins and utrophin in agrin-induced AChR clustering. *Cell* 77(5): 663-674.

Campbell, K.P. 1995. Three muscular dystrophies: Loss of cytoskeleton-extracellular matrix linkage. *Cell* 80: 675-679.

Carafoli, E., and M. Molinari. 1998. Calpain: A protease in search of a function? *Biochemical and Biophysical Research Communications* 247: 193-203.

Chang, W.J., S.T. Iannaccone, K.S. Lau, B.S. Masters, T.J. McCabe, K. McMillan, R.C. Padre, M.J. Spencer, J.G. Tidball, and J.T. Stull. 1996. Neuronal nitric oxide synthase and dystrophin-deficient muscular dystrophy. *Proceedings of the National Academy of Sciences USA* 93(17): 9142-9147.

Chen, Y., P. Zhao, R. Borup, and E.P. Hoffman. 2000. Expression profiling in the muscular dystrophies: Identification of novel aspects of molecular pathophysiology. *Journal of Cell Biology* 151(6): 1321-1336.

Cooney, R.N., S.R. Kimball, and T.C. Vary. 1997. Regulation of skeletal muscle protein turnover during sepsis: Mechanisms and mediators. *Shock* 7: 1-16.

Cozzi, F., M. Cerletti, G.C. Luvoni, R. Lombardo, P.G. Brambilla, S. Faverzani, F. Blasevich, F. Cornelio, O. Pozza, and M. Mora. 2001. Development of muscle pathology in canine X-linked muscular dystrophy. Quantitative characterization of histopathological progression during postnatal skeletal development. *Acta Neuropathologica* 101: 469-478.

Crosbie, R.H. 2001. NO vascular control in Duchenne muscular dystrophy. *Nature Medicine* 7(1): 27-29.

Deconinck, N., J.A. Rafael, G. Beckers-Bleukx, D. Kahn, A.E. Deconinck, K.E. Davies, and J.M. Gillis. 1998. Consequences of the combined deficiency in dystrophin and utrophin on the mechanical properties and myosin composition of some limb and respiratory muscles of the mouse. *Neuromuscular Disorders* 8: 362-370.

Deconinck, N., J. Tinsley, F. De Backer, R. Fisher, D. Kahn, S. Phelps, K. Davies, and J.M. Gillis. 1997. Expression of truncated utrophin leads to major functional improvements in dystrophin-deficient muscles of mice. *Nature Medicine* 3(11): 1216-1221.

Durbeej, M., and K.P. Campbell. 2002. Muscular dystrophies involving the dystrophin-glycoprotein complex: An overview of current mouse models. *Current Opinions in Genetics and Development* 12: 349-361.

Eisner, D.A., G. Isenberg, and K.R. Sipido. 2003. Normal and pathological excitation-contraction coupling in the heart—an overview. *Journal of Physiology* 546(Pt 1): 3-4.

Emery, A.E. 1989. Clinical and molecular studies in Duchenne muscular dystrophy. *Progress in Clinical and Biological Research* 306: 15-28.

Emery, A.E.H. 1993. *Duchenne muscular dystrophy.* 2nd ed. New York: Oxford University Press.

Engel, A.G., and E. Ozawa. 2004. Dystrophinopathies. In *Myology,* 3rd ed., Vol. 2, ed. A.G. Engel and C. Franzini-Armstrong, 961-1025. New York: McGraw-Hill.

Ervasti, J.M. 2003. Costameres: The Achilles' heel of Herculean muscle. *Journal of Biological Chemistry* 278(16): 13591-13594.

Ervasti, J.M. 2007. Dystrophin, its interactions with other proteins, and implications for muscular dystrophy. *Biochimica and Biophysica Acta* 1772: 108-117.

Franco, A.J., and J.B. Lansman. 1990. Calcium entry through stretch-activated ion channels in mdx myotubes. *Nature* 344: 670-673.

Gardner, R.J., M. Bobrow, and R.G. Roberts. 1995. The identification of point mutations in Duchenne muscular dystrophy patients by using reverse-transcription PCR and the protein truncation test. *American Journal of Human Genetics* 57: 311-320.

Grady, R.M., J.P. Merlie, and J.R. Sanes. 1997. Subtle neuromuscular defects in utrophin-deficient mice. *Journal of Cell Biology* 136(4): 871-882.

Grady, R., H. Teng, M. Nichol, J. Cunningham, R. Wilkinson, and J.R. Sanes. 1997. Skeletal and cardiac myopathies in mice lacking utrophin and dystrophin: A model for Duchennes muscular dystrophy. *Cell* 90: 729-738.

Grange, R.W., and J.A. Call. 2007. Recommendations to define exercise prescription for Duchenne muscular dystrophy. *Exercise and Sport Sciences Reviews* 35(1): 12-17.

Grange, R.W., T.G. Gainer, K.M. Marschner, R.J. Talmadge, and J.T. Stull. 2002. Fast-twitch skeletal muscles of dystrophic mouse pups are resistant to injury from acute mechanical stress. *American Journal of Physiology: Cell Physiology* 283(4): c1090-1101.

Grange, R.W., E. Isotani, K.S. Lau, K.E. Kamm, P.L. Huang, and J.T. Stull. 2001. Nitric oxide contributes to vascular smooth muscle relaxation in contracting fast-twitch skeletal muscles. *Physiological Genomics* 5: 35-44.

Hare, J.M. 2003. Nitric oxide and excitation-contraction coupling. *Journal of Molecular and Cellular Cardiology* 35(7): 719-729.

Hoffman, E.P., R.H. Brown, and L.M. Kunkel. 1987. Dystrophin: The protein product of the Duchenne muscular dystrophy locus. *Cell* 51: 919-928.

Hoffman, E.H., A.P. Monaco, C.A. Feener, and L.M. Kunkel. 1987. Conservation of the Duchenne muscular dystrophy gene in mice and humans. *Science* 238: 347-350.

Hopf, F.W., P. Reddy, J. Hong, and R.A. Steinhardt. 1996b. A capacitative calcium current in cultured skeletal muscle cells is mediated by the calcium-specific leak channel and inhibited by dihydropyridine compounds. *Journal of Biological Chemistry* 271(37): 22358-22367.

Hopf, F.W., P.R. Turner, W.F. Denetclaw Jr., P. Reddy, and R.A. Steinhardt. 1996. A critical evaluation of resting intracellular free calcium regulation in dystrophic mdx muscle. *American Journal of Physiology* 271: C1325-1339.

Hutter, O.F. 1992. The membrane hypothesis of Duchenne muscular dystrophy: Quest for functional evidence. *Journal of Inherited Metabolic Disorders* 15: 565-577.

Jejurikar, S.S., and W.M. Kuzon Jr. 2003. Satellite cell depletion in degenerative skeletal muscle. *Apoptosis* 8: 573-578.

Kapsa, R., A.J. Kornberg, and E. Byrne. 2003. Novel therapies for Duchenne muscular dystrophy. *Lancet Neurology* 2: 299-310.

Karol, L.A. 2007. Scoliosis in Duchenne muscular dystrophy. *Journal of Bone and Joint Surgery* (American) 89: 155-162.

Khairallah, M., Khairallah, R., Young, M.E., Dyck, J.R., Petrof, B.J., and Des Rosiers, C. 2007. Metabolic and signaling alterations in dystrophin-deficient hearts precede overt cardiomyopathy. *Journal of Molecular and Cellular Cardiology* 43(2):119-29.

Khurana, T.S., S.C. Watkins, P. Chafey, J. Chelly, E.M. Tome, M. Fardeau, J.C. Kaplan, and L.M. Kunkel. 1991. Immunolocalization and developmental expression of dystrophin related protein in skeletal muscle. *Neuromuscular Disorders* 1(3): 185-194.

Koenig, M., E.P. Hoffman, C.J. Bertelson, A.P. Monaco, C. Feener, and L.M. Kunkel. 1987. Complete cloning of the Duchenne muscular dystrophy (DMD) cDNA and preliminary genomic organization of the DMD gene in normal and affected individuals. *Cell* 50: 509-517.

Kunkel, L.M., A.H. Beggs, and E.P. Hoffman. 1989. Molecular genetics of Duchenne and Becker muscular dystrophy: Emphasis on improved diagnosis. *Clinical Chemistry* 35: B21-24.

Lapidos, K.A., R. Kakkar, and E.M. McNally. 2004. The dystrophin glycoprotein complex: Signaling strength and integrity for the sarcolemma. *Circulation Research* 94(8): 1023-1031.

Lau, K.S., R.W. Grange, W.J. Chang, K.E. Kamm, I. Sarelius, and J.T. Stull. 1998. Skeletal muscle contractions stimulate cGMP formation and attenuate vascular smooth muscle myosin phosphorylation via nitric oxide. *FEBS Letters* 431(1): 71-74.

Lenk, U., R. Hanke, H. Thiele, and A. Speer. 1993. Point mutations at the carboxy terminus of the human dystrophin gene: Implications for an association with mental retardation in DMD patients. *Human Molecular Genetics* 2: 1877-1881.

Love, D.R., B.C. Byth, J.M. Tinsley, D.J. Blake, and K.E. Davies. 1993. Dystrophin and dystrophin-related proteins. *Neuromuscular Disorders* 3: 5-21.

Lowe, D.A., B.O. Williams, D.D. Thomas, and R.W. Grange. 2006. Molecular and cellular contractile dysfunction of dystrophic muscle from young mice. *Muscle and Nerve* 34: 92-100.

Mallouk, N., V. Jacquemond, and B. Allard. 2000. Elevated subsarcolemmal Ca^{2+} in mdx mouse skeletal muscle fibers detected with Ca^{2+}-activated K^+ channels. *Proceedings of the National Academy of Sciences USA* 97: 4950-4955.

Martin, P.T. 2003. Role of transcription factors in skeletal muscle and the potential for pharmacological manipulation. *Current Opinion in Pharmacology* 3: 300-308.

Martinez-Moreno, M., A. Alvarez-Barrientos, F. Roncal, J.P. Albar, G. Gavilanes, S. Lamas, and I. Rodriguez-Crespo. 2005. Direct interaction between the reductase domain of endothelial nitric oxide synthase and the ryanodine receptor. *FEBS Letters* 579(14): 3159-3163.

Matsumura, K., J.M. Ervasti, K. Ohlendieck, S.D. Kahl, and K.P. Campbell. 1992. Association of dystrophin-related protein with dystrophin-associated proteins in mdx mouse muscle. *Nature* 360(6404): 588-591.

McArdle, A., R.H.T. Edwards, and M.J. Jackson. 1994. Time course of changes in plasma membrane permeability in the dystrophin-deficient mdx mouse. *Muscle and Nerve* 17: 1378-1384.

Mendell, J.R., Z. Sahenk, and T.W. Prior. 1995. The childhood muscular dystrophies: Diseases sharing a common pathogenesis of membrane instability. *Journal of Child Neurology* 10: 150-159.

Morrison, J., Q.L. Lu, C. Patoret, T. Partridge, and G. Bou-Gharios. 2000. T-cell-dependent fibrosis in the mdx dystrophic mouse. *Laboratory Investigation* 6: 881-891.

Ohlendieck, K., and K.P. Campbell. 1991. Dystrophin constitutes 5% of membrane cytoskeleton in skeletal muscle. *FEBS Letters* 283: 230-234.

Ohlendieck, K., K. Matsumara, V.V. Ionasescu, J.A. Towbin, E.P. Bosch, S.L. Weinstein, S.W. Sernett, and K.P. Campbell. 1993. Duchenne muscular dystrophy: Deficiency of dystrophin-associated proteins in the sarcolemma. *Neurology* 43: 795-800.

Ozawa, E. 2004. The muscle fiber cytoskeleton: The dystrophin system. In *Myology,* 3rd ed., Vol. 1, ed. A.G. Engel and C. Franzini-Armstrong, 455-470. New York: McGraw Hill.

Pariat, M., C. Salvat, M. Bebien, F. Brockly, E. Altieri, S. Carillo, I. Jariel-Encontre, and M. Piechaczyk. 2000. The sensitivity of c-Jun and c-Fos proteins to calpains depends on conformational determinants of the monomers and not on formation of dimers. *Biochemical Journal* 345(Pt 1): 129-138.

Pessah, I.N., and W. Feng. 2000. Functional role of hyper-reactive sulfhydryl moieties within the ryanodine receptor complex. *Antioxid Redox Signal* 2(1): 17-25.

Petrof, B.J. 1998. The molecular basis of activity-induced muscle injury in Duchenne muscular dystrophy. *Molecular and Cellular Biology* 179: 111-123.

Petrof, B.J., J.B. Shrager, H.H. Stedman, A.M. Kelly, and H.L. Sweeney. 1993. Dystrophin protects the sarcolemma from stresses developed during muscle contraction. *Proceedings of the National Academy of Sciences USA* 90: 3710-1714.

Porter, J.D., S. Khanna, H.J. Kaminski, J.S. Rao, A.P. Merriam, C.R. Richmonds, P. Leahy, J. Li, W. Guo, and F.H. Andrade. 2002. A chronic inflammatory response dominates the skeletal muscle molecular signature in dystrophin-deficient mdx mice. *Human Molecular Genetics* 11(3): 263-272.

Porter, J.D., A.P. Merriam, P. Leahy, B. Gong, J. Feuerman, G. Cheng, and S. Khanna. 2004. Temporal gene expression profiling of dystrophin-deficient (mdx) mouse diaphragm identifies conserved and muscle group-specific mechanisms in the pathogenesis of muscular dystrophy. *Human Molecular Genetics* 13(3): 257-269.

Rando, T.A. 2002. Oxidative stress and the pathogenesis of muscular dystrophies. *American Journal of Physical Medicine and Rehabilitation* 81(11): S175-186.

Renault, V., G. Piron-Hamelin, C. Forestier, S. Didonna, S. Decary, F. Hentati, G. Saillant, G.S. Butler-Browne, and V. Mouly. 2000. Skeletal muscle regeneration and the mitotic clock. *Experimental Gerontology* 35: 711-719.

Rezvani, M., E. Cafarelli, and D.A. Hood. 1995. Performance and excitability of mdx mouse muscle at 2, 5, and 13 wk of age. *Journal of Applied Physiology* 78: 961-967.

Richard, I., O. Broux, V. Allamand, F. Fougerousse, N. Chiannilkulchai, N. Bourg, L. Brenguier, C. Devaud, P. Pasturaud, C. Roudaut et al. 1995. Mutations in the proteolytic enzyme calpain 3 cause limb-girdle muscular dystrophy type 2A. *Cell* 81: 27-40.

Robert, V., M.L. Massimino, V. Tosello, R. Marsault, M. Cantini, V. Sorrentino, and T. Pozzan. 2001. Alteration in calcium handling at the subcellular level in mdx myotubes. *Journal of Biological Chemistry* 276(7): 4647-4651.

Roberts, R.G. 2001. Protein family review (dystrophin). *Genome Biology* 2: 3006.1-3006.5.

Roberts, R.G., R.J. Gardner, and M. Bobrow. 1994. Searching for the 1 in 2,400,000: A review of dystrophin gene point mutations. *Human Mutation* 4: 1-11.

Rybakova, I.N., J.R. Patel, and J.M. Ervasti. 2000. The dystrophin complex forms a mechanically strong link between the sarcolemma and costameric actin. *Journal of Cell Biology* 150: 1209-1214.

Sander, M., B. Chavoshan, S.A. Harris, S.T. Iannaccone, J.T. Stull, G.D. Thomas, and R.G. Victor. 2000. Functional muscle ischemia in neuronal nitric oxide synthase-deficient skeletal muscle of children with Duchenne muscular dystrophy. *Proceedings of the National Academy of Sciences USA* 97(25): 13818-13823.

Sanes, J.R. 2004. The extracellular matrix. In *Myology,* 3rd ed., Vol. 1, ed. A.G. Engel and C. Franzini-Armstrong, 471-487. New York. McGraw-Hill.

Schmalbruch, H. 1984. Regenerated muscle fibers in Duchenne muscular dystrophy: A serial section study. *Neurology* 34(1): 60-65.

Schofield, J.N., D.J. Blake, C. Simmons, G.E. Morris, J.M. Tinsley, K.E. Davies, and Y.H. Edwards. 1994. Apo-dystrophin-1 and apo-dystrophin-2, products of the Duchenne muscular dystrophy locus: Expression during mouse embryogenesis and in cultured cell lines. *Human Molecular Genetics* 3: 1309-1316.

Sicinski, P., Y. Geng, A.S. Ryder-Cook, E.A. Barnard, M.G. Darlison, and P.J. Barnard 1989. The molecular basis of muscular dystrophy in the mdx mouse: A point mutation. *Science* 244: 1578-1580.

Spencer, M.J., Croall, D.E., and Tidball, J.G. 1995. Calpains are activated in necrotic fibers from mdx dystrophic mice. *Journal of Biological Chemistry* 270, 10909-14.

Spencer, M.J., and J.G. Tidball. 2001. Do immune cells promote the pathology of dystrophin-deficient myopathies? *Neuromuscular Disorders* 11: 556-564.

Spencer, M.J., C.M. Walsh, K.A. Dorshkind, E.M. Rodriquez, and J.G. Tidball. 1997. Myonuclear apoptosis in dystrophic mdx muscle occurs by perforin-mediated cytotoxicity. *Journal of Clinical Investigation* 99: 2745-2751.

Stamler, J.S., and G. Meissner. 2001. Physiology of nitric oxide in skeletal muscle. *Physiological Review* 81(1): 209-237.

Stedman, H.H., H.L. Sweeney, J.B. Shrager, H.C. Maguire, R.A. Panettieri, B. Petrof, M. Narusawa, J.M. Leferovich, J.T. Sladky, and A.M. Kelly. 1991. The mdx mouse diaphragm reproduces the degenerative changes of Duchenne muscular dystrophy. *Nature* 352(6335): 536-539.

Stoyanovsky, D., T. Murphy, P.R. Anno, Y.M. Kim, and G. Salama. 1997. Nitric oxide activates skeletal and cardiac ryanodine receptors. *Cell Calcium* 21(1): 19-29.

Straub, V., and K.P. Campbell. 1997. Muscular dystrophies and the dystrophin-glycoprotein complex. *Current Opinions in Neurology* 110: 168-175.

Straub, V., J.A. Rafael, J.S. Chamberlain, and K.P. Campbell. 1997. Animal models for muscular dystrophy show different patterns of sarcolemmal disruption. *Journal of Cell Biology* 139(2): 375-385.

Stupka, N., P. Gregorevic, D.R. Plant, and G.S. Lynch. 2004. The calcineurin signal transduction pathway is essential for successful muscle regeneration in mdx dystrophic mice. *Acta Neuropathology* 107: 299-310.

Thomas, G.D., M. Sander, K.S. Lau, P.L. Huang, J.T. Stull, and R.G. Victor. 1998. Impaired metabolic modulation of alpha-adrenergic vasoconstriction in dystrophin-deficient skeletal muscle. *Proceedings of the National Academy of Sciences USA* 95(25): 15090-15158.

Tidball, J.G., and D.J. Law. 1991. Dystrophin is required for normal thin filament-membrane associations at myotendinous junctions. *American Journal of Pathology* 138(1): 17-21.

Tinsley, J.M., D.J. Blake, A. Roche, U. Fairbrother, J. Riss, B.C. Byth, A.E. Knight, J. Kendrick-Jones, G.K. Suthers, D.R. Love, et al. 1992. Primary structure of dystrophin-related protein. *Nature* 360(6404): 591-593.

Tinsley, J., N. Deconinck, R. Fisher, D. Kahn, S. Phelps, J.M. Gillis, and K.E. Davies. 1998. Expression of full-length utrophin prevents muscular dystrophy in mdx mice. *Nature Medicine* 4(12): 1441-1444.

Tkatchenko, A.V., G. Le Cam, J.J. Leger, and C.A. Dechesne. 2000. Large-scale analysis of differential gene expression in the hindlimb muscles and diaphragm of mdx mouse. *Biochimica et Biophysica Acta* 1500: 17-30.

Tkatchenko, A.V., G. Pietu, N. Cros, L. Gannoun-Zaki, C. Auffray, J.J. Leger, and C.A. Dechesne. 2001. Identification of altered gene expression in skeletal muscles from Duchenne muscular dystrophy patients. *Neuromuscular Disorders* 11: 269-277.

Torres, L.F., and L.W. Duchen. 1987. The mutant mdx:Inherited myopathy in the mouse. Morphological studies of nerves, muscles and end-plates. *Brain* 110(Pt 2): 269-299.

Turner, P.R., P.Y. Fong, W.F. Denetclaw, and R.A. Steinhardt. 1991. Increased calcium influx in dystrophic muscle. *Journal of Cell Biology* 115(6): 1701-1712.

Turner, P.R., R. Schultz, B. Ganguly, and R.A. Steinhardt. 1993. Proteolysis results in altered leak channel kinetics and elevated free calcium in mdx muscle. *Journal of Membrane Biology* 133(3): 243-251.

Turner, P.R., T. Westwood, C.M. Regen, and R.A. Steinhardt. 1988. Increased protein degradation results from elevated free calcium levels found in muscle from mdx mice. *Nature* 335: 735-738.

Tyler, K.L. 2003. Origins and early descriptions of "Duchenne Muscular Dystrophy." *Muscle and Nerve* 28: 402-422.

Vandebrouck, C., D. Martin, M. Colson-Van Schoor, H. Debaix, and P. Gailly. 2002. Involvement of TRPC in the abnormal calcium influx observed in dystrophic (mdx) mouse skeletal muscle fibers. *Journal of Cell Biology* 158: 1089-1096.

Wang, K.K., and P.W. Yuen. 1997. Development and therapeutic potential of calpain inhibitors [review]. *Advances in Pharmacology* 37: 117-152.

Wehling, M., M.J. Spencer, and J.G. Tidball. 2001. A nitric oxide synthase transgene ameliorates muscular dystrophy in mdx mice. *Journal of Cell Biology* 155(1): 123-131.

Whitehead, N.P., E.W. Yeung, and D.G. Allen. 2006. Muscle damage in mdx (dystrophic) mice: Role of calcium and reactive oxygen species. *Clinical and Experimental Pharmacology and Physiology* 33(7): 657-662.

Williams, M.W., and R.J. Bloch. 1999. Differential distribution of dystrophin and beta-spectrin at the sarcolemma of fast twitch skeletal muscle fibers. *Journal of Muscle Research and Cell Motility* 20(4): 383-393.

Wilson, L.A., B.J. Cooper, L. Dux, V. Dubowitz, and C.A. Sewry. 1994. Expression of utrophin (dystrophin-related protein) during regeneration and maturation of skeletal muscle in canine X-linked muscular dystrophy. *Neuropathology and Applied Neurobiology* 20: 359-367.

Wolff, A.V., A.K. Niday, K.A. Voelker, J.A. Call, N.P. Evans, K.P. Granata, and R.W. Grange. 2006. Passive mechanical properties of maturing extensor digitorum longus are not affected by lack of dystrophin. *Muscle and Nerve* 34(3): 304-312.

Yan, Z., S. Choi, X. Liu, M. Zhang, J.J. Schageman, S.Y. Lee, R. Hart, L. Lin, F.A. Thurmond, and R.S. Williams. 2003. Highly coordinated gene regulation in mouse skeletal muscle regeneration. *Journal of Biological Chemistry* 278(10): 8826-8836.

Chapter 10

Amelink GJ, Koot RW, Erich W, Van Gijn J and Bär PR. 1990. Sex-linked variation in creatine kinase release and its dependence on oestradiol can be demonstrated in an in-vitro rat skeletal muscle preparation. *Acta Physiol Scand* 128: 115-122.

Amelink GJ, Van der Waal W, Van Gijn J and Bär PR. 1991. Exercise induced muscle damage in the rat: The effect of vitamin E deficiency. *Pflugers Arch* 412: 417-421.

Anderson JE and Wozniak AC. 2004. Satellite cell activation on fibers: Modeling events in vivo—an invited review. *Can J Physiol Pharmacol* 82: 300-310.

Apple FS, Rogers MA and Casal DC. 1987. Skeletal muscle creatine kinase-MB alterations in women marathon runners. *Eur J Appl Physiol* 56: 49-52.

Ashcroft GS, Greenwell-Wild T, Horan MA, Wahl SM and Ferguson M. 1999. Topical estrogen accelerates cutaneous wound healing in aged humans associated with an altered inflammatory response. *Am J Pathol* 155: 1137-1146.

Ayers S, Baer J and Subbiah MTR. 1998. Exercise-induced increase in lipid peroxidation parameters in amenorrheic female athletes. *Fertility Steril* 69: 73-77.

Bär PR and Amelink GJ. 1997. Protection against muscle damage exerted by oestrogen: Hormonal or antioxidant action? *Biochem Soc Trans* 25: 50-54.

Bär PR, Amelink GJ, Oldenburg B and Blankenstein MA. 1988. Prevention of exercise-induced muscle membrane damage by oestradiol. *Life Sci* 42: 2677-2680.

Beaton LJ, Tarnopolsky M and Phillips SM. 2002. Variability in estimating eccentric contraction-induced muscle damage and inflammation in humans. *Can J Appl Physiol* 27: 516-526.

Belcastro AN, Shewchuk L and Raj DA. 1998. Exercise-induced muscle injury: A calpain hypothesis. *Mol Cell Biochem* 179: 135-145.

Brown MB, Foley A and Ferreria JA. 2005. Ovariectomy, hindlimb unweighting and recovery effects on skeletal muscle in adult rats. *Av Space Environ Med* 76: 1012-1018.

Buckley-Bleiler R, Maughan R and Clarkson PM. 1989. Serum creatine kinase activity after isometric exercise in perimenopausal and postmenopausal women. *Exp Aging Res* 5: 195-198.

Bundy S, Crawley JM and Edwards JH. 1979. Serum creatine kinase levels in pubertal, mature, pregnant and post-menopausal women. *J Med Genet* 16: 117-121.

Chalmers RL and McDermott J. 1996. Molecular basis of skeletal muscle regeneration. *Can J Appl Physiol* 21: 155-184.

Chargé SBP and Rudnicki MA. 2004. Cellular and molecular regulation of muscle regeneration. *Physiol Rev* 84: 209-238.

Chung SC, Goldfarb AH, Jamurtas AZ, Hegde SS and Lee J. 1999. Effect of exercise during the follicular and luteal phases on indices of oxidant stress in healthy women. *Med Sci Sports Exerc* 31: 409-413.

Clarkson PM and Hubal MJ. 2001. Are women less susceptible to exercise-induced muscle damage? *Curr Op Clin Nutr Mebabol Care* 4: 527-531.

Clarkson PM and Sayers SP. 1999a. Etiology of exercise induced muscle damage. *Can J Appl Physiol* 24: 234-248.

Clarkson PM and Sayers SP. 1999b. Gender differences in exercise-induced muscle damage. In Tarnopolsky M. (Ed.); *Gender differences in metabolism.* Boca Raton, FL: CRC Press, pp. 265-299.

D'eon TM, Sharoff C, Chipkin SR, Grow D, Ruby BC and Braun B. 2000. Regulation of exercise carbohydrate metabolism by estrogen and progesterone in women. *Am J Physiol* 283: E1046-E1055.

Dernbach A, Sherman W, Simonsen J, Flowers K and Lamb D. 1993. No evidence of oxidant stress during high-intensity rowing training. *J Appl Physiol* 74: 2140-2145.

Dubey RK and Jackson EK. 2001. Cardiovascular protective effects of 17b-estradiol metabolites. *J Appl Physiol* 91: 1868-1883.

Feng X, Li G and Wand S. 2004. Effects of estrogen on gastrocnemius muscle strain injury and regeneration in female rats. *Acta Pharmacol Sin* 25: 1489-1494.

Harada H, Pavlick KP, Hines IN, Hoffman JM, Bharwani S, Gray L, Wolf RE and Grisham MB. 2001. Effects of gender on reduced-size liver ischemia and reperfusion injury. *J Appl Physiol* 91: 2816-2822.

Hatae J. 2001. Effects of 17b-estradiol on tension responses and fatigue in the skeletal twitch muscle fibers of frog. *Japanese J Physiol* 51: 753-759.

Hawke TJ. 2005. Muscle stem cells and exercise training. *Exerc Sport Sci Rev* 33: 63-68.

Hawke TJ and Garry DJ. 2001. Myogenic satellite cells: Physiology to molecular biology. *J Appl Physiol* 91: 534-551.

Hernandez I, Delgado JL, Diaz J, Quesada T, Teruel MJG, Llanos C and Carbonell LF. 2000. 17b-Estradiol prevents oxidative stress and decreases blood pressure in ovariectomized rats. *Am J Physiol* 279: R1599-R1605.

Joo MH, Maehata E, Adachi T, Ishida A, Murai F and Mesaki N. 2004. The relationship between exercise-induced oxidative stress and the menstrual cycle. *Eur J Appl Physiol* 93: 82-86.

Kadi F, Karlsson C, Larsson B, Eriksson J, Larval M, Billig H and Jonsdottir I. 2002. The effects of physical activity and estrogen treatment on rat fast and slow skeletal muscles following ovariectomy. *J Muscle Res Cell Motil* 23: 335-339.

Kahlert S, Grohe C, Karas RH, Lobber K, Ludwig N and Vetter H. 1997. Effects of estrogen on skeletal myoblast growth. *Biochem Biophys Res Comm* 232: 373-378.

Karas RH, Schulten H, Pare G, Aronovitz MJ, Ohlsson C, Gustafsson JA and Mendelsohn ME. 2001. Effects of estrogen on the vascular injury response in estrogen receptor double knockout mice. *Circ Res* 89: 534-539.

Keans AK, Holbrook MT and Clarkson PM. 2005. Variability of creatine kinase increase and strength loss in men and women following eccentric exercise. *Med Sci Sports Exerc* 37 (Suppl): S318. [abstract]

Kendall B and Eston R. 2002. Exercise-induced muscle damage and the potential protective role of estrogen. *Sports Med* 32: 103-123.

Koh TJ and Tidball JG. 2000. Nitric oxide inhibits calpain-mediated proteolysis of talin in skeletal muscle cells. *Am J Physiol* 279: C806-C812.

Komulainen J, Koskinen S, Kalliokoski R, Takala T and Vihko V. 1999. Gender differences in skeletal muscle fibre damage after eccentrically biased downhill running in rats. *Acta Physiol Scand* 165: 57-63.

Laughlin MH, Welshons WV, Sturek M, Rush JE, Turk JR, Taylor JA, Judy BM, Henderson KK and Ganjam VK. 2003. Gender, exercise training and eNOS expression in porcine skeletal muscle arteries. *J Appl Physiol* 95: 250-264.

Lemoine S, Granier P, Tiffoche C, Rannou-Bekono F, Thieulant ML and Delamarche P. 2003. Estrogen receptor alpha mRNA in human skeletal muscles. *Med Sci Sports Exerc* 35: 439-443.

MacIntyre DL, Reid WD, Lyster DM and McKenzie DC. 2000. Different effects of strenuous eccentric exercise on the accumulation of neutrophils in muscle in women and men. *Eur J Appl Physiol* 81: 47-53.

Malm C, Nyberg P and Endstrom M. 2000. Immunological changes in human skeletal muscle and blood after eccentric exercise and multiple biopsies. *J Physiol* 529: 243-262.

McClung JM, Davis JM and Carson JA. 2007. Ovarian hormone status and skeletal muscle inflammation during recovery from disuse in rats. *Exp Physiol* 92: 219-232.

McCormick KM, Burns KL, Piccone CM, Gosselin LE and Braseau GA. 2004. Effects of ovariectomy and estrogen on skeletal muscle function in growing rats. *J Muscle Res Cell Motil* 25: 21-27.

Mishra DK, Friden J, Schmitz MC and Leiber RL. 1995. Anti-inflammatory medication after muscle injury provides a short term improvement but a long term loss of muscle function. *J Bone Joint Surg* 77A: 1510-1519.

Moosmann, B. and Behl, C. 1999. The antioxidant neuroprotective effects of estrogens and phenolic compounds are independent from their estrogenic properties. *Proceedings of the National Academy of Science* 96: 8867-8872.

Moran AL, Nelson SA, Landisch RM, Warren GL and Lowe DA. 2007. Estradiol replacement reverses ovariectomy-induced muscle contractile and myosin dysfunction in mature female mice. *J Appl Physiol* 102: 1387-1393.

Nadal A, Diaz M and Valverde MA. 2001. The estrogen trinity: Membrane, cytosolic and nuclear effects. *News Physiol Sci* 16: 251-255.

Nilsson, S, Makela, S., Treuter E, Tujague M, Thomsen J, Andersson G, Enmark E, Pettersson K, Warner M, and Gustafson JA. 2001. Mechanisms of estrogen action. *Physiol Rev* 81: 1535-1565.

Node K, Kitakaze M, Kosaka H, Minamino T, Funaya H and Masatsugu H. 1997. Amelioration of ischemia- and reperfusion-induced myocardial injury by 17b-estradiol. *Circulation* 96: 1953-1963.

Ogawa S, Chan J, Gustafsson J, Korach KS and Pfaff DW. 2003. Estrogen increases locomotor activity in mice through estrogen receptor alpha: Specificity for the type of activity. *Endocrinol* 144: 230-239.

Pare G, Krust A, Karas RH, Dupont S, Aronovitz M, Chambon P and Mendelsohn ME. 2002. Estrogen receptor alpha mediates the protective effects of estrogen against vascular injury. *Circ Res* 90: 1087-1092.

Paroo Z, Dipchand ES and Noble EG. 2002. Estrogen attenuates postexercise HSP70 expression in skeletal muscle. *Am J Physiol* 282: C245-C251.

Paroo Z, Tiidus PM and Noble EG. 1999. Estrogen attenuates HSP72 expression in acutely exercised male rodents. *Eur J Appl Physiol* 80: 180-184.

Persky AM, Green PS, Stubley L, Howell CO, Zaulyanov L, Braseau GA and Simpkins JW. 2000. Protective effect of estrogens against oxidative damage to heart and skeletal muscle in vivo and in vitro. *Proc Soc Exp Biol Med* 223: 59-66.

Phaneuf S and Leeuwenburgh C. 2001. Apoptosis and exercise. *Med Sci Sports Exerc* 33: 393-396.

Prorock AJ, Hafezi-Moghadam A, Laubach VE, Liao JK and Ley K. 2003. Vascular protection by estrogen in ischemia-reperfusion injury requires endothelial nitric oxide synthase. *Am J Physiol* 284: H133-H140.

Proske U and Allen TJ. 2005. Damage to skeletal muscle from eccentric exercise. *Exerc Sport Sci Rev* 33: 98-104.

Rinard, J, Clarkson PM, Smith LL and Grossman M. 2000. Response of males and females to high-force eccentric exercise. *J Sports Sci* 18: 229-236.

Roth S, Martel G and Ivy F. 2000. High volume heavy-resistance strength training and muscle damage in young and older women. *J Appl Physiol* 88: 1112-1118.

Salimena MC, Lagrota-Candido J and Quirico-Santos T. 2004. Gender dimorphism influences extracellular matrix expression and regeneration of muscular tissue in mdx dystrophic mice. *Histochem Cell Biol* 122: 435-444.

Salminen A and Kihlstrom M. 1985. Lysosomal changes in mouse skeletal muscle during repair of exercise injuries. *Muscle Nerve* 8: 269-279.

Sayers SP and Clarkson PM. 2001. Force recovery after eccentric exercise in males and females. *Eur J Appl Physiol* 84: 122-126.

Sciote JJ, Horton MJ, Zyman Y and Pascoe G. 2001. Differential effects of diminished oestrogen and androgen levels on development of skeletal muscle fibres in hypogonadal mice. *Acta Physiol Scand* 172: 179-187.

Shumate JB, Brooke M, Carroll J and Davis JE. 1979. Increased serum creatine kinase after exercise: A sex-linked phenomenon. *Neurology* 29: 902-909.

Simoncini T, Hafez-Moghadam A, Brazil DP, Ley K, Chin WW and Liao JK. 2000. Interaction of oestrogen receptor with the regulatory subunit of phosphatidylinositol-3-OH kinase. *Nature* 407: 538-541.

Sipilä S, Taafe DR, Cheng S, Puolakka J, Toivanen J and Suominen H. 2001. Effects of hormone replacement therapy and high-impact physical exercise on skeletal muscle in postmenopausal women: A randomized controlled study. *Clin Sci* 101: 147-157.

Sorensen MB, Rosenfalck AM, Hojgaard L and Ottesen B. 2001. Obesity and sarcopenia after menopause are reversed by sex hormone replacement therapy. *Obes Res* 9: 622-626.

Sorichter S, Mair J and Koller A. 2001. Release of muscle proteins after downhill running in male and female subjects. *Scand J Med Sci Sports* 11: 28-32.

Sotiriadou S, Kyparos A, Mougios V, Trontzos C, Sidiras G and Matziari C. 2003. Estrogen effect on some enzymes in female rats after downhill running. *Physiol Res* 52: 743-748.

Squadrito F, Altavilla D, Squadrito G, Campo G, Arlotta M, Arcoraci V, Minutoli L, Serrano M, Saitta A and Caputi A. 1997. 17b-oestradiol reduces cardiac leukocyte accumulation in myocardial ischaemia reperfusion injury in rat. *Eur J Pharmacol* 335: 185-192.

Sribnick EA, Swapan RK and Banik NL. 2004. Estrogen as a multi-active neuroprotective agent in traumatic injuries. *Neurochem Res* 29: 2007-2014.

St. Pierre-Schneider B, Correia L and Cannon J. 1999. Sex differences in leukocyte invasion in injured murine skeletal muscle. *Res Nurs Health* 22: 243-250.

St. Pierre-Schneider B, Fine JP, Nadolski T and Tiidus PM. 2004. The effects of estradiol and progesterone on plantarflexor muscle fatigue in ovariectomized mice. *Biol Res Nursing* 5: 265-275.

Stupka N, Lowther S, Chorneyko K, Bourgeois JM, Hogben C and Tarnopolsky MA. 2000. Gender differences in muscle inflammation after eccentric exercise. *J Appl Physiol* 89: 2325-2332.

Stupka N, Tarnopolsky MA, Yardley NJ and Phillips SM. 2001. Cellular adaptation to repeated eccentric exercise-induced muscle damage. *J Appl Physiol* 91: 1669-1678.

Stupka N and Tiidus PM. 2001. Effects of ovariectomy and estrogen on ischemia-reperfusion injury in hindlimbs of female rats. *J Appl Physiol* 91: 1828-1835.

Sudo M and Kano Y. 2005. Effects of isometric and eccentric contractions on apoptosis of skeletal muscle in male and female rats. *Med Sci Sports Exerc* 37 (Suppl): S317. [abstract]

Sugioka K, Shimosegawa Y and Nakano M. 1987. Estrogens as natural antioxidants of membrane and phospholipid peroxidation. *FEBS Lett* 210: 37-39.

Tiidus PM. 1995. Can estrogens diminish exercise induced muscle damage? *Can J Appl Physiol* 20: 26-38.

Tiidus PM. 1998. Radical species in inflammation and over-training. *Can J Physiol Pharmacol* 76: 533-538.

Tiidus PM. 2000. Estrogen and gender effects on post-exercise muscle damage, inflammation and oxidative stress. *Can J Appl Physiol* 25: 274-287.

Tiidus PM. 2001. Oestrogen and sex influence on muscle damage and inflammation: Evidence from animal models. *Cur Opin Clin Nutr Metabol Care* 4: 509-513.

Tiidus PM. 2003. Influence of estrogen and gender on muscle damage, inflammation and repair. *Exerc Sport Sci Rev* 31: 40-44.

Tiidus PM. 2005. Can oestrogen influence skeletal muscle damage, inflammation, and repair? *Br J Sports Med* 39: 251-253.

Tiidus PM and Bombardier E. 1999. Oestrogen attenuates post-exercise myeloperoxidase activity in skeletal muscle of male rats. *Acta Physiol Scand* 166: 85-90.

Tiidus PM, Bombardier E, Hidiroglou N and Madere R. 1998. Estrogen administration, post-exercise tissue oxidative stress and vitamin C status in male rats. *Can J Physiol Pharmacol* 76: 952-960.

Tiidus PM, Deller M, Bombardier E, Gul M and Liu XL. 2005. Estrogen supplementation failed to attenuate biochemical indices of neutrophil infiltration or damage in rat skeletal muscles following ischemia. *Biol Res* 38: 213-223.

Tiidus PM, Deller M and Liu XL. 2005. Oestrogen influence on myogenic satellite cells following downhill running in male rats: A preliminary study. *Acta Physiol Scand* 184: 67-72.

Tiidus PM, Holden D, Bombardier E, Zachowski S, Enns D and Belcastro A. 2001. Estrogen effect on post-exercise skeletal muscle neutrophil infiltration and calpain activity. *Can J Physiol Pharmacol* 79: 400-406.

Tiidus PM and Ianuzzo CD. 1983. Effects of intensity and duration of muscular exercise on delayed soreness and serum enzyme activities. *Med Sci Sports Exerc* 15: 461-465.

Van Buren GA, Yang DS and Clark KE. 1992. Estrogen-induced uterine vasodilation is antagonized by L-nitroargenine methyl ester, an inhibitor of nitric oxide synthesis. *Am J Obstet Gynecol* 167: 828-833.

Van der Meulen J, Kuipers H and Drukker J. 1991. Relationship between exercise-induced muscle damage and enzyme release in rats. *J Appl Physiol* 71: 999-1004.

Walden D, McCutchan J, Enquist E, Schwappach J, Shanley P, Reiss O, Terada L, Leff J and Repine JE. 1990. Neutrophils accumulate and contribute to skeletal muscle dysfunction after ischemia-reperfusion. *Am J Physiol* 259: H1809-H1812.

Whiting KP, Restall CJ and Brain PF. 2000. Steroid hormone-induced effects on membrane fluidity and their potential roles in non-genomic mechanisms. *Life Sci* 67: 743-757.

Wiik A, Glenmark B, Ekman M, Esbjornsson-Liljedahl M, Johansson O, Bodin K, Enmark E and Jansson E. 2003. Oestrogen receptor β is expressed in adult human skeletal muscle both at the mRNA and protein level. *Acta Physiol Scand* 179: 381-387.

Wise PM, Dubal DB, Wilson ME, Rau SW and Bottner M. 2001. Minireview: Neuroprotective effects of estrogen—new insights into mechanisms of action. *Endocrinol* 142: 969-973.

Xing D, Miller A, Novak L, Rocha R, Chen YF and Oparil S. 2004. Estradiol and progestins differentially modulate leukocyte infiltration after vascular injury. *Circulation* 109: 234-241.

Xu Y, Armstrong SJ, Arenas IA, Pehowich DJ and Davidge ST. 2004. Cardioprotection by chronic estrogen or superoxide dismutase mimetic treatment in the aged female rat. *Am J Physiol* 287: H165-H175.

Yagi, K., and Komura, S. 1986. Inhibitory effect of female hormones on lipid peroxidation. *Biochemistry International* 13: 1051-1055.

Chapter 11

Achten, J. and Jeukendrup, A.E. 2004. Optimizing fat oxidation through exercise and diet. *Nutrition* 20(7-8): 716-727.

Albright, A., Franz, M., Hornsby, G., Kriska, A., Marrero, D., Ullrich, I. and Verity, L.S. 2000. American College of Sports Medicine position stand: Exercise and type 2 diabetes. *Med Sci Sports Exerc* 32(7): 1345-1360.

Alt, N., Carson, J.A., Alderson, N.L., Wang, Y., Nagai, R., Henle, T., Thorpe, S.R. and Baynes, J.W. 2004. Chemical modification of muscle protein in diabetes. *Arch Biochem Biophys* 425(2): 200-206.

Andersen, J.L., Schjerling, P., Andersen, L.L. and Dela, F. 2003. Resistance training and insulin action in humans: Effects of de-training. *J Physiol* 551(3): 1049-1058.

Antonetti, D.A., Reynet, C. and Kahn, C.R. 1995. Increased expression of mitochondrial-encoded genes in skeletal muscle of humans with diabetes mellitus. *J Clin Invest* 95: 1383-1388.

Aragno, M., Mastrocola, R., Catalano, M.G., Brignardello, E., Danni, O. and Boccuzzi, G. 2004. Oxidative stress impairs skeletal muscle repair in diabetic rats. *Diabetes* 53(4): 1082-1088.

Association, American Diabetes. 2004. Standards of medical care in diabetes. *Diabetes Care* 27(Suppl 1): S15-35.

Bardsley, J. and Want, L. 2004. Overview of diabetes. *Crit Care Nurs Q* 27(2): 106-112.

Betteridge, D.J. 2000. What is oxidative stress? *Metabolism* 49(Suppl 1): 3-8.

Bjornholm, M., Kawano, Y., Lehtihet, M. and Zierath, J.R. 1997. Insulin receptor substrate-1 phosphorylation and phosphatidylinositol 3-kinase activity are decreased in skeletal muscle from NIDDM subjects following in vivo insulin stimulation. *Diabetes* 46: 524-527.

Bjornskov, E.K., Carry, M.R., Katz, F.H., Leflowitz, J. and Ringel, S.P. 1995. Diabetic muscle infarction: A new perspective on pathogenesis and management. *Neuromusc Disord*

5: 39-45.

Blaak, E.E. 2004. Basic disturbances in skeletal muscle fatty acid metabolism in obesity and type 2 diabetes mellitus. *Proc Nutr Soc* 63(2): 323-330.

Blayo, A. and Mandelbrot, L. 2004. Screening and diagnosis of gestational diabetes. *Diabetes Metab* 30(6): 575-580.

Bonen, A., Parolin, M.L., Steinberg, G.R., Calles-Escandon, J., Tandon, N.N., Glatz, J.F., Luiken, J.J., Heigenhauser, G.J. and Dyck, D.J. 2004. Triacylglycerol accumulation in human obesity and type 2 diabetes is associated with increased rates of skeletal muscle fatty acid transport and increased sarcolemmal FAT/CD36. *FASEB J* 18(10): 1144-1146.

Bruce, C.R., Anderson, M.J., Carey, A.L., Newman, D.G., Bonen, A., Kriketos, A.D., Cooney, G.J. and Hawley, J.A. 2003. Muscle oxidative capacity is a better predictor of insulin sensitivity than lipid status. *J Clin Endocrinol Metab* 88(11): 5444-5451.

Bruce, C.R., Carey, A.L., Hawley, J.A. and Febbraio, M.A. 2003. Intramuscular heat shock protein 72 and heme oxygenase-1 mRNA are reduced in patients with type 2 diabetes: Evidence that insulin resistance is associated with a disturbed antioxidant defense mechanism. *Diabetes* 52(9): 2338-2345.

Bruce, C.R. and Hawley, J.A. 2004. Improvements in insulin resistance with aerobic exercise training: A lipocentric approach. *Med Sci Sports Exerc* 36(7): 1196-1201.

Busija, D.W., Miller, A.W., Katakam, P., Simandle, S. and Erdos, B. 2004. Mechanisms of vascular dysfunction in insulin resistance. *Curr Opin Invest Drugs* 5(9): 929-935.

Cantley, L. 2002. The phosphoinositide 3-kinase pathway. *Science* 296(5573): 1655-1657.

Cho, H., Mu, J., Kim, J.K., Thorvaldsen, J.L., Chu, Q., Crenshaw, E.B., Kaestner, K.H., Bartolomei, M.S., Shulman, G.I. and Birnbaum, M.J. 2001. Insulin resistance and a diabetes mellitus-like syndrome in mice lacking the protein kinase Akt2 (PKB beta). *Science* 292: 1728-1731.

Christ-Roberts, C.Y., Pratipanawatr, T., Pratipanawatr, W., Berria, R., Belfort, R., Kashyap, S. and Mandarino, L.J. 2004. Exercise training increases glycogen synthase activity and GLUT4 expression but not insulin signaling in overweight nondiabetic and type 2 diabetic subjects. *Metabolism* 53(9): 1233-1242.

Cohen, N., Rossetti, L., Shlimovich, P., Halberstam, M., Hu, M. and Shamoon, H. 1995. Counterregulation of hypoglycemia: Skeletal muscle glycogen metabolism during three hours of physiological hyperinsulinemia in humans. *Diabetes* 44: 423-430.

Crowther, G.J., Milstein, J.M., Jabria, S.A., Kushmerick, M.J., Gronka, R.K. and Conley, K.E. 2003. Altered energetic properties in skeletal muscle of men with well-controlled insulin-dependent (type 1) diabetes. *Am J Physiol Endocrinol Metab* 284: E655-E662.

De Angelis, K.L., Cestari, I.A., Barp, J., Dall'Ago, P., Fernandes, T.G., de Bittencourt, P.I., Bello-Klein, A., Bello, A.A., Llesuy, S. and Irigoyen, M.C. 2000. Oxidative stress in the latissimus dorsi muscle of diabetic rats. *Braz J Med Biol Res*

33(11): 1363-1368.

Debard, C., Laville, M., Berbe, V., Loizon, E., Guillet, C., Morio-Liondore, B., Boirie, Y. and Vidal, H. 2004. Expression of key genes of fatty acid oxidation, including adiponectin receptors, in skeletal muscle of type 2 diabetic patients. *Diabetologia* 47(5): 917-925.

DeFronzo, R.A., Jacot, E., Jequier, E., Maeder, E., Wahren, J. and Felber, J.P. 1981. The effect of insulin on the disposal of intravenous glucose. Results from indirect calorimetry and hepatic and femoral venous catherization. *Diabetes* 30: 1000-1007.

DeGroot, J. 2004. The AGE of the matrix: Chemistry, consequence and cure. *Curr Opin Pharmacol* 4(3): 301-305.

Dohm, G.L., Tapscott, E.B., Pories, W.J., Dabbs, D.J., Flickinger, E.G., Meelheim, D., Fushiki, T., Atkinson, S.M., Elton, C.W. and Caro, J.F. 1988. An in vitro human skeletal muscle preparation suitable for metabolic studies: Decreased insulin stimulation of glucose transport in muscle from morbidly obese and diabetic subjects. *J Clin Invest* 82: 486-494.

Dunstan, D.W., Daly, R.M., Owen, N., Jolley, D., De Courten, M., Shaw, J. and Zimmet, P. 2002. High-intensity resistance training improves glycemic control in older patients with type 2 diabetes. *Diabetes Care* 25(10): 1729-1736.

Dunstan, D.W., Daly, R.M., Owen, N., Jolley, D., Vulikh, E., Shaw, J. and Zimmet, P. 2005. Home-based resistance training is not sufficient to maintain improved glycemic control following supervised training in older individuals with type 2 diabetes. *Diabetes Care* 28(1): 3-9.

Fushiki, T., Wells, J.A., Tapscott, E.B. and Dohm, G.L. 1989. Changes in glucose transporters in muscle in response to exercise. *Am J Physiol* 256(5 Pt 1): E580-587.

Gerbitz, K.D., Gempel, K. and Brdiczka, D. 1996. Mitochondria and diabetes: Genetic, biochemical, and clinical implications of the cellular energy circuit. *Diabetes* 45: 113-126.

Goodyear, L.J., Giorgino, F., Sherman, L.A., Carey, J., Smith, R.J. and Dohm, G.L. 1995. Insulin receptor phosphorylation, insulin receptor substrate-1 phosphorylation and phosphatidylinositol 3-kinase activity are decreased in intact skeletal muscle strips from obese subjects. *J Clin Invest* 95: 2195-2204.

Grigoriadis, E., Fam, A.G., Starok, A.G. and Cyn-Ang, L. 2000. Skeletal muscle infarction in diabetes mellitus. *J Rheumatol* 27: 1063-1068.

Group, T.D.C.C.T.R. 1993. The effect of intensive treatment of diabetes on the development and progression of long-term complications in insulin-dependent diabetes mellitus. *N Engl J Med* 329: 977-986.

Gumieniczek, A., Hopkala, H., Wojtowicz, Z. and Nieradko, M. 2001. Differences in antioxidant status in skeletal muscle tissue in experimental diabetes. *Clin Chim Acta* 314: 39-45.

Habib, G.S., Nashashibi, M. and Saliba, W. 2003. Diabetic muscular infarction: Emphasis on pathogenesis. *Clin Rheumatol* 22: 450-451.

Hamilton, J.A. and Kamp, F. 1999. How are free fatty acids transported in membranes? Is it by proteins or the free diffusion through the lipids? *Diabetes* s48: 2255-2269.

Holmes, B. and Dohm, G.L. 2004. Regulation of GLUT4 gene expression during exercise. *Med Sci Sports Exerc* 36(7): 1202-1206.

Holten, M.K., Zacho, M., Gaster, M., Juel, C., Wojtaszewski, J.F. and Dela, F. 2004. Strength training increases insulin-mediated glucose uptake, GLUT4 content, and insulin signaling in skeletal muscle in patients with type 2 diabetes. *Diabetes* 53(2): 294-305.

Jack, L.J., Boseman, L. and Vinicor, F. 2004. Aging Americans and diabetes. A public health and clinical response. *Geriatrics* 59(4): 14-17.

James, D. 2005. MUNC-ing around with insulin action. *J Clin Invest* 115(2): 219-221.

Jerums, G., Panagiotopoulos, S., Forbes, J., Osicka, T. and Cooper, M. 2003. Evolving concepts in advanced glycation, diabetic nephropathy, and diabetic vascular disease. *Arch Biochem Biophys* 419: 55-62.

Johansson, B.L., Sundell, J., Ekberg, K., Jonsson, C., Seppanen, M., Raitakari, O., Loutolahti, M., Nuutila, P., Wahren, J. and Knuuti, J. 2004. C-peptide improves adenosine-induced myocardial vasodilation in type 1 diabetes patients. *Am J Physiol Endocrinol Metab* 286(1): E14-19.

Kahn, C. 1978. Insulin resistance, insulin insensitivity, and insulin unresponsiveness: A necessary distinction. *Metabolism* 27(12): 1893-1902.

Kasuga, M., Hedo, J., Yamada, K. and Kahn, C. 1982. The structure of insulin receptor and its subunits. Evidence for multiple nonreduced forms and a 210,000 possible proreceptor. *J Biol Chem* 257(17): 10392-10399.

Kasuga, M., Karlsson, F. and Kahn, C.R. 1982. Insulin stimulates the phosphorylation of the 95,000-dalton subunit of its own receptor. *Science* 215(4529): 185-187.

Kelley, D.E., He, J., Menshikova, E.V. and Ritov, V.B. 2002. Dysfunction of mitochondria in human skeletal muscle in type 2 diabetes. *Diabetes* 51: 2944-2950.

Kelley, D.E. and Simoneau, J.-A. 1994. Impaired free fatty acid utilization by skeletal muscle in non-insulin dependent diabetes mellitus. *J Clin Invest* 94: 2349-2356.

Khan, A. and Pessin, J. 2002. Insulin regulation of glucose uptake: A complex interplay of intracellular signalling pathways. *Diabetologia* 45(11): 1475-1483.

Kim, H.J., Lee, J.S. and Kim, C.K. 2004. Effect of exercise training on muscle glucose transporter 4 protein and intramuscular lipid content in elderly men with impaired glucose tolerance. *Eur J Appl Physiol* 93: 353-358.

Kim, Y.B., Nikoulina, S.E., Ciaraldi, T.P., Henry, R.R. and Kahn, B.B. 1999. Normal insulin-dependent activation of Akt/protein kinase B with diminished activation of phosphoinositide 3-kinase in muscle in type 2 diabetes. *J Clin Invest* 104: 733-741.

Kirwan, J.P. and del Aguila, L.F. 2003. Insulin signalling, exercise and cellular integrity. *Biochem Soc Trans* 31: 1281-1285.

Koistinen, H.A. and Zierath, J.R. 2002. Regulation of glucose transport in human skeletal muscle. *Ann Med* 34(6): 410-418.

Koonen, D.P., Benton, C.R., Arumugam, Y., Tandon, N.N., Calles-Escandon, J., Glatz, J.F., Luiken, J.J. and Bonen, A. 2004. Different mechanisms can alter fatty acid transport when muscle contractile activity is chronically altered. *Am J Physiol Endocrinol Metab* 286(6): E1042-1049.

Koopman, R.J., Mainous, A.G., Diaz, V.A. and Geesey, M.E. 2005. Changes in age at diagnosis of type 2 diabetes in the United States, 1988-2000. *Ann Fam Med* 3(1): 60-63.

Krook, A., Bjornholm, M., Galuska, D., Jiang, X.J., Fahlman, R., Myers, M.G.J., Wallberg-Henriksson, H. and Zierath, J.R. 2000. Characterization of signal transduction and glucose transport in skeletal muscle from type 2 diabetic patients. *Diabetes* 49: 284-292.

Krook, A., Roth, R.A., Jiang, X.J. and Zierath, J.R. 1998. Insulin-stimulated Akt kinase activity is reduced in skeletal muscle from non-insulin dependent diabetic subjects. *Diabetes* 47: 1281-1286.

Krook, A., Wallberg-Henriksson, H. and Zierath, J. 2004. Sending the signal: Molecular mechanisms regulating glucose uptake. *Med Sci Sports Exerc* 36(7): 1212-1217.

Lazar, M.A. 2005. How obesity causes diabetes: not a tall tale. *Science* 307(5708): 373-375.

Lee, S.W., Dai, G., Hu, Z., Wang, X., Du, J. and Mitch, W.E. 2004. Regulation of muscle protein degradation: Coordinated control of apoptotic and ubiquitin-proteasome systems by phosphatidylinositol 3 kinase. *Am Soc Nephrol* 15(6): 1537-1545.

Lowell, B.B. and Shulman, G.I. 2005. Mitochondrial dysfunction and type 2 diabetes. *Science* 307: 384-387.

Matkovics, B., Sasvari, M., Kotorman, M., Varga, I.S., Hai, D.Q. and Varga, C. 1997. Further proof on oxidative stress in alloxan diabetic rat tissues. *Acta Physiol Hung* 85(3): 183-192.

MiSAD, Group. 1997. Prevalence of unrecognized silent myocardial ischaemia and its association with atherosclerotic risk factors in noninsulin-dependent diabetes mellitus. *Am J Cardiol* 79: 134-139.

Nagasawa, T., Tabata, N., Ito, Y., Nishizawa, N., Aiba, Y. and Kitts, D.D. 2003. Inhibition of glycation reaction in tissue protein incubations by water soluble rutin derivative. *Mol Cell Biochem* 249(1-2): 3-10.

Paolisso, G., D'Amore, A., Volpe, C., Balbi, V., Saccomanno, F., Galzerano, D., Gugliano, D., Varricchio, M. and D'Onofrio, F. 1994. Evidence for a relationship between oxidative stress and insulin action in non-insulin-dependent (type II) diabetic patients. *Metabolism* 43: 1426-1429.

Parthiban, A., Vijayalingam, S., Shanmugasundaram, K.R. and Mohan, R. 1995. Oxidative stress and the development of diabetic complications—antioxidants and lipid peroxidation in erythrocytes and cell membrane. *Cell Biol Int* 19: 987-993.

Peltoniemi, P., Yki-Jarvinen, H., Oikonen, V., Oksanen, A., Takala, T.O., Ronnemaa, T., Erkinjuntti, M., Knuuti, J. and Nuutila, P. 2001. Resistance to exercise-induced increase in glucose uptake during hyperinsulinemia in insulin-resistant skeletal muscle of patients with type 1 diabetes. *Diabetes* 50: 1371-1377.

Petersen, A.M. and Pedersen, B.K. 2005. The anti-inflammatory effect of exercise. *J Appl Physiol* 98(4): 1154-1162.

Ramasamy, R., Vannucci, S.J., Shi Du Yan, S., Herold, K., Fang Yan, S. and Marie Schmidt, A. 2005. Advanced glycation end products and RAGE: A common thread in aging, diabetes, neurodegeneration and inflammation. *Glycobiology.* 15(7): 16R-28R.

Reddy, G.K., Stehno-Bittel, L. and Enwemeka, C.S. 2002. Glycation-induced matrix stability in the rabbit achilles tendon. *Arch Biochem Biophys* 399(2): 174-180.

Rudich, A., Tirosh, A., Potashnik, R., Hemi, R., Kanety, H. and Bashan, N. 1998. Prolonged oxidative stress impairs insulin-induced GLUT4 translocation in 3T3-L1 adipocytes. *Diabetes*: 47(10): 1562-1569.

Schrauwen, P. and Hesselink, M. 2004. Oxidative capacity, lipotoxicity, and mitochondrial damage in type 2 diabetes. *Diabetes* 53: 1412-1417.

Searls, Y.M., Smirnova, I.V., Fegley, B.R. and Stehno-Bittel, L. 2004. Exercise attenuates diabetes-induced ultrastructural changes in rat cardiac tissue. *Med Sci Sports Exerc* 36(11): 1863-1870.

Segal, K. 2004. Type 2 diabetes and disease management: Exploring the connections. *Dis Manag* 7(Suppl 1): S11-22.

Simoneau, J.-A., Veerkamp, J.H., Turcotte, L.P. and Kelley, D.E. 1999. Impaired free fatty acid utilization by skeletal muscle in non-insulin dependent diabetes mellitus. *J Clin Invest* 94: 2349-2356.

Stellingwerff, T., Boon, H., Jonkers, R.A., Seden, J.M., Spriet, L.L., Koopman, R., and van Loon, L.J. 2007. Significant intramyocellular lipid use during prolonged cycling in endurance trained males as assessed by three different methodologies. *Am J Physiol Endocrinol Metab* (e-pub)

Tirosh, A., Potashnik, R., Bashan, N. and Rudich, A. 1999. Oxidative stress disrupts insulin-induced cellular redistribution of insulin receptor substrate-1 and phosphatidylinositol 3-kinase in 3T3-L1 adipocytes. A putative cellular mechanism for impaired protein kinase B activation and GLUT4 translocation. *J Biol Chem* 274: 10595-10602.

Tisch, R. and Vedevit, H. 1996. Insulin-dependent diabetes mellitus. *Cell* 85: 291-297.

Trujillo-Santos, A.J. 2003. Diabetic muscle infarction. *Diabetes Care* 26(1): 211-215.

Walder, K., Kerr-Bayles, L., Civitarese, A., Jowett, J., Curran, J., Elliott, K., Trevaskis, J., Bishara, N., Zimmet, P., Mandarino, L., Ravussin, E., Blangero, J., Kissebah, A. and Collier, G.R. 2005. The mitochondrial rhomboid protease PSARL is a new candidate gene for type 2 diabetes. *Diabetologia* 48(3): 459-468.

Zorzano, A., Palacin, M. and Guma, A. 2005. Mechanisms regulating GLUT4 glucose transporter expression and glucose transport in skeletal muscle. *Acta Physiol Scand* 183(1): 43-58.

Chapter 12

Alberta Human Resources and Employment. 2006. Occupational Injuries and Diseases in Alberta: Lost-Time Claims and Claim Rates 2005 Summary, released July 2006, 15-16 (http://employment.alberta.ca/cps).

Allen D.L., R.R. Roland, and V.R. Edgerton. 1999. Myonuclear domains in muscle adaptation and disease. *Muscle and Nerve* 22: 1350-1360.

Almekinders L.C., and J.A. Gilbert. 1986. Healing of experimental muscle strains and the effects of nonsteroidal anti-inflammatory medication. *American Journal of Sports Medicine* 14(4): 303-308.

Al-Shatti T., A.E. Barr, F. Safadi, M. Amin, and M.F. Barbe. 2005. Increase in pro- and anti-inflammatory cytokines in median nerves in a rat model of repetitive motion injury. *Journal of Neuroimmunology* 167: 13-22.

Aubert A., C. Vega, R. Dantzer, and G. Goodall. 1995. Pyrogens specifically disrupt the acquisition of a task involving cognitive processing in the rat. *Brain Behavior and Immunity* 9: 129-148.

Babenko V., T. Graven-Nielsen, P. Svensson, A.M. Drewes, T.S. Jensen, and L. Arendt-Nielsen. 1999. Experimental human muscle pain and muscular hyperalgesia induced by combinations of serotonin and bradykinin. *Pain* 82: 1-8.

Banasik J. 2000. Cell injury, aging and death. In *Pathophysiology: Biological and behavioral perspectives,* ed. C. Banasik. Philadelphia: Saunders.

Barbe M.F., A.E. Barr, I. Gorzelany, M. Amin, J.P. Gaughan, and F.F. Safadi. 2003. Repetitive motion causes local injury, systemic inflammation and reach pattern decrements in rats. *Journal of Orthopaedic Research* 21: 167-176.

Barr A.E., M. Amin, and M.F. Barbe. 2002. Dose-response relationship between reach repetition and indicators of inflammation and movement dysfunction in a rat model of work-related musculoskeletal disorder. *Proceedings of the HFES 46th annual meeting,* 1486-1490. Human Factors and Ergonomics Society, Santa Monica, CA.

Barr A.E., and M.F. Barbe. 2002. Pathophysiological tissue changes associated with repetitive movement: A review of the evidence. *Physical Therapy* 82: 173-187.

Barr A.E., and M.F. Barbe. 2004. Inflammation reduces physiological tissue tolerance in the development of work related musculoskeletal disorders. *Journal of Electromyography and Kinesiology* 14: 77-85.

Barr A.E., M.F. Barbe, and B.D. Clark. 2004a. Systemic inflammatory mediators may contribute to widespread symptoms in work-related musculoskeletal disorders. *Exercise and Sport Sciences Reviews* 32: 135-142.

Barr A.E., M.F. Barbe, and B.D. Clark. 2004b. Work-related musculoskeletal disorders of the hand and wrist: epidemiology, pathophysiology, and sensorimotor changes. *Journal of Orthopaedic and Sports Physical Therapy* 34: 610-627.

Barr A.E., F.F. Safadi, R.P. Garvin, S.N. Popoff, and M.F. Barbe. 2000. Evidence of progressive tissue pathophysiology and motor behavior degradation in a rat model of work related musculoskeletal disease. *Proceedings of the IEA 2000/HFES 2000 Congress* 5: 584-587.

Barr A.E., F.F. Safadi, I. Gorzelany, M. Amin, S.N. Popoff, and M.F. Barbe. 2003. Repetitive, negligible force reaching in rats induces pathological overloading of upper extremity bones. *Journal of Bone and Mineral Research* 18: 2023-2032.

Bureau of Labor Statistics. 2006. Nonfatal Occupational Injuries and Illnesses Requiring Days Away from Work, 2005, United States Department of Labor 06-1982, Washington, DC (www.bls.gov/iif/home.htm).

Capuron L., and R. Dantzer. 2003. Cytokines and depression: The need for a new paradigm. *Brain Behavior and Immunity* 17: S119-S124.

Carp S.J., M.F. Barbe, K.A. Winter, M. Amin, and A.E. Barr. 2007. Cytokines and C-reactive protein in human serum are associated with severity of musculoskeletal disorders from overuse. *Clinical Science* 112: 305-314.

Clark B.D., T.A. Al-Shatti, A.E. Barr, M. Amin, and M.F. Barbe. 2004. Performance of a high-repetition, high-force task induces carpal tunnel syndrome in rats. *Journal of Orthopaedic and Sports Physical Therapy* 34: 244-254.

Clark B.D., A.E. Barr, F.F. Safadi, L. Beitman, T. Al-Shatti, and M.F. Barbe. 2003. Median nerve trauma in a rat model of work-related musculoskeletal disorder. *Journal of Neurotrauma* 20: 681-695.

Dantzer R. 2004. Cytokine-induced sickness behavior: A neuroimmune response to activation of innate immunity. *European Journal of Pharmacology* 500: 399-411.

Dennett X., and H.J.H. Fry. 1998. Overuse syndrome: A muscle biopsy study. *Lancet* 23: 905-908.

Dupre D. 2001. Work-related health problems in the EU 1998-1999. European Communities Cat. No. KS-NK-01-017-EN-I, ISSN 1024-4352, European Communities (http://epp.eurostat.ec.europa.eu/cache/ITY_OFFPUB/KS-NK-01-017/EN/KS-NK-01-017-EN.PDF).

Fritz V.K., and W.T. Stauber. 1988. Characterization of muscles injured by forced lengthening. II. Proteoglycans. *Medicine and Science in Sports and Exercise* 20: 354-361.

Gibala M.J., J.D. MacDougall, M.A. Tarnopolsky, and W.T. Stauber. 1995. Changes in human skeletal muscle ultrastructure and force production after acute exercise. *Journal of Applied Physiology* 78: 702-708.

Goehler L.E., J.K. Relton, D. Cripps, R. Kiechle, N. Tartaglia, S.F. Maier, and L.R. Watkins. 1997. Vagal paraganglia bind biotinylated interleukin-1 receptor agonist: A possible mechanism for immune-to-brain communication. *Brain Research Bulletin* 43: 357-364.

Gute D.C., T. Ishida, K. Yarimizu, and R.J. Korthuis. 1998. Inflammatory responses to ischemia and reperfusion in skeletal muscle. *Molecular and Cellular Biochemistry* 179: 169-187.

Hagg G. 2000. Human muscle fibre abnormalities related to occupational load. *European Journal of Applied Physiology* 83: 159-165.

Hesselink M.K.C., H. Kuipers, P. Guerten, and H. Van Straaten. 1996. Structural muscle damage and muscle strength after incremental number of isometric forced lengthening contractions. *Journal of Muscle Research and Cell Motility* 17: 335-341.

Kadi F., G. Hagg, R. Hakansson, S. Holmner, G.S. Butler-Browne, and L.-E. Thornell. 1998. Structural changes in male trapezius muscle with work-related myalgia. *Acta Neuropathologica* 95: 352-360.

Kadi F., K. Waling, C. Ahlgren, G. Sundlein, S. Holmner, G.S. Butler-Browne, and L.-E. Thornell. 1998. Pathological mechanisms implicated in localized female trapezius myalgia. *Pain* 78: 191-196.

Kanemaki T., H. Kitade, M. Kaibori, K. Sakitani, Y. Hiramatsu, Y. Kamiyama, S. Ito, and T. Okumura. 1998. Interleukin 1beta and interleukin 6, but not tumor necrosis factor alpha, inhibit insulin-stimulated glycogen synthesis in rat hepatocytes. *Hepatology* 27: 1296-1303.

Kelley K.W., R.M. Bluthe, R. Dantzer, J.H. Zhou, W.H. Chen, R.W. Johnson, and S.R. Broussard. 2003. Cytokine-induced sickness behavior. *Brain Behavior and Immunity* 17: S112-S118.

Keogh J.P., I. Nuwayhid, J.L. Gordon, and P. Gucer. 2000. The impact of occupational injury on injured worker and family: Outcomes of upper extremity cumulative trauma disorders in Maryland workers. *American Journal of Industrial Medicine* 38: 498-506.

Lapointe B.M., J. Frenette, and C.H. Côté. 2002. Lengthening contraction-induced inflammation is linked to secondary damage but devoid of neutrophil invasion. *Journal of Applied Physiology* 92: 1995-2004.

Larsson B., J. Björk, J. Elert, and B. Gerdle. 2000. Mechanical performance and electromyography during repeated maximal isokinetic shoulder forward flexions in female cleaners with and without myalgia of the trapezius muscle and in healthy controls. *European Journal of Applied Physiology* 83: 257-267.

Larsson B., J. Björk, J. Elert, R. Lindman, and B. Gerdle. 2001. Fibre type proportion and fibre size in trapezius muscle biopsies from cleaners with and without myalgia and its correlation with ragged red fibres, cytochrome-c-oxidase-negative fibres, biomechanical output, perception of fatigue, and surface electromyography during repetitive forward flexions. *European Journal of Applied Physiology* 84: 492-502.

Larsson B., J. Björk, K.-G. Henriksson, B. Gerdle, and R. Lindman. 2000. The prevalences of cytochrome c oxidase negative and superpositive fibres and ragged-red fibres in the trapezius muscle of female cleaners with and without myalgia and of female healthy controls. *Pain* 84: 379-387.

Larsson B., J. Björk, F. Kadi, R. Lindman, and B. Gerdle. 2004. Blood supply and oxidative metabolism in muscle biopsies of female cleaners with and without myalgia. *Clinical Journal of Pain* 20: 440-446.

Larsson S.E., L. Bodegard, K.G. Kenriksson, and P.A. Oberg. 1990. Chronic trapezius myalgia: Morphology and blood flow studies in 17 patients. *Acta Orthopaedica Scandinavica* 61: 394-398.

Leclerc A., M.F. Landre, J.F. Chastang, I. Niedhammer, and Y. Roquelaure. 2001. Upper limb disorders in repetitive work. *Scandinavian Journal of Work, Environment and Health* 27: 268-278.

LeMay L.G., A.J. Vander, and M.J. Kluger. 1990. The effects of psychological stress on plasma interleukin-6 activity in rats. *Physiology and Behavior* 47: 957-961.

Lindman R., M. Hagberg, K.A. Angqvist, K. Soderlund, E. Hultman, and L.E. Thornell. 1991. Changes in muscle morphology in chronic trapezius myalgia. *Scandinavian Journal of Work, Environment and Health* 17: 347-355.

Ljung B.O., R.L. Lieber, and J. Friden. 1999. Wrist extensor muscle pathology in lateral epicondylitis. *Journal of Hand Surgery* 24B: 177-183.

Lyngso D., L. Simonsen, and J. Bulow. 2002. Metabolic effects of interleukin-6 in human splanchnic and adipose tissue. *Journal of Physiology* 543: 379-386.

Mathis L.B., R.J. Gatchel, P.B. Polatin, H.J. Boulas, and R.K. Kinney. 1994. Prevalence of psychopathology in carpal tunnel syndrome patients. *Journal of Occupational Rehabilitation* 4: 199-210.

Mishra K.D., J. Friden, M.C. Schmitz, and R.L. Lieber. 1995. Antiinflammatory medication after muscle injury: A treatment resulting in short-term improvement but subsequent loss of muscle function. *Journal of Bone and Joint Surgery (American)* 77: 1510-1519.

National Research Council and Institute of Medicine. 2001. *Musculoskeletal disorders and the workplace: Low back and upper extremities*. Washington, DC: National Academy Press.

Nemet D., S. Hong, P.J. Mills, M.G. Ziegler, M. Hill, and D.M. Cooper. 2002. Systemic vs. local cytokine and leukocyte responses to unilateral wrist flexion exercise. *Journal of Applied Physiology* 93: 546-554.

Nikolaou P.K., B.L. MacDonald, R.R. Glisson, A.V. Seaber, and W.E. Garrett. 1987. Biomechanical and histological evaluation of muscle after controlled strain injury. *American Journal of Sports Medicine* 15(1): 9-14.

Rosendal L., A.K. Blangsted, J. Kristiansen, K. Sogaard, H. Langberg, G. Sjogaard, and M. Kjaer. 2004. Interstitial muscle lactate, pyruvate and potassium dynamics in the trapezius muscle during repetitive low-force arm movements, measured with microdialysis. *Acta Physiologica Scandinavica* 182: 379-388.

Rosendal L., B. Larsson, J. Kristiansen, M. Peolsson, K. Sogaard, M. Kjaer, J. Sorensen, and B. Gerdle. 2004. Increase in muscle nociceptive substances and anaerobic metabolism in patients with trapezius myalgia: Microdialysis in rest and during exercise. *Pain* 112: 324-334.

Rosendal L., K. Sogaard, M. Kjaer, G. Sjogaard, H. Langberg, and J. Kristiansen. 2004. Increase in interstitial interleukin-6 of human skeletal muscle with repetitive low-force exercise. *Journal of Applied Physiology* 98: 477-481.

Starkie R., R. Ostrowski, S. Jauffred, M. Febbraio, and B.K. Pedersen. 2003. Exercise and IL-6 infusion inhibit endotoxin-induced TNF-α production in humans. *FASEB Journal* 17: 884-886.

Stauber W.T., P.M. Clarkson, V.K. Fritz, and W.J. Evans. 1990. Extracellular matrix disruption and pain after eccentric muscle action. *Journal of Applied Physiology* 69: 868-874.

Stauber W.T., K.K. Knack, G.R. Miller, and J.G. Grimmett. 1996. Fibrosis and intercellular collagen connections from four weeks of muscle strains. *Muscle and Nerve* 19: 423-430.

Stauber W.T., G.R. Miller, J.G. Grimmett, and K.K. Knack. 1994. Adaptation of rat soleus muscle to 4 wk of intermittent strain. *Journal of Applied Physiology* 77: 58-62.

Stauber W.T., C.A. Smith, G.R. Miller, and M.A. Stauber. 2000. Recovery from 6 weeks of repeated strain injury to rat soleus muscles. *Muscle and Nerve* 23: 1819-1825.

Stauber W.T., and M.E. Willems. 2002. Prevention of histopathic changes from 30 repeated stretches of active rat skeletal muscles by long inter-stretch times. *European Journal of Applied Physiology* 88: 94-99.

Stupka N., and P.M. Tiidus. 2001. Effects of ovariectomy and estrogen on ischemia-reperfusion injury in hindlimbs of female rats. *Journal of Applied Physiology* 91: 1828-1835.

Tiidus P.M. 2003. Influence of estrogen on skeletal muscle damage, inflammation, and repair. *Exercise and Sport Sciences Reviews* 31: 40-44.

Watkins L.R., and S.F. Maier. 1999. Implications of immune-to-brain communication for sickness and pain. *Proceedings of the National Academy of Sciences USA* 96: 7710-7713.

Weigert B.J., A.A. Rodriguez, R.G. Radwin, and J. Sherman. 1999. Neuromuscular and psychological characteristics in subjects with work-related forearm pain. *American Journal of Physical Medicine and Rehabilitation* 78: 545-551.

Winkelstein B.A. 2004. Mechanisms of central sensitization, neuroimmunology and injury biomechanics in persistent pain: Implications for musculoskeletal disorders. *Journal of Electromyography and Kinesiology* 14: 87-93.

Zimmermann M., and T. Herdegen. 1996. Plasticity of the nervous system at the systemic, cellular and molecular levels: A mechanism of chronic pain and hyperalgesia. *Progress in Brain Research* 110: 233-259.

Chapter 13

Andriacchi, T.P. 1998. Practical and theoretical considerations in the application in the development of clinical gait analysis. *Biomedical Materials and Engineering* 8(3-4):137-143.

Arsenault, A.B., D.A. Winter, and R.G. Marteniuk. 1986. Is there a normal profile of EMG activity in gait? *Medical and Biological Engineering and Computing* 24(4):337-343.

Braune, W., and G. Fisher. 1889. The center of gravity of the human body as related to the equipment of the German infantry. *Treatises of the Mathematical-Physical Class of the Royal Academy of Sciences of Saxony*, VII:1-130.

Bresler, B., and J.P. Frankel. 1950. The forces and moments in the leg during level walking. *American Society of Mechanical Engineers* 48-A-62:27-35.

Crowninshield, R.D. 1978. Use of optimization techniques to predict muscle forces. *Journal of Biomechanical Engineering—Transactions of the ASME* 100(2):88-92.

Crowninshield, R.D. 1983. A physiologically based criterion for muscle force predictions on locomotion. *Bulletin of the Hospital for Joint Diseases Orthopaedic Institute* 43(2):164-170.

Davis, R.B. 1997. Reflections on clinical gait analysis. *Journal of Electromyography and Kinesiology* 7(4):251-257.

Edman, K.A.P., G. Elzinga, and M.I.M. Noble. 1978, August. Enhancement of mechanical performance by stretch during tetanic contractions of vertebrate skeletal-muscle fibers. *Journal of Physiology* (London) 281:139-155.

Elftman, H. 1938. Forces and energy changes in the leg during walking. *American Journal of Physiology* 125:339-356.

Elftman, H. 1939. The function of the muscles in locomotion. *American Journal of Physiology* 125:-357-366.

Eng, J.J., and D.A. Winter. 1995. Kinetic-analysis of the lower-limbs during walking—what information can be gained from a 3-dimensional model? *Journal of Biomechanics* 28(6):753-758.

Eston, R.G., J. Mickleborough, and V. Baltzopoulos. 1995. Eccentric activation and muscle damage—biomechanical and physiological considerations during downhill running. *British Journal of Sports Medicine* 29(2):89-94.

Frigo, C., M. Rabuffetti, D.C. Kerrigan, L.C. Deming, and A. Pedotti. 1998. Functionally oriented and clinically feasible quantitative gait analysis method. *Medical and Biological Engineering and Computing* 36(2):179-185.

Gates, D.H., J.L. Su, and J.B. Dingwell. 2007. Possible bio-mechanical origins of the long-range correlations in stride intervals of walking. *Physica A* 380:259-270.

Gordon, A.M., A.F. Huxley, and F.J. Julian. 1966. The variation in isometric tension with sarcomere length in vertebrate muscle fibers. *Journal of Rehabilitation Research and Development* 184:170-192.

Hatze, H. 1981. *Myocybernetic control models of skeletal muscles: Characteristics and applications.* University of South Africa, Pretoria.

Herzog, W. 1987. Individual muscle force estimations using a nonlinear optimal-design. *Journal of Neuroscience Methods* 21(2-4):167-179.

Hill, A.V. 1938. The heat of shortening and the dynamic constants of muscle. *Proceedings of the Royal Society* (B) 126:136-195.

Huxley, A.F., and R.M. Simmons. 1971. Proposed mechanism of force generation in striated muscle. *Nature* 233:533-538.

Inman, V.T., H.J. Ralston, and F. Todd. 1981. *Human walking.* Baltimore: Williams & Wilkins.

Kepple, T., K. Siegel, and S. Stanhope. 1997. Relative contributions of the lower extremity joint moments to forward progression and support during gait. *Gait and Posture* 6:1-8.

Kepple, T., K. Siegel, and S. Stanhope. 2000. Modeling the relative compensatory ability of lower extremity muscle groups during normal walking. *Gait and Posture* 9:119-120.

Kopf, A., S. Pawelka, and A. Kranzl. 1998. Clinical gait analysis—methods, limitations, and indications. *Acta Medica Austriaca* 25(1):27-32.

Mulder, T., B. Nienhuis, and J. Pauwels. 1998. Clinical gait analysis in a rehabilitation context: some controversial issues. *Clinical Rehabilitation* 12(2):99-106.

Pathokinesiology Service and Physical Therapy Department. 2001. *Observational gait analysis handbook.* Downey, CA: Los Amigos Research and Education Institute, Inc.

Perry, J. 1992. *Gait analysis: Normal and pathological function.* Thorofare, NJ: SLACK Inc.

Pierrynowski, M.R., P.M. Tiidus, and V. Galea. 2005. Women with fibromyalgia walk with an altered muscle synergy. *Gait and Posture* 22: 210-218.

Pollack, G.H. 1983. The cross-bridge theory. *Physiological Reviews* 63(3):1049-1113.

Rozendal, R.H. 1991. Clinical gait analysis—problems and solutions. *Human Movement Science* 10(5):555-564.

Sadeghi, H., P. Allard, F. Barbier, S. Sadeghi, S. Hinse, R. Perrault, and H. Labelle. 2002. Main functional roles of knee flexors/extensors in able-bodied gait using principal component analysis (I). *Knee* 9(1):47-53.

Sutherland, D.H. 2001. The evolution of clinical gait analysis part I: Kinesiological EMG. *Gait and Posture* 14(1):61-70.

White, S.C., and D.A. Winter. 1992. Predicting muscle forces in gait from EMG signals and musculotendon kinematics. *Journal of Electromyography and Kinesiology* 2(4):217-231.

Whittle, M.W. 1996. Clinical gait analysis: A review. *Human Movement Science* 15(3):369-387.

Winter, D.A. 1980. Overall principle of lower-limb support during stance phase of gait. *Journal of Biomechanics* 13(11):923-927.

Winter, D.A. 1984. Kinematic and kinetic patterns in human gait—variability and compensating effects. *Human Movement Science* 3(1-2):51-76.

Winter, D.A. 1989. Biomechanics of normal and pathological gait—implications for understanding human locomotor control. *Journal of Motor Behavior* 21(4):337-355.

Winter, D.A., and P. Eng. 1995. Kinetics—our window into the goals and strategies of the central-nervous-system. *Behavioural Brain Research* 67(2):111-120.

Winter, D.A., C.D. Mackinnon, G.K. Ruder, and C. Wieman. 1993. An integrated EMG/biomechanical model of upper-body balance and posture during human gait. *Progress in Brain Research* 97:359-367.

Winter, D.A., S.J. Olney, J. Conrad, S.C. White, S. Ounpuu, and J.R. Gage. 1990. Adaptability of motor patterns in pathological gait. In *Multiple muscle systems: Biomechanics and movement organization,* ed. J.M. Winters and S.L-Y. Woo. New York: Springer-Verlag.

Winter, D.A., and S.H. Scott. 1991. Technique for interpretation of electromyography for concentric and eccentric contractions in gait. *Journal of Electromyography and Kinesiology* 1(4):263-269.

Winters, J.M. 1990. Hill-based muscle models: A systems engineering perspective. In *Multiple muscle systems: Biomechanics and movement organization.* New York: Springer-Verlag.

Yamaguchi, G.T., and F.E. Zajac. 1989. A planar model of the knee-joint to characterize the knee extensor mechanism. *Journal of Biomechanics* 22(1):1-10.

Zahalak, G.I. 1990. Modeling muscle mechanics (and energetics). In *Multiple muscle systems: Biomechanics and movement organization.* New York: Springer-Verlag.

Zajac, F.E., and M.E. Gordon. 1989. Determining muscles force and action in multi-articular movement. *Exercise and Sport Sciences Reviews* 17:187-230.

Zajac, F.E., and J.M. Winters. 1990. Modeling musculoskeletal movement systems: Joint and body segmental dynamics, musculoskeletal actuation, and neuromuscular control. In *Multiple muscle systems: Biomechanics and movement organization.* New York: Springer-Verlag.

Chapter 14

Adams, G.R. 1998. Role of insulin-like growth factor-I in the regulation of skeletal muscle adaptation to increased loading. *Exercise and Sport Sciences Reviews* 26: 31-60.

Andersen, J.L., and P. Aagaard. 2000. Myosin heavy chain IIX overshoot in human skeletal muscle. *Muscle and Nerve* 23: 1095-1104.

Baldwin, K.M., and F. Haddad. 2002. Skeletal muscle plasticity: Cellular and molecular responses to altered physical activity paradigms. *American Journal of Physiology and Medicine Rehabilitation* 81 (11 Suppl): S 40-51.

Bigard, A.X., H. Sanchez, O. Biort, and B. Serrurier. 2000. Myosin heavy chain composition of skeletal muscles in young rats growing under hypobaric hypoxia conditions. *Journal of Applied Physiology*: 88 (2): 479-486.

Booth, F.W., B.S. Tseng, M. Flück, and J.A. Carson. 1998. Molecular and cellular adaptation of muscle in response to physical training. *Acta Physiologica Scandinavica* 343-350.

Bottinelli, R. 2001. Functional heterogeneity of mammalian single muscle fibres: Do myosin isoforms tell the whole story? *European Journal of Physiology* 443: 6-17.

Demirel, H.A., S.K. Powers, H. Naito, M. Hughes, and J.S. Coombes. 1999. Exercise-induced alterations in skeletal muscle myosin heavy chain phenotype: Dose-response relationship. *Journal of Applied Physiology* 86: 1002-1008.

Foster, C., J.T. Daniels, and S. Seiler. 1999. Perspectives on correct approaches to training. In: *Overload, performance incompetence, and regeneration in sport,* ed. M. Lehmann, C. Foster, U. Gastmann, H. Keizer, and J.M. Steinacker, 27-41. New York, : Kluwer Academic/Plenum Press.

Foster, C., A. Synder, and R. Welsh. 1999. Monitoring of training, warm up, and performance in athletes. In: *Overload, performance incompetence, and regeneration in sport,* ed. M. Lehmann, C. Foster, U. Gastmann, H. Keizer, and J.M. Steinacker, 43-51. New York, Boston, Dordrecht, London, Moscow: Kluwer Academic/Plenum Press.

Goldspink, G. 1999. Changes in muscle mass and phenotype and the expression of autocrine and systemic growth factors by muscle in response to stretch and overload. *Journal of Anatomy* 194: 323-334.

Goldspink, G. 2000. Cloning of local growth factors involved in the determination of muscle mass. *British Journal of Sports Medicine* 34: 159-160.

Gosker, H.R., E.F. Wouters, G.J. van der Vusse, and A.M. Schols. 2000. Skeletal muscle dysfunction in chronic obstructive pulmonary disease and chronic heart failure: Underlying mechanisms and therapy perspectives. *American Journal of Clinical Nutrition* 71: 1033-1047.

Guezennec, C., Y. Gilson, and B. Serrurier. 1990. Comparative effects of hindlimb suspension and exercise on skeletal muscle myosin isozymes in rats. *European Journal of Applied Physiology* 60: 430-435.

Hayashibara, T., and T. Miyanishi. 1994. Binding of the aminoterminal region of myosin alkali 1 light chain to actin and its effect on actin-myosin interaction. *Biochemistry* 33: 12821-12827.

Hofmann, P.A., M.L. Greaser, and R.L. Moss. 1991. C-protein limits shortening velocity of rabbit skeletal muscle fibres at low levels of Ca^{2+} activation. *Journal of Physiology* (London) 439: 701-715.

Holloszy, J.O., and F.W. Booth. 1976. Biochemical adaptation to endurance exercise in muscle. *Annual Review of Physiology* 38: 273-291.

Hooper, S.B., L.T. Mackinnon, A. Howard, R.D. Gordon, and A.W. Bachmann. 1995. Markers for monitoring overtraining and recovery. *Medicine and Science in Sports and Exercise* 27: 106-112.

Hussar, Ü., T. Seene, and M. Umnova. 1992. Changes in the mast cell number and the degree of its degranulation in different skeletal muscle fibres and lymphoid organs of rat after administration of glucocorticoids. In: *Tissue biology,* 7-9. Tartu University Press. Tartu, Estonia.

Itoh, H., T. Ohkuwa, T. Yamamoto, Y. Sato, M. Miyamura, and M. Naoi. 1998. Effects of training on hydroxyl radical generation in rat tissues. *Life Sciences* 63: 1921-1929.

Kibler, W.B., and T.J. Chandler. 1998. Musculoskeletal and orthopedic considerations. In: *Overtraining in sport,* ed. R.B. Kreider, A.C. Fray, and M.L. O'Toole, 169-190. Champaign, IL: Human Kinetics.

Lambert, M.I., A. St.C. Gibson, W. Derman, and T.D. Noakes. 1999. Regeneration after ultra-endurance exercise. In: *Overload, performance incompetence, and regeneration in sport,* ed. M. Lehmann, C. Foster, U. Gastmann, H. Keizer, and J.M. Steinacker, 163-172. New York, Kluwer Academic/Plenum Press.

Lehmann, M., C. Foster, N. Netzer, W. Lormes, J.M. Steinacker, Y. Liu, A. Opitz-Gress, and U. Gastmann. 1998. Physiological responses to short- and long-term overtraining in endurance athletes. In: *Overtraining in sport,* ed. R.B. Kreider, A.C. Fry, and M.L. O'Toole, 19-46. Champaign, IL: Human Kinetics.

Lehmann, M., C. Foster, J. Steinacker, W. Lormes, A. Opitz-Gress, J. Keul, and U. Gastmann. 1997. Training and overtraining: Overview and experimental results. *Journal of Sports Medicine and Physical Fitness* 37: 7-17.

Lehmann, M., U. Gastmann, S. Bauer, Y. Liu, W. Lormes, A. Opitz-Gress, S. Reissnecker, C. Simsch, and J.M. Steinacker. 1999. Selected parameters and mechanisms of peripheral and central fatigue and regeneration in overtrained athletes. In: *Overload, performance incompetence, and regeneration in sport*, ed. M. Lehmann, C. Foster, U. Gastmann, H. Keizer, and J.M. Steinacker, 7-25. New York, Kluwer Academic/Plenum Press.

Leiber, R.L., and J. Friden. 1993. Muscle damage is not a function of muscle force but active muscle strain. *Journal of Applied Physiology* 74: 520-526.

Liu, Y., and J.M. Steinacker. 2001. Changes in skeletal muscle heat shock proteins: Pathological significance. *Frontiers in Bioscience* 6: 12-25.

MacLennan, D.H. 2000. Ca^{2+} signalling and muscle disease. *European Journal of Biochemistry* 267: 5291-5297.

Noakes, T.D. 2000. Physiological models to understand exercise fatigue and the adaptations that predict or enhance athletic performance. *Scandinavian Journal of Medicine and Science in Sports* 10: 123-145.

O'Toole, M.L. 1998. Overreaching and overtraining in endurance athletes. In: *Overtraining in sport*, ed. R.B. Kreider, A.C. Fry, and M.L. O'Toole, 3-17. Champaign, IL: Human Kinetics.

Packer, L. 1986. Oxygen radicals and antioxidants in endurance exercise. In: *Biochemical aspects of physical exercise*, 73-92. New York: Elsevier.

Pansarasa, O., G. D'Antona, M.R. Gualea, B. Marzani, M.A. Pellegrino, and F. Marzatico. 2002. "Oxidative stress": effects of mild endurance training and testosterone treatment on rat gastrocnemius muscle. *European Journal of Applied Physiology* 87: 550-555.

Pette, D. 2001. Historical perspectives: Plasticity of mammalian skeletal muscle. *Journal of Applied Physiology* 90: 1119-1124.

Pette, D., and R.S. Staron. 2000. Myosin isoforms, muscle fiber types, and transitions. *Microscopy Research and Technique* 50: 500-509.

Seene, T. 1994. Turnover of skeletal muscle contractile proteins in glucocorticoid myopathy. *Journal of Steroid Biochemistry and Molecular Biology* 50 (1/2): 1-4.

Seene, T., and K. Alev. 1991. Effect of muscular activity on the turnover rate of actin and myosin heavy and light chains in different types of skeletal muscle. *International Journal of Sports Medicine* 12: 204-207.

Seene, T., K. Alev, P. Kaasik, and A. Pehme. 2007. Changes in fast-twitch muscle oxidative capacity and myosin isoforms modulation during endurance training. *Journal of Sports Medicine and Physical Fitness* 47:124-132.

Seene, T., K. Alev, P. Kaasik, A. Pehme, and A.-M. Parring. 2005. Endurance training: Volume-dependent adaptational changes in myosin. *International Journal of Sports Medicine* 26: 815-821.

Seene, T., P. Kaasik, K. Alev, A. Pehme, and E.M. Riso. 2004. Composition and turnover of contractile proteins in volume-

overtrained skeletal muscle. *International Journal of Sports Medicine* 25: 438-445.

Seene, T., P. Kaasik, A. Pehme, K. Alev, and E.-M. Riso. 2003. The effect of glucocorticoids on the myosin heavy chain isoforms' turnover in skeletal muscle. *Journal of Steroid Biochemistry and Molecular Biology* 86: 201-206.

Seene, T., and M. Umnova. 1992. Relations between the changes in the turnover rate of contractile proteins, activation of satellite cells and ultra-structural response of neuromuscular junctions in the fast-oxidative-glycolytic muscle fibers in endurance trained rats. *Basic and Applied Myology* 2: 34-46.

Seene, T., and M. Umnova. 1996. The muscle spindle of the rat: Peculiarities of motor innervation and ultrastructure and effect of increased activity. *Scandinavian Journal of Laboratory Animal Science* 23 (1): 19-26.

Seene, T., M. Umnova, K. Alev, and A. Pehme. 1988. Effect of glucocorticoids on contractile apparatus of rat skeletal muscle. *Journal of Steroid Biochemistry* 29 (3): 313-317.

Seene, T., M. Umnova, K. Alev, R. Puhke, P. Kaasik, J. Järva, and A. Pehme. 1995. Effect of overtraining on skeletal muscle at the ultrastructural and molecular level. In: *The way to win*, ed. J. T. Viitasalo and U. Kujala, 187-189. Helsinki: Hakapaino OY.

Seene, T., M. Umnova, and P. Kaasik. 1999. The exercise myopathy. In: *Overload, performance incompetence, and regeneration in sport,* ed. M. Lehmann, C. Foster, U. Gastmann, H. Keizer, and J.M. Steinacker, 119-130. New York: Kluwer Academic/Plenum Press.

Seene, T., and A. Viru. 1982. The catabolic effect of glucocorticoids on different types of skeletal muscle fibers and its dependence upon muscle activity in interaction with anabolic steroids. *Journal of Steroid Biochemistry* 16: 349-352.

Steinacker, J.M., M. Kellmann, B.O. Böhn, Y. Liu, A. Opitz-Gress, K.W. Kallus, M. Lehmann, D. Altenburg, and W. Lormes. 1999. Clinical findings and parameters of stress and regeneration in rowers before world championships. In: *Overload, performance incompetence, and regeneration in sport,* ed. M. Lehmann, C. Foster, U. Gastmann, H. Keizer, and J.M. Steinacker, 71-80. New York: Kluwer Academic/Plenum Press.

Steinacker, J.M., and Y. Liu. 2002. Stress proteins and applied exercise physiology. In: *Exercise and stress response: the role of stress proteins*, ed. M. Locke and E.G. Noble, 197-216. Boca Raton, FL: CRC Press.

Thayer, R., J. Collins, E.G. Noble, and A.W. Taylor. 2000. A decade of aerobic endurance training: Histological evidence for fibre type transformation. *Journal of Sports Medicine and Physical Fitness* 40: 284-289.

Umnova, M.M., and T.P. Seene. 1991. The effect of increased functional load on the activation of satellite cells in the skeletal muscle of adult rats. *International Journal of Sports Medicine* 12: 501-504.

Venditti, P., and S. Di Meo. 1997. Effects of training on antioxidant capacity, tissue damage and endurance of adult male rats. *International Journal of Sports Medicine* 18 (7): 497-502.

Vescovo, G., R. Zennaro, and M. Sandri. 1998. Apoptosis of skeletal muscle myofibers and interstitial cells in experimental heart failure. *Journal of Medicine and Cell Cardiology* 30: 2449-2459.

Wahrmann, J.P., R. Winand, and M. Rieu. 2001. Plasticity of skeletal myosin in endurance-trained rats (I): A quantitative study. *European Journal of Applied Physiology* 84: 367-372.

Chapter 15

Baumeister, F.A., M. Gross, D.R. Wagner, D. Pongratz, and R. Eife. 1993. Myoadenylate deaminase deficiency with severe rhabdomyolysis. *European Journal of Pediatrics* 152(6): 513-515.

Braseth, N.R., E.J. Allison, Jr., and J.E. Gough. 2001. Exertional rhabdomyolysis in a body builder abusing anabolic androgenic steroids. *European Journal of Emergency Medicine* 8(2): 155-157.

Brown, T.P. 2004. Exertional rhabdomyolysis. *Physician and Sportsmedicine* 32(4), online.

Centers for Disease Control. 1990. Exertional rhabdomyolysis and acute renal impairment—New York City and Massachusetts, 1988. *Morbidity and Mortality Weekly Report* 39(42): 751-756.

Chen, T.C. 2006. Variability in muscle damage after eccentric exercise and the repeated bout effect. *Research Quarterly for Exercise and Sport* 77(3): 362-371.

Clarkson, P.M. 1993. Worst case scenarios: Exertional rhabdomyolysis and acute renal failure. *Sports Science Exchange* 4(41): 1-5.

Clarkson, P.M., J.M. Devaney, H. Gordish-Dressman, P.D. Thompson, M.J. Hubal, M. Urso, T.B. Price, T.J. Angelopoulos, P.M. Gordon, N.M. Moyna, L.S. Pescatello, P.S. Visich, R.F. Zoeller, R.L. Seip, and E.P. Hoffman. 2005. ACTN3 genotype is associated with increases in muscle strength and response to resistance training in women. *Journal of Applied Physiology*. 99(1):154-163, 2005.

Clarkson, P.M., E.P. Hoffman, E. Zambraski, H. Gordish-Dressman, A. Kearns, M. Hubal, B. Harmon, and J. Devaney. 2005. ACTN3 and MLCK genotype associations with exertional muscle damage. *Journal of Applied Physiology*. 99(2):564-569, 2005.

Clarkson, P.M., A.K. Kearns, P. Rouzier, R. Rubin, and P.D. Thompson. 2005. Serum creatine kinase levels and renal function measures in exertional muscle damage. *Medicine and Science in Sports and Exercise*. 38(4):623-7, 2006.

Clarkson, P.M., K. Nosaka, and B. Braun. 1992. Muscle function after exercise-induced muscle damage and rapid adaptation. *Medicine and Science in Sports and Exercise* 24(5): 512-520.

Davis, M., R. Brown, A. Dickson, H. Horton, D. James, N. Laing, R. Marston, M. Norgate, D. Perlman, N. Pollock, and K. Stowell. 2002. Malignant hyperthermia associated with exercise-induced rhabdomyolysis or congenital abnormalities and a novel RYR1 mutation in New Zealand and Australian pedigrees. *British Journal of Anaesthesiology* 88(4): 508-515.

Demos, M.A., E.L. Gitin, and L.J. Kagen. 1974. Exercise myoglobinemia and acute exertional rhabdomyolysis. *Archives of Internal Medicine* 134(4): 669-673.

Figarella-Branger, D., A.M. Baeta Machado, G.A. Putzu, P. Malzac, M.A. Voelckel, and J.F. Pellissier. 1997. Exertional rhabdomyolysis and exercise intolerance revealing dystrophinopathies. *Acta Neuropathology* (Berlin) 94(1): 48-53.

Grange, R.W., C.R. Cory, R. Vandenboom, and M.E. Houston. 1995. Myosin phosphorylation augments force-displacement and force-velocity relationships of mouse fast muscle. *American Journal of Physiology* 269(3 Pt 1): C713-724.

Greenberg, J., and L. Arneson. 1967. Exertional rhabdomyolysis with myoglobinuria in a large group of military trainees. *Neurology* 17(3): 216-222.

Gulbin, J.P., and P.T. Gaffney. 2002. Identical twins are discordant for markers of eccentric exercise-induced muscle damage. *International Journal of Sports Medicine* 23(7): 471-476.

Hackl, W., M. Winkler, W. Mauritz, P. Sporn, and K. Steinbereithner. 1991. Muscle biopsy for diagnosis of malignant hyperthermia susceptibility in two patients with severe exercise-induced myolysis. *British Journal of Anaesthesiology* 66(1): 138-140.

Hassanein, T., J.A. Perper, L. Tepperman, T.E. Starzl, and D.H. Van Thiel. 1991. Liver failure occurring as a component of exertional heatstroke. *Gastroenterology* 100(5 Pt 1): 1442-1447.

Hoffman, E.P., and G.A. Nader. 2004. Balancing muscle hypertrophy and atrophy. *Nature Medicine* 10(6): 584-585.

Kamber, M., N. Baume, M. Saugy, and L. Rivier. 2001. Nutritional supplements as a source for positive doping cases? *International Journal of Sport Nutrition and Exercise Metabolism* 11(2): 258-263.

Katzir, Z., B. Hochman, A. Biro, D.I. Rubinger, D. Feigel, J. Silver, M.M. Friedlaender, M.M. Popovtzer, and S. Smetana. 1996. Carnitine palmitoyltransferase deficiency: An underdiagnosed condition? *American Journal of Nephrology* 16(2): 162-166.

Knochel, J.P. 1982. Rhabdomyolysis and myoglobinuria. *Annual Review of Medicine* 33: 435-443.

Knochel, J.P. 1990. Catastrophic medical events with exhaustive exercise: "White collar rhabdomyolysis." *Kidney International* 38(4): 709-719.

Knochel, J.P. 1992. Hypophosphatemia and rhabdomyolysis. *American Journal of Medicine* 92(5): 455-457.

Knochel, J.P. 2000. Hemangioma steal syndrome: Another cause of exertional rhabdomyolysis. *American Journal of Medicine* 108(7): 594-595.

Kojima, Y., S. Oku, K. Takahashi, and K. Mukaida. 1997. Susceptibility to malignant hyperthermia manifested as delayed return of increased serum creatine kinase activity and episodic rhabdomyolysis after exercise. *Anesthesiology* 87(6): 1565-1567.

Legros, P., P. Jehenson, J.P. Gascard, and G. Kozak-Reiss. 1992. Long-term relationship between acute rhabdomyolysis and abnormal high-energy phosphate metabolism potentiated by ischemic exercise. *Medicine and Science in Sports and Exercise* 24(3): 298-302.

Lieber, R.L., and J. Friden. 1993. Muscle damage is not a function of muscle force but active muscle strain. *Journal of Applied Physiology* 74(2): 520-526.

Mantz, J., C. Hindelang, J.M. Mantz, and M.E. Stoeckel. 1992. Vascular and myofibrillar lesions in acute myoglobinuria associated with carnitine-palmityl-transferase deficiency. *Virchows Archiv* 421(1): 57-64.

Marinella, M.A. 1998. Exertional rhabdomyolysis after recent coxsackie B virus infection. *Southern Medical Journal* 91(11): 1057-1059.

Milne, C.J. 1988. Rhabdomyolysis, myoglobinuria and exercise. *Sports Medicine* 6(2): 93-106.

Mongini, T., C. Doriguzzi, I. Bosone, L. Chiado-Piat, E.P. Hoffman, and L. Palmucci. 2002. Alpha-sarcoglycan deficiency featuring exercise intolerance and myoglobinuria. *Neuropediatrics* 33(2): 109-111.

Newham, D.J., D.A. Jones, and R.H. Edwards. 1983. Large delayed plasma creatine kinase changes after stepping exercise. *Muscle and Nerve* 6(5): 380-385.

Nosaka, K., and P.M. Clarkson. 1996a. Changes in indicators of inflammation after eccentric exercise of the elbow flexors. *Medicine and Science in Sports and Exercise* 28(8): 953-961.

Nosaka, K., and P.M. Clarkson. 1996b. Variability in serum creatine kinase response after eccentric exercise of the elbow flexors. *International Journal of Sports Medicine* 17(2): 120-127.

Nosaka, K., P.M. Clarkson, and F.S. Apple. 1992. Time course of serum protein changes after strenuous exercise of the forearm flexors. *Journal of Laboratory and Clinical Medicine* 119(2): 183-188.

Olerud, J.E., L.D. Homer, and H.W. Carroll. 1976. Incidence of acute exertional rhabdomyolysis. Serum myoglobin and enzyme levels as indicators of muscle injury. *Archives of Internal Medicine* 136(6): 692-697.

Poels, P.J., R.A. Wevers, J.P. Braakhekke, A.A. Benders, J.H. Veerkamp, and E.M. Joosten. 1993. Exertional rhabdomyolysis in a patient with calcium adenosine triphosphatase deficiency. *Journal of Neurological and Neurosurgical Psychiatry* 56(7): 823-826.

Pretzlaff, R.K. 2002. Death of an adolescent athlete with sickle cell trait caused by exertional heat stroke. *Pediatric Critical Care Medicine* 3(3): 308-310.

Rawson, E.S., B. Gunn, and P.M. Clarkson. 2001. The effects of creatine supplementation on exercise-induced muscle damage. *Journal of Strength and Conditioning Research* 15(2): 178-184.

Sayers, S.P., and P.M. Clarkson. 2001. Force recovery after eccentric exercise in males and females. *European Journal of Applied Physiology* 84(1-2): 122-126.

Sayers, S.P., P. Clarkson, and J.J. Patel. 2002. Metabolic response to light exercise after exercise-induced rhabdomyolysis. *European Journal Applied Physiology* 86(3): 280-282.

Sayers, S.P., P.M. Clarkson, P.A. Rouzier, and G. Kamen. 1999. Adverse events associated with eccentric exercise protocols: Six case studies. *Medicine and Science in Sports and Exercise* 31(12): 1697-1702.

Schulze, V.E., Jr. 1982. Rhabdomyolysis as a cause of acute renal failure. *Postgraduate Medicine* 72(6): 145-147, 150-148.

Shumate, J.B., M.H. Brooke, J.E. Carroll, and J.E. Davis. 1979. Increased serum creatine kinase after exercise: a sex-linked phenomenon. *Neurology* 29(6): 902-904.

Springer, B.L., and P.M. Clarkson. 2003. Two cases of exertional rhabdomyolysis precipitated by personal trainers. *Medicine and Science in Sports and Exercise* 35(9): 1499-1502.

Sweeney, H.L., B.F. Bowman, and J.T. Stull. 1993. Myosin light chain phosphorylation in vertebrate striated muscle: Regulation and function. *American Journal of Physiology* 264(5 Pt 1): C1085-1095.

Sweeney, H.L., and J.T. Stull. 1990. Alteration of cross-bridge kinetics by myosin light chain phosphorylation in rabbit skeletal muscle: Implications for regulation of actin-myosin interaction. *Proceedings of the National Academy of Science USA* 87(1): 414-418.

Szczesna, D., J. Zhao, M. Jones, G. Zhi, J. Stull, and J.D. Potter. 2002. Phosphorylation of the regulatory light chains of myosin affects Ca^{2+} sensitivity of skeletal muscle contraction. *Journal of Applied Physiology* 92(4): 1661-1670.

Thompson, P.D., N. Moyna, R. Seip, T. Price, P. Clarkson, T. Angelopoulos, P. Gordon, L. Pescatello, P. Visich, R. Zoeller, J.M. Devaney, H. Gordish, S. Bilbie, and E.P. Hoffman. 2004. Functional polymorphisms associated with human muscle size and strength. *Medicine and Science in Sports and Exercise* 36(7): 1132-1139.

Tiidus, P.M. 2001. Oestrogen and sex influence on muscle damage and inflammation: Evidence from animal models. *Current Opinions in Clinical Nutrition and Metabolic Care* 4(6): 509-513.

Tiidus, P.M. 2003. Influence of estrogen on skeletal muscle damage, inflammation, and repair. *Exercise and Sport Sciences Reviews* 31(1): 40-44.

Tiidus, P.M., D. Holden, E. Bombardier, S. Zajchowski, D. Enns, and A. Belcastro. 2001. Estrogen effect on post-exercise skeletal muscle neutrophil infiltration and calpain activity. *Canadian Journal of Physiology and Pharmacology* 79(5): 400-406.

Trimarchi, H.M., A. Muryan, J. Schropp, O. Colombo, A. Garcia, H. Pereyra, B. Sarachian, and E.A. Freixas. 2000. Focal exertional rhabdomyolysis associated with a hemangioma steal syndrome. *American Journal of Medicine* 108(7): 577-580.

Vertel, R.M., and J.P. Knochel. 1967. Acute renal failure due to heat injury. An analysis of ten cases associated with a high incidence of myoglobinuria. *American Journal of Medicine* 43(3): 435-451.

Volek, J.S., S.A. Mazzetti, W.B. Farquhar, B.R. Barnes, A.L. Gomez, and W.J. Kraemer. 2001. Physiological responses to short-term exercise in the heat after creatine loading. *Medicine and Science in Sports and Exercise* 33(7): 1101-1108.

Warren, G.L., D.A. Lowe, and R.B. Armstrong. 1999. Measurement tools used in the study of eccentric contraction-induced injury. *Sports Medicine* 27(1): 43-59.

Wirthwein, D.P., S.D. Spotswood, J.J. Barnard, and J.A. Prahlow. 2001. Death due to microvascular occlusion in sickle-cell trait following physical exertion. *Journal of Forensic Science* 46(2): 399-401.

Chapter 16

Armstrong, R.B. 1984. Mechanisms of exercise-induced delayed onset muscular soreness. A brief review. *Medicine and Science in Sports and Exercise* 16: 529-538.

Blaha, J., and I. Pondelicek. 1997. Prevention and therapy of postburn scars. *Acta Chiropidi Plastica* 39: 17-21.

Cafarelli, E., and F. Flint. 1992. The role of massage in preparation for and recovery from exercise. *Sports Medicine* 14: 1-9.

Callaghan, M.J. 1993. The role of massage in the management of the athlete: A review. *British Journal of Sports Medicine* 27: 28-33.

Cheung, K., P. Hume, and L. Maxwell. 2003. Delayed onset muscle soreness: Treatment strategies and performance factors. *Sports Medicine* 33: 145-164.

Clarkson, P.M., K. Nosaka, and B. Braun. 1992. Muscle function after exercise-induced muscle damage and rapid adaptation. *Medicine and Science in Sports and Exercise* 24: 512-520.

Clarkson, P.M., and S. Sayers. 1999. Etiology of exercise-induced muscle damage. *Canadian Journal of Applied Physiology* 24: 234-248.

Dawson, L.G., K.A. Dawson, and P.M. Tiidus. 2004. Evaluating the influence of massage on leg strength, swelling and pain following a half-marathon. *Journal of Sports Science and Medicine* 3 (YISI 1): 37-43.

Drews, T., R.B. Kreider, B. Drinkard, C.W. Cotres, C. Lester, C. Somma, L. Shall, and M. Woodhouse. 1991. Effects of post-event massage therapy on repeated ultra-endurance cycling. *International Journal of Sports Medicine* 11: 407. [abstract]

Durst, B., G. Atkinson, W. Gregson, D. French, and D. Binningsley. 2003. The effects of massage on intra muscular temperature in the vastus lateralis muscle in humans. *International Journal of Sports Medicine* 24: 395-399.

Ebel, A., and L.H. Wisham. 1952. Effect of massage on muscle temperature radiosome clearance. *Archives of Physical Medicine* 34: 31-39.

Elkins, E.C., J. Herrick, J. Grindlay, F. Mann, and R.E. Deforest. 1953. Effect of various procedures on the flow of lymph. *Archives of Physical Medicine* 33: 399-405.

Ernst, E. 1998. Does post-exercise massage treatment reduce delayed onset muscle soreness? A systematic review. *British Journal of Sports Medicine* 32: 212-214.

Ernst, E. 2004. Manual therapies for pain control: Chiropractic and massage. *Clinical Journal of Pain* 20: 8-12.

Farr, T., C. Nottle, K. Nosaka, and P. Sacco. 2002. The effects of therapeutic massage on delayed onset muscle soreness and muscle function following downhill walking. *Journal of Science and Medicine in Sport* 5: 297-306.

Galloway, S.D., and J.M. Watt. 2004. Massage provision by physiotherapists at major athletic events between 1987 and 1998. *British Journal of Sports Medicine* 38: 235-237.

Gam, A.N., S. Warming, L.H. Larsen, B. Jensen, O. Hoydalsmo, I. Allon, B. Andersen, N. Gotzche, M. Petersen, and B. Mathiesen. 1998. Treatment of myofascial trigger-points with ultrasound combined with massage and exercise—a randomized controlled study. *Pain* 77: 73-79.

Goats, G.C. 1994a. Massage—the scientific basis of an ancient art: Part 1. The techniques. *British Journal of Sports Medicine* 28: 149-152.

Goats, G.C. 1994b. Massage—the scientific basis of an ancient art: Part 2. Physiological and therapeutic effects. *British Journal of Sports Medicine* 28: 153-156.

Gregory, M.A., and M. Mars. 2005. Compressed air massage causes capillary dilation in untraumatized skeletal muscle: a morphometric and ultrastructural study. *Physiotherapy* 91: 131-137.

Gupta, S., A. Goswami, A.K. Sadhukhan, and D.N. Mathur. 1996. Comparative study of lactate removal in short term massage of extremities, active recovery and passive recovery period after supramaximal exercise sessions. *International Journal of Sports Medicine* 17: 106-110.

Hart, J.M., C.B. Swanik, and R.T. Tierney. 2005. Effects of sport massage on limb girth and discomfort associated with eccentric exercise. *Journal of Athletic Training* 40: 181-185.

Hemmings, B., M. Smith, J. Graydon, and R. Dyson. 2000. Effects of massage on physiological restoration, perceived recovery and repeated sports performance. *British Journal of Sports Medicine* 34: 109-115.

Hilbert, J.E., G.A. Sforzo, and T. Swensen. 2003. The effects of massage on delayed onset muscle soreness. *British Journal of Sports Medicine* 37: 72-75.

Hinds, T., I. McEwan, J. Perkes, E. Dawson, D. Ball, and K. George. 2004. Effects of massage on limb and skin blood flow after quadriceps exercise. *Medicine and Science in Sports and Exercise* 36: 1308-1313.

Hopper, D., M. Conneely, F. Chromiak, E. Canini, J. Berggren, and K. Briffa. 2005a. Evaluating the effect of two massage techniques on hamstring muscle length in competitive female hockey players. *Physical Therapy in Sport* 6: 137-145.

Hopper, D., S. Deacon, S. Das, A. Jain, D. Riddell, T. Hall, and K. Briffa. 2005b. Dynamic soft tissue mobilization increases hamstring flexibility in healthy male subjects. *British Journal of Sports Medicine* 39: 594-598.

Hovind, H., and S.L. Nielson. 1974. Effect of massage on blood flow in skeletal muscle. *Scandinavian Journal of Rehabilitation Medicine* 6: 74-77.

Howell, J.N., G. Chleboun, and R. Conatser. 1993. Muscle stiffness, strength loss, swelling and soreness following exercise-induced injury in humans. *Journal of Physiology* 464: 183-196.

Huard, J., Y. Li, and F. Fu. 2002. Muscle injury and repair: Current trends in research. *Journal of Bone and Joint Surgery of America* 84: 822-832.

Huguenin, L.K. 2004. Myofascial trigger points: The current evidence. *Physical Therapy in Sport* 5: 2-12.

Jonhagen, S., P. Ackermann, T. Eriksson, T. Saartok, and P.A. Renstrom. 2004. Sports massage after eccentric exercise. *American Journal of Sports Medicine* 32: 1499-1503.

Lewis, M., and M.I. Johnson. 2006. The clinical effectiveness of therapeutic massage for musculoskeletal pain: A systematic review. *Physiotherapy* 92: 146-158.

Longworth, J.C. 1982. Psychophysiological effects of slow stroke back massage on normotensive females. *Advances in Nursing Science* 4: 44-61.

McNeely, M.L., D.J. Magee, A. Lees, K. Bagnall, M. Haykowski, and J. Hanson. 2004. The addition of manual lymph drainage to compression therapy for breast cancer related lymphedema: A randomized control trial. *Breast Cancer Research and Treatment* 86: 95-106.

Mok, E., and C.P. Woo. 2004. The effect of slow-stroke back massage on anxiety and shoulder pain in elderly stroke patients. *Complementary Therapies in Nursing and Midwifery* 10: 209-216.

Monedero, J., and B. Donne. 2000. Effect of recovery interventions on lactate removal and subsequent performance. *International Journal of Sports Medicine* 21: 593-597.

Moraska, A. 2005. Sports massage, a comprehensive review. *Journal of Sports Medicine and Physical Fitness* 45: 370-380.

Morelli, M., D.E. Seaborne, and J. Sullivan. 1990. Changes in H-reflex amplitude during massage of triceps surae in healthy subjects. *Journal of Orthopaedic and Sports Physical Therapy* 12: 55-59.

Mori, H., H. Ohsawa, T. Tanaka, E. Taniwaki, G. Leisman, and K. Nishijo. 2004. Effect of massage on blood flow and muscle fatigue following isometric lumbar exercise. *Medical Science Monitor* 10: CR173-CR178.

Moyer, C.A., J. Rounds, and J.W. Hannum. 2004. A meta-analysis of massage therapy research. *Psychological Bulletin* 130: 3-18.

Patino, O., C. Novek, A. Merlo, and F. Benaim. 1999. Massage in hypertrophic scars. *Journal of Burn Care and Rehabilitation* 20: 189-193.

Proske, U., and T.J. Allen. 2005. Damage to skeletal muscle from eccentric exercise. *Exercise and Sport Sciences Reviews* 33: 98-104.

Ramey, D.W., and P.M. Tiidus. 2002. Massage therapy in horses: Assessing its effectiveness from empirical data in humans and animals. *Compendium on Continuing Education for the Practicing Veterinarian* 24: 418-423.

Robertson, A., J.M. Watt, and S. Galloway. 2004. Effects of leg massage on recovery from high intensity cycling exercise. *British Journal of Sports Medicine* 38: 173-176.

Sato, K., Y. Li, W. Foster, K. Fukushima, N. Badlani, N. Adachi, A. Usas, F. Fu, and J. Huard. 2003. Improvement in muscle healing through enhancement of muscle regeneration and prevention of fibrosis. *Muscle and Nerve* 28: 365-272.

Saxon, J.M., and E.A. Donnelly. 1995. Light concentric exercise during recovery from exercise-induced muscle damage. *International Journal of Sports Medicine* 16: 347-351.

Shoemaker, J.K., P.M. Tiidus, and R. Mader. 1997. Failure of manual massage to alter limb blood flow: Measures by Doppler ultrasound. *Medicine and Science in Sports and Exercise* 29: 610-614.

Smith, L.L., M.N. Keating, D. Holbert, D. Spratt, M.R. McCammon, S.S. Smith, and R.G. Israel. 1994. The effects of athletic massage on delayed onset muscle soreness, creatine kinase and neutrophil count: A preliminary report. *Journal of Orthopaedic and Sports Physical Therapy* 19: 93-99.

Sullivan, J., L.R. Williams, and D.E. Seaborne. 1991. Effects of massage on alpha motoneuron excitability. *Physical Therapy* 71: 555-560.

Tiidus, P.M. 1997. Manual massage and recovery of muscle function following exercise: A literature review. *Journal of Orthopaedic and Sports Physical Therapy* 25: 107-112.

Tiidus, P.M. 1999. Massage and ultrasound as therapeutic modalities in exercise-induced muscle damage. *Canadian Journal of Applied Physiology* 24: 267-278.

Tiidus, P.M. 2002. Massage and the athlete: Often used, rarely researched. *Acta Academiae Olympiquae Estoniae* 10 (2): 5-15.

Tiidus, P.M., and J.K. Shoemaker. 1995. Effleurage massage, muscle blood flow and long term post-exercise strength recovery. *International Journal of Sports Medicine* 16: 478-483.

Vickers, A. 1996. *Massage and aromatherapy: A guide for health professionals.* London: Chapman and Hall.

Wakim, K.G., G.M. Martin, J.C. Terrier, E.C. Elkins, and F.H. Krusen. 1949. The effects of massage on the circulation in normal and paralyzed extremities. *Archives of Physical Medicine* 30: 135.

Warren, G.L., C.P. Ingalls, D.A. Lowe, and R.B. Armstrong. 2001. Excitation-contraction uncoupling. Major role in contraction-induced muscle injury. *Exercise and Sport Sciences Reviews* 29: 82-87.

Warren, G.L., D.A. Lowe, and R.B. Armstrong. 1999. Measurement tools used in the study of eccentric contraction-induced injury. *Sports Medicine* 27: 43-59.

Weber, M.D., F.J. Servedio, and W.R. Woodall. 1994. The effects of three modalities on delayed onset muscle soreness. *Journal of Orthopaedic and Sports Physical Therapy* 20: 236-242.

Weerapong, P., P.A. Hume, and G.S. Kolt. 2005. The mechanisms of massage and effects on performance, muscle recovery and injury prevention. *Sports Medicine* 35: 235-256.

Zainuddin, Z., M. Newton, P. Sacco, and K. Nosaka. 2005. Effects of massage on delayed-onset muscle soreness, swelling and recovery of muscle function. *Journal of Athletic Training* 40: 174-180.

Chapter 17

Baker, K.G., V.J. Robertson, and F.A. Duck. 2001. A review of therapeutic ultrasound: Biophysical effects. *Physical Therapy* 81(7): 1351-1358.

Balmaseda, M.T., M.T. Fatehi, S.H. Koozekanani, and A.L. Lee. 1986. Ultrasound therapy: A comparative study of different coupling media. *Archives of Physical Medicine and Rehabilitation* 67: 147-150.

Belanger, A-Y. 2002. *Evidence-based guide to therapeutic physical agents*. Philadelphia: Lippincott, Williams & Wilkins.

Benson, H.A., and J.C. McElnay. 1987. A high performance liquid chromatography assay for the measurement of benzydamine hydrochloride in topical pharmaceutical preparations. *Journal of Chromatography* 394: 395-399.

Benson, H.A., J.C. McElnay, and R. Harland. 1989. Use of ultrasound to enhance percutaneous absorption of Benzydamine. *Physical Therapy* 69(2): 113-118.

Binder, A., G. Hodge, A.M. Greenwood, B.L. Hazleman, and Page Thomas DP. 1985. Is therapeutic ultrasound effective in treating soft tissue lesions? *British Medical Journal* 290: 512-514.

Bommannan, D., G.K. Menon, H. Okuyama, P.M. Elias, and R.H. Guy. 1992. Sonophoresis II. Examination of the mechanisms of ultrasound enhanced transdermal drug delivery. *Pharmaceutical Research* 9(8): 1043-1047.

Bommannan, D., H. Okuyama, P. Stauffer, and R.H. Guy. 1992. Sonophoresis I. The use of high-frequency ultrasound to enhance transdermal drug delivery. *Pharmaceutical Research* 9(4): 559-564.

Byl, N.N. 1995. The use of ultrasound as an enhancer for transcutaneous drug delivery: Phonophoresis. *Physical Therapy* 75(6): 539-353.

Byl, N., A.L. McKenzie, B. Halliday, T. Wong, and J. O'Connell. 1993. The effects of phonophoresis with corticosteroids: A controlled pilot study. *Journal of Orthopedic and Sports Physical Therapy* 18(5): 590-600.

Cameron, M., and L. Monroe. 1992. Relative transmission of ultrasound by media customarily used for phonophoresis. *Physical Therapy* 72(2): 142-148.

Cameron, M.H. 2003. *Physical agents in rehabilitation*. 2nd ed., 185-214. St. Louis: Saunders.

Ciccone, C.D., B.G. Leggin, and J. Callamaro. 1991. Effects of ultrasound and trolamine salicylate phonophoresis on delayed-onset muscle soreness. *Physical Therapy* 71: 666-678.

Craig, J.A., J. Bradley, D.M. Walsh, D. Baxter, and J.M. Allen. 1999. Delayed onset muscle soreness: Lack of effect of therapeutic ultrasound in humans. *Archives of Physical Medicine and Rehabilitation* 80: 318-323.

Davick, J.P., R.K. Martin, and J.P. Albright. 1988. Distribution and deposition of tritiated cortisol using phonophoresis. *Physical Therapy* 68: 1672-1675.

Dellagatto, E.M., and E.C. Thompson. 1999. The effect of an ester of nicotinic acid and ultrasound on cutaneous circulation. Platform presentation at APTA Combined Sections Meeting, Seattle, February.

Draper, D.O. 2002. Don't disregard ultrasound yet—the jury is still out. *Physical Therapy* 82(2): 190-191.

Draper, D.O., J.C. Castel, and D. Castel. 1995. Rate of temperature increase in human tissue during 1 and 3 MHz continuous ultrasound. *Journal of Orthopedic and Sports Physical Therapy* 22(4): 142-150.

Draper, D.O., S.T. Harris, S. Schulthies, E. Durrant, K. Knight, and M. Ricard. 1998. Hot-pack and 1-MHz ultrasound treatments have an additive effect on muscle temperature increase. *Journal of Athletic Training* 33: 21-24.

Draper, D.O., C. Hatheway, and D. Fowler. 1991. Methods of applying underwater ultrasound: Science versus folklore. *Journal of Athletic Training* 26: 152-154.

Draper, D.O., C. Mahaffey, D.A. Kaiser, D.L. Eggett, J. Jarman. 2007. Therapeutic ultrasound softens trigger points in upper trapezius muscles. *Journal of Athletic Training* 42(2-supplement): S-40.

Draper, D.O., and W.E. Prentice. 1998. Therapeutic ultrasound. In *Therapeutic modalities for allied health professionals,* ed. W.E. Prentice, 263-298. New York: McGraw-Hill.

Draper, D.O., and M.D. Ricard. 1995. Rate of temperature decay in human muscle following 3 MHz ultrasound: The stretching window revealed. *Journal of Athletic Training* 30: 304-307.

Draper, D.O., S. Schulthies, P. Sorvisto, and A-M. Hautala. 1995. Temperature changes in deep muscle of humans during ice and ultrasound therapies: an in vivo study. *Journal of Orthopedic and Sports Physical Therapy* 21(3): 153-157.

Draper, D.O., and S. Sunderland. 1993. Examination of the law of Grotthus-Draper: Does ultrasound penetrate subcutaneous fat in humans? *Journal of Athletic Training* 28: 246-250.

Draper, D.O., S. Sunderland, D.T. Kirkendall, and M. Ricard. 1993. A comparison of temperature rise in human calf muscles following application of underwater and topical gel ultrasound. *Journal of Orthopedic and Sports Physical Therapy* 17(5): 247-251.

Drez, D. 1990. *Therapeutic modalities for sports injuries.* Chicago: Yearbook Medical.

Fabrizio, P.A., J.A. Schmidt, F.R. Clemente, L.A. Lankiewicz, and Z.A. Levine. 1996. Acute effects of therapeutic ultrasound delivered at varying parameters on the blood flow velocity in a muscular distribution artery. *Journal of Orthopedic and Sports Physical Therapy* 24: 294-302.

Fellinger, K., and J. Schmid. 1954. *Klinik und Therapie des chronischen Gelenkheumatismum,* 549-522. Vienna: Maudrich.

Fisher, B.D., S.M. Hiller, and S.G. Rennie. 2003. A comparison of continuous ultrasound and pulsed ultrasound on soft tissue injury markers in the rat. *Journal of Physical Therapy Science* 15: 65-70.

Forrest, G., and K. Rosen. 1992. Ultrasound treatments in degassed water. *Journal of Sport Rehabilitation* 1: 284-289.

Fyfe, M.C., and L.A. Chahl. 1982. Mast cell degranulation: A possible mechanism of action of therapeutic ultrasound. *Ultrasound in Medicine and Biology* 8 (Suppl. 1): 62.

Gatto, J., I.F. Kimura, D.T. Gulick, C. Mattacola, M.R. Sitler, and Z. Kendrick. 1999. Effects of beam nonuniformity ratio of three ultrasound machines on tissue phantom temperature. Poster presentation at NATA Annual Conference, June. (Abstract published in 1999, April-June, *NATA Journal* 34(2): S-69.)

Gersten, J.W. 1958. Effect of metallic objects on temperature rise produced in tissue by ultrasound. *American Journal of Physical Medicine* 37(2): 75-82.

Griffin, J., and T. Karselis. 1982. *Physical agents for physical therapists.* Springfield, IL: Charles C Thomas.

Gulick, D.T., J. Barsky, M. Bersheim, K. Katz, and M. Lescallette. 2001. Effect of ultrasound on pain associated with myofascial trigger points. Platform presentation at APTA Combined Sections Meeting, February. (Abstract published in 2001, January, *Journal of Orthopedic and Sports Physical Therapy).*

Gulick, D.T., N. Ingram, T. Krammes, and C. Wilds. 2004. Comparison of tissue heating using 3 MHz ultrasound with methyl nicotinate versus aquasonic gel. Presented at NATA Annual Meeting and Clinical Symposium, Baltimore, June

Gulick, D.T., J.J. Nevulis, J. Fagnani, M. Long, and K. Morris. 2006. Ultrasound treatment may not be a contraindication for joint arthroplasty. *Orthopedic Physical Therapy Practice* 18(4): 20-22.

Harvey, W., M. Dyson, J.B. Pond, and R. Graham. 1975. The stimulation of protein synthesis in human fibroblasts by therapeutic ultrasound. *Rheumatologic Rehabilitation* 14: 237.

Hasson, S., W. Mundorf, J. Barnes, J. Williams, and M. Fujii. 1990. Effects of pulsed ultrasound versus placebo on muscle soreness perception and muscular performance. *Scandinavian Journal of Rehabilitative Medicine* 22: 199-205.

Hecox, B., T.A. Mehreteab, J. Weisberg, and J. Sanko. 2006. *Integrating physical agents in rehabilitation.* 2nd ed. Upper Saddle River, NJ: Prentice-Hall.

Henley, E.J. 1990. Iontophoresis and phonophoresis transcutaneous drug delivery. Presented at APTA Annual Conference, Anaheim, CA, June 24.

Holland, C.K., C.X. Deng, R.E. Apfel J.L. Alderman, L.A. Fernandez, and K.J. Taylor. 1996. Direct evidence of cavitation in vivo from diagnostic ultrasound. *Ultrasound Medicine and Biology.* 22: 917-925.

Kimura, I.F., D.T. Gulick, J. Shelly, and M.C. Ziskin. 1998. Effects of two ultrasound machines and angle of application on the temperature of tissue mimicking material. *Journal of Orthopedics and Sports Physical Therapy* 27: 27-31.

Klaiman, M.D., J.A. Shrader, J.V. Danoff, J.E. Hicks, W.J. Pesce, and J. Ferland. 1998. Phonophoresis versus ultrasound in the treatment of common musculoskeletal conditions. *Medicine and Science in Sport and Exercise* 30(9): 1349-1355.

Kleinkort, J.A., and A.F. Wood. 1975. Phonophoresis with 1% versus 10% hydrocortisone. *Physical Therapy* 55: 1320-1324.

Klucinec, B. 1996. The effectiveness of the aquaflex gel pad in the transmission of acoustic energy. *Journal of Athletic Training* 31: 313-317.

Klucinec, B., M. Scheidler, C. Denegar, E. Domholdt, and S. Burgess. 2000. Transmissivity of coupling agents used to deliver ultrasound through indirect methods. *Journal of Orthopedic and Sports Physical Therapy* 30(5): 263-269.

Kramer, J.F. 1984. Ultrasound: Evaluation of its mechanical and thermal effects. *Archives of Physical Medicine and Rehabilitation* 65(5): 223-227.

Lehmann, J., A. Masock, C. Warren, and J.N. Koblanski. 1970. Effects of therapeutic temperatures on tendon extensibility. *Archives of Physical Medicine and Rehabilitation* 51: 481-487.

Lehmann, J., J.B. Stonebridge, B.J. DeLateur, C.G. Warren, and E. Halar. 1978. Temperatures in human thighs after hot pack treatment followed by ultrasound. *Archives of Physical Medicine and Rehabilitation* 59: 472-475.

Lentell, G., T. Hetherington, and J. Eagan. 1992. The use of thermal agents to influence the effectiveness of low-load prolonged stretch. *Journal of Orthopedic and Sports Physical Therapy* 16(5): 200-207.

Majlesi, J., and H. Unalan. 2004. High-power pain threshold ultrasound technique in the treatment of active myofascial trigger points: A randomized, double-blind, case-control study. *Archives of Physical Medicine and Rehabilitation* 85: 833-836.

McElnay, J.C., T.A. Kennedy, and R. Harland. 1987. The influence of ultrasound on the percutaneous absorption of fluocinolone acetonide. *International Journal of Pharmaceutics* 40: 105-110.

McLeod, D.R., and S.B. Fowlow. 1989. Multiple malformations and exposure to therapeutic ultrasound during organogenesis. *American Journal of Medical Genetics* 34: 317-319.

McNeill, S.C., R.O. Potts, and M.L. Francoer. 1992. Local enhanced topical drug delivery of drugs: Does it truly exist? *Homological Research* 9: 1422-1427.

Menon, G.K., D.B. Bommannan, and P.M. Elias. 1994. High frequency sonophoresis: Permeation pathways and structural basis for enhanced permeability. *Skin Pharmacology* 7: 130-139.

Merrick, M.A., M.R. Mihalyov, J.L. Roethemeier, M.L. Cordova, and C.D. Ingersoll. 2002. A comparison of intramuscular temperatures during ultrasound treatments with coupling gel or gel pads. *Journal of Orthopedic and Sports Physical Therapy* 32(5): 216-220.

Michlovitz, S. 1996. *Thermal agents in rehabilitation,* 141-173. Philadelphia: Davis.

Moll, M.J. 1979. A new approach to pain: Lidocaine and Decadron with ultrasound. *USAF Medical Service Digest* 30: 8-11.

Mortimer, A.J., and M. Dyson. 1988. The effect of therapeutic ultrasound on calcium uptake by fibroblasts. *Ultrasound in Medicine and Biology* 14(6): 499-506.

Penderghest, C., I.F. Kimura, and D.T. Gulick. 1998. Double blind clinical efficacy study of pulsed phonophoresis on perceived pain associated with symptomatic tendonitis. *Journal of Sport Rehabilitation* 7(1): 9-19.

Plaskett, C., P.M. Tiidus, and L. Livingston. 1999. Ultrasound treatment does not affect postexercise muscle strength recovery or soreness. *Journal of Sport Rehabilitation* 8: 1-9.

Ramirez, A., J.A. Schwane, C. McFarland, and B. Starcher. 1997. The effect of ultrasound on collagen synthesis and fibroblast proliferation in vitro. *Medicine and Science in Sports and Exercise* 29: 326-332.

Rantanen, J., O. Thorsson, P. Wollmer, T. Hurme, and H. Kallimo. 1999. Effects of therapeutic ultrasound on the regeneration of skeletal myofibers after experimental muscle injury. *American Journal of Sports Medicine* 27: 54-59.

Rennie, G.A., and S.L. Michlovitz. 1996. Biophysical principles of heating and superficial heating agents. In *Thermal agents in rehabilitation,* ed. S.L. Michlovitz. Philadelphia: Davis.

Reid, D.C., and G.E. Cummings. 1977. Efficiency of ultrasound coupling agents. *Physiotherapy* 63(8): 255-257.

Rimington, S., D.O. Draper, E. Durrant, and G. Fellingham. 1994. A common practice scrutinized: Is a 15-minute ice pack treatment prior to ultrasound application wise? Presented at NATA Conference, Dallas, June 11.

Robertson, V.J. 2002. Dosage and treatment response in randomized clinical trials of therapeutic ultrasound. *Physical Therapy in Sport* 3: 124-133.

Robertson, V.J., and K.G. Baker. 2001. A review of therapeutic ultrasound: Effectiveness studies. *Physical Therapy* 81(7): 1339-1350.

Rose, S., D.O. Draper, S.S. Schulthies, and E. Durrant. 1996. The stretching window part two: Rate of thermal decay in deep muscle following 1 MHz ultrasound. *Journal of Athletic Training* 31: 139-143.

Sicard-Rosenbaum, L., D. Lord, J.V. Danoff, A.K. Thom, and M.A. Eckhaus. 1995. Effects of continuous therapeutic ultrasound on growth and metastasis of subcutaneous murine tumors. *Physical Therapy* 75: 3-13.

Smith, W., F. Winn, and R. Parette. 1986. Comparative study using four modalities in shin splints treatments. *Journal of Orthopedic and Sports Physical Therapy* 8: 77-80.

Sparrow, K.J. 2005. Therapeutic ultrasound. In *Modalities for therapeutic intervention,* ed. S.L. Michlovitz and T. Nolan. Philadelphia: Davis.

Stay, J.C., M.D. Ricard, D.O. Draper, S.S. Schulthies, and E. Durrant. 1998. Pulsed ultrasound fails to diminish delayed-onset muscle soreness symptoms. *Journal of Athletic Training* 33(4): 341-346.

Stewart, H.F., J.L. Absug, and G.R. Harris. 1980. Considerations in ultrasound therapy and equipment performance. *Physical Therapy* 60: 425.

Tiidus, P.M. 1999. Massage and ultrasound as therapeutic modalities in exercise-induced muscle damage. *Canadian Journal of Applied Physiology* 24: 267-278.

Tiidus, P.M., J. Cort, S.J. Woodruff, and P. Bryden. 2002. Ultrasound treatment and recovery from eccentric-exercise-induced muscle damage. *Journal of Sport Rehabilitation* 11: 305-314.

Vore, M.E., J. Kotmel, and E. Wenzler. 2006. Transmission of hydrocortisone through unembalmed cadaver skin using clinical parameters for phonophoresis. Presented at APTA Combined Sections Meeting, San Diego, February.

Wessling, K.C., D.A. DeVane, and C.R. Hylton. 1987. Effects of static stretch versus static stretch and ultrasound combined on triceps surae muscle extensibility in healthy women. *Physical Therapy* 67(5): 674-679.

Wilkin, L.D., M.A. Merrick, T.E. Kirby, and S.T. Devor. 2004. Influence of therapeutic ultrasound on skeletal muscle regeneration following blunt contusion. *International Journal of Sports Medicine* 25: 73-77.

Williams, A.R. 1983. *Ultrasound: Biological effects and potential hazards.* London: Academic Press.

Young, R., I.F. Kimura, and D.T. Gulick. 1999. Accuracy of intensity output, beam nonuniformity ratio, and effective radiating area of four therapeutic ultrasound machines. Poster presentation at NATA Annual Conference, June. (Abstract published in 1999, April-June, *NATA Journal* 34(2): S-69.)

Young, S.R., and M. Dyson. 1990. Macrophage responsiveness to therapeutic ultrasound. *Ultrasound in Medicine and Biology* 16(8): 809-816.

Chapter 18

Allen, D.L., Linderman, J.K., Roy, R.R., Bigbee, A.J., Grindeland, R.E., Mukku, V., and Edgerton, V.R. (1997) Apoptosis: A mechanism contributing to remodeling of skeletal muscle in response to hindlimb unweighting. *Am J Physiol* 273, C579-C587.

Almekinders, L.C. and Gilbert, J.A. (1986) Healing of experimental muscle strains and the effects of nonsteroidal antiinflammatory medication. *Am J Sports Med* 14, 303-308.

Armstrong, R.B., Ogilvie, R.W., and Schwane, J.A. (1983) Eccentric exercise-induced injury to rat skeletal muscle. *J Appl Physiol* 54, 80-93.

Arrington, E.D. and Miller, M.D. (1995) Skeletal muscle injuries. *Orthop Clin North Am* 26, 411-422.

Bandy, W.D. and Irion, J.M. (1994) The effect of time on static stretch on the flexibility of the hamstring muscles. *Phys Ther* 74, 845-850.

Banga, A.K., Bose, S., and Ghosh, T.K. (1999) Iontophoresis and electroporation: Comparisons and contrasts. *Int J Pharm* 179, 1-19.

Bansal, D., Miyake, K., Vogel, S.S., Groh, S., Chen, C.C., Williamson, R., McNeil, P.L., and Campbell, K.P. (2003) Defective membrane repair in dysferlin-deficient muscular dystrophy. *Nature* 423, 168-172.

Barash, I.A., Mathew, L., Ryan, A.F., Chen, J., and Lieber, R.L. (2004) Rapid muscle-specific gene expression changes after a single bout of eccentric contractions in the mouse. *Am J Physiol Cell Physiol* 286, C355-C364.

Bartleson, J.D. (2001) Low back pain. *Curr Treat Options Neurol* 3(2), 159-168.

Beiner, J.M. and Jokl, P. (2001) Muscle contusion injuries: Current treatment options. *J Am Acad Orthop Surg* 9, 227-237.

Beiner, J.M. and Jokl, P. (2002) Muscle contusion injury and myositis ossificans traumatica. *Clin Orthop* S110-S119.

Beiner, J.M., Jokl, P., Cholewicki, J., and Panjabi, M.M. (1999) The effect of anabolic steroids and corticosteroids on healing of muscle contusion injury. *Am J Sports Med* 27, 2-9.

Berg, E. (2000) Deep muscle contusion complicated by myositis ossificans (a.k.a. heterotopic bone). *Orthop Nurs* 19, 66-67.

Best, T.M. and Hunter, K.D. (2000) Muscle injury and repair. *Phys Med Rehabil Clin N Am* 11, 251-266.

Best, T.M., Loitz-Ramage, B., Corr, D.T., and Vanderby, R. (1998) Hyperbaric oxygen in the treatment of acute muscle stretch injuries. Results in an animal model. *Am J Sports Med* 26, 367-372.

Bischoff, R. (1989) Analysis of muscle regeneration using single myofibers in culture. *Med Sci Sports Exerc* 21, S164-S172.

Black, J.D., Freeman, M., and Stevens, E.D. (2002) A 2 week routine stretching programme did not prevent contraction-induced injury in mouse muscle. *J Physiol* 544, 137-147.

Black, J.D. and Stevens, E.D. (2001) Passive stretching does not protect against acute contraction-induced injury in mouse EDL muscle. *J Muscle Res Cell Motil* 22, 301-310.

Blankenbaker, D.G. and De Smet, A.A. (2004) MR imaging of muscle injuries. *Applied Radiology* 33(4), 14-26.

Bolin, D. and Goforth, M. (2004) Electric delivery. Generally well tolerated by patients, iontophoresis has many uses for the rehab clinician. *Rehab Manag* 17, 18-21, 54.

Bondesen, B.A., Mills, S.T., Kegley, K.M., and Pavlath, G.K. (2004) The COX-2 pathway is essential during early stages of skeletal muscle regeneration. *Am J Physiol Cell Physiol* 287, C475-C483.

Bonnar, B.P., Deivert, R.G., and Gould, T.E. (2004) The relationship between isometric contraction durations during hold-relax stretching and improvement of hamstring flexibility. *J Sports Med Phys Fitness* 44, 258-261.

Bonutti, P.M., Windau, J.E., Ables, B.A., and Miller, B.G. (1994) Static progressive stretch to reestablish elbow range of motion. *Clin Orthop Relat Res* 303, 128-134.

Brockett, C.L., Morgan, D.L., and Proske, U. (2001) Human hamstring muscles adapt to eccentric exercise by changing optimum length. *Med Sci Sports Exerc* 33, 783-790.

Brooks, S.V., Zerba, E., and Faulkner, J.A. (1995) Injury to muscle fibres after single stretches of passive and maximally stimulated muscles in mice. *J Physiol* 488 (Pt 2), 459-469.

Bullock, M.P., Foster, N.E., and Wright, C.C. (2005) Shoulder impingement: The effect of sitting posture on shoulder pain and range of motion. *Man Ther* 10, 28-37.

Carrino, J.A., Chandnanni, V.P., Mitchell, D.B., Choi-Chinn, K., DeBerardino, T.M., and Miller, M.D. (2000) Pectoralis major muscle and tendon tears: Diagnosis and grading using magnetic resonance imaging. *Skeletal Radiol* 29, 305-313.

Cavagna, G.A. (1977) Storage and utilization of elastic energy in skeletal muscle. *Exerc Sport Sci Rev* 5, 89-129.

Charge, S.B. and Rudnicki, M.A. (2004) Cellular and molecular regulation of muscle regeneration. *Physiol Rev* 84, 209-238.

Chazaud, B., Sonnet, C., Lafuste, P., Bassez, G., Rimaniol, A.C., Poron, F., Authier, F.J., Dreyfus, P.A., and Gherardi, R.K. (2003) Satellite cells attract monocytes and use macrophages as a support to escape apoptosis and enhance muscle growth. *J Cell Biol* 163, 1133-1143.

Chopard, A., Pons, F., and Marini, J.F. (2001) Cytoskeletal protein contents before and after hindlimb suspension in a fast and slow rat skeletal muscle. *Am J Physiol Regul Integr Comp Physiol* 280, R323-R330.

Cook, H.A., Morales, M., La Rosa, E.M., Dean, J., Donnelly, M.K., McHugh, P., Otradovec, A., Wright, K.S., Kula, T., and Tepper, S.H. (1994) Effects of electrical stimulation on lymphatic flow and limb volume in the rat. *Phys Ther* 74, 1040-1046.

Cosgrove, K.A., Alon, G., Bell, S.F., Fischer, S.R., Fowler, N.R., Jones, T.L., Myaing, J.C., Crouse, T.M., and Seaman, L.J. (1992) The electrical effect of two commonly used clinical stimulators on traumatic edema in rats. *Phys Ther* 72, 227-233.

Costello, C.T. and Jeske, A.H. (1995) Iontophoresis: Applications in transdermal medication delivery. *Phys Ther* 75, 554-563.

Crisco, J.J., Hentel, K.D., Jackson, W.O., Goehner, K., and Jokl, P. (1996) Maximal contraction lessens impact response in a muscle contusion model. *J Biomech* 29, 1291-1296.

Croisier, J.L., Forthomme, B., Namurois, M.H., Vanderthommen, M., and Crielaard, J.M. (2002) Hamstring muscle strain recurrence and strength performance disorders. *Am J Sports Med* 30, 199-203.

Curl, W.W., Smith, B.P., Marr, A., Rosencrance, E., Holden, M., and Smith, T.L. (1997) The effect of contusion and cryotherapy on skeletal muscle microcirculation. *J Sports Med Phys Fitness* 37, 279-286.

Cushner, F.D. and Morwessel, R.M. (1992) Myositis ossificans traumatica. *Orthop Rev* 21, 1319-1326.

Decoster, L.C., Scanlon, R.L., Horn, K.D., and Cleland, J. (2004) Standing and supine hamstring stretching are equally effective. *J Athl Train* 39, 330-334.

De Deyne, P.G. (2001) Application of passive stretch and its implications for muscle fibers. *Phys Ther* 81, 819-827.

Dolan, M.G., Mychaskiw, A.M., and Mendel, F.C. (2003) Cool-water immersion and high-voltage electric stimulation curb edema formation in rats. *J Athl Train* 38, 225-230.

Donnelly, A.E., Clarkson, P.M., and Maughan, R.J. (1992) Exercise-induced muscle damage: Effects of light exercise on damaged muscle. *Eur J Appl Physiol Occup Physiol* 64, 350-353.

Dudkiewicz, I., Salai, M., and Chechik, A. (2001) A young athlete with myositis ossificans of the neck presenting as a soft-tissue tumour. *Arch Orthop Trauma Surg* 121, 234-237.

Fish, D.R., Mendel, F.C., Schultz, A.M., and Gottstein-Yerke, L.M. (1991) Effect of anodal high voltage pulsed current on edema formation in frog hind limbs. *Phys Ther* 71, 724-730.

Friden, J. and Lieber, R.L. (2001) Eccentric exercise-induced injuries to contractile and cytoskeletal muscle fibre components. *Acta Physiol Scand* 171, 321-326.

Friden, J., Sjostrom, M., and Ekblom, B. (1981) A morphological study of delayed muscle soreness. *Experientia* 37, 506-507.

Fu, F.H., Cen, H.W., and Eston, R.G. (1997) The effects of cryotherapy on muscle damage in rats subjected to endurance training. *Scand J Med Sci Sports* 7, 358-362.

Garrett, W.E., Jr. (1996) Muscle strain injuries. *Am J Sports Med* 24, S2-S8.

Germain, G., Delaney, J., Moore, G., Lee, P., Lacroix, V., and Montgomery, D. (2003) Effect of hyperbaric oxygen therapy on exercise-induced muscle soreness. *Undersea Hyperb Med* 30, 135-145.

Gierer, P., Mittlmeier, T., Bordel, R., Schaser, K.D., Gradl, G., and Vollmar, B. (2005) Selective cyclooxygenase-2 inhibition reverses microcirculatory and inflammatory sequelae of closed soft-tissue trauma in an animal model. *J Bone Joint Surg Am* 87-A, 153-160.

Glass, J.B. (1979) Acute lumbar strain: Clinical signs and prognosis. *Practitioner* 222, 821-825.

Goddard, G., Karibe, H., McNeill, C., and Villafuerte, E. (2002) Acupuncture and sham acupuncture reduce muscle pain in myofascial pain patients. *J Orofac Pain* 16, 71-76.

Gokoglu, F., Fndkoglu, G., Yorgancoglu, Z.R., Okumus, M., Ceceli, E., and Kocaoglu, S. (2005) Evaluation of iontophoresis and local corticosteroid injection in the treatment of carpal tunnel syndrome. *Am J Phys Med Rehabil* 84, 92-96.

Greenfield, B., Catlin, P.A., Coats, P.W., Green, E., McDonald, J.J., and North, C. (1995) Posture in patients with shoulder overuse injuries and healthy individuals. *J Orthop Sports Phys Ther* 21, 287-295.

Gregorevic, P., Lynch, G.S., and Williams, D.A. (2000) Hyperbaric oxygen improves contractile function of regenerating rat skeletal muscle after myotoxic injury. *J Appl Physiol* 89, 1477-1482.

Grounds, M.D. (1991) Towards understanding skeletal muscle regeneration. *Pathol Res Pract* 187, 1-22.

Halbertsma, J.P., van Bolhuis, A.I., and Goeken, L.N. (1996) Sport stretching: Effect on passive muscle stiffness of short hamstrings. *Arch Phys Med Rehabil* 77, 688-692.

Hamer, P.W., McGeachie, J.M., Davies, M.J., and Grounds, M.D. (2002) Evans Blue Dye as an in vivo marker of myofibre damage: Optimising parameters for detecting initial myofibre membrane permeability. *J Anat* 200, 69-79.

Harrison, B.C., Robinson, D., Davison, B.J., Foley, B., Seda, E., and Byrnes, W.C. (2001) Treatment of exercise-induced muscle injury via hyperbaric oxygen therapy. *Med Sci Sports Exerc* 33, 36-42.

Hartig, D.E. and Henderson, J.M. (1999) Increasing hamstring flexibility decreases lower extremity overuse injuries in military basic trainees. *Am J Sports Med* 27, 173-176.

Hasselman, C.T., Best, T.M., Seaber, A.V., and Garrett, W.E., Jr. (1995) A threshold and continuum of injury during active stretch of rabbit skeletal muscle. *Am J Sports Med* 23, 65-73.

Hasson, S.M., Wible, C.L., Reich, M., Barnes, W.S., and Williams, J.H. (1992) Dexamethasone iontophoresis: Effect on delayed muscle soreness and muscle function. *Can J Sport Sci* 17, 8-13.

Herbert, R.D. and Gabriel, M. (2002) Effects of stretching before and after exercising on muscle soreness and risk of injury: Systematic review. *BMJ* 325, 468.

Hierton, C. (1983) Regional blood flow in experimental myositis ossificans. A microsphere study in conscious rabbits. *Acta Orthop Scand* 54, 58-63.

High, D.M., Howley, E.T., and Franks, B.D. (1989) The effects of static stretching and warm-up on prevention of delayed-onset muscle soreness. *Res Q Exerc Sport* 60, 357-361.

Ho, K.W., Roy, R.R., Tweedle, C.D., Heusner, W.W., Van Huss, W.D., and Carrow, R.E. (1980) Skeletal muscle fiber splitting with weight-lifting exercise in rats. *Am J Anat* 157, 433-440.

Howatson, G. and Van Someren, K.A. (2003) Ice massage. Effects on exercise-induced muscle damage. *J Sports Med Phys Fitness* 43, 500-505.

Huard, J., Li, Y., and Fu, F.H. (2002) Muscle injuries and repair: Current trends in research. *J Bone Joint Surg Am* 84-A, 822-832.

Hubbard, T.J. and Denegar, C.R. (2004) Does cryotherapy improve outcomes with soft tissue injury? *J Athl Train* 39, 278-279.

Hultman, E., Sjoholm, H., Jaderholm-Ek, I., and Krynicki, J. (1983) Evaluation of methods for electrical stimulation of human skeletal muscle in situ. *Pflugers Arch* 398, 139-141.

Hunter, K.D. and Faulkner, J.A. (1997) Pliometric contraction-induced injury of mouse skeletal muscle: Effect of initial length. *J Appl Physiol* 82, 278-283.

Irintchev, A., Langer, M., Zweyer, M., Theisen, R., and Wernig, A. (1997) Functional improvement of damaged adult mouse muscle by implantation of primary myoblasts. *J Physiol* 500 (Pt 3), 775-785.

Jackson, D.W. and Feagin, J.A. (1973) Quadriceps contusions in young athletes. Relation of severity of injury to treatment and prognosis. *J Bone Joint Surg Am* 55, 95-105.

Jarvinen, M. (1977) Immobilization effect on the tensile properties of striated muscle: An experimental study in the rat. *Arch Phys Med Rehabil* 58, 123-127.

Jarvinen, M.J., Einola, S.A., and Virtanen, E.O. (1992) Effect of the position of immobilization upon the tensile properties of the rat gastrocnemius muscle. *Arch Phys Med Rehabil* 73, 253-257.

Jarvinen, M.J. and Lehto, M.U. (1993) The effects of early mobilisation and immobilisation on the healing process following muscle injuries. *Sports Med* 15, 78-89.

Jayaraman, R.C., Reid, R.W., Foley, J.M., Prior, B.M., Dudley, G.A., Weingand, K.W., and Meyer, R.A. (2004) MRI evaluation of topical heat and static stretching as therapeutic modalities for the treatment of eccentric exercise-induced muscle damage. *Eur J Appl Physiol* 93, 30-38.

Jensen, K. and Di Fabio, R.P. (1989) Evaluation of eccentric exercise in treatment of patellar tendinitis. *Phys Ther* 69, 211-216.

Julius, A., Lees, R., Dilley, A., and Lynn, B. (2004) Shoulder posture and median nerve sliding. *BMC Musculoskelet Disord* 5, 23.

Kauhanen, S., von Boguslawsky, K., Michelsson, J.E., and Leivo, I. (1998) Satellite cell proliferation in rabbit hindlimb muscle following immobilization and remobilization: An immunohistochemical study using MIB 1 antibody. *Acta Neuropathol* (Berl) 95, 165-170.

Kernell, D., Eerbeek, O., Verhey, B.A., and Donselaar, Y. (1987) Effects of physiological amounts of high- and low-rate chronic stimulation on fast-twitch muscle of the cat hindlimb. I. Speed- and force-related properties. *J Neurophysiol* 58, 598-613.

King, J.B. (1998) Post-traumatic ectopic calcification in the muscles of athletes: A review. *Br J Sports Med* 32, 287-290.

Kirkendall, D.T. and Garrett, W.E., Jr. (2002) Clinical perspectives regarding eccentric muscle injury. *Clin Orthop Relat Res* 403, S81-S89.

Koh, T.J. and Brooks, S.V. (2001) Lengthening contractions are not required to induce protection from contraction-induced muscle injury. *Am J Physiol Regul Integr Comp Physiol* 281, R155-R161.

LaStayo, P.C., Reich, T.E., Urquhart, M., Hoppeler, H., and Lindstedt, S.L. (1999) Chronic eccentric exercise: Improvements in muscle strength can occur with little demand for oxygen. *Am J Physiol* 276, R611-R615.

LaStayo, P.C., Woolf, J.M., Lewek, M.D., Snyder-Mackler, L., Reich, T., and Lindstedt, S.L. (2003) Eccentric muscle contractions: Their contribution to injury, prevention, rehabilitation, and sport. *J Orthop Sports Phys Ther* 33, 557-571.

Lehti, T.M., Kalliokoski, R., Komulainen, J. (2007) Repeated bout effect on the cytoskeletal proteins titin, desmin, and dystrophin in rat skeletal muscle. *J Muscle Res Cell Motil* 28(1), 39-47.

Levine, W.N., Bergfeld, J.A., Tessendorf, W., and Moorman, C.T., III (2000) Intramuscular corticosteroid injection for hamstring injuries. A 13-year experience in the National Football League. *Am J Sports Med* 28, 297-300.

Lewis, J.S., Wright, C., and Green, A. (2005) Subacromial impingement syndrome: The effect of changing posture on shoulder range of movement. *J Orthop Sports Phys Ther* 35, 72-87.

Lieber, R.L. and Friden, J. (1993) Muscle damage is not a function of muscle force but active muscle strain. *J Appl Physiol* 74, 520-526.

Lieber, R.L. and Kelly, M.J. (1991) Factors influencing quadriceps femoris muscle torque using transcutaneous neuromuscular electrical stimulation. *Phys Ther* 71, 715-721.

Lieber, R.L., Thornell, L.E., and Friden, J. (1996) Muscle cytoskeletal disruption occurs within the first 15 min of cyclic eccentric contraction. *J Appl Physiol* 80, 278-284.

Lindh, M. (1979) Increase of muscle strength from isometric quadriceps exercises at different knee angles. *Scand J Rehabil Med* 11, 33-36.

Lindstedt, S.L., LaStayo, P.C., and Reich, T.E. (2001) When active muscles lengthen: Properties and consequences of eccentric contractions. *News Physiol Sci* 16, 256-261.

Lipscomb, A.B., Thomas, E.D., and Johnston, R.K. (1976) Treatment of myositis ossificans traumatica in athletes. *Am J Sports Med* 4, 111-120.

Lovering, R.M. and De Deyne, P.G. (2004) Contractile function, sarcolemma integrity, and the loss of dystrophin after skeletal muscle eccentric contraction-induced injury. *Am J Physiol Cell Physiol* 286, C230-C238.

Lovering, R.M., Hakim, M., Moorman, C.T., and De Deyne, P.G. (2005) The contribution of contractile pre-activation to loss of function after a single lengthening contraction. *J Biomech* 38, 1501-1507.

Lund, H. (2003) Stretching before or after exercising has no effect on muscle soreness or risk of injury. *Aust J Physiother* 49, 73.

Lynch, G.S. and Faulkner, J.A. (1998) Contraction-induced injury to single muscle fibers: Velocity of stretch does not influence the force deficit. *Am J Physiol* 275, C1548-C1554.

MacIntyre, D.L., Reid, W.D., Lyster, D.M., Szasz, I.J., and McKenzie, D.C. (1996) Presence of WBC, decreased strength, and delayed soreness in muscle after eccentric exercise. *J Appl Physiol* 80, 1006-1013.

Mafi, N., Lorentzon, R., and Alfredson, H. (2001) Superior short-term results with eccentric calf muscle training compared to concentric training in a randomized prospective multicenter study on patients with chronic Achilles tendinosis. *Knee Surg Sports Traumatol Arthrosc* 9, 42-47.

Maier, F. and Bornemann, A. (1999) Comparison of the muscle fiber diameter and satellite cell frequency in human muscle biopsies. *Muscle Nerve* 22, 578-583.

Mair, S.D., Seaber, A.V., Glisson, R.R., and Garrett, W.E., Jr. (1996) The role of fatigue in susceptibility to acute muscle strain injury. *Am J Sports Med* 24, 137-143.

McHugh, M.P., Connolly, D.A., Eston, R.G., and Gleim, G.W. (1999) Exercise-induced muscle damage and potential mechanisms for the repeated bout effect. *Sports Med* 27, 157-170.

McNeil, P.L. and Khakee, R. (1992) Disruptions of muscle fiber plasma membranes. Role in exercise-induced damage. *Am J Pathol* 140, 1097-1109.

McNeil, P.L. and Terasaki, M. (2001) Coping with the inevitable: How cells repair a torn surface membrane. *Nat Cell Biol* 3, E124-E129.

Melzack, R. and Wall, P.D. (1965) Pain mechanisms: A new theory. *Science* 150, 971-979.

Mendias, C.L., Tatsumi, R., and Allen, R.E. (2004) Role of cyclooxygenase-1 and -2 in satellite cell proliferation, differentiation, and fusion. *Muscle Nerve* 30, 497-500.

Menth-Chiari, W.A., Curl, W.W., Paterson-Smith, B., and Smith, T.L. (1999) [Microcirculation of striated muscle in closed soft tissue injury: Effect on tissue perfusion, inflammatory cellular response and mechanisms of cryotherapy. A study in rat by means of laser Doppler flow-measurements and intravital microscopy]. *Unfallchirurg* 102, 691-699.

Merrick, M.A., Rankin, J.M., Andres, F.A., and Hinman, C.L. (1999) A preliminary examination of cryotherapy and secondary injury in skeletal muscle. *Med Sci Sports Exerc* 31, 1516-1521.

Mishra, D.K., Friden, J., Schmitz, M.C., and Lieber, R.L. (1995) Anti-inflammatory medication after muscle injury. A treatment resulting in short-term improvement but subsequent loss of muscle function. *J Bone Joint Surg Am* 77, 1510-1519.

Moffroid, M.T. and Whipple, R.H. (1970) Specificity of speed of exercise. *Phys Ther* 50, 1692-1700.

Mohr, M., Krustrup, P., Nybo, L., Nielsen, J.J., and Bangsbo, J. (2004) Muscle temperature and sprint performance during soccer matches—beneficial effect of re-warm-up at half-time. *Scand J Med Sci Sports* 14, 156-162.

Myrer, W.J., Myrer, K.A., Measom, G.J., Fellingham, G.W., and Evers, S.L. (2001) Muscle temperature is affected by overlying adipose when cryotherapy is administered. *J Athl Train* 36, 32-36.

Nelson, A.G., Wolf, E.G., Jr., and Li, B. (1994) Influence of delayed hyperbaric oxygenation on recovery from mechanically induced damage. *Undersea Hyperb Med* 21, 185-191.

Noonan, T.J., Best, T.M., Seaber, A.V., and Garrett, W.E., Jr. (1993) Thermal effects on skeletal muscle tensile behavior. *Am J Sports Med* 21, 517-522.

Norman, A. and Dorfman, H.D. (1970) Juxtacortical circumscribed myositis ossificans: Evolution and radiographic features. *Radiology* 96, 301-306.

Nosaka, K., Sakamoto, K., Newton, M., and Sacco, P. (2001a) How long does the protective effect on eccentric exercise-induced muscle damage last? *Med Sci Sports Exerc* 33, 1490-1495.

Nosaka, K., Sakamoto, K., Newton, M., and Sacco, P. (2001b) The repeated bout effect of reduced-load eccentric exercise on elbow flexor muscle damage. *Eur J Appl Physiol* 85, 34-40.

Obremsky, W.T., Seaber, A.V., Ribbeck, B.M., and Garrett, W.E., Jr. (1994) Biomechanical and histologic assessment of a controlled muscle strain injury treated with piroxicam. *Am J Sports Med* 22, 558-561.

Otte, J.W., Merrick, M.A., Ingersoll, C.D., and Cordova, M.L. (2002) Subcutaneous adipose tissue thickness alters cooling time during cryotherapy. *Arch Phys Med Rehabil* 83, 1501-1505.

Paddon-Jones, D.J. and Quigley, B.M. (1997) Effect of cryotherapy on muscle soreness and strength following eccentric exercise. *Int J Sports Med* 18, 588-593.

Papadimitriou, J.M., Robertson, T.A., Mitchell, C.A., and Grounds, M.D. (1990) The process of new plasmalemma formation in focally injured skeletal muscle fibers. *J Struct Biol* 103, 124-134.

Pizza, F.X., Peterson, J.M., Baas, J.H., and Koh, T.J. (2005) Neutrophils contribute to muscle injury and impair its resolution after lengthening contractions in mice. *J Physiol* 562, 899-913.

Poppele, R.E. and Quick, D.C. (1981) Stretch-induced contraction of intrafusal muscle in cat muscle spindle. *J Neurosci* 1, 1069-1074.

Prausnitz, M.R., Bose, V.G., Langer, R., and Weaver, J.C. (1993) Electroporation of mammalian skin: A mechanism to enhance transdermal drug delivery. *Proc Natl Acad Sci USA* 90, 10504-10508.

Rahusen, F.T., Weinhold, P.S., and Almekinders, L.C. (2004) Nonsteroidal anti-inflammatory drugs and acetaminophen in the treatment of an acute muscle injury. *Am J Sports Med* 32, 1856-1859.

Rodenburg, J.B., Steenbeek, D., Schiereck, P., and Bar, P.R. (1994) Warm-up, stretching and massage diminish harmful effects of eccentric exercise. *Int J Sports Med* 15, 414-419.

Ryan, J.B., Wheeler, J.H., Hopkinson, W.J., Arciero, R.A., and Kolakowski, K.R. (1991) Quadriceps contusions. West Point update. *Am J Sports Med* 19, 299-304.

Sachs, F. (1997) Mechanical transduction by ion channels: How forces reach the channel. *Soc Gen Physiol Ser* 52, 209-218.

Safran, M.R., Garrett, W.E., Jr., Seaber, A.V., Glisson, R.R., and Ribbeck, B.M. (1988) The role of warmup in muscular injury prevention. *Am J Sports Med* 16, 123-129.

Sandberg, M., Lundeberg, T., Lindberg, L.G., and Gerdle, B. (2003) Effects of acupuncture on skin and muscle blood flow in healthy subjects. *Eur J Appl Physiol* 90, 114-119.

Sawyer, P.C., Uhl, T.L., Mattacola, C.G., Johnson, D.L., and Yates, J.W. (2003) Effects of moist heat on hamstring flexibility and muscle temperature. *J Strength Cond Res* 17, 285-290.

Sayers, S.P., Clarkson, P.M., and Lee, J. (2000a) Activity and immobilization after eccentric exercise: I. Recovery of muscle function. *Med Sci Sports Exerc* 32, 1587-1592.

Sayers, S.P., Clarkson, P.M., and Lee, J. (2000b) Activity and immobilization after eccentric exercise: II. Serum CK. *Med Sci Sports Exerc* 32, 1593-1597.

Schiaffino, S., Bormioli, S.P., and Aloisi, M. (1976) The fate of newly formed satellite cells during compensatory muscle hypertrophy. *Virchows Arch B Cell Pathol* 21, 113-118.

Schuldt, K., Ekholm, J., Harms-Ringdahl, K., Nemeth, G., and Arborelius, U.P. (1986) Effects of changes in sitting work posture on static neck and shoulder muscle activity. *Ergonomics* 29, 1525-1537.

Sherry, M.A. and Best, T.M. (2004) A comparison of 2 rehabilitation programs in the treatment of acute hamstring strains. *J Orthop Sports Phys Ther* 34, 116-125.

Singh, J. and Bhatia, K.S. (1996) Topical iontophoretic drug delivery: Pathways, principles, factors, and skin irritation. *Med Res Rev* 16, 285-296.

Smith, T.L., Curl, W.W., Smith, B.P., Holden, M.B., Wise, T., Marr, A., and Koman, L.A. (1993) New skeletal muscle model for the longitudinal study of alterations in microcirculation following contusion and cryotherapy. *Microsurgery* 14, 487-493.

Snyder-Mackler, L., Delitto, A., Bailey, S.L., and Stralka, S.W. (1995) Strength of the quadriceps femoris muscle and functional recovery after reconstruction of the anterior cruciate ligament. A prospective, randomized clinical trial of electrical stimulation. *J Bone Joint Surg Am* 77, 1166-1173.

Sorichter, S., Koller, A., Haid, C., Wicke, K., Judmaier, W., Werner, P., and Raas, E. (1995) Light concentric exercise and heavy eccentric muscle loading: Effects on CK, MRI and markers of inflammation. *Int J Sports Med* 16, 288-292.

Speer, K.P., Lohnes, J., and Garrett, W.E., Jr. (1993) Radiographic imaging of muscle strain injury. *Am J Sports Med* 21, 89-95.

Staples, J.R., Clement, D.B., Taunton, J.E., and McKenzie, D.C. (1999) Effects of hyperbaric oxygen on a human model of injury. *Am J Sports Med* 27, 600-605.

Stauber, W.T., Clarkson, P.M., Fritz, V.K., and Evans, W.J. (1990) Extracellular matrix disruption and pain after eccentric muscle action. *J Appl Physiol* 69, 868-874.

Stewart, D., Macaluso, A., and De Vito, G. (2003) The effect of an active warm-up on surface EMG and muscle performance in healthy humans. *Eur J Appl Physiol* 89, 509-513.

Strickler, T., Malone, T., and Garrett, W.E. (1990) The effects of passive warming on muscle injury. *Am J Sports Med* 18, 141-145.

Svernlov, B. and Adolfsson, L. (2001) Non-operative treatment regime including eccentric training for lateral humeral epicondylalgia. *Scand J Med Sci Sports* 11, 328-334.

Talbot, J.A. and Morgan, D.L. (1998) The effects of stretch parameters on eccentric exercise-induced damage to toad skeletal muscle. *J Muscle Res Cell Motil* 19, 237-245.

Taylor, D.C., Dalton, J.D., Jr., Seaber, A.V., and Garrett, W.E., Jr. (1990) Viscoelastic properties of muscle-tendon units. The biomechanical effects of stretching. *Am J Sports Med* 18, 300-309.

Taylor, D.C., Dalton, J.D., Jr., Seaber, A.V., and Garrett, W.E., Jr. (1993) Experimental muscle strain injury. Early functional and structural deficits and the increased risk for reinjury. *Am J Sports Med* 21, 190-194.

Thorsson, O., Rantanen, J., Hurme, T., and Kalimo, H. (1998) Effects of nonsteroidal antiinflammatory medication on satellite cell proliferation during muscle regeneration. *Am J Sports Med* 26, 172-176.

Tidball, J.G. (1995) Inflammatory cell response to acute muscle injury. *Med Sci Sports Exerc* 27, 1022-1032.

Tidball, J.G. (2005) Inflammatory processes in muscle injury and repair. *Am J Physiol Regul Integr Comp Physiol* 288, R345-R353.

Tiidus, P.M. (1999) Massage and ultrasound as therapeutic modalities in exercise-induced muscle damage. *Can J Appl Physiol* 24, 267-278.

Trappe, T.A., Carrithers, J.A., White, F., Lambert, C.P., Evans, W.J., and Dennis, R.A. (2002) Titin and nebulin content in human skeletal muscle following eccentric resistance exercise. *Muscle Nerve* 25, 289-292.

Vierck, J., O'Reilly, B., Hossner, K., Antonio, J., Byrne, K., Bucci, L., and Dodson, M. (2000) Satellite cell regulation following myotrauma caused by resistance exercise. *Cell Biol Int* 24, 263-272.

Wall, P.D. and Sweet, W.H. (1967) Temporary abolition of pain in man. *Science* 155, 108-109.

Warren, G.L., Hermann, K.M., Ingalls, C.P., Masselli, M.R., and Armstrong, R.B. (2000) Decreased EMG median frequency during a second bout of eccentric contractions. *Med Sci Sports Exerc* 32, 820-829.

Warren, G.L., Ingalls, C.P., Lowe, D.A., and Armstrong, R.B. (2002) What mechanisms contribute to the strength loss that occurs during and in the recovery from skeletal muscle injury? *J Orthop Sports Phys Ther* 32, 58-64.

Webster, A.L., Syrotuik, D.G., Bell, G.J., Jones, R.L., and Hanstock, C.C. (2002) Effects of hyperbaric oxygen on recovery from exercise-induced muscle damage in humans. *Clin J Sport Med* 12, 139-150.

Willems, M.E. and Stauber, W.T. (2000) Force output during and following active stretches of rat plantar flexor muscles: Effect of velocity of ankle rotation. *J Biomech* 33, 1035-1038.

Williams, G.R., Jr. (1997) Painful shoulder after surgery for rotator cuff disease. *J Am Acad Orthop Surg* 5, 97-108.

Yackzan, L., Adams, C., and Francis, K.T. (1984) The effects of ice massage on delayed muscle soreness. *Am J Sports Med* 12, 159-165.

Yang, S., Alnaqeeb, M., Simpson, H., and Goldspink, G. (1996) Cloning and characterization of an IGF-1 isoform expressed in skeletal muscle subjected to stretch. *J Muscle Res Cell Motil* 17, 487-495.

Yoshimura, K. and Harii, K. (1999) A regenerative change during muscle adaptation to denervation in rats. *J Surg Res* 81, 139-146.

Zarins, B. and Ciullo, J.V. (1983) Acute muscle and tendon injuries in athletes. *Clin Sports Med* 2, 167-182.

Chapter 19

Alessio, H.M., A.H. Goldfarb, and G. Cao. 1997. Exercise-induced oxidative stress before and after vitamin C supplementation. *International Journal of Sports Nutrition* 7: 1-9.

Aoi, W., Y. Naito, Y. Takanami, Y. Kawai, K. Sakuma, H. Ichikawa, N. Yoshida, and T. Yoshikawa. 2004. Oxidative stress and delayed-onset muscle damage after exercise. *Free Radical Biology and Medicine* 37(4): 480-487.

Avery, N.G., J.L. Kaiser, M.J. Sharman, T.P. Scheett, D.M. Barnes, A.L. Gomez, W.J. Kraemer, and J.S. Volek. 2003. Effects of vitamin E supplementation on recovery from repeated bouts of resistance exercise. *Journal of Strength and Conditioning Research* 17(4): 801-809.

Baskin, C.R., K.W. Hinchcliff, R.A. DiSilvestro, G.A. Reinhart, M.G. Hayek, B.P. Chew, J.R. Burr, and R.A. Swenson. 2000. Effects of dietary antioxidant supplementation on oxidative damage and resistance to oxidative damage during prolonged exercise in sled dogs. *American Journal of Veterinary Research* 61: 886-891.

Beaton, L.J., D.A. Allan, M.A. Tarnopolsky, P.M. Tiidus, and S.M. Phillips. 2002. Contraction-induced muscle damage is unaffected by vitamin E supplementation. *Medicine and Science in Sports and Exercise* 34: 798-805.

Best, T.M., R. Fiebig, D.T. Corr, S. Brickson, and L. Ji. 1999. Free radical activity, antioxidant enzyme, and glutathione changes with muscle stretch injury in rabbits. *Journal of Applied Physiology* 87: 74-82.

Bhagavan, H.N., and R.K. Chopra. 2006. Coenzyme Q10: absorption, tissue uptake, metabolism and pharmacokinetics. *Free Radical Research* 40: 445-453.

Bloomer, R.J., A.H. Goldfarb, and M.J. McKenzie. 2006. Oxidative stress response to aerobic exercise: Comparison of antioxidant supplements. *Medicine and Science in Sports and Exercise* 38(6): 1098-1105.

Bloomer, R.J., A.H. Goldfarb, M.J. McKenzie, T. You, and L. Nguyen. 2004. Effects of antioxidant therapy in women exposed to eccentric exercise. *International Journal of Sport Nutrition and Exercise Metabolism* 14: 377-388.

Bowles, D.K., C.E. Torgan, S. Ebner, J.P. Keher, J.L. Ivy, and J.W. Starnes. 1991. Effects of acute submaximal exercise on skeletal muscle vitamin E. *Free Radical Research Communications* 14: 139-143.

Boyer, B.T., A.H. Goldfarb, and A.Z. Jamurtas. 1996. Relationship of prostaglandin E_2, leukotriene B_4, creatine kinase, lactic acid and DOMS. *Medicine and Science in Sports and Exercise* 28: S154.

Braun, B., P.M. Clarkson, P.S. Freedson, and R.L. Kohl. 1991. Effects of coenzyme Q10 supplementation on exercise performance, VO_2max, and lipid peroxidation in trained cyclists. *International Journal of Sports Nutrition* 1: 353-365.

Bryant, R.J., J. Ryder, P. Martino, J. Kim, and B.W. Craig. 2003. Effects of vitamin E and C supplementation either alone or in combination on exercise-induced lipid peroxidation in trained cyclists. *Journal of Strength and Conditioning Research* 17(4): 792-800.

Bryer, S.C., and A.H. Goldfarb. 2006. Effect of high dose vitamin C supplementation on muscle soreness, damage, function, and oxidative stress to eccentric exercise. *International Journal of Sport Nutrition and Exercise Metabolism* 16(3): 270-280.

Cai, Q., and H. Wei. 1996. Effect of dietary genistein on antioxidant enzyme activities in SENCAR mice. *Nutritional Cancer* 25: 1-7.

Cakatay, U., R. Kayal, A. Sivas, and F. Tekeli. 2005. Prooxidant activities of alpha-lipoic acid on oxidative protein damage in the aging rat heart muscle. *Archives of Gerontology and Geriatrics* 40: 231-240.

Cakatay, U., T. Telci, R. Kayali, A. Sivas, and T. Akcay. 2000. Effect of alpha-lipoic acid supplementation on oxidative protein damage in the streptozotocin-diabetic rat. *Research in Experimental Medicine* (Berlin) 199: 243-251.

Cannon, J.G., and J.B. Blumberg. 2000. Acute phase immune responses in exercise. In: *Handbook of oxidants and antioxidants in exercise,* ed. C.K. Sen, L. Packer, and O. Hanninen, 177-193. Amsterdam: Elsevier.

Cannon, J.G., S.N. Meydani, R.A. Fielding, M.A. Fiatrone, M. Meydani, M. Farhangmehr, S.F. Orencole, J.B. Blumberg, and W.J. Evans. 1991. Acute phase response in exercise. II. Associations between vitamin E, cytokines, and muscle proteolysis. *American Journal of Physiology* 260: R1235-R1240.

Cannon, J.G., S.F. Orencole, R.A. Fielding, M. Meydani, S.N. Meydani, M.A. Fiatrone, J.B. Blumberg, and W.J. Evans. 1990. Acute phase response in exercise: interaction of age and vitamin E on neutrophils and muscle enzyme release. *American Journal of Physiology* 259: R1214-R1219.

Cantini, M., E. Giurisato, C. Radu, S. Tiozzo, F. Pampinella, D. Senigaglia, G. Zaniolo, F. Mazzoleni, and L. Vitiello. 2002. Macrophage-secreted myogenic factors: A promising tool for greatly enhancing the proliferative capacity of myoblasts in vitro and in vivo. *Neurological Science* 23: 189-194.

Cassatella, M.A. 1999. *Neutrophil-derived proteins: Selling cytokines by the pound.* New York: Academic Press.

Chen, C., and R.M. Bakhiet. 1998. The effect of acute strenuous exercise on the activities of antioxidant enzymes and plasma genistein concentration in rats fed a genistein supplemented diet. *Federation of Allied Societies in Experimental Biology Journal* 12(5), A560. [abstract]

Chen, C.Y., R.M. Bakhiet, V. Hart, and G. Holtzman. 2005. Isoflavones improve plasma homocysteine status and antioxidant defense system in healthy young men at rest but do not ameliorate oxidative stress induced by 80% VO(2)pk exercise. *Annals of Nutrition and Metabolism* 25(49): 33-41.

Childs, A., C. Jacobs, T. Kaminski, B. Halliwell, and C. Leeuwenburgh. 2001. Supplementation with vitamin C and N-acetyl-cysteine increases oxidative stress in humans after an acute muscle injury induced by eccentric exercise. *Free Radical Biology and Medicine* 31: 745-753.

Close, G.L., T. Ashton, T. Cable, D. Doran, and D.P. MacLaren. 2004. Eccentric exercise, isokinetic muscle torque and delayed onset muscle soreness: The role of reactive oxygen species. *European Journal of Applied Physiology* 91: 615-621.

Dawson, B., G.J. Henry, C. Goodman, I. Gillam, J.R. Beilby, S. Ching, V. Fabian, D. Dasig, P. Morling, and B.A. Kakulus. 2002. Effect of vitamin C and E supplementation on biochemical and ultrastructural indices of muscle damage after a 21 km run. *International Journal of Sports Medicine* 23: 10-15.

Dincer, Y., A. Telci, R. Kayali, I.A. Yilmaz, U. Cakatay, and T. Akcay. 2002. Effect of alpha-lipoic acid on lipid peroxidation and anti-oxidant enzyme activities in diabetic rats. *Clinical and Experimental Pharmacology and Physiology* 29: 281-284.

Essig, D.A., and T.M. Nosek. 1997. Muscle fatigue and induction of stress protein genes: A dual function of reactive oxygen species? *Canadian Journal of Applied Physiology* 22: 409-428.

Faff, J., and A. Frankiewicz-Jozko. 1997. Effect of ubiquinone on exercise-induced lipid peroxidation in rat tissues. *European Journal of Applied Physiology and Occupational Physiology* 75: 413-417.

Frei, B., L. England, and B.N. Ames. 1989. Ascorbate is an outstanding antioxidant in human blood plasma. *Proceedings of the National Academy of Sciences USA* 86: 6377-6381.

Gohil, K., L. Rothfuss, J. Lang, and L. Packer. 1987. Effect of exercise training on tissue vitamin E and ubiquinone content. *Journal of Applied Physiology* 63(4): 1638-1641.

Goldfarb, A.H. 1999. Nutritional antioxidants as therapeutic and preventive modalities in exercise-induced muscle damage. *Canadian Journal of Applied Physiology* 24(3): 249-266.

Goldfarb, A.H., R.J. Bloomer, and M.J. McKenzie. 2005a. Combined antioxidant treatment effects on blood oxidative stress after eccentric exercise. *Medicine and Science in Sports and Exercise* 37: 234-239.

Goldfarb, A.H., S.W. Patrick, S.C. Bryer, and T. You. 2005b. Vitamin C supplementation affects oxidative-stress blood markers in response to a thirty minute run at 75% VO_2 max. *International Journal of Sport Nutrition and Exercise Metabolism* 15: 279-290.

Goldfarb, A.H., M. K. McIntosh, B.T. Boyer, and J. Fatouros. 1994. Vitamin E effects on indexes of lipid peroxidation in muscle from DHEA-treated and exercised rats. *Journal of Applied Physiology* 76(4):1630-1635.

Hagen, T.M., R.T. Ingersoll, J. Lykkesfeldt, J. Liu, C.M. Wehr, V. Vinarsky, J.C. Bartholomew, and A.B. Ames. 1999. (R)-alpha-lipoic acid-supplemented old rats have improved mitochondrial function, decreased oxidative damage, and increased metabolic rate. *Federation of Allied Societies in Experimental Biology Journal* 13: 411-418.

Halliwell, B., and J.M.C. Gutteridge. 1985. *Free radicals in biology and medicine.* Oxford: Clarendon Press.

Han, D., G.J. Handelman, and L. Packer. 1995. Analysis of reduced and oxidized lipoic acid in biological samples by high-performance liquid chromatography. *Methods in Enzymology* 251: 315-325.

Hollman, P.C., and M.B. Katan. 1998. Bioavailability and health effects of dietary flavonols in man. *Archives of Toxicology* Suppl. 20: 237-248.

Jackson, M.J. 2000. Exercise and oxygen radical production by muscle. In: *Handbook of oxidants and antioxidants in exercise,* ed. C.K. Sen, L. Packer, and O. Hanninen, 57-68. Amsterdam: Elsevier.

Jackson, M.J., D.A. Jones, and R.H. Edwards. 1983. Vitamin E and skeletal muscle. *Ciba Found Symposium* 101:224-239.

Jakeman, P., and S. Maxwell. 1993. Effect of antioxidant vitamin supplementation on muscle function after eccentric exercise. *European Journal of Applied Physiology* 67: 426-430.

Jenkins, R.R., and B. Halliwell. 1994. Metal binding agents: Possible role in exercise. In: *Exercise and oxygen toxicity,* ed. C.K. Sen, L. Packer, and O. Hanninen, 59-76. Amsterdam: Elsevier.

Jha, H.C., G. Von Recklinghausen, and F. Killiken. 1985. Inhibition of in vitro microsomal lipid peroxidation by isoflavonoids. *Biochemical Pharmacology* 34: 1367-1369.

Kaikkonen, J., L. Kosonen, K. Nyyssonen, E. Porkkala-Saratho, R. Salonen, H. Korpela, and J.T. Salonen. 1998. Effect of combined coenzyme Q10 and d-alpha-tocopheryl; acetate supplementation on exercise-induced lipid peroxidation and muscular damage: A placebo-controlled double-blind study in marathon runners. *Free Radical Research* 29: 85-92.

Kaminski, M., and R. Boal. 1992. An effect of ascorbic acid on delayed-onset muscle soreness. *Pain* 50: 317-321.

Kamzalov, S., N. Sumien, M.J. Forster, and R.S. Sohal. 2003. Coenzyme Q intake elevates the mitochondrial and tissue levels of coenzyme Q and alpha-tocopherol in young mice. *Journal of Nutrition* 133: 3175-3180.

Kanter, M.M., L.A. Nolte, and J.O. Holloszy. 1993. Effects of an antioxidant vitamin mixture on lipid peroxidation at rest and postexercise. *Journal of Applied Physiology* 74: 965-969.

Karlsson, J. 1997. *Antioxidants and exercise.* Champaign, IL: Human Kinetics

Keith, R.E. 1997. Ascorbic acid. In: *Sports nutrition: Vitamins and trace elements,* ed. I. Wolinsky and J.A. Driskell, 29-45 and 119-131. New York: CRC Press.

Khanna, S., M. Atalay, D.E. Laaksonen, M. Gul, S. Roy, and C.K. Sen. 1999. α-Lipoic acid supplementation: Tissue glutathione homeostasis at rest and after exercise. *Journal of Applied Physiology* 86: 1191-1196.

Klebanoff, S.J. 1982. Oxygen-dependent cytotoxic mechanisms of phagocytes. In: *Advances in host defense mechanisms,* ed. J.I. Gallin and A.S. Fauci, Vol. 1, 111-162. New York: Raven Press.

Laaksonen, R., M. Fogelholm, J.J. Himberg, J. Laako, and Y. Salorinne. 1995. Ubiquinone supplementation and exercise capacity in trained young and older men. *European Journal of Applied Physiology* 72: 95-100.

Langen, R.C., A.M. Schols, M.C. Kelders, J.L. Van Der Velden, E.F. Wouters, and Y.M. Janssen-Heininger. 2002. Tumor necrosis factor-alpha inhibits myogenesis through redox-dependent and -independent pathways. *American Journal of Physiology: Cell Physiology* 283: C714-C721.

Lee, J., A.H. Goldfarb, M.H. Rescino, S. Hegde, S. Patrick, and K. Apperson. 2002. Eccentric exercise effect on blood oxidative-stress markers and delayed onset of muscle soreness. *Medicine and Science in Sports and Exercise* 34: 443-448.

Lenn, J., T. Uhl, C. Mattacola, G. Boissonneault, J. Yates, W. Ibrahim, and G. Bruckner. 2002. The effects of fish oil and isoflavones on delayed onset muscle soreness. *Medicine and Science in Sports and Exercise* 34: 1605-1613.

Levine, M., C. Conry-Cantilena, Y. Wang, R.W. Welch, P.W. Washko, K.R. Dhariwal, J.B. Park, A. Lazarev, J.F. Graumlich, J. King, and L.R. Cantilena. 1996. Vitamin C pharmacokinetics in healthy volunteers: Evidence for a recommended dietary allowance. *Proceedings of the National Academy of Sciences USA* 93: 3704-3709.

Lonnrot, K., P. Holm, A. Lagerstedt, H. Huhtala, and H. Alho. 1998. The effects of lifelong ubiquinone Q10 supplementation on the Q9 and Q10 tissue concentrations and life span of male rats and mice. *Biochemistry and Molecular Biology International* 44: 727-737.

Madani, S., J. Prost, and J. Belleville. 2000. Dietary protein level and origin (casein and highly purified soybean protein) affect hepatic storage, plasma lipid transport, and antioxidative defense status in the rat. *Nutrition* 16: 368-375.

Malm, C., M. Svensson, B. Ekblom, and B. Sjodin. 1997a. Effects of ubiquinone-10 supplementation and high intensity training on physical performance in humans. *Acta Physiologica Scandinavia* 161: 379-384.

Malm, C., M. Svensson, B. Sjoberg, B. Ekblom, and B. Sjodin. 1997b. Supplementation with ubiquinone-10 causes cellular damage during intense exercise. *Acta Physiologica Scandinavia* 157: 511-512.

McArdle, A., R.H.T. Edwards, and M.J. Jackson. 1991. Effects of contractile activity on muscle damage in the dystrophin-deficient mouse. *Clinical Science* 80: 367-371.

McArdle, A., R.H.T. Edwards, and M.J. Jackson. 1992. Accumulation of calcium by normal and dystrophin-deficient mouse muscle during contractile activity in vitro. *Clinical Science* 82: 455-459.

McHugh, M.P., D.A.J. Connolly, R.G. Eston, and G.W. Gleim. 1999. Exercise-induced muscle damage and potential mechanisms for the repeated bout effect. *Sports Medicine* 27: 157-170.

McLoughlin, T.J., E. Mylona, T.A. Hornberger, K.A. Esser, and F.X. Pizza. 2003. Inflammatory cells in rat skeletal muscle are elevated after electrically stimulated contractions. *Journal of Applied Physiology* 94: 876-882.

Meydani, M., W.J. Evans, G. Handelman, L. Biddle, R.A. Fielding, S.N. Meydani, J. Burrill, M.A. Fiatrone, J.B. Blumberg, and J.G. Cannon. 1993. Protective effect of vitamin E on exercise-induced oxidative damage in young and older adults. *American Journal of Physiology* 33: R992-R998.

Neiman, D., D. Henson, D. Butterworth, B. Warren, J. Davis, O. Fagoaga, and S. Nehlsen-Cannarella. 1997. Vitamin C supplementation does not alter the immune response to 2.5 hours of running. *International Journal of Sports Nutrition* 7: 173-184.

Nikawa, T., M. Ikemoto, T. Sakai, M. Kano, T. Kitano, T. Kawahara, S. Teshima, K. Rokutan, and K. Kishi. 2002. Effects of a soy protein diet on exercise-induced muscle protein catabolism in rats. *Nutrition* 18: 490-495.

Niki, E., T. Saito, A. Kawakami, and Y. Kamiya. 1984. Inhibition of oxidation of methyl linoleate in solution by vitamin E and vitamin C. *Journal of Biological Chemistry* 259: 4177-4182.

Packer, L., A.L. Almada, L.M. Rothfuss, and D.S. Wilson. 1989. Modulation of tissue vitamin E levels by physical exercise. *Annals of the New York Academy of Sciences* 570: 311-321.

Packer, L., E.H. Witt, and H.J. Tritschler. 1995. Alpha-lipoic acid as a biological antioxidant. *Free Radical Biology and Medicine* 19: 227-250.

Phillips, T., A.C. Childs, D.M. Dreon, S. Phinney, and C. Leeuwenburgh. 2003. A dietary supplement attenuates IL-6 and CRP after eccentric exercise in untrained males. *Medicine and Science in Sports and Exercise* 35: 2032-2037.

Pizza, F.X., T.J. Koh, S.J. McGregor, and S.V. Brooks. 2002. Muscle inflammatory cells after passive stretches, isometric contractions, and lengthening contractions. *Journal of Applied Physiology* 92(5):1873-1878.

Pizza, F.X., T.J. McLoughlin, S.J. McGregor, E.P. Calomeni, and W.T. Gunning. 2001. Neutrophils injure cultured skeletal myotubes. *American Journal of Physiology: Cell Physiology* 281: C335-C341.

Pizza, F.X., J.M. Peterson, J.H. Baas, and T.J. Koh. 2005. Neutrophils contribute to muscle injury and impair its resolution after lengthening contractions in mice. *Journal of Physiology* 562(3): 899-913.

Resnick, A.Z., E. Witt, M. Matsumato, and L. Packer. 1992. Vitamin E inhibits protein oxidation in skeletal muscle of resting and exercised rats. *Biochemical and Biophysical Research Communications* 189: 801-806.

Rokitski, L., E. Logemann, G. Huber, E. Keck, and J. Keul. 1994a. Alpha-tocopherol supplementation in racing cyclists during extreme endurance training. *International Journal of Sports Nutrition* 4: 253-264.

Rokitski, L., E. Logemann, A.N. Sagredos, M. Murphy, W. Wetzel-Roth, and J. Keul. 1994b. Lipid peroxidation and antioxidative vitamins under extreme endurance stress. *Acta Physiologica Scandinavia* 151: 149-158.

Rosenfeldt, F., D. Hilton, S. Pepe, and H. Krum. 2003. Systematic review of effect of coenzyme Q10 in physical exercise, hypertension and heart failure. *Biofactors* 18: 91-100.

Rossi, A., R.A. DiSilvestro, and A. Blostein-Fujii. 2000. Effects of soy consumption on exercise-induced acute muscle damage and oxidative stress in young adult males. *Journal of Nutraceuticals, Functional and Medical Foods* 3: 1-5.

Sacheck, J.M., P.E. Milbury, J.G. Cannon, R. Roubenoff, and J.B. Blumberg. 2003. Effect of vitamin E and eccentric exercise on selected biomarkers of oxidative stress in young and elderly men. *Free Radical Biology and Medicine* 34: 1575-1588.

Saengsirisuwan, V., T.R. Kinnick, M.B. Schmit, and E.J. Henriksen. 2001. Interactions of exercise training and lipoic acid on skeletal muscle glucose transport in obese Zucker rats. *Journal of Applied Physiology* 91: 145-153.

Saengsirisuwan, V., F.R. Perez, J.A. Sloniger, T. Maier, and E.J. Henriksen. 2004. Interactions of exercise training and alpha-lipoic acid on insulin signaling in skeletal muscle of obese Zucker rats. *American Journal of Physiology: Endocrinology and Metabolism* 287: E529-E536.

Saxton, J.M., A.E. Donnelly, and H.P. Roper. 1994. Indices of free-radical-mediated damage following maximal voluntary eccentric and concentric muscular work. *European Journal of Applied Physiology* 68: 189-193.

Sen, C.K., M. Ataly, J. Agren, D.E. Laaksonen, S. Roy, and O. Hänninen. 1997. Fish oil and vitamin E supplementation in oxidative stress at rest and after physical exercise. *Journal of Applied Physiology* 83(1):189-195.

Shafat, A., P. Butler, R.L. Jensen, and A.E. Donnelly. 2004. Effects of dietary supplementation with vitamins C and E on muscle function during and after eccentric contractions in humans. *European Journal of Applied Physiology* 93: 196-202.

Shimomura, Y., M. Suzuki, S. Sugiyama, Y. Hanaki, and T. Ozawa. 1991. Protective effect of coenzyme Q10 on exercise-induced muscular injury. *Biochemical and Biophysical Research Communications* 176: 349-355.

Smith, L.L., J.M. Wells, J.A. Houmard, S.T. Smith, R.G. Israel, T.C. Chenier, and S.N. Pennington. 1993. Increases in plasma prostaglandin E$_2$ after eccentric exercise. *Hormone and Metabolism Research* 25: 451-452.

Suh, J.H., E.T. Shigeno, J.D. Morrow, B. Cox, A.E. Rocha, B. Frei, and T.M. Hagen. 2001. Oxidative stress in the aging rat heart is reversed by dietary supplementation with (R)-(alpha)-lipoic acid. *Federation of Allied Societies in Experimental Biology Journal* 15: 700-706.

Sumida, S., K. Tanaka, H. Kitao, and F. Nakadomo. 1989. Exercise-induced lipid peroxidation and leakage of enzymes before and after vitamin E supplementation. *International Journal of Biochemistry* 21(8):835-838.

Svensson, M., C. Malm, M. Tonkonogi, B. Ekblom, B. Sjodin, and K. Sahlin. 1999. Effect of Q10 supplementation on tissue Q10 levels and adenine nucleotide catabolism during high-intensity exercise. *International Journal of Sports Nutrition* 9: 166-180.

Telford, R.D., E.A. Catchpole, V. Deakin, A.C. McLeay, and A.W. Plank. 1992. The effect of 7-8 months of vitamin/mineral supplementation on the vitamin and mineral status of athletes. *International Journal of Sports Nutrition* 2: 132-134.

Thompson, D., D.M. Bailey, J, Hill, T. Hurst, J.R. Powell, and C. Williams. 2004. Prolonged vitamin C supplementation and recovery from eccentric exercise. *European Journal of Applied Physiology* 92: 133-138.

Thompson, D., C. Williams, P. Garcia-Roves, S.J. McGregor, F. McArdle, and M.J. Jackson. 2003. Post-exercise vitamin C supplementation and recovery from demanding exercise. *European Journal of Applied Physiology* 89: 393-400.

Thompson, D., C. Williams, M. Kingsley, C. Nicholas, H. Lakomy, F. McArdle, and M.J. Jackson. 2001a. Muscle soreness and damage parameters after prolonged intermittent shuttle-running following acute vitamin C supplementation. *International Journal of Sports Medicine* 22: 68-75.

Thompson, D., C. Williams, S.J. McGregor, C. Nicholas, F. McArdle, M.J. Jackson, and J.R. Powell. 2001b. Prolonged vitamin C supplementation and recovery from demanding exercise. *International Journal of Sports Nutrition* 11: 466-481.

Tiidus, P.M., W.A. Behrens, R. Madere, and M.E. Houston. 1993. Muscle vitamin E levels following acute exercise in female rats. *Acta Physiologica Scandinavica* 147: 249-250.

Tikkanen, M.J., K. Wahala, S. Ojala, V. Vihma, and H. Adlecreutz. 1998. Effect of soybean phytoestrogen intake on low density lipoprotein oxidation resistance. *Proceedings of the National Academy of Sciences USA* 95: 3106-3110.

Tsivitse, S.K., T.J. McLoughlin, J.M. Peterson, E. Mylona, S.J. McGregor, and F.X. Pizza. 2003. Downhill running in rats: Influence on neutrophils, macrophages, and MyoD+ cells in skeletal muscle. *European Journal of Applied Physiology* 90: 633-638.

Van der Meulen, J.H., A. McArdel, M.J. Jackson, and J.A. Faulkner. 1997. Contraction-induced injury to the extensor digitorum longus muscle of rats: The role of vitamin E. *Journal of Applied Physiology* 83: 817-823.

Vasankari, T., U. Kujala, S. Sarna, and M. Ahotupa. 1998. Effects of ascorbic acid and carbohydrate ingestion on exercise induced oxidative stress. *Journal of Sports Medicine and Physical Fitness* 38: 281-285.

Viguie, C., L. Packer, and G.A. Brooks. 1989. Antioxidant supplementation affects indices of muscle trauma and oxidant stress in human blood during exercise. *Medicine and Science in Sports and Exercise* 21: S16.

Warren, J.A., R.R. Jenkins, L. Packer, E.H. Witt, and R.B. Armstrong. 1992. Elevated muscle vitamin E does not attenuate eccentric exercise-induced muscle injury. *Journal of Applied Physiology* 72: 2168-2175.

Wedworth, S.M., and S. Lynch. 1995. Dietary flavonoids in atherosclerosis prevention. *Annals of Pharmacotherapy* 29: 627-628.

Wei, H., Q. Cai, and R.O. Rahn. 1996. Inhibition of UV light- and Fenton-reaction-induced oxidative DNA damage by the soybean isoflavone genistein. *Carcinogenesis* 17: 73-77.

You, T., A.H. Goldfarb, R.J. Bloomer, L. Nguyen, X. Sha, and M.J. McKenzie. 2005. Oxidative stress response in normal and antioxidant supplemented rats to a downhill run: Changes in blood and skeletal muscles. *Canadian Journal of Applied Physiology* 30(6): 677-689.

Zerba, E., T.E. Komorowski, and J.A. Faulkner. 1990. Free radical injury to skeletal muscles of young, adult, and old mice. *American Journal of Physiology* 258: C429-C435.

Chapter 20

Altun, L., Bingul, U., Aykuc, M., and Yurtkuran, M. (2005) Investigation of the effects of GaAs laser therapy on cervical myofascial pain syndrome. *Rheumatology International,* 25, 23-27.

Bartels, E., and Dannieskold-Samsoe, B. (1986) Histological abnormalities in muscle from patients with certain types of fibrosistis. *Lancet,* 8484, 755-757.

Bendtsen, L., Jensen, R., and Olesen, J. (1996) Qualitatively altered nociception in chronic myofascial pain. *Pain,* 65, 259-264.

Bengtsson, A., Henrikkson, K., and Larsson, J. (1986) Reduced high energy phosphate levels in the painful muscles of patients with primary fibromyalgia. *Arthritis and Rheumatism,* 29, 817-821.

Benoit, P. (1978) Reversible skeletal muscle damage after administration of local anaesthetics with and without epinephrine. *Journal of Oral Surgery,* 36, 198-201.

Bialla, G., Sotgiu, M., Pellegata, G., Paulesu, E., Castiglioni, I., and Fazio, F. (2001) Acupuncture produces central activations in pain regions. *Neuroimage,* 14, 60-66.

Button, M. (1940) Muscular rheumatism—local injection treatment as a means to rapid restoration of function. *British Medical Journal,* August, 183-185.

Byrn, C., Olsson, I., Falkheden, L., Lindh, M., Hosteroy, U., Fogelberg, M., Linder, L.-E., and Bunketorp, O. (1993) Subcutaneous sterile water injections for chronic neck and shoulder pain following whiplash injuries. *Lancet,* 341, 449-452.

Carrasco, A., Wescoat, L., and Roman, A. (2003) A retrospective review of botulinum toxin type A compared with standard therapy in the treatment of lumbar myofascial back pain patients. *Pain Clinic,* 15, 205-211.

Ceccherelli, F., Altafini, L., Lo Castro, G., Avila, A., Ambrosio, F., and Giron, G.P. (1989) Diode laser in cervical myofascial pain: A double-blind study versus placebo. *Clinical Journal of Pain,* 5, 301-304.

Ceylan, Y., Hizmeti, S., and Silig, Y. (2004) The effects of infrared laser and medical treatments on pain and serotonin degradation products in patients with myofascial pain syndrome. A controlled trial. *Rheumatology International,* 24, 260-263.

Chen, J.-T., Chen, S.-M., Kuan, T.-S., Chung, K.-C., and Hong, C.-Z. (1998) Phentolamine effect on the spontaneous electrical activity of active loci in a myofascial trigger spot of rabbit skeletal muscle. *Archives of Physical Medicine and Rehabilitation,* 79, 789-794.

Chen, J., Chung, K., Hou, C., Kuan, C., Chen, S., and Hong, C. (2001) Inhibitory effect of dry needling on the spontaneous electrical activity recorded from myofascial trigger spots of rabbit skeletal muscle. *American Journal of Physical Medicine and Rehabilitation,* 80, 729-735.

Cheshire, W., Abashian, S., and Mann, J. (1994) Botulinum toxin in the treatment of myofascial pain syndrome. *Pain,* 59, 65-69.

Diakow, P. (1988) Thermographic imaging of myofascial trigger points. *Journal of Manipulative and Physiological Therapeutics,* 11, 114-117.

Diakow, P. (1992) Differentiation of active and latent trigger points by thermography. *Journal of Manipulative and Physiological Therapeutics,* 15, 439.

Ernberg, M., Lundeberg, T., and Kopp, S. (2000) Pain and allodynia/hyperalgesia induced by intramuscular injection of serotonin in patients with fibromyalgia and healthy individuals. *Pain,* 85, 31-39.

Ernberg, M., Lundeberg, T., and Kopp, S. (2003) Effects on muscle pain by intramuscular injection of granisetron in patients with fibromyalgia. *Pain,* 101, 275-282.

Fine, P.G., Milano, R., and Hare, B.D. (1988) The effects of myofascial trigger point injections are naloxone reversible. *Pain,* 32, 15-20.

Foster, A., and Carlson, B. (1980) Myotoxicity of local anaesthetics and regeneration of the damaged muscle fibres. *Anesthesia and Analgesia,* 59, 727-736.

Franz, M., and Mense, S. (1975) Muscle receptors with group IV afferent fibres responding to application of bradykinin. *Brain Research,* 92, 369-383.

Frost, A. (1986) Diclofenac versus lidocaine as injection therapy in myofascial pain. *Scandinavian Journal of Rheumatology,* 15, 153-156.

Frost, F., Jessen, B., and Siggard-Andersen, J. (1980) A control, double blind comparison of mepivicaine injection versus saline injection for myofascial pain. *Lancet,* March, 499-500.

Furlan, A., van Tulder, M., Cherkin, D., Tsukayama, H., Lao, L., Koes, B., and Berman, B. (2005) Acupuncture and dry needling for low back pain. *The Cochrane Database of Systematic Reviews,* Art. no. CD001351.pub2.DOI:10.1002/14651858. CD00135.pub2.

Gam, A., Warming, S., Larsen, L.H., Jensen, B., Hoydalsmo, O., Allon, I., Andersen, B., Gotzsche, N., Petersen, M., and Mathiesen, B. (1998) Treatment of myofascial trigger points with ultrasound combined with massage and exercise—a randomised controlled trial. *Pain,* 77, 73-79.

Garvey, T.A., Marks, M.R., and Wiesel, S.W. (1989) A prospective, randomised, double blind evaluation of trigger point therapy for lower back pain. *Spine,* 14, 962-964.

Gerwin, R. (1994) Neurobiology of the myofascial trigger point. *Bailliere's Clinical Rheumatology,* 8, 747-762.

Gerwin, R., Shannon, S., Hong, C.-Z., Hubbard, D., and Gevirtz, R. (1997) Interrater reliability in myofascial trigger point examination. *Pain,* 69, 65-73.

Graff-Redford, S.B., Reeves, J.L., Baker, R.L., and Chiu, D. (1989) Effects of transcutaneous electrical nerve stimulation on myofascial pain and trigger point sensitivity. *Pain,* 37, 1-5.

Graven-Nielsen, T., and Mense, S. (2001) The peripheral apparatus of muscle pain: Evidence from animal and human studies. *Clinical Journal of Pain,* 17, 2-9.

Grevert, P., Albert, L., and Goldstein, A. (1983) Partial antagonism of placebo analgesia by naloxone. *Pain,* 16, 129-143.

Gunn, C. (1997) Radiculopathic pain: Diagnosis and treatment of segmental irritation or sensitisation. *Journal of Musculoskeletal Pain,* 5, 119-134.

Hameroff, S.R., Crago, B.R., Blitt, C.D., Womble, J., and Kanel, J. (1981) Comparison of bupivacaine, etidocaine and saline for trigger point therapy. *Anesthesia and Analgesia,* 60, 752-755.

Hanten, W., and Chandler, S. (1994) Effects of myofascial release leg pull and sagittal plane isometric contract-relax techniques on passive straight-leg raise angle. *Journal of Orthopaedic and Sports Physical Therapy,* 20, 138-144.

Harden, R., Breuhl, S., Gass, S., Niemiec, C., and Barbick, B. (2000) Signs and symptoms of the myofascial pain syndrome: A national survey of pain management providers. *Clinical Journal of Pain,* 16, 64-72.

Hedenberg-Magnusson, B., Ernberg, M., Alstergren, P., and Kopp, S. (2001) Pain mediation by prostaglandin E2 and leukotriene b4 in the human masseter muscle. *Acta Odontologica Scandinavia,* 59, 348-355.

Henry, J. (1989) Concepts of pain sensation and its modulation. *Journal of Rheumatology,* 16, 104-112.

Hoheisal, U., Mense, S., Simons, D., and Yu, X.-M. (1993) Appearance of new receptive fields in rat dorsal horn neurons following noxious stimulation of skeletal muscle: A model for referral of muscle pain? *Neuroscience Letters,* 153, 9-12.

Hong, C-Z. (1994) Lidocaine injection versus dry needling to myofascial trigger point. The importance of the local twitch response. *American Journal of Physical Medicine and Rehabilitation,* 73, 256-263.

Hong, C-Z. (1996) Pathophysiology of myofascial trigger point. *Journal of the Formosan Medical Association,* 95, 93-104.

Hong, C-Z., and Hsueh, T.C. (1996) Difference in pain relief after trigger point injections in myofascial pain patients with and without fibromyalgia. *Archives of Physical Medicine and Rehabilitation,* 77, 1161-1165.

Hong, C-Z., Torigoe, Y., and Yu, J. (1995) The localised twitch responses in responsive taut bands of rabbit skeletal muscle fibres are related to the reflexes at a spinal cord level. *Journal of Musculoskeletal Pain,* 3, 15-34.

Howard, R. (1941) The use of local anaesthesia in the relief of chronic pain. *Medical Journal of Australia,* March, 298-299.

Hsieh, C.-Y., Hong, C.-Z., Adams, A., Platt, K., Danielson, C., Hoehler, F., and Tobis, J. (2000) Interexaminer reliability of the palpation of trigger points in the trunk and lower limb muscles. *Archives of Physical Medicine and Rehabilitation,* 81, 258-264.

Hsueh, T.C., Cheng, P.T., Kuan, T.S., and Hong, C.Z. (1997) The immediate effectiveness of electrical nerve stimulation and electrical muscle stimulation on myofascial trigger points. *American Journal of Physical Medicine and Rehabilitation,* 76, 471-476.

Hubbard, D., and Berkhoff, G. (1993) Myofascial trigger points show spontaneous needle EMG activity. *Spine,* 18, 1803-1807.

Huguenin, L., Brukner, P., McCrory, P., Smith, P., Wajswelner, H., and Bennell, K. (2005) Effect of dry needling of gluteal muscles on straight leg raise: A randomised, placebo controlled, double blind trial. *British Journal of Sports Medicine,* 39, 84-90.

Irnich, D., Behrens, N., Gleditsch, J., Stor, W., Schreiber, M., Schops, P., Vickers, A., and Beyer, A. (2002) Immediate effects of dry needling and acupuncture at distant points in chronic neck pain: Results of a randomised, double blind, sham-controlled crossover trial. *Pain,* 99, 83-89.

Jaeger, B., and Reeves, L J. (1986) Quantification of changes in myofascial trigger point sensitivity with the pressure algometer following passive stretch. *Pain,* 27, 203-210.

Kamanli, A., Kaya, A., Ardicoglu, O., Ozgocmen, S., Zengin, F., and Bayik, Y. (2005) Comparison of lidocaine injection, botulinum toxin injection, dry needling to trigger points in myofascial pain syndrome. *Rheumatology International,* 25, 604-611.

Karakurum, B., Karaalin, O., Coskun, O., Dora, B., Ucler, S., and Inan, L. (2001) The "dry needle technique": Intramuscular stimulation in tension-type headache. *Cephalalgia,* 21, 813-817.

Kellgren, J. (1938a) Observations on referred pain arising from muscle. *Clinical Science,* 3, 175-190.

Kellgren, J. (1938b) A preliminary account of referred pains arising from muscle. *British Medical Journal,* February, 325-327.

Koltyn, K. (2002) Exercise-induced hypoalgesia and intensity of exercise. *Sports Medicine,* 32, 477-487.

Kruse, R., and Christiansen, J. (1992) Thermographic imaging of myofascial trigger points: A follow up study. *Archives of Physical Medicine and Rehabilitation,* 73, 819-823.

Kushmerick, M. (1989) Muscle energy metabolism, nuclear magnetic resonance spectroscopy and their potential in the study of fibromyalgia. *Journal of Rheumatology,* 16, 40-46.

Lew, P., Lewis, J., and Story, I. (1997) Inter-therapist reliability in locating latent myofascial trigger points using palpation. *Manual Therapy,* 2, 87-90.

Lewis, J., and Tehan, P. (1999) A blinded pilot study investigating the use of diagnostic ultrasound for detecting active myofascial trigger points. *Pain,* 79, 39-44.

Lund, E., Kendall, S., Janerot-Sjoberg, B., and Bengtsson, A. (2003) Muscle metabolism in fibromyalgia studied by P-31 magnetic resonance spectroscopy during aerobic and anaerobic exercise. *Scandinavian Journal of Rheumatology,* 32, 138-145.

Lund, N., Bengtsson, A., and Thorborg, P. (1986) Muscle tissue oxygen pressure in primary fibromyalgia. *Scandinavian Journal of Rheumatology,* 15, 165-173.

Mathur, A., Gatter, R., Bank, W., and Schumacher, H. (1988) Abnormal 31-P-NMR spectroscopy of painful muscles of patients with fibromyalgia [abstract]. *Arthritis and Rheumatism,* 31, 337.

McMillan, A., Nolan, A., and Kelly, P. (1997) The efficacy of dry needling and procaine in the treatment of myofascial pain in the jaw muscles. *Journal of Orofacial Pain,* 11, 307-314.

McNulty, W., Gervirtz, R., Hubbard, D., and Berkhoff, G. (1994) Needle electromyographic evaluation of trigger point response to a psychological stressor. *Psychophysiology,* 31, 313-316.

Melzack, R. (1982) Recent concepts of pain. *Journal of Medicine,* 13, 147-160.

Melzack, R., Stillwell, D., and Fox, E. (1977) Trigger points and acupuncture points for pain: Correlations and implications. *Pain,* 3, 3-23.

Mense, S. (1993) Nociception from skeletal muscle in relation to clinical muscle pain. *Pain,* 54, 241-289.

Mense, S. (1996) Biochemical pathogenesis of myofascial pain. *Journal of Musculoskeletal Pain,* 1, 145-162.

Njoo, K., and Van der Does, E. (1994) The occurrence and interrater reliability of myofascial trigger points in the quadratus lumborum and gluteus medius: A prospective study in non-specific low back pain patients and controls in general practice. *Pain,* 58, 317-324.

Olavi, A., Pekka, R., Pertii, K., and Pekka, P. (1989) Effects of the infra red laser therapy at treated and non-treated trigger points. *Acupuncture and Electrotherapeutics Research,* 14, 9-14.

Quintner, J., and Cohen, M. (1994) Referred pain of peripheral nerve origin: An alternative to the "myofascial pain" construct. *Clinical Journal of Pain,* 10, 243-251.

Ray, M. (1941) Isotonic glucose solution in the treatment of fibrositis [correspondence]. *British Medical Journal,* December, 850.

Russell, I. (1997) A reason to biopsy myofascial trigger points [editorial]. *Journal of Musculoskeletal Pain,* 5, 1-3.

Simone, D., Marchinetti, P., Caputi, G., and Ochoa, J. (1994) Identification of muscle afferents subserving sensation of deep pain in humans. *Journal of Neurophysiology,* 72, 883-889.

Simons, D. (1996) Clinical and etiological update of myofascial pain from trigger points. *Journal of Musculoskeletal Pain,* 4, 93-121.

Simons, D. (1997) Myofascial pain syndromes—trigger points. *Journal of Musculoskeletal Pain,* 5, 105-107.

Simons, D. (2001) Do endplate noise and spikes arise from normal motor endplates? *American Journal of Physical Medicine and Rehabilitation,* 80, 134-140.

Simons, D., Hong, C., and Simons, L. (1995) Prevalence of spontaneous electrical activity at trigger spots and at control sites in rabbit skeletal muscle. *Journal of Musculoskeletal Pain,* 3, 35-48.

Simons, D., Hong, C.-Z., and Simons, L. (2002) Endplate potentials are common to midfiber myofascial trigger points. *American Journal of Physical Medicine and Rehabilitation,* 81, 212-222.

Simons, D., Travell, J., and Simons, L. (1998) *Travell & Simons' myofascial pain and dysfunction: The trigger point manual.* Baltimore: Williams & Wilkins.

Snyder-Mackler, L., Bork, C., Bourbon, B., and Trumbore, D. (1986) Effect of helium-neon laser on musculoskeletal trigger points. *Physical Therapy,* 66, 1087-1090.

Souttar, H. (1923) Acute lumbago treated by the injection of quinine and urea. *British Medical Journal,* November, 915-916.

Steinbrocker, O. (1944) Therapeutic injections in painful musculoskeletal disorders. *Journal of the American Medical Association,* 125, 397-401.

Swerdlow, B., and Dieter, J. (1992) An evaluation of the sensitivity and specificity of medical thermography for the documentation of myofascial trigger points. *Pain,* 48, 205-213.

Thorsen, H., Gam, A., Svensson, B., Jess, M., Jensen, M., Piculell, I., Schack, L., and Skjott, K. (1992) Low level laser therapy for myofascial pain in the neck and shoulder girdle. A double-blind, cross-over study. *Scandinavian Journal of Rheumatology,* 21, 139-141.

Vaeroy, H., Sakurada, T., Forre, O., Kass, E., and Terenius, L. (1989) Modulation of pain in fibromyalgia (fibrositis syndrome): Cerebrospinal fluid (CSF) investigation of pain related neuropeptides with special reference to calcitonin gene related peptide (CGRP). *Journal of Rheumatology,* 16, 94-97.

Waylonis, G., Wilke, S., O'Toole, D., Waylonis, D., and Waylonis, D. (1988) Chronic myofascial pain: Management by low-output helium-neon laser therapy. *Archives of Physical Medicine and Rehabilitation,* 69, 1017-1020.

Wheeler, A., Goolkasian, P., and Gretz, S. (1998) A randomized, double blind, prospective pilot study of botulinum toxin injection for refractory, unilateral, cervicothoracic, paraspinal, myofascial pain syndrome. *Spine,* 23, 1662-1667.

Witting, N., Svensson, P., Gottrup, H., Arendt-Nielsen, L., and Jensen, T. (2000) Intramuscular and intradermal injection of capsaicin: A comparison of local and referred pain. *Pain,* 84, 407-412.

Wolfe, F., Simons, D., Bennett, R., Goldenberg, D., Gerwin, R., Hathaway, D., McCain, G., Russell, I., Sanders, H., and Skootsky, S. (1992) The fibromyalgia and myofascial pain syndromes: A preliminary study of tender points and trigger points in persons with fibromyalgia, myofascial pain syndrome and no disease. *Journal of Rheumatology,* 19, 944-951.

Wu, M.-T., Hsieh, J.-C., Xiong, J., Yang, C.-F., Pan, H.-B., Chen, Y.-C.I., Tsai, G., Rosen, B., and Kwong, K. (1999) Central nervous pathway for acupuncture stimulation: Localisation of processing with functional MR imaging of the brain—preliminary experience. *Radiology,* 212, 133-141.

Xian-Min, Y., Hoheisel, U., and Mense, S. (1992) Effect of a novel piperazine derivative (CGP 29030A) on nociceptive dorsal horn neurons in the rat. *Drugs and Experimental Clinical Research,* 17, 447-459.

Yunus, M.B., and Kalyan-Raman, U.P. (1986) Muscle biopsy findings in primary fibromyalgia and other forms of non-articular rheumatism. In: *The fibromyalgia syndrome,* ed. Danneskiold-Somsoe, 115-131. Portland, OR: Hawood Medical Press.

Yunus, M.B., Kalyan-Raman, U.P., Kalyan-Raman, K., and Masi, A.T. (1986) Pathologic changes in muscle in primary fibromyalgia syndrome. *American Journal of Medicine,* 81, 38-42.

Chapter 21

Babul, S., E.C. Rhodes, J.E. Taunton and M. Lepawski. 2004. Effects of intermittent exposure to hyperbaric oxygen for the treatment of an acute soft tissue injury. *Clinical Journal of Sports Medicine* 13: 138-147.

Beitzel, F., P. Gregorevic, J.G. Ryall, D.R. Plant, M.N. Sillence and G.S. Lynch. 2004. β_2-Adrenoceptor agonist fenoterol enhances functional repair of regenerating rat skeletal muscle after injury. *Journal of Applied Physiology* 96: 1385-1392.

Best, T.M., B. Loitz-Ramage, D.T. Corr and R. Vanderby. 1998. Hyperbaric oxygen in the treatment of acute muscle stretch injuries: Results in an animal model. *American Journal of Sports Medicine* 26: 367-372.

Bouachour, G., P. Cronler and J.P. Gouello. 1996. Hyperbaric oxygen therapy in the management of crush injuries: A randomized double-blind controlled clinical trial. *Journal of Trauma* 41: 333-339.

Carlson, B.M. and J.A. Faulkner. 1996. The regeneration of noninnnervated muscle grafts and marcaine-treated muscles in young and old rats. *Journals of Gerontology Series A: Biological Science and Medical Science* 51: B43-B49.

Chan, Y.S., Y. Li, W. Foster, T. Horaguchi, G. Somogyi, F.H. Fu and J. Huard. 2003. Antifibrotic effects of suramin in injured skeletal muscle after laceration. *Journal of Applied Physiology* 95: 771-780.

Cianci, P. and R. Sato. 1994. Adjunctive hyperbaric oxygen therapy in the treatment of thermal burns: A review. *Burns* 20: 5-14.

Fukuyama, R., M. Ohta, K. Ohta, S. Saiwaki, S. Fushiki, and A. Awaya. 2000. A synthesized pyrimidine compound, MS-818, promotes walking function recovery from crush injury of the sciatic nerve through its indirect stimulation of Schwann cells. *Restorative Neurology and Neuroscience* 17: 9-16.

Glasgow, P.D., I.D. Hill, A.M. McKevitt, A.S. Lowe and D. Baxter. 2001. Low intensity monochromatic infrared therapy: a preliminary study of the effects of a novel treatment unit upon experimental muscle soreness. *Lasers in Surgery and Medicine* 28: 33-39.

Gregorevic, P., G.S. Lynch and D.A. Williams. 2000. Hyperbaric oxygen improves contractile function of regenerating rat skeletal muscle after myotoxic injury. *Journal of Applied Physiology* 89: 1477-1482.

Haapaniemi, T., G. Nylander, A. Sirsjo and J. Larsson. 1996. Hyperbaric oxygen reduces ischemia-induced skeletal muscle injury. *Plastic and Reconstructive Surgery* 97: 602-607.

Harrison, B.C., D. Robinson, B.J. Davison, B. Foley, E. Seda and W.C. Byrnes. 2001. Treatment of exercise-induced muscle injury via hyperbaric oxygen therapy. *Medicine and Science in Sports and Exercise* 33: 36-42.

Huard, J., Y. Li and F.H. Fu. 2002. Muscle injury and repair: Current trends in research. *Journal of Bone and Joint Surgery* (American) 84: 822-832.

James, P.B., B. Scott and M.W. Allen. 1993. Hyperbaric oxygen therapy in sports injuries: A preliminary study. *Physiotherapy* 79: 571-572.

Leach, R.E. 1998. Hyperbaric oxygen therapy in sports. *American Journal of Sports Medicine* 26: 489-490.

Li, Y. and J. Huard. 2002. Differentiation of muscle-derived cells into myofibroblasts in injured skeletal muscle. *American Journal of Pathology* 161: 895-907.

Mekjavic, I.B., J.A. Exner, P.A. Tesch and O. Eiken. 2000. Hyperbaric oxygen therapy does not affect recovery from delayed onset muscle soreness. *Medicine and Science in Sports and Exercise* 32: 558-563.

Navegantes, L., R.H. Migliorini and I.C. Kettelhut. 2002. Adrenergic control of protein metabolism in skeletal muscle. *Current Opinion in Clinical Nutrition and Metabolic Care* 5: 281-286.

Ryall, J.G., P. Gregorevic, D.R. Plant, M.N. Sillence and G.S. Lynch. 2002. β_2-Agonist fenoterol has greater effects on contractile function of rat skeletal muscles than clenbuterol. *American Journal of Physiology: Regulatory, Integrative, and Comparative Physiology* 283: R1386-R1394.

Sato, K., Y. Li, W. Foster, K. Fukushima, N. Badlani, N. Adachi, A. Usas, F.H. Fu and J. Huard. 2003. Improvement of muscle healing through enhancement of muscle regeneration and prevention of fibrosis. *Muscle and Nerve* 28: 365-372.

Staples, J.R., D.B. Clement, J.E. Taunton and D.C. McKenzie. 1999. Effects of hyperbaric oxygen on a human model of injury. *American Journal of Sports Medicine* 27: 600-605.

Sugiyama, N., A. Yoshimura, C. Fujitsuka, H. Iwata, A. Awaya, S. Mori, H. Yoshizato and N. Fujitsuka. 2002. Acceleration by MS-818 of early muscle regeneration and enhanced muscle recovery after surgical transection. *Muscle and Nerve* 25: 218-229.

Tabrah, F.L., R. Tanner, R. Vega and S. Batkin. 1994. Baromedicine today—rational uses of hyperbaric oxygen therapy. *Hawaii Medical Journal* 53: 112-115.

Thorsson, O., J. Rantanen, T. Hurme and H. Kalimo. 1998. Effects of nonsteroidal antiinflammatory medication on satellite cell proliferation during muscle regeneration. *American Journal of Sports Medicine* 26: 172-176.

Van Heest, J.L., J. Stoppani, J. Scheett, V. Collins, M. Roti, J. Anderson, G.J. Allen, J. Hoffman, W.J. Kraemer and C.M. Maresh. 2002. Effects of ibuprofen and Vicoprofen on physical performance after exercise-induced muscle damage. *Journal of Sports Rehabilitation* 11: 224-234.

Vignaud, A., J. Cerian, I. Martelly, J.P. Caruelle and A. Ferry. 2005. Effect of anti-inflammatory and antioxidant drugs on the long-term repair of severely injured mouse skeletal muscle. *Experimental Physiology* 90: 487-495.

Warren, G.L., D.A. Lowe and R.B. Armstrong. 1999. Measurement tools used in the study of eccentric contraction-induced injury. *Sports Medicine* 27: 43-59.

Note: The italicized *f* and *t* following page numbers refer to figures and tables, respectively.

Photo courtesy of Wilfrid Laurier University

Peter M. Tiidus, PhD, is a professor and the department chair of the department of kinesiology and physical education at Wilfrid Laurier University in Waterloo, Ontario, Canada. For more than 20 years he has focused his research on the physiological mechanisms of and practical interventions in muscle damage and repair, employing both animal models and human subjects.

Tiidus has authored more than 60 publications and presented his research in more than 80 lectures and conference presentations on the topics of estrogen and muscle damage, inflammation, and repair and the influence of treatment interventions on muscle recovery from damage and physiological responses. He currently serves as an editorial board member for *Medicine and Science in Sports and Exercise*. He is also a former member of the board of directors of the Canadian Society for Exercise Physiology.

A former competitive swimmer and swim coach, Tiidus continues to enjoy swimming as well as cycling, skiing, and reading. He and his wife, Ann, have two sons and live in Waterloo, Ontario, Canada.